Ordinary Differential Equations

Victor Henner • Alexander Nepomnyashchy
Tatyana Belozerova • Mikhail Khenner

Ordinary Differential Equations

Analytical Methods and Applications

 Springer

Victor Henner
Department of Physics
Perm State University
Perm, Russia

Alexander Nepomnyashchy
Department of Mathematics
Israel Institute of Technology
Haifa, Israel

Tatyana Belozerova
Department of Mathematics
Perm State University
Perm, Russia

Mikhail Khenner
Department of Mathematics
Western Kentucky University
Bowling Green, KY, USA

ISBN 978-3-031-25132-0   ISBN 978-3-031-25130-6   (eBook)
https://doi.org/10.1007/978-3-031-25130-6

Mathematics Subject Classification: 33-01; 34-01; 45-01

This Springer imprint is published by the registered company Springer Nature Switzerland AG
The registered company address is: Gewerbestrasse 11, 6330 Cham, Switzerland

Preface

The primary goal of the book is the development of practical skills in solutions and qualitative investigations of ODEs and teaching how to apply these skills to solve real-world problems. The book rests on "teaching-by-examples" framework, thus we provide numerous carefully chosen examples that guide step-by-step learning of concepts and techniques. Quite a few of these examples come from mechanics, physics, and biology.

The book includes all basic topics that form the core of a typical applied ODE course. In addition to standard techniques presented in undergraduate-level ODE textbooks, e.g., in the book by W.E. Boyce and R.C. DiPrima [1], this textbook introduces other useful approaches, such as qualitative methods and stability of ODE solutions, integral equations, calculus of variations, etc. Among these are the topics that prepare the readers for studying PDEs (boundary value problems for second-order ODEs, Green's function, Sturm-Liouville problem, Fourier series, generalized Fourier series, Fourier transform, and special functions). In contradistinction to the popular textbook of E.A. Coddington [2], which contains primarily a mathematical theory, our book is more driven by applications. In that aspect, it is similar to the textbook by M. Braun [3]. However, that textbook is written for students in applied mathematics, whereas our textbook is better suited for physics and engineering students.

Despite that we focus on solutions and applications, we keep all rigor of mathematics (de-emphasizing proofs of theorems). The text is written in such a way that, depending on a course syllabus, an instructor can easily adopt certain chapters and omit other chapters. Numerous examples, applications, and graphical material make the book useful for students taking a distance learning course or studying the subject independently. We include the suggested syllabi.

The book structure is as follows:

The first four chapters cover the basic topics of a standard ODE course for undergraduate students: general definitions (Chap. 1), first-order ODEs (Chap. 2), linear ODEs of order two and higher (Chap. 3), and systems of linear ODEs (Chap. 4). In addition to presenting general theory and solution techniques, we provide, wherever possible, the applications to real science problems. Chapter 5 contains an introduction to the qualitative theory and qualitative techniques for ODEs that allow to analyze the behavior of the entire set of solutions and to investigate the stability of solutions.

The following chapters of the book introduce additional tools for solving ODEs: power series solutions (Chap. 6), method of Laplace transform (Chap. 7), and Fourier series and Fourier transform (Chap. 8). Chapter 9 contains the formulations and analyses of boundary value problems for second-order ODEs. These techniques include the generalized Fourier series and the method of Green's function. Chapter 10 discusses the properties of basic special functions.

The topics in Chaps. 11, 12, and 13 typically are not included in the standard ODE courses, but they are important for mathematical formulations of real-world problems and understanding the laws of nature. Chapter 11 is devoted to integral equations, Chap. 12 presents the calculus of variations, and Chap. 13 describes the application of ODEs to solutions of three linear PDEs, e.g., the heat equation, the wave equation, and the Laplace equation. Also, we briefly discuss the Helmholtz equation and the Schrödinger equation. The last Chap. 14 contains an introduction to numerical methods for ODEs. Finally, some auxiliary topics are relegated to the Appendices.

Optional user-friendly laboratory-type software is included with the textbook (there is NO requirement for using it for any purpose, although suggestions for use are provided). Familiarity with a programming language or a computer algebra system is not a prerequisite for using a software. Software allows to "play" with the problem setup (primarily, by entering new sets of parameters via clicking on menu items and typing) and solution and thus better understand an application. Real-world problems form the core of pre-set problem library for the laboratory. A student can also formulate new problems by modifying the library problems and then follow by their analysis in the software.

Perm, Russia   Victor Henner
Haifa, Israel   Alexander Nepomnyashchy
Perm, Russia   Tatyana Belozerova
Bowling Green, KY, USA   Mikhail Khenner

Contents

Chapter 1
Introduction

The main subject of this book is *ordinary differential equations*.

"Equation" means that something unknown has to be found. In contradistinction to algebraic equations, where *unknown numbers* have to be found, an ordinary differential equation (ODE) contains an *unknown function of one variable*, e.g., $y = y(x)$, and its *derivatives*: $y'(x)$, $y''(x)$, ..., $y^{(n)}(x)$. The general form of an ODE is

$$f\left(x, y(x), y'(x), \ldots, y^{(n)}(x)\right) = 0, \tag{1.1}$$

where f is a certain function of its arguments. Any function satisfying Eq. (1.1) is its *solution*. The *order of the ODE* is the order of its highest derivative, $n \geq 1$.

As the first example, consider a population of animals. The number $N(t)$ of animals changes with time due to their births and deaths, hence,

$$N'(t) = B(t) - D(t), \tag{1.2}$$

where $B(t)$ and $D(t)$ are the rates of births and deaths. The simplest assumption is that both rates are just proportional to $N(t)$:

$$B(t) = bN(t), \quad D(t) = dN(t),$$

where b and d are positive numbers. Then, the population evolution is governed by the first-order ODE:

$$N'(t) = rN(t), \quad r = b - d. \tag{1.3}$$

In order to solve Eq. (1.3), we may use the inverse function $t = t(N)$. It is known from the calculus that

© The Author(s), under exclusive license to Springer Nature Switzerland AG 2023
V. Henner et al., *Ordinary Differential Equations*,
https://doi.org/10.1007/978-3-031-25130-6_1

$$t'(N) = \frac{1}{N'(t)},$$

hence Eq. (1.3) can be rewritten as

$$t'(N) = \frac{1}{rN}. \tag{1.4}$$

Notice that N now is the independent variable. Integrating both sides of Eq. (1.4) with respect to N, we find

$$t(N) = \frac{1}{r} \ln N + C,$$

hence

$$N(t) = e^{r(t-C)}, \tag{1.5}$$

where C is an arbitrary number. We see that solutions of the first-order ODE (1.3) form a one-parameter family. The full set of solutions (1.5), is called the *general solution* (or the *general integral*) of the ODE.

To predict the evolution of a particular population, it is necessary to count the number of animals at a certain time, $t = t_0$. Let us assume that the result of that measurement is

$$N(t_0) = N_0. \tag{1.6}$$

Relation (1.6) is called the *initial condition*; the ODE (1.3) and the initial condition (1.6) form together an *initial value problem* (IVP). Substituting (1.5) into (1.6), we obtain the C value:

$$N_0 = e^{r(t_0 - C)}, \quad \text{hence} \quad e^{-rC} = N_0 e^{-rt_0}.$$

Thus, the *particular solution* of Eq. (1.3) satisfying the initial condition (1.6), i.e., the solution of the initial value problem (1.3) and (1.6) is unique, and it is

$$N(t) = N_0 e^{r(t - t_0)}. \tag{1.7}$$

If $N_0 = 0$, then $N(t) = 0$ for all t. Otherwise, the population either decays exponentially (if $r < 0$) or grows exponentially (if $r > 0$). For each set (t_0, N_0), we obtain a curve (1.7) called the *integral curve* in the plane (t, N) which passes through the point $t = t_0$, $N = N_0$ (see Fig. 1.1).

The unbounded growth of the population in the case $r > 0$, being applied to the humankind, led *Thomas Malthus*, the prominent English economist and demographer, to alarmist predictions (1798). Model (1.3) is, however, unrealistic: the growth

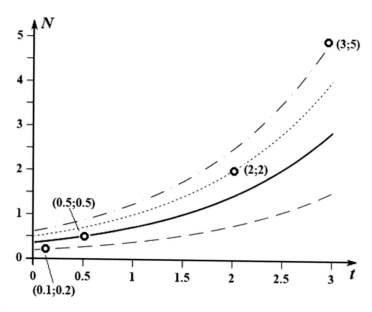

Fig. 1.1 Solutions of Eq. (1.3), $r = 0.7$. The initial conditions are indicated by circles. Solid line, $N_0 = 0.5$, $t_0 = 0.5$; dashed line, $N_0 = 0.2$, $t_0 = 0.1$; dotted line, $N_0 = 2$, $t_0 = 2$; dashed-dotted line $N_0 = 5$, $t_0 = 3$

rate is not a constant but depends on the number of animals in the population: $r = r(N)$; hence, the model has to be improved as follows:

$$N'(t) = r(N(t))N(t). \tag{1.8}$$

For the inverse function, we write the equation:

$$t'(N) = \frac{1}{r(N)N},$$

and obtain the general solution of ODE (1.8) in the *implicit form*:

$$t(N) = \int \frac{dN}{r(N)N} + C. \tag{1.9}$$

It is quite natural to expect that often the growth rate of an animal population decreases with N due to an insufficient food supply. *Pierre-François Verhulst*, a Belgian mathematician, suggested (1836) a linear dependence between r and N:

$$r(N) = r_0 \left(1 - \frac{N}{N_\infty} \right), \tag{1.10}$$

where r_0 and N_∞ are positive numbers. In the framework of that model, the growth rate $r(N) > 0$ only if $N < N_\infty$; otherwise, $r(N) < 0$. Substituting (1.10) into (1.9), we find

$$t(N) = \frac{N_\infty}{r_0} \int \frac{dN}{N(N_\infty - N)} = \frac{1}{r_0} \int dN \left(\frac{1}{N} + \frac{1}{N_\infty - N} \right) = \frac{1}{r_0} \ln \frac{N}{|N_\infty - N|} + C,$$

thus, the general solution is

$$N(t) = \frac{N_\infty}{1 \pm e^{-r_0(t - C)}}, \tag{1.11}$$

where C is an arbitrary real number. If the initial condition is

$$N(t_0) = N_0, \tag{1.12}$$

we find that

$$\pm e^{r_0(C - t_0)} = \frac{N_\infty - N_0}{N_0},$$

thus, the solution of the initial value problem (1.8), (1.10), and (1.12) is

$$N(t) = N_\infty \frac{N_0}{N_0 + e^{-r_0(t - t_0)}(N_\infty - N_0)}. \tag{1.13}$$

The behavior of solutions of the Verhulst model (1.13) (also called the logistic model) is very different from that of solutions of the Malthus model (1.7) (see Fig. 1.2), except for the trivial case $N_0 = 0$. For any $N_0 \neq 0$, $\lim_{t \to \infty} N(t) = N_\infty$, i.e., the population size tends to a definite *equilibrium* value.

The basic difference between the Malthus model,

$$f_M(t, N, N') = N' - rN = 0,$$

and that of Verhulst,

$$f_V(t, N, N') = N' - rN \left(1 - \frac{N}{N_\infty} \right) = 0,$$

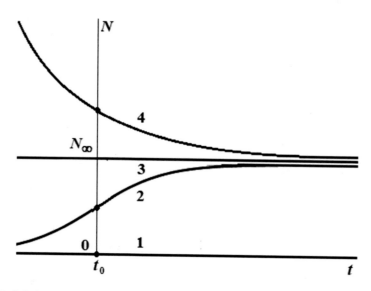

Fig. 1.2 Solution of Eq. (1.10). Line 1, $N_0 = 0$; line 2, $N_0 < N_\infty$; line 3, $N_0 = N_\infty$; line 4, $N_0 > N_\infty$

is as follows: function f_M is a linear function of N and N', while f_V is a not a linear function. In the former case, the ODE is *linear*; in the latter case, it is *nonlinear*. As we have seen, the behavior of solutions in both cases is quite different.

Let us now formulate the general definition of the linear ODE. Equation (1.1) is linear, if f is a linear function of y, y', ..., $y^{(n)}$ (but it can be an arbitrary function of x). The general form of a linear equation of the n-th -order is

$$f\left(x, y(x), \ldots, y^{(n)}(x)\right) = a_0(x)y^{(n)}(x) + a_1(x)y^{(n-1)}(x) + \ldots$$
$$+ a_n(x)y(x) + b(x) = 0. \tag{1.14}$$

As an example of a linear second-order equation, let us recall the high-school mechanical problem: an object moving in the vertical direction under the action of the gravity force. The motion of the object is described by the function $z = z(t)$, where t is time and z is the coordinate of the object (see Fig. 1.3).

The derivatives of the function $z = z(t)$ have a clear meaning: $v(t) = z'(t)$ is the velocity and $a(t) = z''(t)$ is the acceleration. The second law of Newton says that

$$mz''(t) = F(t, z(t), z'(t)), \tag{1.15}$$

where m is the mass of the object and F is the force acting on the object. Obviously, Eq. (1.15) is the second-order ODE. If the friction is neglected, the only force acting on the object is the constant gravity force:

Fig. 1.3 An object moving
in the vertical direction

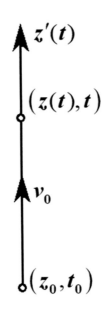

$$F = -mg, \tag{1.16}$$

where g is the acceleration of gravity; hence, we obtain a linear second-order
equation:

$$z''(t) = -g. \tag{1.17}$$

In order to solve this equation, we rewrite it as the first-order equation for the
velocity:

$$v'(t) = -g \tag{1.18}$$

and integrate both sides of Eq. (1.18) with respect to t. We find:

$$v(t) = z'(t) = -gt + C_1, \tag{1.19}$$

where C_1 is an arbitrary constant. Integrating both sides of (1.19), we find:

$$z(t) = -\frac{gt^2}{2} + C_1 t + C_2, \tag{1.20}$$

where C_2 is an arbitrary constant. Thus, the general solution (1.20) of the second
order equation is a *two-parameter* family of solutions, $z = z(t; C_1, C_2)$. To predict the
actual motion of the object, it is necessary to impose *two initial conditions*, the initial
position of the object $z(t_0)$ and its initial velocity $v(t_0) = z'(t_0)$. Let us choose $t_0 = 0$
and assume

Fig. 1.4 Parallelogram of forces: the weight force P, the tension force T, and the net force F; $P = mg$, $F = mg \sin \theta$

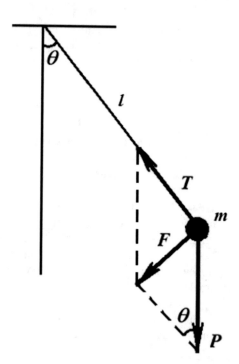

$$z(0) = z_0, \quad z'(0) = v_0. \tag{1.21}$$

Substituting (1.20) into these initial conditions, we find that $C_1 = v_0$ and $C_2 = z_0$. As the solution of the IVP (1.17) and (1.21), we recover the high-school formula:

$$z(t) = z_0 + v_0 t - \frac{gt^2}{2}. \tag{1.22}$$

As a physical example of a nonlinear second-order equation, let us consider "the mathematical pendulum," which is a point-like object of mass m attached to a weightless thread of the length l and swinging in the gravity field (see Fig. 1.4).

Applying the equation for the rotational moment, we find:

$$ml^2 \theta''(t) = -mgl \sin \theta(t),$$

where $\theta(t)$ is the angle. Thus, we obtain the following nonlinear second-order ODE governing pendulum's swinging:

$$\theta'' + \frac{g}{l} \sin \theta(t) = 0. \tag{1.23}$$

We are not ready yet to solve this equation, and we shall return to it in Sect. 2.8. However, it is intuitively clear that the pendulum motion depends on two quantities, its initial position, $\theta(t_0)$, and its initial angular velocity, $\theta'(t_0)$. Thus, the general solution of the second-order ODE (1.23) depends on two parameters.

One can guess that the general solution of Eq. (1.1), which is of order n, contains n arbitrary constants:

$$y = y(x; C_1, C_2, \ldots, C_n),$$

that can be determined by n initial conditions:

$$y(x_0) = y_0, y'(x_0) = y_1, \ldots, y^{(n-1)}(x_0) = y_{n-1}.$$

We shall prove this in Chap. 3.

In conclusion, note that according to the general definition (1.1) of the ODE, the function f can explicitly depend on the independent variable. However, in examples (1.3), (1.8), (1.17), and (1.23), the equations do not include the variable t explicitly. This means that the evolution of the system is not influenced by a time-dependent external factor but is determined solely by its internal dynamics. An equation of that kind, $f(x), y'(x), \ldots, y^{(n)}(x)) = 0$, is called an *autonomous equation* (otherwise, the equation is nonautonomous). Autonomous systems can be studied by means of certain special techniques that we consider in Sect. 2.7 and Chap. 5.

Problems
For the following equations, determine their order and type (linear or nonlinear), and solve them by integration:

1. $y'(x) = \sin x$; $y(0) = 0$.
2. $y''(x) = \sin x$; $y(0) = 0, y'(0) = 1$.
3. $y'(x) = -y(x)$; $y(0) = 1$. *Hint*: use the inverse function $x = x(y)$.
4. $y'(x) = y^2(x)$; $y(0) = 1$.
5. $y'(x) = 1/y(x)$; $y(1) = 1$.
6. $y''(x) + y'(x) = 0$; $y(0) = 1, y'(0) = -1$.
7. $y''(x) + 2y'(x) + y(x) = 0$; $y(0) = 1, y'(0) = -1$.
   Hint: use the auxiliary variable $z = y' + y$.
8. $y''(x) = 2yy'(x)$; $y(0) = 1, y'(0) = 1$.

Chapter 2
First-Order Differential Equations

In this chapter we systematically develop solution techniques for first-order ODEs.

2.1 Existence and Uniqueness of a Solution

First-order equations have either the form

$$F(x, y, y') = 0 \tag{2.1}$$

or the form

$$y' = f(x, y). \tag{2.2}$$

An explicit form of a general solution contains one arbitrary constant: $y = y(x, C)$. Quite often a general solution is *implicit*:

$$\Phi(x, y, C) = 0.$$

A particular solution emerges when C takes on a certain numerical value. In order to find a particular solution, a single initial condition

$$y(x_0) = y_0 \tag{2.3}$$

is needed. The problem (2.1), (2.3), (or (2.2), (2.3)) is the *initial value problem* (IVP), also called *Cauchy problem* (Augustin-Louis Cauchy (1789–1857), a French mathematician) for the first-order ODE. To be specific, we will assume in what follows the IVP (2.2) and (2.3).

© The Author(s), under exclusive license to Springer Nature Switzerland AG 2023
V. Henner et al., *Ordinary Differential Equations*,
https://doi.org/10.1007/978-3-031-25130-6_2

The question that we now should be asking is the following one: under what conditions on a function $f(x, y)$ does there exist a unique solution of the IVP? The following Theorem provides the answer.

Theorem (*Existence and Uniqueness of a Solution*)
   Let a function f be continuous in a rectangle D: $x_0 - \alpha \leq x \leq x_0 + \alpha$, $y_0 - \beta \leq y \leq y_0 + \beta$ that contains the point (x_0, y_0). Also let the partial derivative $\partial f/\partial y$ exists and be bounded in D. Then, the solution of IVP (2.2) and (2.3) exists and is unique in the vicinity of (x_0, y_0).

The condition of finiteness of the partial derivative $\partial f/\partial y$ can be replaced by a weaker *Lipschitz condition* (Rudolf Otto Sigismund Lipschitz (1832–1903), a German mathematician) in y, i.e., there exists a constant $K > 0$ independent of x such that for any two points (x, y_1) and (x, y_2) in D:

$$|f(x, y_1) - f(x, y_2)| \leq K|y_1 - y_2|.$$

The proof of the Existence and Uniqueness Theorem is presented in Appendix A. If function $f(x, y)$ is linear in y, Eq. (2.2) can be put in the form

$$y' + p(x)y = g(x), \tag{2.4}$$

which is called a *linear equation*. If functions $p(x)$ and $g(x)$ are continuous, the conditions of the Existence and Uniqueness Theorem are obviously satisfied.

Notice that even when the conditions of the theorem are satisfied in a rectangle D, the theorem does not guarantee the existence of the solution in the whole interval $[x_0 - \alpha, x_0 + \alpha]$.

Example 2.1 Discuss the initial value problems: (a) $y' = y^2$, $y(1) = 1$; (b) $y' = y^{1/3}$, $y(0) = 0$.

(a) The function $f(y) = y^2$ at the right-hand side of the equation is continuous and differentiable everywhere, and the derivative $df/dy = 2y$ is bounded in any rectangle containing the point $(1,1)$. Thus, the conditions of the Existence and Uniqueness Theorem are satisfied. Therefore, there exists a certain region of x around the point $x_0 = 1$, where the solution exists.
   Now let us solve the problem. We can rewrite the equation as $dx/dy = y^{-2}$. Integrating both sides in y, we find $x(y) = -1/y + C$, i.e., the general solution of the equation is $y(x) = 1/(C - x)$, where C is an arbitrary constant. To satisfy the initial condition, we have to take $C = 2$, i.e., the solution of the initial value problem is $y = -\frac{1}{x-2}$. This solution, existing "in the vicinity of the point $(1, 1)$," does not exist for $x = 2$, and thus it cannot be extended into the domain $x > 2$.

(b) The function $f(y) = y^{1/3}$ is not differentiable at $y = 0$; therefore, the conditions of the Existence and Uniqueness Theorem are not satisfied. To understand the consequences, let us first rewrite the equation as $dx/dy = y^{-1/3}$. Integrating both parts of the equation, we find $x(y) = (3/2)y^{2/3} + C$. To satisfy the initial

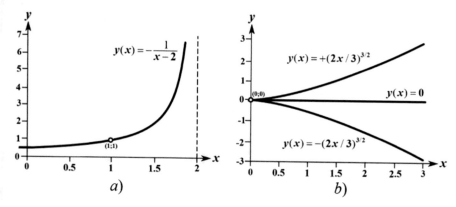

Fig. 2.1 Illustration for Example 2.1

condition, we must take $C = 0$. That gives us two solutions for $x > 0$: $y(x) = + (2x/3)^{3/2}$ and $y(x) = - (2x/3)^{3/2}$. Moreover, it is obvious that there exists one more solution, $y(x) = 0$ for any x. We see that in this example, the violation of the conditions of the Existence and Uniqueness Theorem leads to nonuniqueness of the IVP solution. Note that in both situations (a) and (b), the functions $f(x, y)$ are nonlinear with respect to y. Figure 2.1 shows the solutions.

Contrary to the solution considered in this example, for a linear Eq. (2.4), a solution of an IVP exists and is unique for *any* x where $p(x)$ and $g(x)$ are continuous. For instance, consider equation $y' = y + 1$ with the initial condition $y(1) = 0$. Let us rewrite the equation as $dx/dy = (y + 1)^{-1}$. Integrating in y both sides of that equation, we find $x(y) = \ln |1 + y| + C$, where C is an arbitrary constant. To satisfy the initial condition, we have to take $C = 1$. Thus, $|1 + y| = e^{x-1}$. For $x = 1$, $y(x) = 0$; hence, $1 + y(x) > 0$. Obviously, the value of $1 + y(x)$ cannot change the sign at any finite x; therefore, $1 + y(x) > 0$ for any x. We find that the solution of the initial value problem is $y(x) = - 1 + e^{x-1}$ – this function satisfies both the differential equation and the initial condition. This solution exists and is unique for any x. This result agrees with the fact that the linear function $f(y) = y + 1$ is continuous and differentiable everywhere, and the derivative $df/dy = 1$ is bounded.

Conditions of the Existence and Uniqueness Theorem can be illustrated by the simple method of the numerical integration of the ODE $y' = f(x, y)$ on the interval $[x_0, b]$. Let $[x_0, b]$ be divided into n subintervals $[x_0, x_1], [x_1, x_2], \ldots, [x_{n-1}, b]$ of equal length h, where $h = \frac{b - x_0}{n}$. The quantity h is the step size of the computation (see Fig. 2.2). The idea of the method is to approximate the integral curve by the set of the straight-line segments on the intervals $[x_i, x_{i+1}]$, such that each segment is tangent to the solution curve at one of the endpoints of the subintervals. For these purposes, the function $f(x, y)$ should be smooth in the vicinity of the point (x_0, y_0).

First, consider the interval $[x_0, x_0 + h]$ and find the value y_1 from equation $y' = f(x, y)$, as follows. Suppose h is small, then the derivative $y' \approx \frac{\Delta y}{\Delta x} = \frac{y_1 - y_0}{h}$, and thus $y_1 \approx y_0 + y'h$. Also, if h is small and the function $f(x, y)$ is continuous

Fig. 2.2 Illustration of
Euler's method

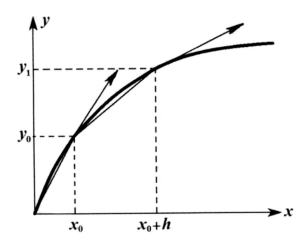

and changes slowly, then y' can be replaced by $f(x_0, y_0)$ on this interval. Then,
$y_1 \approx y_0 + f(x_0, y_0)h$.

Next, let us repeat this construction on the interval $[x_0 + h, x_0 + 2h]$ (which is
$[x_1, x_2]$). Taking

$$y' \approx \frac{y_2 - y_1}{h} \approx f(x_1, y_1),$$

we find $y_2 \approx y_1 + hf(x_1, y_1)$. Then, repeating the construction for other subintervals,
we finally arrive to *Euler's method* (Leonhard Euler (1707–1783), a genius of
mathematics and physics):

$$y_{i+1} = y_i + hf(x_i, y_i), \quad i = 0, 1, \ldots, n - 1 \tag{2.5}$$

(this algorithm can be applied also when $b < x_0$. In that case the step size $h < 0$.)
As $|h|$ decreases, the line segments become shorter, and they trace the integral curve
better – thus the accuracy of the approximate solution increases as $h \to 0$. The
integral curve that passes through the point (x_0, y_0) represents the particular solution.

Another important question arises: what will happen with the solution if the initial
conditions slightly change? Will the solution change also slightly, or in other words,
is the solution stable? This question is important from a practical standpoint, because
in effect we ask: how an error in initial conditions (which often are obtained from
experimental data, or from calculations performed with some limited precision) can
affect a solution of a Cauchy problem? The answer to this question is that it can be
shown that if the conditions of the Existence and Uniqueness Theorem are satisfied,
then a solution continuously depends on the initial conditions. It also can be shown
that if the equation contains a parameter λ:

$$y' = f(x, y, \lambda), \tag{2.6}$$

then a solution continuously depends on λ if a function f is a continuous function of λ. The problem of a *solution stability* that we only touched here will be discussed in detail in Chap. 5.

To end this section, let us consider the inverse problem, i.e., the determination of a differential equation that has a given solution $\Phi(x, y, C) = 0$. Here one has to differentiate a function Φ, considering y as a function of x (thus using the Chain Rule), and then eliminate C with the help of equation $\Phi(x, y, C) = 0$.

For example, let the general solution of an equation be $y = Cx^3$, and we wish to find the equation. Differentiation of the general solution gives $y' = 3Cx^2$. Substitution into this expression $C = y/x^3$ from the general solution results in the differential equation $y' = 3y/x$.

Problems

In the following problems, (i) check whether the conditions of the Existence and Uniqueness Theorem are satisfied; (ii) find the solution by integrating both sides of the equation; (iii) analyze the properties of the solution; and (iv) find the domain in which the solution exists and plot the solution graph:

1. $yy' = 1$, $y(0) = 1$.
2. $y' = -2y$, $y(0) = 2$.
3. $y' = 2\sqrt{y}$, $y(0) = 0$.
4. $y' = \sqrt[3]{y^2}$, $y(0) = 1$.
5. $y' = -5y^2$, $y(0) = -1$.
6. $y' = y^3$, $y(0) = 3$.
7. $4y' = y^{3/4}$, $y(0) = 0$.

In the following problems, construct differential equations describing given families of solutions:

8. $y = (x - C)^2$.
9. $x^3 + Cy^2 = 5y$.
10. $y^3 + Cx = x^2$.
11. $y = e^{Cx + 5}$.
12. $y = Ce^x - 3x + 2$.
13. $\ln y = Cx + y$.
14. $y = C \cos x + \sin x$.
15. $e^{-y} = x + Cx^2$.
16. $y^2 + Cx = x^3$.
17. $x = Cy^2 + y^3$.

2.2   Integral Curves and Isoclines

A differential equation $\frac{dy}{dx} = f(x, y)$ can be seen as a formula that provides a connection between the Cartesian coordinates of a point, (x, y), and the slope (i.e., the tangent of the inclination angle) of the integral curve, $\frac{dy}{dx}$, at this point.

To visualize this slope, we can draw at a point (x, y) a short line segment (called the *slope mark*) that has the slope $f(x, y)$. Repeating this for some other points in the

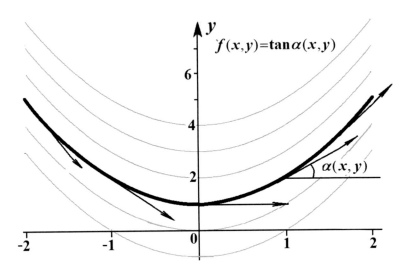

Fig. 2.3 Integral curves $y = x^2 + C$ of equation $y' = 2x$ on the (x, y) plane. For the bold line $C = 1$

Cartesian plane gives the *direction field*. (As a rule, tens or even hundreds of points are required for the construction of a quality direction field.) Thus, from a geometrical perspective, to solve a differential equation $y' = f(x, y)$ means to find curves that are tangent to the direction field at each point. These solution curves are called the integral curves. Figure 2.3 shows several integral curves of equation $y' = 2x$ and the tangent vectors for one chosen curve.

When the analytical solution is unavailable, the construction of the direction field often can be made easier by first drawing lines of a constant slope, $y' = k$. Such lines are called *isoclines*. Since $y' = f(x, y)$, the equation of an isocline is $f(x, y) = k$. This means that an isocline is a *level curve* of the function $f(x, y)$.

Example 2.2 Consider equation $dy/dx = y$. Here, the isoclines are just straight lines $y = k$ parallel to the x-axis (see Fig. 2.4). Starting at any point (x_0, y_0), we can sketch the plot of the solution. The solution is monotonically increasing if $y_0 > 0$ and monotonically decreasing if $y_0 < 0$. For $y_0 = 0$, $y(x)$ is a constant (zero). The integral curves become steeper when x grows and flatter when x decreases. All curves approach the x-axis as $x \to -\infty$.

The analytical solution of this equation is given in Chap. 1 (with different notations of the variables).

Example 2.3 Consider equation $\frac{dy}{dx} = \frac{y}{x}$. We notice that at any point (x, y), the slope of the integral curve, y/x, is also the slope of the straight lines leaving the origin $(0, 0)$ (in the right half-plane), and entering the origin (in the left half-plane). The direction field is shown in Fig. 2.5 with the short black arrows. (Since the function $f(x, y) = y/x$ is not defined for $x = 0$, the equation does not have a solution corresponding to the y-axis, but the direction field can be plotted everywhere in the plane except the origin.

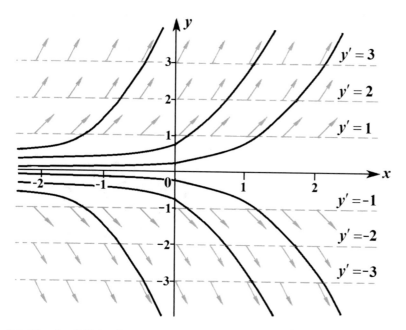

Fig. 2.4 Direction field, isoclines, and integral curves for equation $y' = y$

Fig. 2.5 Direction field
and isoclines for equation
$y' = y/x$

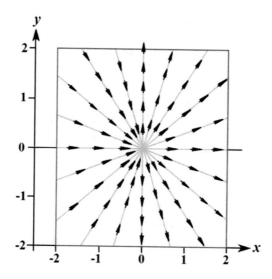

Note that along the y-axis the slope is infinite.) The isoclines are given by equation
$y/x = k$. Thus, in this example the isoclines are the straight lines coinciding with the
direction field. In Fig. 2.5 the isoclines are shown by gray lines. Dark gray rectangle
is the chosen domain for the graphical representation of the solution.

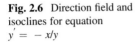

Fig. 2.6 Direction field and isoclines for equation $y' = -x/y$

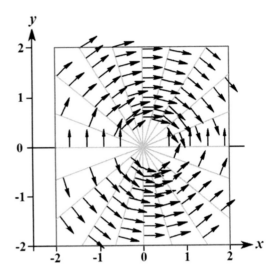

In Figs. 2.5–2.7, the isoclines and the direction fields are plotted with the software *ODE 1st order*; its description is in Appendix B.

Example 2.4 Consider equation $\frac{dy}{dx} = -\frac{x}{y}$. Isocline equation $-x/y = k$ gives straight lines shown in gray in Fig. 2.6. Along each isocline the value of y' does not change. This means that all slope marks (black arrows) along a particular isocline are parallel. Connecting the slope marks on the neighboring isoclines, we can plot integral curves – which are obviously the circles centered at the origin: $x^2 + y^2 = C^2$. In other words, the general solution has the form $y = \sqrt{C^2 - x^2}$ ($y = -\sqrt{C^2 - x^2}$) in the upper (lower) half-plane. For the particular solution, starting from the point (x_0, y_0), we obtain the value of C (the circle radius): $C = \sqrt{x_0^2 + y_0^2}$. Function $f(x, y) = -x/y$ is not defined at $x = y = 0$; therefore, the direction field can be plotted everywhere in the plane (x, y) except the point $(0, 0)$, where $f(x, y)$ is not defined and the limit depends on the approach direction.

So, the solution of the IVP satisfying the initial condition $y(x_0) = y_0$ with y_0 different from 0 exists only inside the interval $|x| < C = \sqrt{x_0^2 + y_0^2}$. Toward the end of this interval, $y(x)$ tends to zero, so that the right-hand side of the equation tends to positive or negative infinity. The situation when the solution exists only in a bounded domain is quite common for nonlinear IVPs.

Example 2.5 Sketch several solution curves for equation $y' = x + y - 1$ using isoclines.

The isocline equation is $x + y - 1 = k$, or $y = -x + (k + 1)$. First, using several k values, we plot isoclines. Then, we plot the direction marks (that have slope k) along each isocline. Next, starting at a particular initial point (the initial condition), we connect the neighboring slope marks by a smooth curve in such a way that a curve's slope at each point is shown by these marks. If we use a reasonably large

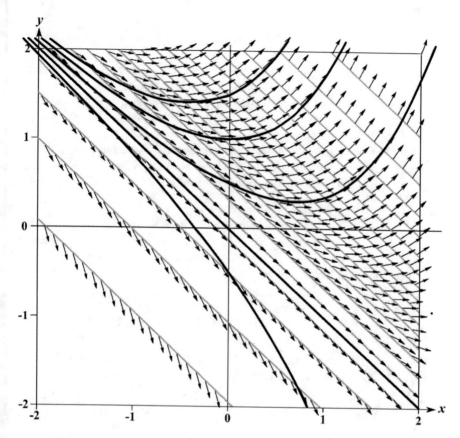

Fig. 2.7 Direction field, isoclines, and five integral curves for equation $y' = x + y - 1$

number of isoclines, then this process results in a qualitative plot of the integral curve. Another initial point gives another integral curve, and so on. This approach can be useful in some situations, such as when the analytical solution is unknown and the integral curves are relatively simple. Figure 2.7 shows the sketch of the isoclines and the direction field, as well as the five integral curves plotted for the initial points $(0, -0.5)$, $(0, 0)$, $(0, 0.5)$, $(0, 1.0)$, $(0, 1.5)$.

In many problems, especially those of a physical or a geometrical nature, the variables x and y are indistinguishable in the dependent/independent sense (i.e., y may be considered independent and x dependent, or vice versa). Thus, if such problem is described by a differential equation

$$\frac{dy}{dx} = f(x, y) \tag{2.7}$$

that has a solution $y = y(x)$, then it is natural to also consider the equation:

$$\frac{dx}{dy} = \frac{1}{f(x,y)},\tag{2.8}$$

that has the solution $x = x(y)$. It is clear from (2.7) and (2.8) that these solutions are equivalent, and thus their graphs (the integral curves) coincide.

In the situations without obvious dependent and independent variables and when one of Eq. (2.7) or (2.8) does not make sense at a certain point or points (thus the right-hand side is undefined), it is natural to replace the equation at such points by its counterpart. For instance, the right-hand side of equation $\frac{dy}{dx} = \frac{y}{x}$ is undefined at $x = 0$. The solution of equation, considering $x \neq 0$, is $y = Cx$. When this equation is replaced by $\frac{dx}{dy} = \frac{x}{y}$, one finds that the latter equation has the trivial solution $x = 0$.

Problems
Plot the isoclines and the direction field for the given differential equation in the given domain $[a, b; c, d]$. With the help of isoclines, make a sketch of two particular solutions of this equation passing through the points (a) $y_1 = y(x_1)$; (b) $y_2 = y(x_2)$ (this can be done with the help of program ODE 1st order):

1. $y' = y - x^2$, $[-2,2;-2,2]$, (a) $y(-1) = 1$;   (b) $y(0) = -1$.
2. $y' = 2x - e^y$, $[-2,2;-2,2]$, (a) $y(-1) = 1$;   (b) $y(0) = 0.5$.
3. $y' = \frac{5x}{x^2+y^2}$, $[1,5;1,5]$,   (a) $y(1) = 2$;   (b) $y(4) = 3$.
4. $y' = x + 1 - 2y$, $[-1,1;-1,1]$, (a) $y(-0.5) = -1$;   (b) $y(0) = 0.5$.
5. $y' = \frac{y-3x}{x+3y}$, $[1,3;1,3]$,   (a) $y(1.5) = 2.5$;   (b) $y(2) = 1.5$.
6. $y' = 2x^2 - y$, $[-2,2;-2,2]$, (a) $y(-1) = 1$;   (b) $y(0) = -1.5$.
7. $y' = \sin x \sin y$, $[-3,3;-3,3]$, (a) $y(-2) = -2$;   (b) $y(2) = 1$.
8. $y' = x^2 + \sin y$, $[-2,2;-2,2]$, (a) $y(-1) = -1$;   (b) $y(1) = 1$.
9. $y' = x(x - y)^2$, $[-1,1;-1,1]$, (a) $y(-1) = -0.5$;   (b) $y(-0.5) = 0$.
10. $y' = 3\cos^2 x + y$, $[-2,2;-2,2]$, (a) $y(-2) = 0$;   (b) $y(2) = 1$.

There is no general solution method applicable to all differential equations, and it is possible to solve analytically only the first-order equations of certain types. In the next sections, we introduce these types and the solution methods.

2.3 Separable Equations

The first-order equation

$$\frac{dy}{dx} = f(x,y)\tag{2.9}$$

can be written in the equivalent form:

$$dy = f(x, y)dx. \tag{2.10}$$

If a function $f(x, y)$ can be represented as a quotient, $f_2(x)/f_1(y)$, then Eq. (2.10) takes the form:

$$f_1(y)dy = f_2(x)dx. \tag{2.11}$$

In Eq. (2.11) variables x and y are separated to the right and to the left of the equality sign. Equations that allow a transformation from (2.9) to (2.11) are called *separable*, and the process of a transformation is called *separation of variables*. By integrating both sides of (2.11), one obtains

$$F_1(y) = F_2(x) + C. \tag{2.12}$$

where

$$F_1(y) = \int f_1(y)dy, \quad F_2(x) = \int f_2(x)dx.$$

If Eq. (2.12) can be explicitly resolved with respect to y, one obtains the general solution of the differential equation in an explicit form, as a function $y = F(x, C)$. Otherwise, (2.12) determines the general solution in an implicit form, as

$$\Phi(x, y, C) = F_1(y) - F_2(x) - C.$$

Notice, even if the integrals in (2.12) cannot be evaluated in elementary functions, it is customary to declare that the solution of a separable equation has been determined.

The solution of an IVP can be obtained by substituting the general solution into the initial condition $y(x_0) = y_0$ (or vice versa), which allows to determine C value. The corresponding particular solution can be written also as

$$\int_{y_0}^{y} f_1(r)dr = \int_{x_0}^{x} f_2(r)dr, \tag{2.13}$$

where no arbitrary constant is present. Notice that r is the dummy integration variable. Obviously, formula (2.13) gives the equality $0 = 0$ at the point (x_0, y_0); thus, this particular solution satisfies the initial condition $y(x_0) = y_0$.

Example 2.6 Consider again the differential equation from Example 2.4, $\frac{dy}{dx} = -\frac{x}{y}$. Multiplication of the equation by ydx gives $ydy = -xdx$. In the latter equation, variables are separated. Next, integrate $\int ydy = -\int xdx + c$. Evaluation of the integrals gives the general solution $x^2 + y^2 = 2c$, which is a circle of radius $C = \sqrt{2c}$.

Let the initial condition be $y(3) = 4$. Substituting the initial condition in the general solution, we obtain the particular solution $x^2 + y^2 = 25$.

It is instructive to solve some problems using the program *ODE 1st order*. The program offers a simple and convenient graphical interface to the numerical solution procedures for first-order IVPs given in the form $y' = f(x, y)$, $y(x_0) = y_0$. Numerical solution may be difficult if a function $f(x, y)$ does not satisfy the conditions of the Existence and Uniqueness Theorem at some points. Thus, in *ODE 1st order*, a region D for the integration of a differential equation must be chosen so that it does not contain such points. For instance, in Example 2.6 a domain D must include the point (3,4) (the initial condition), but it should not include any point with the zero y-coordinate. Since the particular solution is the circle of radius 5 centered at the origin, the interval of x-values for the numerical solution has to be $-5 < x < 5$. For the initial condition $y(3) = 4$, the numerical solution gives the upper semicircle $y = +\sqrt{25 - x^2}$, and for the initial condition $y(3) = -4$, the numerical solution gives the lower semicircle $y = -\sqrt{25 - x^2}$.

Often, in order to transform (4.2) into (4.3), it is necessary to divide both sides of (4.2) by a function of x, y. Zero(s) of such function may be the solution(s) of (4.1), which are lost when the division is performed. Thus, we must take a note of these solutions before division. This situation occurs, for instance, in problems 13, 16, and 17 given at the end of this section.

Example 2.7 Next, we consider the differential equation from Example 2.3: $y' = \frac{y}{x}$. Dividing both sides by y and multiplying by dx, the equation becomes $\frac{dy}{y} = \frac{dx}{x}$. Dividing by y, we may have lost the solution $y = 0$. Indeed, plugging $y = 0$ into both sides of the original equation $y' = \frac{y}{x}$, we see that equation becomes the identity $0 = 0$. Thus, $y = 0$ indeed is the solution. To find other solutions, we integrate $\frac{dy}{y} = \frac{dx}{x}$, and obtain: $\ln|y| = \ln|x| + c$. Exponentiation gives $|y(x)| = e^{\ln|x|}e^c = C|x|$ with $C = e^c$. Notice that $C > 0$. Now obviously, $y(x) = \pm Cx$. The trivial solution $y = 0$ can be included in the solution $y(x) = Cx$ by allowing $C = 0$. Thus, finally, the general solution is $y = Cx$, where C is *arbitrary constant* (can be either positive, or negative, or zero).

To find particular solutions, any initial condition $y(x_0) = y_0$ is allowed, except x_0 equal to zero. This is because the conditions of Existence and Uniqueness Theorem are violated when $x_0 = 0$: the function $f(x, y) = y/x$ is unbounded. When $x_0 \neq 0$, the particular solution is $y = y_0 x/x_0$.

Solutions are always defined in a certain domain of x. In this example the line $x = 0$ (the y-axis) is the line of discontinuity. Therefore, for the initial condition taken at $x_0 > 0$, we can solve the equation in the domain $0 < x \leq a$, if $x_0 < 0$ – in the domain $a \leq x < 0$.

Example 2.8 Consider a differential equation $\frac{dy}{dx} = -\frac{y}{x}$. (Note that this equation differs from Example 2.7 only in that the right-hand side is multiplied by -1). Functions $f(x, y) = -y/x$ and $f_y(x, y) = -1/x$ are continuous at $x \neq 0$; thus, equation

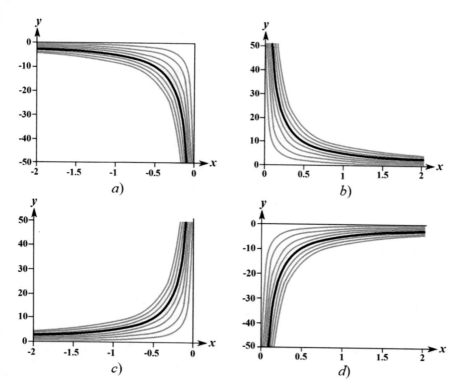

Fig. 2.8 Integral curves for the equation $\frac{dy}{dx} = -\frac{y}{x}$. General solution is $y = C/x$. In panels (**a**, **b**), the curves for $C = 1, 2, \ldots, 9$ are shown; the bold black curve is the particular solution $y = 5/x$. In panels (**c**, **d**), the curves for $C = -9, -8, \ldots, -1$ are shown; the bold black curve is the particular solution $y = -5/x$

satisfies the conditions of the Existence and Uniqueness Theorem in the entire (x, y) plane, except again at the y-axis. Separating the variables we have $dy/y = -dx/x$ and after the integration obtain $\ln|y| = \ln C - \ln|x|$ (here we choose the arbitrary constant in the form $\ln C$, $C > 0$), which gives $|y| = C/|x|$. Then, $y = \pm C/x$ and the general solution can be finally written as $y = C/x$, where C is arbitrary constant ($C = 0$ is also allowed because $y = 0$ is the trivial solution of the given equation). This solution holds in the domains $x < 0$ and $x > 0$. The integral curves are the hyperbolas shown in Fig. 2.8. An initial condition $y(x_0) = y_0$ taken at a point (x_0, y_0) in each of these four quadrants determines only one integral curve corresponding to the particular solution of the differential equation.

Example 2.9   Solve equation

$$y' = ky,$$

where k is a constant. We already considered this equation in the discussion of "The Malthus Model" in Chap. 1, where it was solved using a different approach.

Fig. 2.9 Illustration to
Example 2.10

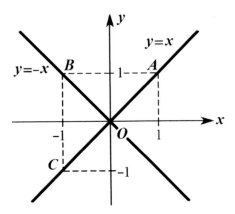

Division by y and multiplication by dx gives $\frac{dy}{y} = k\,dx$. Integrating, $\ln|y| = kx + C$. Next, exponentiation gives $|y| = Ce^{kx}$, $C > 0$, or equivalently $y = Ce^{kx}$, where C is arbitrary and nonzero. By allowing $C = 0$, this formula again includes the trivial solution $y = 0$ which has been lost when the equation was divided by y. Thus, $y(x) = Ce^{kx}$ is the general solution, where C is arbitrary constant.

Since the right-hand side of the equation $y' = ky$ satisfies the conditions of Existence and Uniqueness Theorem everywhere in the plane, there are no limitations on the initial condition (any x_0 and y_0 can be used). Substituting the initial condition $y(x_0) = y_0$ in the general solution, we obtain $y_0 = Ce^{kx_0}$; thus, $C = y_0e^{-kx_0}$. Substituting this value of C into the general solution, we find that the IVP solution takes the form $y = y_0e^{k(x-x_0)}$.

Example 2.10 Solve equation

$$x\left(1 + y^2\right)dx = y\left(1 + x^2\right)dy.$$

Division of both sides by $(1 + y^2)(1 + x^2)$ separates variables. This product obviously is never zero; thus, there is no loss of solutions. Integrating,

$$\int \frac{x\,dx}{x^2 + 1} = \int \frac{y\,dy}{y^2 + 1} + C,$$

and then, $\ln(1 + x^2) = \ln(1 + y^2) + c$. Thus, the general solution is $x^2 + 1 = C(y^2 + 1)$, where $C = e^c > 0$.

Let the initial condition be $y(1) = 1$. Substitution in the general solution gives $C = 1$. Thus, the particular solution satisfying the initial condition is $x^2 = y^2$. Writing it in the form $y = \pm x$, it becomes clear that only the function $y = x$ satisfies the initial condition (see Fig. 2.9).

Next, let us try to apply the initial condition at the origin: $y(0) = 0$. From the general solution, we still get $C = 1$. Now both solutions $y = \pm x$ match the initial

Fig. 2.10 Two families of solutions for Example 2.24

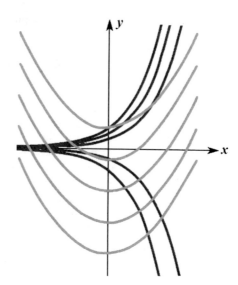

condition. This seems to contradict the uniqueness property stated in the Existence and Uniqueness Theorem. But from the equation in the form $y' = f(x, y)$ with

$$f(x, y) = \frac{x(1 + y^2)}{y(1 + x^2)}$$

it is clear that at $y = 0$, the function $f(x, y)$ has an infinite discontinuity – thus, the condition of the Existence and Uniqueness Theorem is violated, and there is no guarantee of the uniqueness of the particular solution near the origin.

In practice this means that in the case of a nonunique solution, when an integral curve is traced through the discontinuity line, one can slip from this curve to an adjacent one. For instance, let us solve our example equation on the interval $[-1, 1]$ with the initial condition $y(1) = 1$, i.e., start at point A in Fig. 2.10. Moving along the solution $y = x$ toward the origin, at the destination point $(0, 0)$, we have two possibilities for the following motion: to continue along the line $y = x$ toward the final point C$(-1, -1)$ or to continue along the line $y = -x$ toward the final point B $(-1, 1)$. This demonstrates that when a solution is not unique and one of the solution curves is traced, we may slip to an adjacent curve. Problems of this kind are especially difficult for numerical solutions of differential equations, when there is a danger of missing a discontinuity point and not find several branches of a solution. The choice of a proper integral curve in the case of a nonunique solution can be based on a real situation described by the differential equation. If, for instance, it is known that for $x < 0$ the values of a function y, describing a particular process, are positive, then on the interval $[-1, 1]$ the line AOB should be taken for the solution; if it is known that for $x < 0$ the values of a function y are negative, then the line AOC should be taken. Often the choice can be made based on the knowledge of the asymptotes at $x \rightarrow \infty$ or $x \rightarrow -\infty$ of a function modeled by a differential equation. The example discussed above demonstrates the importance of the analysis of the properties of a differential equation before attempting a solution.

To summarize, separable equations can be always solved analytically. Note that some types of equations can be transformed into separable equations using the change of a variable. Two such types of equations are the equations $y' = f(ax + by + c)$ and $y' = f(y/x)$.

1. Consider equation $y' = f(ax + by + c)$, where a, b, and c are constants.

   Let

   $$z = ax + by + c \qquad (2.14)$$

   be a new variable. Differentiating z with respect to x, we obtain $\frac{dz}{dx} = a + b\frac{dy}{dx}$. Notice that $\frac{dy}{dx} = f(z)$ (Eq. 2.14); thus,

   $$\frac{dz}{dx} = a + bf(z). \qquad (2.15)$$

   Now variables separate:

   $$\frac{dz}{a + bf(z)} = dx.$$

   Integration gives:

   $$x = \int \frac{dz}{a + bf(z)} + C.$$

   After the integration at the right-hand side is completed, replacing the variable z by $ax + by + c$ gives the general solution of (2.14).

Example 2.11   Solve $\frac{dy}{dx} = 2x - 3y - 5$.

Using a new variable $z = 2x - 3y - 5$, we find $z' = 2 - 3y' = 2 - 3z$. Separating variables, $\frac{dz}{2-3z} = dx$. Integration gives:

$$x + c = -\frac{1}{3}\ln|2 - 3z|, \text{ or } |2 - 3z| = e^{-3x+c}$$

This can be written as $2 - 3z = Ce^{-3x}$ with C either positive or negative. Substituting $z = 2x - 3y - 5$ gives $2 - 6x + 9y + 15 = Ce^{-3x}$. Thus, the general solution is

$$y = Ce^{-3x} + 2x/3 - 17/9.$$

It is easy to check that for $C = 0$, the function $y = 2x/3 - 17/9$ also satisfies the given equation; thus, C in the general solution is an arbitrary constant.

Example 2.12 Solve $y' = \frac{2}{x-y} - 3$.

Let $z = x - y$ be the new variable. Differentiation gives $z' = 1 - y'$. Since $y' = \frac{2}{z} - 3$, the equation for the function z is $1 - z' = \frac{2}{z} - 3$.

Separation of variables gives $dx = \frac{zdz}{4z-2}$. The right-hand side can be algebraically transformed: $\frac{zdz}{4z-2} = \frac{1}{4}\frac{4z-2+2}{4z-2}dz = \frac{1}{4}\left(1 + \frac{2}{4z-2}\right)dz$. Integration of the latter equation gives $x + c = \frac{1}{4}z + \frac{1}{8}\ln|4z-2|$. Multiplying by 8, and denoting $8c = C$, finally gives the general solution in the implicit form:

$$8x + C = 2(x-y) + \ln|4x-4y-2|.$$

2. Consider equation:

$$y' = f\left(\frac{y}{x}\right), \qquad (2.16)$$

in which a function $f(x, y)$ contains variables x and y in the combination y/x. Note that an equation $y' = f(x, y)$ with a function $f(x, y)$ satisfying the property $f(cx, cy) = f(x, y)$ can be transformed to the form (2.16).

Let

$$z = \frac{y}{x} \qquad (2.17)$$

be the new variable. Then, $y = zx$. Differentiating this in x, one obtains $y' = z + xz'$, or

$$\frac{dy}{dx} = z + x\frac{dz}{dx}.$$

Thus

$$f(z) = z + x\frac{dz}{dx}. \qquad (2.18)$$

Separating variables $\frac{dx}{x} = \frac{dz}{f(z)-z}$, and then integration gives:

$$\ln|x| + c = \int \frac{dz}{f(z) - z},$$

or

$$x = C \exp\left[\int \frac{dz}{f(z) - z}\right]$$

with arbitrary C. Finally, replacement of z by y/x in the solution results in the general solution of (2.16).

Example 2.13 Solve equation $(x^2 + y^2)dx - 2xydy = 0$.

Division by $2xy\, dx$ gives $\frac{dy}{dx} = (x^2 + y^2)/2xy$. Function $f(x, y) = (x^2 + y^2)/2xy$ does not change if both x and y are multiplied by a constant. With the new variable $z = \frac{y}{x}$, we have $y' = z + xz'$. Equating this to $f(z) = x^2(1 + z^2)/2x^2z = (1 + z^2)/2z$ gives $z + xz' = \frac{1+z^2}{2z}$. Next, separate variables, $\frac{dx}{x} = \frac{2zdz}{1 - z^2}$, and integrate. This gives $|x| = \frac{c}{|z^2 - 1|}$, $c > 0$. Thus, $x = \frac{C}{1 - z^2}$, where C is arbitrary constant. Substitution $z = \frac{y}{x}$ gives $x = \frac{Cx^2}{x^2 - y^2}$. Finally, $y^2 = x^2 - Cx$ is the general solution.

When we divided the original equation by $2xy$, we lost the solution $x(y) = 0$. Thus, the solution $x = 0$ complements the general solution and together they form the solution set.

Example 2.14 Solve equation $y' = y/x + \tan(y/x)$.

First, introduce new variable z: $y = xz$, $y' = z + xz'$. Then, the equation becomes $x\frac{dz}{dx} + z = z + \tan z$. Next, separate variables $\frac{dx}{x} = \frac{\cos z}{\sin z}dz$, and integrate. This gives $\ln|\sin z| = \ln |x| + \ln c$, $c > 0$. Thus, $\sin z = Cx$, or $\sin(y/x) = Cx$, where C is arbitrary constant.

Problems

1. $2y'y = 1$.
2. $(x + 1)^3 dy - (y - 2)^2 dx = 0$.
3. $e^{x + y} - 3e^{-x - 2y}y' = 0$.
4. $y'\left(\sqrt{x} + \sqrt{xy}\right) = y$.
5. $y'x^2 + y^2 = 0$.
6. $y'y + x = 1$.
7. $y'x^2y^2 - y + 1 = 0$.
8. $(x^2 + 1)y^3 dx = (y^2 - 1)x^3 dy$.
9. $xydx + (x + 1)dy = 0$.
10. $x\frac{dx}{dt} + t = 1$.
11. $\sqrt{1 + y^2}dx = xydy$.
12. $(x^2 - 1)y' + 2xy^2 = 0$.
13. $y' = -\frac{xy}{x+2}$.
14. $\sqrt{y^2 + 5}dx = 2xydy$.
15. $2x^2yy' + y^2 = 5$.
16. $xy' + y = y^2y' - 2xy^2 = 3xy$.
17. $xy' + y = y^2$.
18. $2x\sqrt{1 + y^2}dx + ydy = 0$.
19. $y' = \frac{\sqrt{y}}{x}$.
20. $\sqrt{1 - y^2}dx + \sqrt{1 - x^2}dy = 0$.
21. $(2x + y + 1)dx - (4x + 2y - 3)dy = 0$.
22. $x - y - 1 - (y - x + 2)y' = 0$.

23. $y' = (y + x + 3)(y + x + 1)$.
25. $y' = (y - 4x + 3)^2$.
27. $y' = \frac{1}{x+2y}$.
29. $y' = \frac{x+3y}{2x+6y-5}$.
31. $xy' = y + \sqrt{x^2 + y^2}$.
33. $xy' - y = (x + y) \ln \frac{x+y}{x}$.
35. $y\,dx + \left(2\sqrt{xy} - x\right)dy = 0$.
37. $(y^2 - 2xy)dx + x^2 dy = 0$.

24. $y' = \sqrt{2x - y + 3} + 2$.
26. $y' - y = 3x - 5$.
28. $y' = \sqrt{4x + 2y - 1}$.
30. $(x + 2y)dx - xdy = 0$.
32. $xy' = y - xe^{y/x}$.
34. $xdy = \left(\sqrt{xy} + y\right)dx$.
36. $2x^3 dy - y(2x^2 - y^2)dx = 0$.
38. $y^2 + x^2 y' = xyy'$.

In the problems of this set, the reader can choose the initial condition and find the particular solution.

Next, the reader can run the program *ODE 1st order* and compare the particular solution to the program output.

2.4 Linear First-Order Differential Equations

Equation that is linear in y and y'

$$y' + p(x)y = f(x) \qquad (2.19)$$

is called a *linear first-order differential equation*. We assume that $p(x)$ and $f(x)$ are continuous in a domain where the solution is sought. The coefficient of y' in (2.19) is taken equal to one simply for convenience – if it is not equal to one, then the equation can be divided by this coefficient prior to the solution, which would cast the equation in the form (2.19). Linearity of the expression $y' + p(x)y$ in y and y' means that when $y(x)$ is replaced by $C_1 y_1(x) + C_2 y_2(x)$, the left-hand side of Eq. (2.19) transforms into $C_1 (y_1' + py_1) + C_2 (y_2' + py_2)$. For instance, equations $y'y + 2y = x$ and $y' - \sin y = 1$ are nonlinear: the first equation due to the term $y'y$ and the second one – due to the term $\sin y$. An example of a linear equation may be $x^2 y' + e^x y = 5 \sin x$, and the assumption that $p(x)$ and $f(x)$ are continuous in a domain where the solution is sought means that we may consider either the domain $x < 0$ or $x > 0$.

If a function $f(x)$ is not identically zero, the first-order Eq. (2.19) is called a *linear nonhomogeneous equation*.

2.4.1 Homogeneous Linear Equations

When $f(x) \equiv 0$, equation

$$y' + p(x)y = 0 \tag{2.20}$$

is called a *linear homogeneous equation*. In such equation variables can be separated:

$$\frac{dy}{y} = -p(x)dx.$$

Integration gives

$$\ln|y| = -\int p(x)dx + c;$$

hence, $|y| = e^c e^{-\int p(x)dx}$, where c is arbitrary number.

An additional solution lost by the transformation of the original equation is $y(x) = 0$. Taking into account the absolute value, we obtain the *general solution of the linear homogeneous equation*:

$$y(x) = Ce^{-\int p(x)dx}, \tag{2.21}$$

where C is arbitrary. (Note that here and below, the notation $\int p(x)dx$ means a definite function, that is, an arbitrary constant of integration is omitted.)

Example 2.15 Solve $y' + 5y = 0$.

Separating variables, $dy/y = -5dx$, and integration gives $\ln|x| = -5x + c$; hence, $|y| = e^c e^{-5x}$, where c is arbitrary. While e^c in the expression for $|y|$ is positive, for y we obtain $y = Ce^{-5x}$ with $C \equiv \pm e^c$ of arbitrary sign. Also, $y = 0$ is a solution of the equation. Thus, the general solution of this homogeneous equation is $y = Ce^{-5x}$ with arbitrary C.

2.4.2   Nonhomogeneous Linear Equations: Method of a Parameter Variation

Recall that our goal is to determine the *general* solution of the *nonhomogeneous* Eq. (2.19). Now that the homogeneous equation has been solved, this can be accomplished by using the *method of a parameter variation*. The idea of the method is that the solution of (2.19) is sought in the form (2.21), *where the constant C is replaced by a function u(x)*:

$$y(x) = u(x)e^{-\int p(x)dx}. \tag{2.22}$$

To find $u(x)$, we substitute (2.22) in (2.19). First, find the derivative y':

$$y' = u'(x)e^{-\int p(x)dx} - u(x)p(x)e^{-\int p(x)dx},$$

and substitute y' and y (see (2.22)) in (2.19):

$$y' + p(x)y = u'(x)e^{-\int p(x)dx} - u(x)p(x)e^{-\int p(x)dx} + u(x)p(x)e^{-\int p(x)dx} = f(x).$$

Terms with $u(x)$ cancel, and we obtain:

$$u'(x) = f(x)e^{\int p(x)dx}.$$

Integrating, we find:

$$u(x) = \int f(x)e^{\int p(x)dx}dx + C.$$

Substituting this $u(x)$ in (2.22), we finally obtain *the general solution of* (2.19):

$$y(x) = u(x)e^{-\int p(x)dx} = Ce^{-\int p(x)dx} + e^{-\int p(x)dx}\int f(x)e^{\int p(x)dx}dx. \tag{2.23}$$

Notice that the first term at the right-hand side is the general solution (2.21) of the homogeneous equation.

The second term in (2.23) is the *particular solution* of the nonhomogeneous Eq. (2.19) (this can be checked by substituting the second term in (2.23) for y in (2.19), and substituting its derivative for y'; also note that the particular solution does not involve an arbitrary constant).

To summarize, *the general solution of the nonhomogeneous equation is the sum of the general solution of the homogeneous equation and the particular solution of the nonhomogeneous equation.* In this textbook we denote these solutions by $Y(x)$ and $\bar{y}(x)$, respectively. Thus, the general solution of (2.19) is written as

$$y(x) = Y(x) + \bar{y}(x). \tag{2.24}$$

When solving problems, the complicated formula (2.23) usually is not used. Instead, all steps of the solution scheme are applied as described above.

Example 2.16 Solve $y' + 5y = e^x$.

In Example 2.15 we found that the general solution of the corresponding homogeneous equation $y' + 5y = 0$ is $y = Ce^{-5x}$.

Next, we proceed to find the general solution of the nonhomogeneous equation using the variation of the parameter:

$$y = u(x)e^{-5x}.$$

The derivative is $y' = u'(x)e^{-5x} - 5u(x)e^{-5x}$. Substitution of y and y' in $y' + 5y = e^x$ gives

$$u'(x)e^{-5x} + 5u(x)e^{-5x} - 5u(x)e^{-5x} = e^x,$$

or $u'(x) = e^{6x}$. Integration results in

$$u(x) = e^{6x}/6 + C.$$

The general solution of the nonhomogeneous equation is

$$y(x) = u(x)e^{-5x} = Ce^{-5x} + e^x/6.$$

It is easy to check that $e^x/6$ satisfies the given nonhomogeneous equation.

Example 2.17 Solve $y' - \frac{y}{x} = x$.

First, solve the homogeneous equation:

$$y' - y/x = 0.$$

Separating variables, we obtain $\frac{dy}{y} = \frac{dx}{x}$. Integration gives $\ln|y| = \ln|x| + c$. Using the arguments presented in the previous example, we come to conclusion that the general solution of the homogeneous equation is $y = Cx$, where C is arbitrary.

The general solution of the original nonhomogeneous equation is sought in the form:

$$y = u(x)x.$$

Substitution of y and its derivative $y' = u'(x)x + u(x)$ into $y' - y/x = x$ gives $u'(x)x = x$, and then $u(x) = x + C$. Finally, the general solution of the problem is $y = Cx + x^2$.

2.4.3   *Nonhomogeneous Linear Equations: Method of Integrating Factor*

Another widely used method for the solution of (2.19) is based on using an *integrating factor*. The integrating factor for a linear differential equation is the function:

$$\mu(x) = e^{\int p(x)dx}.$$ (2.25)

When (2.19) is multiplied by $\mu(x)$, the equation becomes

$$\frac{d}{dx}[\mu(x)y] = \mu(x)f(x).$$ (2.26)

(You can check that the left-hand side in (2.26) equals $\mu y' + \mu p y$.)
Multiplication of (2.26) by dx, integration and division by $\mu(x)$ gives the solution:

$$y(x) = \frac{1}{\mu(x)}\left[\int \mu(x)f(x)dx + C\right].$$ (2.27)

Substitution of $\mu(x)$ from (2.25) shows that (2.27) coincides with (2.23), which was obtained by the method of variation of the parameter.

Example 2.18   Solve equation $y' + 5y = e^x$ from Example 2.16 using the integrating factor.

Here $\mu(x) = e^{\int 5dx} = e^{5x}$ (the integration constant in $\int 5dx = 5x + c$ is omitted because it is sufficient to find one integrating factor), and the solution is

$$y(x) = \frac{1}{e^{5x}}\left[\int e^{5x}e^x dx + C\right] = \frac{1}{e^{5x}}\left[\frac{1}{6}e^{6x} + C\right] = Ce^{-5x} + \frac{e^x}{6}.$$

2.4.4 Nonlinear Equations That Can Be Transformed into Linear Equations

(a) *Equations with Hidden Linearity*

Some equations that are not linear for the function $y(x)$ can be transformed into linear equations if we consider them as equations for a function $x(y)$. Such transformation is based on the relationship between the derivatives: $x'(y) = dx/dy = 1/y'(x)$.

Example 2.19 Solve $(2e^y - x)y' = 1$.

Let us write this equation in the form $2e^y - x = \frac{dx}{dy}$, which is the linear nonhomogeneous equation for $x(y)$. The solution of the homogeneous equation $dx/dy + x = 0$ is $x = Ce^{-y}$. Then, seek the general solution of the nonhomogeneous equation in the form $x = u(y)e^{-y}$. Substitution in $2e^y - x = x'$ gives $u'(y) = 2e^{2y}$, and the integration gives $u(y) = e^{2y} + C$. Thus, the general solution of the nonhomogeneous equation is $x = Ce^{-y} + e^y$.

(b) The *Bernoulli equation* (Jacob Bernoulli (1654–1705)) is

$$y' + p(x)y = f(x)y^n, \qquad (2.28)$$

where n is arbitrary real number (for $n = 0, 1$ this equation is linear). Let us show how for $n \neq 0, 1$ the Bernoulli equation can be transformed into a linear equation.
First, we divide (2.28) by y^n:

$$y'y^{-n} + p(x)y^{1-n} = f(x). \qquad (2.29)$$

Let

$$z(x) = y^{1-n}, \qquad (2.30)$$

and find the derivative $z'(x)$ by the chain rule:

$$z'(x) = (1-n)y^{-n}y'.$$

Substitution of $z(x)$ and $z'(x)$ in (2.29) gives the linear nonhomogeneous equation for $z(x)$:

$$\frac{1}{1-n}\frac{dz}{dx} + p(x)z = f(x).$$

The general solution of the latter equation can be found by the methods just described above, and then $y(x)$ is determined from (2.30).

Notice that for $n \geq 1$, Eq. (2.28) also has the solution $y(x) = 0$ which was lost when we divided (2.28) by y^n.

Example 2.20 Solve $y' - y = \frac{3}{y}$.

This is the Bernoulli equation with $n = -1$. Division by y^{-1} (i.e., the multiplication by y) gives $yy' - y^2 = 3$. Let $z = y^2$, then $z' = 2yy'$ and the equation becomes $\frac{1}{2}z' - z = 3$. First, we solve the homogeneous equation $\frac{1}{2}z' - z = 0$. Its general solution is $z = Ce^{2x}$. Using variation of the parameter, we seek the solution of the inhomogeneous equation in the form $z = u(x)e^{2x}$, which gives $u'(x) = 6e^{-2x}$. Integration gives $u(x) = -3e^{-2x} + C$; thus, the nonhomogeneous equation has the general solution:
$z = Ce^{2x} - 3$. Returning to the variable $y(x)$, the solution of the original equation is obtained: $y^2 = Ce^{2x} - 3$.

(c) The *Riccati equation* (Jacopo Francesco Riccati (1676–1754), a Venetian mathematician) is

$$y' + p(x)y + q(x)y^2 = f(x). \tag{2.31}$$

This equation cannot be solved analytically in the general case, but it can be transformed into the Bernoulli equation if one particular solution, $y_1(x)$, of Eq. (2.31) is known. In this case let $y = y_1 + w$; then substitute y in (2.31):

$$y_1' + w' + p(x)(y_1 + w) + q(x)(y_1 + w)^2 = f(x).$$

Next, because $y_1' + p(x)y_1 + q(x)y_1^2 = f(x)$, for function $z(x)$, we obtain the Bernoulli equation:

$$w' + [p(x) + 2q(x)y_1]w + q(x)w^2 = 0.$$

Adding its general solution to the function $y_1(x)$ gives a general solution of Eq. (2.31).

Example 2.21 Solve $\frac{dy}{dx} = y^2 - \frac{2}{x^2}$.

A particular solution of this equation is easy to guess: $y_1 = \frac{1}{x}$. Then, $y = w + \frac{1}{x}$ and $y' = w' - \frac{1}{x^2}$. Substituting y and y' into equation, we obtain $w' - \frac{1}{x^2} = \left(w + \frac{1}{x}\right)^2 - \frac{2}{x^2}$, or $w' = w^2 + 2\frac{w}{x}$. To solve this Bernoulli equation, make the substitution $z = w^{-1}$, which yields the linear equation $\frac{dz}{dx} = -\frac{2z}{x} - 1$.

The general solution of the corresponding homogeneous equation is $\ln|z| = -2\ln|x| + c$; thus, $z = \frac{C}{x^2}$ (here C is arbitrary).

Next, we seek the general solution of the nonhomogeneous equation in the form $z = \frac{u(x)}{x^2}$. This gives $\frac{u'(x)}{x^2} = -1$; then, $u(x) = -\frac{x^3}{3} + C$ and $z = \frac{C}{x^2} - \frac{x}{3}$. Since $z = w^{-1}$, we get $\frac{1}{w} = \frac{C}{x^2} - \frac{x}{3}$, then $\frac{1}{y - 1/x} = \frac{C}{x^2} - \frac{x}{3}$, and the general solution of the given equation is (with $3C$ replaced by C):

$$y = \frac{1}{x} + \frac{3x^2}{C - x^3}.$$

Problems

1. $xy' + x^2 + xy - y = 0.$ 2. $(2x + 1)y' = 4x + 2y.$

3. $x^2y' + xy + 1 = 0.$ 4. $y' \cos x + y \sin x - 1 = 0.$

5. $y' = \frac{2y + \ln x}{x \ln x}.$ 6. $y' - y = xe^x.$

7. $y' - 2y = x.$ 8. $y' + 2xy = xe^{-x^2}.$

9. $y' + (2y + 1)\operatorname{ctg} x = 0.$ 10. $xy' = y + 2x^3.$

11. $y' + \frac{y}{x} - 2e^{x^2} = 0.$ 12. $y' + y \tan x = \frac{1}{\cos x}.$

13. $y' - y \tan x = \sin x.$ 14. $y' = 2x(x^2 + y).$

15. $xy' + (x + 1)y = 3x^2e^{-x}.$ 16. $y' + \frac{3}{x}y = \frac{2}{x^2}.$

17. $(x + y^2)dy = ydx.$ 18. $y'(y^3 + 2x) = y.$

19. $y' + 2y = y^2e^x.$ 20. $xy' + 2y + x^5y^3e^x = 0.$

21. $xy^2y' = x^2 + y^3.$ 22. $xy' + 2x^2\sqrt{y} = 4y.$

23. $xydy = (y^2 + x)dx.$ 24. $y' = \frac{y}{-2x + y - 4\ln y}.$

25. $(3x - y^2)dy = ydx.$ 26. $(1 - 2xy)y' = y(y - 1).$

27. $2y' - \frac{x}{y} = \frac{xy}{x^2 - 1}.$ 28. $y'(2x^2y \ln y - x) = y.$

29. $y' = \frac{x}{x^2 - 2y + 1}.$ 30. $x^2y' + xy + x^2y^2 = 4.$

31. $3y' + y^2 + \frac{2}{x^2} = 0.$ 32. $xy' - (2x + 1)y + y^2 = -x^2.$

33. $y' - 2xy + y^2 = 5 - x^2.$ 34. $y' + 2ye^x - y^2 = e^{2x} + e^x.$

35. $y' = \frac{1}{2}y^2 + \frac{1}{2x^2}.$

2.5 Exact Equations

A first-order equation $y' = f(x, y)$ always can be written as

$$P(x, y)dx + Q(x, y)dy = 0, \tag{2.32}$$

where $P(x, y)$ and $Q(x, y)$ are some functions. If the left-hand side of (2.32) is a total differential of some function $F(x, y)$,

$$dF(x, y) = P(x, y)dx + Q(x, y)dy, \qquad (2.33)$$

then Eq. (2.32) is written as

$$dF(x, y) = 0. \qquad (2.34)$$

Such equations are called *exact*. Solution of (2.34) is

$$F(x, y) = c. \qquad (2.35)$$

Recall from calculus that for $P(x, y)dx + Q(x, y)dy$ to be a total differential, it is necessary and sufficient that

$$\frac{\partial P}{\partial y} = \frac{\partial Q}{\partial x}. \qquad (2.36)$$

Now write $dF(x, y)$ as

$$dF(x, y) = \frac{\partial F}{\partial x}dx + \frac{\partial F}{\partial y}dy. \qquad (2.37)$$

Comparison of (2.37) and (2.33) gives (since dx and dy are arbitrary)

$$\frac{\partial F(x, y)}{\partial x} = P(x, y), \qquad (2.38)$$

$$\frac{\partial F(x, y)}{\partial y} = Q(x, y). \qquad (2.39)$$

Next, integrate (2.38) in x to obtain:

$$F(x, y) = \int P(x, y)dx + C(y), \qquad (2.40)$$

where $C(y)$ is an arbitrary function of y.
Substitution of (2.40) in (2.39) gives

$$\frac{\partial}{\partial y}\left(\int P(x, y)dx \right) + C'(y) = Q(x, y),$$

and thus $C'(y) = Q(x, y) - \frac{\partial}{\partial y} \int P(x, y)dx$. From the last equation, $C(y)$ is determined by integration:

$$C(y) = \int \left[Q(x,y) - \frac{\partial}{\partial x} \left(\int P(x,y)dx \right) \right] dy + C_1.$$

Substitution of this $C(y)$ into (2.40) gives $F(x,y)$. Next, equating this $F(x,y)$ to a constant (see (2.35)) gives the final solution of (2.32). Note that the sum of c and C_1 should be combined into a single constant.

For the method of variation of the parameter, in practice it is better not to use the derived formulas for the solution. Instead, the solution scheme as described should be applied to each equation to be solved.

Example 2.22 Equation $(3x^2y^2 + 7)dx + 2x^3ydy = 0$ is exact because $\frac{d}{dy}(3x^2y^2 + 7) = \frac{d}{dx}(2x^3y) = 6x^2y$. This equation is equivalent to the following equation: $dF(x,y) = 0$, which has the solution $F(x,y) = c$.

From $dF(x,y) = \frac{\partial F}{\partial x}dx + \frac{\partial F}{\partial y}dy$, it follows that $\frac{\partial F(x,y)}{\partial x} = 3x^2y^2 + 7$; thus, $F(x,y) = \int (3x^2y^2 + 7)dx + C(y)$. Since y is considered constant in the integrand, $F(x,y) = x^3y^2 + 7x + C(y)$.

Substitution of $F(x,y)$ in $\frac{\partial F(x,y)}{\partial y} = 2x^3y$ gives $2x^3y + C'(y) = 2x^3y$, from which $C'(y) = 0$. Thus, $C(y) = C_1$, and now $F(x,y) = x^3y^2 + 7x + C_1$. Equating $F(x,y)$ to c gives the (implicit) solution of the problem:

$$x^3y^2 + 7x = C.$$

When the left-hand side of (2.32) is not a total differential (and thus the equation is not exact), in some cases it is possible to find an auxiliary function $\mu(x,y)$ that can serve as *the integrating factor*. When (2.32) is multiplied by such function, its left-hand side becomes a total differential. That is,

$$dF(x,y) = \mu(x,y)P(x,y)dx + \mu(x,y)Q(x,y)dy.$$

Unfortunately, there is no universally applicable method for the determination of the integrating factor.

As *Reading Exercise*, show that the linear nonhomogeneous equation

$$y' + p(x)y = f(x),$$

or $[p(x)y - f(x)]dx + dy = 0$, is transformed into an exact equation when it is multiplied by $\mu = e^{\int p(x)dx}$. It must be taken into consideration that the multiplication by $\mu(x,y)$ can lead to extraneous solutions that must be excluded from the final solution. When such extraneous solutions exist, they are simultaneously the solutions of equation $\mu(x,y) = 0$.

Example 2.23 Equation $[x + x^2(x^2 + y^2)]dx + ydy = 0$ is not exact, but after it is multiplied by $\mu = 1/(x^2 + y^2)$, it takes the form:

$$\frac{xdx + ydy}{x^2 + y^2} + x^2 dx = 0,$$

where the left-hand side is a total differential:

$$dF(x, y) = d\left[\frac{1}{2}\ln(x^2 + y^2) + \frac{x^3}{3}\right].$$

Thus, now we are dealing with the equation $dF(x, y) = 0$, which has the solution $F(x, y) = \ln C$, where for convenience we adopt the integration constant as $\ln C$. Multiplying $F(x, y) = \ln C$ by 2 and choosing to write the arbitrary constant $2 \ln C$ as $\ln C$, we obtain

$$\ln(x^2 + y^2) + \frac{2x^3}{3} = \ln C,$$

and then

$$(x^2 + y^2)e^{2x^3/3} = C \quad (C > 0).$$

Note that $\mu = 1/(x^2 + y^2) \neq 0$; thus, there is no extraneous solution.

Remark Eq. (2.38)

$$\frac{\partial F(x, y)}{\partial x} = P(x, y),$$

where $P(x, y)$ is given and $F(x, y)$ is the unknown function to be determined, is the example of the first-order *partial differential equation* (PDE). Solution of this equation

$$F(x, y) = \int P(x, y)dx + C(y),$$

contains $C(y)$ – an arbitrary function of y. The general solution of a second-order PDE would contain two arbitrary functions, and so on – that is, the number of arbitrary functions in the general solution is n, where n is the order of the equation.

Problems

1. $2xydx + (x^2 - y^2)dy = 0$.

3. $3x^2(1 + \ln y)dx = (2y - x^3/y)dy$.

5. $e^{-y}dx - (2y + xe^{-y})dy = 0$.

2. $(2 - 9xy^2)xdx + (4y^2 - 6x^3)ydy = 0$.

4. $2x\left(1 + \sqrt{x^2 - y}\right)dx - \sqrt{x^2 - y}\,ydy = 0$.

6. $(y \cos x - x^2)dx + (\sin x + y)dy = 0$.

7. $(e^y + 2xy)dx + (e^y + x)xdy = 0.$

8. $(1 + y^2 \sin 2x)dx - 2y\cos^2 xdy = 0.$

9. $\frac{y}{x}dx + (y^3 + \ln x)dy = 0.$

10. $(2x - y + 1)dx + (2y - x - 1)dy = 0.$

11. $(3x^2 - 3y^2 + 4x)dx - (6xy + 4y)dy = 0.$

12. $\dfrac{2x(1 - e^y)dx}{(1 + x^2)^2} + \dfrac{e^y dy}{1 + x^2} = 0.$

13. $\left(\dfrac{\sin 2x}{y} + x\right)dx + \left(y - \dfrac{\sin^2 x}{y^2}\right)dy = 0.$

14. $\left(1 + e^{x/y}\right)dx + e^{x/y}\left(1 - \dfrac{x}{y}\right)dy = 0.$

15. $(6x^2 - y + 3)dx + (3y^2 - x - 2)dy = 0.$

16. $e^y dx + (2y + xe^y)dy = 0.$

17. $(2xy - \sin x)dx + (x^2 - \cos y)dy = 0.$

18. $\left(1 + 2x\sqrt{x^2 - y^2}\right)dx + 2y\sqrt{x^2 - y^2}dy = 0.$

2.6 Equations Unresolved with Respect to a Derivative

2.6.1 Regular and Irregular Solutions

Consider a first-order equation in its general form:

$$F(x, y, y') = 0. \tag{2.41}$$

If this equation can be solved for y', we obtain one or several equations:

$$y'_i = f_i(x, y), \quad i = \overline{1, m}. \tag{2.42}$$

Integrating these equations, we obtain the solution of Eq. (2.41). The complication comparing to the Eq. (2.41) is that Eq. (2.42) determines several direction fields.

The Cauchy problem for Eq. (2.41) is formulated in the same way as for equation $y' = f(x, y)$: find a particular solution of Eq. (2.41) satisfying the initial condition:

$$y(x_0) = y_0.$$

If the number of solutions of this problem is the same as the number of functions, i.e., m, one says that the Cauchy problem has a unique solution. In other words, not more than one integral curve of Eq. (2.41) passes through the point (x_0, y_0) in the given direction (determined by each of the values $f_i(x_0, y_0)$).

Such solutions are called *regular*. If at each point of the solution the uniqueness is not valid, such a solution is called *irregular*.

Consider equations quadratic with respect to y':

$$(y')^2 + 2P(x, y)y' + Q(x, y) = 0. \tag{2.43}$$

Solve (2.43) for y':

$$y' = -P(x,y) \pm \sqrt{P^2(x,y) - Q(x,y)}. \tag{2.44}$$

This expression is defined for $P^2 - Q \geq 0$. Integrating (2.44) we find a general integral of Eq. (2.43).

An irregular solution could be only the curve

$$P^2(x,y) - Q(x,y) = 0,$$

which also should satisfy the system of equations for y':

$$\begin{cases} (y')^2 + 2P(x,y)y' + Q(x,y) = 0, \\ y' + P(x,y) = 0. \end{cases} \tag{2.45}$$

Example 2.24 Find a general solution of equation

$$(y')^2 - 2(x+y)y' + 4xy = 0.$$

Solving this quadratic equation for y', we obtain $y' = (x+y) \pm (x-y)$; thus, $y' = 2x$ and $y' = 2y$. Integrating each of these equations, we find

$$y = x^2 + C, \text{ and } y = Ce^{2x}.$$

Both families of solutions (the general solutions) satisfy the initial equation. Obviously, these solutions are regular. A possible irregular solution could arise if $(x-y) = 0$, but $y = x$ does not satisfy the given differential equation.

Example 2.25 Find a general solution of equation

$$(y')^2 - 4x^2 = 0,$$

and then find the integral curves passing through the points (a) $M(1,1)$, (b) $O(0,0)$.

From the equation one finds $y' = 2x$, $y' = -2x$. Note that these equations are incompatible except at the points with $x = 0$, where the integral curves of each equation are tangent to each other. Integration gives

$$y = x^2 + C, y = -x^2 + C.$$

These general solutions are two families of parabolas and there are no irregular solutions.

Next, we solve two Cauchy problems.

(a) First, let us impose the initial condition at a point with $x \neq 0$, where the slopes of the integral curves corresponding to two families of solutions are different. For instance, substituting the initial condition $x_0 = 1$, $y_0 = 1$ into the general solution $y = x^2 + C$, we obtain $C = 0$; thus, $y = x^2$; substituting the initial condition into $y = -x^2 + C$, we obtain $C = -2$; thus, $y = -x^2 + 2$. Therefore, *two* integral curves, $y = x^2$ and $y = -x^2 + 2$, pass through point $M(1, 1)$.

(b) Impose now the initial condition at $x = 0$, where the slopes of both integral curves coincide, $y' = 0$. For instance, substitution of the initial condition $x_0 = 0$, $y_0 = 0$ into the general solutions gives $y = x^2$ and $y = -x^2$.

Besides, the solutions are

$$y = \begin{cases} x^2, & x \leq 0, \\ -x^2, & x \geq 0 \end{cases} \quad \text{and} \quad y = \begin{cases} -x^2, & x \leq 0, \\ x^2, & x \geq 0. \end{cases}$$

The uniqueness of the solution at the point $O(0,0)$ is violated because the direction fields of both families of solutions at this point are the same: $y'(0) = 0$. At the point $x = 0$, the solution can "switch" from one family to another one.

Example 2.26 Find a general solution of equation

$$e^{y'} + y' = x.$$

This is the equation of the type $F(x, y') = 0$, but it is resolved with respect of x. In such cases it is useful to introduce a parameter in the following way: $y' = t$. For this example, we obtain $x = e^t + t$. As the result, we have a parametric representation of the given equation in the form

$$x = e^t + t, \, y' = t.$$

Next,

$$dy = y' dx = t(e^t + 1) dt, \quad y = \int t(e^t + 1) dt + C = e^t(t - 1) + t^2/2 + C.$$

Thus, the general solution in parametric form is

$$x = e^t + t, \quad y = e^t(t - 1) + t^2/2 + C.$$

Fig. 2.11 Solution curves for Example 2.27

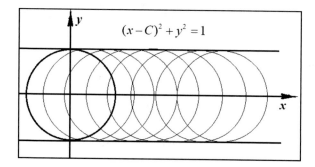

$$(x-C)^2 + y^2 = 1$$

Example 2.27 Find a general solution of equation

$$y^2\left[(y')^2 + 1\right] = 1.$$

The standard way of solution gives

$$\frac{dy}{dx} = \pm \frac{\sqrt{1-y^2}}{y}.$$

The next step is to write:

$$\frac{dx}{dy} = \pm \frac{y}{\sqrt{1-y^2}}, \quad y^2 \neq 1. \tag{2.46}$$

Integration gives a family of circles:

$$x = \mp \sqrt{1-y^2} + C \text{ or } (x - C)^2 + y^2 = 1.$$

The irregular solution $y = \pm 1$ corresponds to the envelopes of that family (see Fig. 2.11).

2.6.2 Lagrange's Equation

Let us consider equation

$$y = f(y')x + g(y'), \tag{2.47}$$

which is called *Lagrange's equation* (Joseph-Louis Lagrange (1736–1813), one of the greatest mathematicians in history).

As in Example 2.26, denote $t(x) + y'(x)$, hence

$$y = f(t)x + g(t).\tag{2.48}$$

Differentiating (2.48) with respect to x, we obtain the equation:

$$t = f'(t)t'(x)x + f(t) + g'(t)t'(x)\tag{2.49}$$

that does not contain y. Next, rewrite (2.49) as

$$(f'(t)x + g'(t))t'(x) = t - f(t).\tag{2.50}$$

First, consider the general case, where $f(t)$ is not *identical* to t (but $f(t)$ can be *equal* to t at some values of t). Using the inverse function $x(t)$ with

$$x'(t) = \frac{1}{t'(x)},$$

we obtain the *linear equation*:

$$x'(t) = \frac{f'(t)x + g'(t)}{t - f(t)}.\tag{2.51}$$

Solving Eq. (2.51) by means of one of the approaches described in Sect. 2.4, we obtain the family of solutions:

$$x = x(t; C).\tag{2.52}$$

Equations (2.48) and (2.52) describe the solution in the parametric form.

Assume now that at a certain point $t = t_0$, $f(t_0) = t_0$. Solving equation $y'(x)$ and using (2.47), we obtain a *special solution*:

$$y = t_0 x + g(t_0).$$

2.6.3 Clairaut's Equation

Consider now the case, where $f(t) \equiv t$ identically, i.e.,

$$y = tx + g(t),\tag{2.53}$$

Fig. 2.12 Solutions of
Clairaut's equation

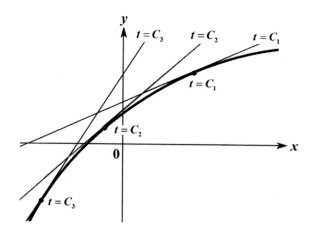

where $t(x) = y'(x)$. Equation (2.53) is *Clairaut's equation* (Alexis Claude Clairaut (1713–1765), a French mathematician, astronomer, and geophysicist). Equation (2.50) becomes:

$$(x + g'(t))t' = 0.$$

Thus, either $t' = 0$ or $x + g'(t) = 0$.
Equation $t' = 0$ gives a family of linear functions:

$$y = Cx + g(C), \tag{2.54}$$

where C is an arbitrary real number (in the domain of a function $g(C)$).
    Another solution originates from the equation

$$x = -g'(t). \tag{2.55}$$

Substituting (2.55) into (2.53), we find

$$y = -tg'(t) + g(t). \tag{2.56}$$

Equations (2.55) and (2.56) describe that special solution in a parametric way.
    Note that for any particular value of $t = t_0$, the point (x, y), according to (2.53), satisfies the relation $y = t_0 x + g(t_0)$, i.e., that point belongs to the straight line (2.54) with $C = t_0$. This means that each point of the curve determined by relations (2.55) and (2.56) is common to a certain line of the family (2.54), i.e., that curve is the *envelope* of that family (see Fig. 2.12).

Problems

1. $(y')^2 - 2xy' = 8x^2$. 2. $y'(2y - y') = y^2 \sin^2 x$.
3. $xy'(xy' + y) = 2y^2$. 4. $(y')^2 + 2yy' \cot x - y^2 = 0$.
5. $(y')^3 + 1 = 0$. 6. $(y')^4 - 1 = 0$.
7. $x = (y')^3 + 1 = 0$. 8. $x(y')^3 = 1 + y'$.

Find solutions of Clairaut's and Lagrange's equations:

Lagrange's equation
9. $y = 2xy' - 3y'^2$. 10. $2y - 4xy' - \ln y' = 0$.
11. $2y(y' + 2) = xy'^2$. 12. $y = xy'^2 + y'^2$.
13. $y = 2xy' + \frac{1}{y'^2}$.

Clairaut's equation
14. $y = xy' + y'^2$.   15. $y = xy' - a\sqrt{1 + y'^2}$.
16. $y = xy' - e^{y'}$.   17. $y = xy' - \sin y'$.
18. $y = xy' - \sqrt{y'}$.

2.7 Qualitative Approach for Autonomous First-Order Equations: Equilibrium Solutions and Phase Lines

In this section, we consider equation

$$dy/dt = f(y), \tag{2.57}$$

where the right-hand side is independent of t. Such equations are called autonomous equations (AE). If t is the time, then the system governed by Eq. (2.57) is driven by its internal dynamics rather than by the change of the external conditions. Some examples of autonomous equations are given in Chap. 1. More examples are presented in Sect. 2.7.

Obviously, Eq. (2.57) is separable; thus, it can be easily solved (see Sect. 2.3). But now our goal is to understand the solutions of (2.57) without even trying to obtain the exact formula for solutions.

In applications, it may be important to determine the qualitative behavior of a solution rather than to find it exactly. Assume that y in (2.57) is a temperature in a chemical reactor, and $f(y)$ characterizes the heat release due to some chemical reactions. From the point of view of an engineer, only the behavior of the solution at large t is significant. One must know whether the solution tends to a constant value corresponding to a stable synthesis of a product or it tends to infinity, which can mean an explosion. The details of a solution $y(t)$ are of minor importance.

In this section we describe the technique called *the phase line* approach that allows to qualitatively understand the solution's behavior without solving an IVP. This method works equally well for linear and nonlinear AEs. It may be fruitful to use the phase line before attempting analytical calculation or computation.

The phase line is most useful, and even indispensable, for the analysis of a nonlinear equation, since such equations rarely permit analytical solution. Thus, one often resorts to solving the equation numerically – but this has a disadvantage of computing only one solution at a time. In the numerical approach, there is always an inherent risk of not choosing the initial condition that results in the solution that is qualitatively different from other solutions, and thus failing to completely determine the solution set. The phase line, on the other hand, allows to determine (qualitatively) all solutions in a single step.

Suppose that equation can be solved analytically, but the solution formula is rather complicated. In such cases, qualitative methods can help to understand the behavior of all solutions in the set. In this section we are applying the phase line method to rather simple equations. When a comparison is possible, we compare the predictions of this qualitative technique to analytical solutions.

We begin the discussion by stating the notion of equilibrium solution of AE. Such solution is the solution that does not change (is a constant). Thus, if $y = C$ is equilibrium solution, then $y' = 0$, and the equation $y' = f(y)$ now reads $f(y) = 0$. Thus, to find all equilibrium solutions means to solve the algebraic equation $f(y) = 0$. Equilibrium solutions are also often called equilibrium points or critical points.

As a side note, an autonomous first-order equation is separable – its solutions (in fact, the inverse solution function $t = F(y)$) are given by the integral of $1/f(y)$. However, for many nonlinear functions $f(y)$, there are no analytical methods to calculate the integral in elementary functions. Even when the integral can be calculated using some special functions, the exact solution can be too cumbersome for understanding its qualitative behavior.

Why are equilibrium solutions useful? Since they are constants, we can plot them on the real number line. Then, we mark the sign of $f(y)$ in the intervals between equilibrium solutions. It is convenient to position the line vertically and use the up (down) arrow for the positive (negative) sign. Notice that since $y' = f(y)$, we are actually marking the sign of y'. Thus, if $f(y) > 0$ (<0) on an interval, then the solution is increasing (decreasing) on that interval. Now, the completed phase line allows to determine the qualitative behavior of time-dependent (nonequilibrium) solutions. For instance, when there are two equilibrium solutions $y = y_e^{(1)} < 0$ and $y = y_e^{(2)} > 0$, the phase line and the solutions may look like in Fig. 2.13. The three examples below demonstrate in detail how this works.

Example 2.28   For the IVP $\frac{dy}{dt} = ky, y(t_0) = y_0$:

(a) Find equilibrium solutions.
(b) Use a phase line to determine qualitative behavior of nontrivial solutions.

Equilibrium solutions are determined from the equation $f(y) = 0$. Thus, $ky = 0$, and the equilibrium solution is $y = 0$. Clearly, for $y < 0, f(y) < 0(>0)$ if $k > 0(<0)$.

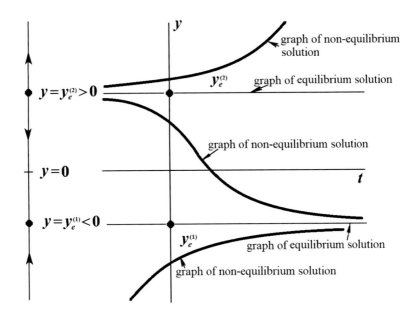

Fig. 2.13 Example phase line (left) and sketch of corresponding time-dependent (nonequilibrium) solutions of ODE $y' = f(y), f\left(y_e^{(1)}\right) = f\left(y_e^{(2)}\right) = 0$ (right)

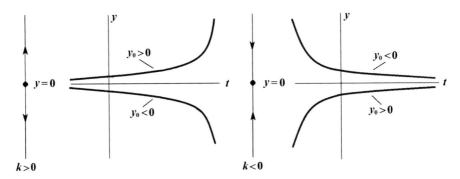

Fig. 2.14 Phase lines and two characteristic solutions of IVP $y' = ky$, $y(t_0) = y_0$. Upward pointing arrows mark increasing solutions ($y' > 0$) and downward pointing arrows decreasing solutions ($y' < 0$). Solutions that correspond to other values of y_0 have same trends as the two solutions shown. Thus, they qualitatively do not differ from the shown solutions

But, for $y > 0$, $f(y) > 0(<0)$ if $k > 0(<0)$. The phase lines for $k > 0$ and $k < 0$ are shown in Fig. 2.14. Next to the phase lines we show the sketches of some solutions. The first phase line indicates that a positive initial condition y_0 gives rise to an increasing solution, and a negative initial condition to a decreasing one. This is certainly correct, since the exact solution of this IVP by separation of variables is $y = y_0 e^{kt}$, which is increasing and positive (decreasing and negative) for $k > 0$, $y_0 > 0$ ($k > 0$, $y_0 < 0$). Similarly, the second line indicates that a positive initial condition y_0

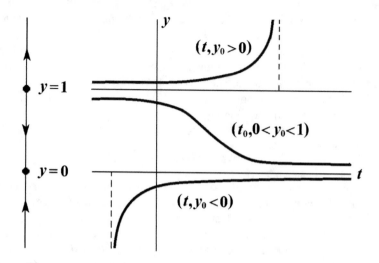

Fig. 2.15 Phase line and three characteristic solutions of IVP

gives rise to a decreasing solution and a negative initial condition to an increasing one. Again, this is correct since the analytical solution $y = y_0 e^{kt}$ is decreasing and positive (increasing and negative) for $k < 0$, $y_0 > 0$ ($k < 0$, $y_0 < 0$). Next, notice that the uniqueness theorem prohibits crossing or merging of the solution curves that correspond to different initial conditions; this includes intersections or merging with the equilibrium solution line $y = 0$. Taken together with the solution increasing/ decreasing analysis just performed, this leaves only one option for the shape of a solution curve when $y > 0$ or $y < 0$, as shown next to the phase line. That is, the time-dependent solutions of $y' = ky$ (integral curves) *are asymptotic to the equilibrium solution.* Certainly, this asymptotic property matches the behavior of the exact solution $y = y_0 e^{kt}$, $y_0 \neq 0$ as $t \to \infty$ ($k < 0$), or $t \to -\infty$ ($k > 0$).

Since the method is qualitative, we are not concerned with the exact shapes of the solution curves. It is sufficient here to determine that the solutions are increasing or decreasing in the certain intervals of y values shown on the phase line. We can sketch more solution curves that follow the solution trends already shown. In the sketch we should only care that the solution curves do not intersect and do not show a tendency toward intersecting, since intersections are not allowed due to the uniqueness property of solutions.

Example 2.29 For the $\frac{dy}{dt} = y(y-1)$, $y(t_0) = y_0$:

(a) Find equilibrium solutions.
(b) Use a phase line to determine the qualitative behavior of nontrivial solutions.

Equilibrium solutions are determined from the equation $f(y) = 0$. Thus, $y(y - 1) = 0$, and the equilibrium solutions are $y = 0$ and $y = 1$. The phase line and the sketch of the solutions are shown in Fig. 2.15. According to the phase line, all initial conditions above the line $y = 1$ in the sketch give rise to increasing

solutions; all initial conditions between this line and the line $y = 0$ give rise to the
decreasing solutions; and all initial conditions below the dashed line $y = 0$ give rise
to increasing solutions. The time-dependent integral curves must be asymptotic to
the equilibrium solutions due to the uniqueness theorem, which results in the unique
trend of IVP solution curves in the sketch. Note that two vertical asymptotes (dashed
lines in Fig. 2.15) are the result of the analytical solution of the IVP; they cannot be
obtained from the qualitative reasoning.

$$y' = y(y-1), y(t_0) = y_0$$

In Example 2.29, the equilibrium solution $y = 1$ can be called a *repelled*, because
all nearby initial conditions produce solutions that diverge (tend away) from $y = 1$ as
t increases. It is commonplace to call such equilibrium solution *a source, or unstable
node*. Similarly, the equilibrium solution $y = 0$ "attracts" its vicinity, and it presents
an example of an *attractor*. The attractor of this kind is called *a sink, or stable node*
since it is stable – all nearby initial conditions produce solutions that converge to
$y = 0$ as t increases. In Example 2.29, $y = 0$ is a source for $k > 0$, and a sink for $k < 0$.
We will discuss stability of solutions and classification of equilibrium solutions in
Sect. 5.3.

Exact analytical solution of IVP 2.28 is $y = 1/(1 \pm Ce^t)$, where $C = e^{-t_0}|1 - 1/y_0|$. It can be easily obtained using separation of variables, partial fraction
decomposition, and integration. To make sure that the phase line gives a correct
qualitative information about the solutions, you can plot a few analytical solution
curves for the initial conditions as in Fig. 2.15.

Example 2.30 For the IVP $\frac{dy}{dt} = ky^2, y(t_0) = y_0$:

(a) Find equilibrium solutions.
(b) Use a phase line to determine qualitative behavior of nontrivial solutions.
(c) Confirm your analysis by solving IVP exactly using separation of variables.

Equilibrium solutions are determined from the equation $f(y) = 0$. Thus, $ky^2 = 0$,
and the equilibrium solution is $y = 0$. The phase line and the sketch of solutions are
shown in Fig. 2.16. According to the phase line, all initial conditions for $k > 0(<0)$
give rise to increasing (decreasing) solutions. The time-dependent integral curves
must be asymptotic to the equilibrium solutions due to the uniqueness theorem,
which results in the unique trend of the IVP solution curves in the sketch.

The exact solution of the IVP obtained by using separation of variables is
$y = 1/\left[\left(y_0^{-1} + kt_0 \right) - kt \right]$. Let k, y_0, $t_0 > 0$. As t increases from t_0, the denominator
decreases, and as long as it is positive, the solution is positive and increases. This
matches the solution trend in Fig. 2.16 (left sketch) for $y > 0$. However, the exact
solution ceases to exist at $t = (ky_0)^{-1} + t_0$ (vertical asymptote there, solution growth
rate tends to infinity as $t \to (ky_0)^{-1} + t_0$), which the qualitative analysis is incapable
of predicting. Other cases of k, y_0, t_0 can be analyzed similarity to this case.

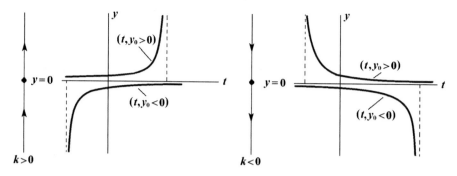

Fig. 2.16 Phase line and two characteristic solutions of IVP $y' = ky^2$, $y(t_0) = y_0$

The equilibrium solution $y = 0$ of Example 2.30 is called *a degenerate node*, and it is unstable.

Lastly, we provide the following analytical criteria for a sink, a source, and a degenerate node. Let $y = y_*$ be:

$$A \text{ degenerate node, then } f'(y_*) = 0.$$
$$A \text{ sink, then } f'(y_*) < 0. \qquad (2.58)$$
$$A \text{ source, then } f'(y_*) > 0.$$

In cases when the derivative $f'(y)$ can be easily calculated, these conditions are convenient for confirming the equilibrium point type. For instance, in Example 2.30 $f'(y) = 2ky$, $y_* = 0$, and $f'(y_*) = 0$. Importantly, the condition $f'(y_*) = 0$ can be useful when one wishes to *find* degenerate nodes. However, caution is needed, as the following example shows.

Example 2.31 For the IVP $\frac{dy}{dt} = ky^3$, $y(t_0) = y_0$:

(a) Find equilibrium solutions.
(b) Use a phase line to determine qualitative behavior of nontrivial solutions.

The equilibrium solution is again $y = 0$. The phase line and the sketch of solutions coincide with the one in Fig. 2.14. The equilibrium solution $y = 0$ is unstable for $k < 0$ and stable for $k > 0$, but notice that $f'(0) = 0$. Thus, it can be concluded that the phase-line analysis is more powerful than the application of criterion (2.28). On has to always supplement the calculation of the derivative at the equilibrium point with the analysis of the sign of $f(y)$ at the neighboring points. (As exercise, obtain the analytical solution: $y = \pm 1/\sqrt{C - 2kt}$, where $C = y_0^{-4} + 2kt_0$.)

A phase line helps in the studies of *bifurcations* in autonomous equations containing a parameter. A bifurcation is an event, such as the parameter attaining a critical value, which results in a drastic (qualitative) change in the solution behavior. For instance, when the parameter k in Example 2.28 changes from a negative value to a positive one (thus k passes through zero), the sink bifurcates into the source. Value $k = 0$ is called the bifurcation value.

We shall discuss bifurcations in more detail in Chap. 5.

In problems below, draw a phase line, sketch solutions next to a line, and classify equilibrium points into sources, sinks, and degenerate nodes. Also, in problem 1 obtain exact solutions using separation of variables and compare the qualitative analysis to these solutions.

Problems

1. $y' = ky^3$, $y(t_0) = y_0$.   2. $y' = y^3(y-2)(y+1)$, $y(t_0) = y_0$.
3. $y' = (y+3) \ln |y|$, $y(t_0) = y_0$.   4. $y' = 2ye^y \sqrt[3]{y}$, $y(t_0) = y_0$.
5. $y' = \frac{-3y}{(y-1)(y-9)}$, $y(t_0) = y_0$.   6. $y' = \sin y$, $y(t_0) = y_0$.
7. $y' = -1 + \cos y$, $y(t_0) = y_0$.   8. $y' = -y \sin y$, $y(t_0) = y_0$.

2.8   Examples of Problems Leading to First-Order Differential Equations

In this section we present some examples of the real-world problems described by first-order ODEs.

Example 2.32 Motion of an object under the action of a time-dependent force: slowing down by a friction force that is proportional to velocity.

Newton's law of a linear motion of an object of mass m is the second-order ordinary differential equation for the object's coordinate $x(t)$:

$$m\frac{dx^2(t)}{dt^2} = F(t,x,x').$$
(2.59)

Consider the situation when F is a friction force proportional to the object's velocity, that is, $F = -kv(t)$, where k is a positive coefficient. Negative sign in this relation indicates that force and velocity have opposite directions.

Since $x'(t) = v(t)$, Eq. (2.59) can be stated as the first-order ODE for $v(t)$:

$$m\frac{dv(t)}{dt} = -kv(t).$$

Since velocity decreases, the derivative $dv/dt < 0$. In the last equation, variables can be separated:

$$\frac{dv}{v} = -\frac{k}{m}dt.$$

Integration gives the general solution:

$$v(t) = Ce^{-kt/m}.$$

Let $t_0 = 0$ be the time instant when the action of the force starts, and let v_0 be the velocity at $t = 0$. Then, the initial condition is $v(0) = v_0$. Substitution in the general solution gives $C = v_0$, and the particular solution thus is

$$v(t) = v_0 e^{-kt/m}, \tag{2.60}$$

i.e., velocity decreases exponentially with time, and it does so faster if k is larger and m is smaller.

The coordinate can be obtained integrating $v(t)$:

$$x(t) = \int_0^t v(t)dt = \frac{v_0 m}{k}\left(1 - e^{-kt/m}\right). \tag{2.61}$$

From there it is seen that the distance covered before the mass comes to rest is $v_0 m/k$.

Example 2.33 Rocket motion

The Tsiolkovsky equation (Konstantin Tsiolkovsky (1857–1935), a Russian rocket scientist who pioneered astronautic theory) for the rocket is

$$m(t)\frac{dv}{dt} = -v_e\frac{dm(t)}{dt}, \tag{2.62}$$

where $v(t)$ is the rocket velocity, v_e is the velocity of the exhaust in the rocket frame, and $m(t)$ is the total rocket mass decreasing with time. The initial total mass with the propellant (the wet mass) is m_0, the final total mass without the propellant (the dry mass) is m_f, and the propellant mass is $m_p = m_0 - m_f$. This equation can be derived using the conservation of the momentum (assuming that gravity and drag forces are neglected) of the rocket and the exhaust.

Integrating Eq. (2.62), we obtain $\int_{v_0}^{v} dv = -v_e \int_{m_0}^{m_f}\frac{dm}{m}$, or

$$\Delta v = v - v_0 = v_e \ln\frac{m_0}{m_f} = v_e \ln\left(1 + \frac{m_p}{m_f}\right). \tag{2.63}$$

For instance, for $m_p/m_f = 6$ and $v_e = 3$ km/s, the velocity is $\Delta v = 5.85$ km/s. It is less than the escape velocity from Earth. The necessary escape velocity can be achieved with several stage rockets (do problems 8 and 9 below).

Fig. 2.17 An object
cooling down in the
surrounding medium

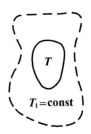

Example 2.34 Cooling of an object

Let the initial object temperature be T_0, and the temperature of the surrounding medium be $T_1 = $ const. (Fig. 2.17). The task is to determine the dependence of the object's temperature $T(t)$ on time.

From the experiments it is known that the speed of cooling, dT/dt, is proportional to a difference $T - T_1$ of the object's temperature and the medium's temperature:

$$\frac{dT}{dt} = -k(T - T_1). \tag{2.64}$$

Here $k > 0$ is a coefficient, and the negative sign in (2.64) is due to the decreasing temperature – the derivative $dT/dt < 0$.

In Eq. (2.64) variables are separated:

$$\frac{dT}{(T - T_1)} = -kdt.$$

Integration gives the general solution:

$$T - T_1 = Ce^{-kt}.$$

Substituting the initial condition $T(0) = T_0$, we find $C = T_0 - T_1$. Substituting this constant in the general solution gives the particular solution:

$$T(t) = T_1 + e^{-kt}(T_0 - T_1). \tag{2.65}$$

Note that formula (2.65) is valid also when the object is heated up, $T_1 > T_0$.

Equation (2.64) also describes diffusion into a surrounding medium, where concentration of the diffusing substance is constant.

Example 2.35 Radioactive decay

In radioactive decay, the mass m of a radioactive substance decreases with time, and the speed of decay, $\frac{dm(t)}{dt}$, is proportional to a mass that has not yet disintegrated. Thus, the radioactive decay is described by the equation:

$$\frac{dm(t)}{dt} = -km. \tag{2.66}$$

Coefficient $k > 0$ depends on the type of radioactive material, and the negative sign corresponds to the decrease of a mass with time.

Separating variables in (2.66) and integrating, we find the general solution of this equation:

$$m(t) = Ce^{-kt}.$$

Let the initial condition be $m(0) = m_0$, where m_0 is the initial mass of a material. Substitution in the general solution gives $C = m_0$; thus, the particular solution is

$$m(t) = m_0 e^{-kt}. \tag{2.67}$$

Half-life is the interval of time it takes to decrease the mass by half in radioactive decay. Denoting this time as $T_{1/2}$, we obtain from (2.67):

$$\frac{1}{2} m_0 = m_0 e^{-kT_{1/2}} \text{ and } T_{1/2} = \frac{1}{k} \ln 2.$$

If $T_{1/2}$ is known (or another, shorter time during which a certain quantity of a material decays), one can find the decay constant k.

If Eq. (2.66) has a positive sign at the right-hand side, then its solution is the expression (2.67) with the positive sign in the exponent. This solution describes, for instance, the exponential growth of the number of neutrons in the nuclear reactions, or the exponential growth of a bacterial colony in the situation when the growth rate of a colony is proportional to its size (see Chap. 1).

Example 2.36 Fluids' mixture
A water tank initially contains 150 L of seawater with a salt concentration of 4%. Then, fresh (unsalted) water is pumped to the tank at the rate of 4 L/min, and the mixture flows out at the rate of 6 L/min. What will be the concentration of salt in the tank in 1 hour?

Solution Let $V(t)$ be the salt volume (in liters) in the tank at time t (in minutes). Since every minute 2 L more water leaves the tank than goes into it, the water volume in the tank at time t is $(150 - 2t)$, and the salt concentration in the water remaining in the tank is $V(t)/(150 - 2t)$. During time dt, $6dt$ liters of water leave the tank. Thus, we arrive to the equation:

$$\frac{V(t)}{(150 - 2t)} \cdot 6dt = -dV.$$

Its general solution is (check as *Reading Exercise*)

$$V(t) = C(75 - t)^3, \quad t < 75,$$

and with the initial condition $V(0) = 6$, we have $C = 6/75^3$ and $V(t) = 6\left(\frac{t}{75} - 1\right)^3$. After 60 min, $V(60) = 6/5^3$, and the salt concentration is

$$\frac{6}{5^3(150 - 120)} = 5^{-4} = 0.0016, \text{ or } 0.16\%.$$

Example 2.37 Mixture of gases
A 50 L tank contains air with the composition of 80% nitrogen and 20% oxygen. 0.2 L of nitrogen per second enters the tank and continuously mixes inside. Same amount of the mixture per second leaves the tank. How long will it take to achieve a 90% nitrogen content in the tank?

Solution Let $V(t)$ be a volume of nitrogen at time t. During time dt, $0.2dt$ liters of nitrogen enter the tank, and $V(t) \cdot 0.2dt/50$ liters of nitrogen leave the tank; thus, we have the equation:

$$V'(t) = 0.2 - 0.004 \cdot V(t).$$

Its general solution is $V(t) = 50 - Ce^{-0.004t}$. With the initial condition, $V(0) = 40$ L, we find the particular solution $V(t) = 50 - 10e^{-0.004t}$. Next, we substitute in this formula $V(t) = 45$ L and find the time $t = 250 \ln 2 \approx 173.3$ s.

Answer 2 min and 53 s

Example 2.38 Atmospheric pressure
Find the atmospheric pressure, $P(h)$, at a height h above the sea level.

Solution Consider a column of air with a height dh and area A. Its weight, $\rho g A dh$, is balanced by the force $A dP$ acting upward; thus, we have the equation $\rho g A dh = -A dP$. At constant temperature, the Boyle-Mariotte law reads $PV = const.$, thus, $\rho = \alpha P$ (α is a constant), and we arrive to the first-order ODE:

$$dP/P = -\alpha g dh,$$

which has the general solution $P = Ce^{-\alpha g h}$. Denoting $P(0) = P_0$ the atmospheric pressure at the sea level (this is the initial condition), we have the particular solution:

$$P(h) = P_0 e^{-\alpha g h},$$

which is called *the barometric formula*.

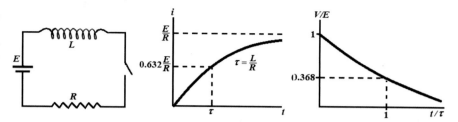

Fig. 2.18 Illustration for Example 2.41. Left panel, *RL* circuit; middle panel, graph $i(t)$; right panel, graph $V(t)$

Example 2.39 Banking interest

A bank client took out a $100,000, 15-year loan at 6% interest per year. What sum does he need to pay back, if the interest is accumulating: (a) at the end of each year and (b) continuously?

Solution

(a) Using the formula for *complex interest* (which can be obtained as the sum of the geometric series), after 15 years the loan balance is $10^5(1 + 0.06)^{15} \approx \$239,656$.
(b) In this situation we have DE $s' = 0.06\,s$ (this equation is similar to the equation describing the exponential growth of a bacterial colony; see Example 2.35), where $s(t)$ is the loan balance and t is the time in years. Its particular solution is $s(t) = 10^5 e^{0.06t}$. At $t = 15$, it gives $s(15) \approx \$245,960$.

Example 2.40 Inflation

Determine the number of years necessary to double the price of certain goods, if the yearly inflation is $p\%$.

Solution   Since during inflation the growth of prices is continuous, we need to use the formula for continuous compound interest. Let $s(t)$ be the price at time t, then $s' = 0.01p \cdot s$, i.e., $s(t) = Ce^{0.01pt}$, where a constant C is the initial price level. When the prices double, $s(t) = 2C$, and $e^{0.01pt} = 2$. Thus $t = \frac{\ln 2}{0.01p} = \frac{100 \cdot \ln 2}{p} \approx \frac{70}{p}$ years. For instance, for $p = 7\%$, $t = \frac{\ln 2}{0.01 \cdot 7} \approx 10$ years.

This *rule of 70* also approximates how long it will take for the size of an economy to double. The number of years it takes for a country's economy to double in size is equal to 70 divided by the growth rate, in percent.

More generally, the rule of 70 is a way to estimate the time it takes to double a value based on its growth rate.

Example 2.41 *RL* circuit

This is an electric circuit containing a resistor R and an inductor L. Assume that the circuit also contains a battery with a voltage E and a switch, like in Fig. 2.18. At time $t = 0$ the switch is closed, and current begins to increase, but the inductor produces an *emf* (electromotive force) that opposes the increasing current. As the result, the current cannot change from zero to its maximum value E/R instantaneously. The voltage across the resistor is Ri, and the voltage across the inductor is

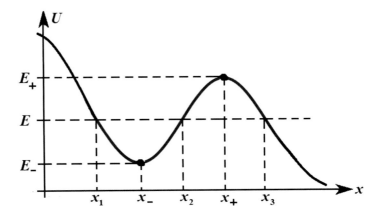

Fig. 2.19 Illustration for Example 2.42

Ldi/dt, then *Kirchhoff's law* (Gustav Robert Kirchhoff (1824–1887), a German physicist) for the electric circuit gives

$$L\frac{di}{dt} + Ri = E. \tag{2.68}$$

With the initial condition, $I(t = 0) = 0$, the solution to this IVP is (find it as an Exercise):

$$i(t) = \frac{E}{R}\left(1 - e^{-t/\tau}\right), \tag{2.69}$$

where $\tau = L/R$ is called the time constant of the circuit. The voltage across the resistor is $V(t) = di/dt = Ee^{-t/\tau}/L$. At $t = \tau$, the current reaches approximately 0.632 of its maximum value, E/R; the voltage drops to 0.368 from its initial value of E. Notice that the solution (2.69) gives an infinite amount of time required for the current to reach its maximum value. This is an artifact of assuming current is composed of moving charges that are infinitesimal.

More sophisticated case of *RLC* circuit will be considered in Sect. 3.10.2.

Example 2.42 Motion of a particle in a potential field
General Case
Let us consider a particle of mass m moving without friction on a curved surface, which is described by equation $z = h(x)$, under the action of gravity (see Fig. 2.19). The kinetic energy of the particle is

$$K = \frac{m}{2}\left(\frac{dx}{dt}\right)^2$$

(we assume that the vertical component of the velocity is negligible), while its potential energy

$$U(x) = mgh(x),$$

where g is the acceleration of gravity. Because there is no friction, the total energy $E = K + U$ is conserved:

$$\frac{m}{2}\left(\frac{dx}{dt}\right)^2 + U(x) = E. \tag{2.70}$$

This first-order differential equation can be written also as

$$\frac{dx}{dt} = \pm\sqrt{\frac{2}{m}(E - U(x))}, \tag{2.71}$$

which is a separable equation. Its general (implicit) solution is

$$t = \pm\sqrt{\frac{m}{2}}\int \frac{dx}{\sqrt{E - U(x)}} + C. \tag{2.72}$$

One has to use the upper sign when the particle moves to the right and the lower sign when it moves to the left. For a given E, the motion can take place only in the regions $U(x) < E$. At the points where $U(x) = E$, called *return points*, the velocity of the particle dx/dt vanishes (because its kinetic energy $K = E - U(x) = 0$).

As an example, let us consider in more detail the shape of the surface shown in Fig. 2.19. For the value of E indicated in the figure, the motion is possible either in the finite interval $x_1(E) \le x \le x_2(E)$ or in the semi-infinite interval $x \ge x_3(E)$.

Assume that at the initial time instant $t = 0$, $x(0) = x_0$, $x_1(E) < x_0 < x_2(E)$, and its initial velocity is negative. Then, the implicit solution of Eq. (2.69) satisfying the initial condition is

$$t = -\sqrt{\frac{m}{2}}\int_{x_0}^{x} \frac{dx}{\sqrt{E - U(x)}} = \sqrt{\frac{m}{2}}\int_{x}^{x_0} \frac{dx}{\sqrt{E - U(x)}}. \tag{2.73}$$

Formula (2.73) is valid until the particle reaches the *turning point* $x = x_1$ at the time instant:

$$t_1 = \sqrt{\frac{m}{2}} \int\limits_{x_1}^{x_0} \frac{dx}{\sqrt{E - U(x)}}.$$

In the turning point, the particle stops. Then, it starts moving in the opposite direction; hence, its motion is governed by Eq. (2.72) with "+" sign:

$$t = t_1 + \sqrt{\frac{m}{2}} \int\limits_{x_1}^{x} \frac{dx}{\sqrt{E - U(x)}},$$

etc.

The particle oscillates periodically between the points $x = x_1(E)$ and $x = x_2(E)$, with the period:

$$T = \sqrt{\frac{m}{2}} \int\limits_{x_1(E)}^{x_2(E)} \frac{dx}{\sqrt{E - U(x)}} - \frac{m}{2} \int\limits_{x_2(E)}^{x_1(E)} \frac{dx}{\sqrt{E - U(x)}} = \sqrt{2m} \int\limits_{x_2(E)}^{x_1(E)} \frac{dx}{\sqrt{E - U(x)}}.$$

When $E \to E_+ = \max U(x)$ (see Fig. 2.19), the period of oscillations tends to infinity; hence, for $E = E_+$, there is only one oscillation with $x(\pm\infty) = x_+ = x_2(E_+)$.

Also, for $E = E_+$ there exists a constant solution (*equilibrium point*) $x = x_+$. For $E = E_- = \min U(x)$, the only solution is the constant one, $x = x_1(E_-) = x_2(E_-) = x_-$. It is clear that the solution $x = x_-$ corresponding to the minimum of the potential energy is *stable* (i.e., a small deviation of the particle position leads to a small-amplitude oscillation), while the solution $x = x_+$ is *unstable* (an arbitrary small initial deviation of the particle position leads to a finite deviation of the solution). We shall present the exact definition of stability and the tools for its analysis in Chap. 5.

If $x_0 \geq x_3$, the rightward motion of the particle is unbounded, and the particle goes to infinity.

The analysis given above is valid actually for a one-dimensional motion of any physical nature when the force F acting on the object depends only on x; hence, the second Newton's law is

$$m \frac{d^2 x}{dt^2} = F(x). \tag{2.74}$$

Defining the potential $U(x)$ by the relation

$$F(x) = - \frac{dU(x)}{dx},$$

and multiplying both parts of (2.72) by dx/dt, we find:

Fig. 2.20 Harmonic
oscillator

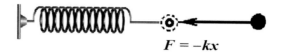

$$F = -kx$$

$$m\frac{dx}{dt}\frac{d^2x}{dt^2} = -\frac{dU(x)}{dx}\frac{dx}{dt},$$

or

$$\frac{d}{dt}\left[\frac{m}{2}\left(\frac{dx}{dt}\right)^2 + U(x)\right] = 0.$$

Therefore, Eq. (2.70) is valid.

Harmonic Oscillator

Let us consider the oscillations of a particle in the potential of quadratic shape:

$$U(x) = \frac{kx^2}{2},$$

that corresponds to the action of a linear quadratic force:

$$F(x) = -\frac{dU}{dx} = -kx.$$

The corresponding physical system, shown in Fig. 2.20, is called *a harmonic oscillator*. The conservation law (2.70) gives:

$$\frac{m}{2}\left(\frac{dx}{dt}\right)^2 + \frac{kx^2}{2} = E. \tag{2.75}$$

For $E = 0$, the only solution is $x = 0$. For $E > 0$, the oscillations occur between the turning points:

$$x_\pm = \pm\sqrt{\frac{2E}{k}}.$$

It is convenient to introduce the variable φ as

$$x = \sqrt{\frac{2E}{k}}\sin\varphi. \tag{2.76}$$

Substituting (2.76) into (2.75), we find:

$$\frac{mE}{k}\cos^2\varphi\left(\frac{d\varphi}{dt}\right)^2 = E\cos^2\varphi,$$

hence

$$\frac{d\varphi}{dt} = \pm\sqrt{\frac{k}{m}}.$$

Unlike the variable x, which oscillates, the variable φ changes monotonically in time.

Taking the upper sign, we get the general solution:

$$\varphi = \sqrt{\frac{k}{m}}t + C,$$

i.e.,

$$x(t) = \sqrt{\frac{2E}{k}}\sin\left(\sqrt{\frac{k}{m}}t + C\right), \tag{2.77}$$

where C is the arbitrary constant.

Taking the lower sign, we obtain:

$$x(t) = \sqrt{\frac{2E}{k}}\sin\left(-\sqrt{\frac{k}{m}}t + C_1\right) = \sqrt{\frac{2E}{k}}\sin\left(\sqrt{\frac{k}{m}}t - C_1 + \pi\right).$$

Thus, we do not get a new solution, but just reproduce solution (2.77) with $C = -C_1 + \pi$.

Problems

1. For certain radioactive substance, 50% of atoms decays within 30 days. When will only 1% of the initial atoms remain?
   Answer: $t \approx 200$ days

2. According to experiments, during a year from each gram of radium, a portion of 0.44 mg decays. After what time does one-half of a gram decay?
   Answer: $T_{1/2} = \frac{\ln 0.5}{\ln(1 - 0.00044)} \approx 1570$ years

3. A measurement of the concentration of the carbon isotope ^{14}C with the half-life of about 5714 years is used for determining the age of an object containing organic materials. When an animal or a plant dies, it stops exchanging carbon with its environment, and thereafter the amount of ^{14}C it contains begins to decrease. What is the age of the object that contains only 5% of the initial ^{14}C?

4. A boat begins to slow down due to water resistance, which is proportional to the boat's velocity. Initial velocity of the boat is 1.5 m/s; after 4 s velocity decreased

to 1 m/s. After it begins to slow down, (a) when will the boat's velocity become 0.5 m/s? (b) What distance will the boat travel during 4 s? (*Hint*: find first the coefficient of friction.)

Answer: (a) 10.8 s, 4.9 m

5. The object cooled down from 100 to 60 °C in 10 min. Temperature of the surrounding air is constant 20 °C. when will the object cool down to 25 °C?

Answer: $t = 40$ min.

6. Let m_0 be the initial mass of the salt dissolving in water of mass M. The rate of dissolution, $\frac{dm(t)}{dt}$, is proportional to a mass of the salt that has not yet dissolved at time t, and the difference between the saturation concentration, $\overline{m}/M$, and the actual concentration, $(m_0 - m)/M$. Thus

$$\frac{dm}{dt} = -km\left(\frac{\overline{m}}{M} - \frac{m_0 - m}{M}\right).$$

Find a solution to this Cauchy problem.

7. In bimolecular chemical reactions, the substances A and B form molecules of type C. If a and b are the original concentrations of A and B, respectively, and x is the concentration of C at time t, then

$$\frac{dx}{dt} = k(a-x)(b-x).$$

Find a solution to this Cauchy problem.

8. Formulate and solve the Tsiolkovsky equation for a two-stage rocket.
9. Formulate and solve the Tsiolkovsky equation for a three-stage rocket.
10. Find the dependence of the oscillation period on the energy for a particle in the potential field: (*i*) $U(x) = A|x|^n$; (*ii*) $U(x) = -A\,\mathrm{sech}^2(ax)$; (*iii*) $U(x) = A\tan^2(ax)$.
11. Find a current in a RL circuit with $R = 2\,\Omega$ (ohms) and $L = 0.4$ H (henry), driven by an *emf* $E = 100 \sin 50t$.

Hint: The equation for the current is $0.4i' + 2i = 100 \sin 50t$. The general solution of this nonhomogeneous linear equation is the sum of the solution of the associated homogeneous equation, which exponentially decays in time, and the particular solution of the nonhomogeneous equation, which oscillates with the period $T = 2\pi/50$. The steady-state regime is reached almost instantaneously.

12. There is the mathematical similarity between Eq. (2.68) and equation for RC circuit with a capacitor instead of an inductor

$$\frac{1}{C}\frac{di}{dt} + Ri = E.$$

Solve this equation with the initial condition $i(t=0) = E/R$ which physically corresponds to the initially uncharged capacitor, $q(t=0) = 0$; notice that $i = dq/dt$. The time constant for a RC circuit is $\tau = RC$.

13. Find the atmospheric pressure at the height 500 m above the sea level. (Ignore the changes of the air temperature at that height.) Use the following values of the atmospheric pressure and air density at the sea level (in SI units):

$$P_0 = 1.013 \times 10^5 \text{ Pa}, \rho = 1.225 \text{ kg/m}^3.$$

Answer: $P_0 = 0.954 \times 10^5$ Pa

Chapter 3
Differential Equations of Order $n > 1$

3.1 General Considerations

In this chapter we consider the general case of the n-th order differential equation, which is written in the form resolved for the highest derivative of the unknown function $y(x)$:

$$y^{(n)} = f\left(x, y, y', \ldots, y^{(n-1)}\right). \tag{3.1}$$

Obviously, a general solution (a general integral) of Eq. (3.1) depends on n arbitrary constants. In the simplest case of Eq. (3.1):

$$y^{(n)} = f(x),$$

a general solution obtained by integrating the equation n times contains n arbitrary constants, which appear as the coefficients of the polynomial of the order $n - 1$:

$$y(x) = \int dx \int dx \ldots \int f(x)dx + \sum_{i=0}^{n-1} C_i x^i.$$

As an example, consider Newton's second law for a one-dimensional motion of an object of mass m, moving under the action of a force F: $m\frac{d^2x}{dt^2} = F$. Integrating, we obtain $x'(t) = \int (F/m)dt + C_1$. Next, assuming constant values of F and m gives $x(t) = Ft^2/2m + C_1 t + C_2$. This is the *general solution* which gives the answer to the problem, i.e., $x(t)$ is a quadratic function of the time t. One can see that the general solution describes the general features of the entire set of solutions.

The constants C_1 and C_2 have a clear physical meaning. The general solution can be presented as $x(t) = at^2/2 + v_0 t + x_0$, where $a = d^2x/dt^2 = F/m$ is the acceleration, and two arbitrary constants are denoted as v_0 and x_0. Obviously, these constants are the position, $x_0 = x(0)$, and velocity, $v_0 = x'(0)$, at time $t = 0$. For the given particular values of $x(0)$ and $x'(0)$, we obtain the *particular solution* of the corresponding initial value problem.

Generally, the Cauchy problem (or IVP) for the n-th order Eq. (3.1) assumes n *initial conditions* at a certain point x_0:

$$y(x_0) = y_0, \quad y'(x_0) = y'_0, \quad y''(x_0) = y''_0, \quad \ldots, \quad y^{(n-1)}(x_0) = y_0^{(n-1)}, \tag{3.2}$$

where $y_0, y'_0, y''_0, \ldots, y_0^{(n-1)}$ are n real numbers – the values of the function $y(x)$ and its $n - 1$ derivatives at x_0. Equation (3.1), along with the initial conditions (3.2), constitutes the IVP.

The following theorem is the primary theoretical result.

Theorem (*Existence and Uniqueness of a Solution to the n-th order differential equation*).

Let f be a continuous function of all its arguments in the region R

$$|x - x_0| < a, \quad |y - y_0| < b_0, \quad |y' - y'_0| < b_1, \ldots, \left| y^{(n-1)} - y_0^{(n-1)} \right| < b_{n-1},$$

which contains the initial point $\left(x_0, y_0, y'_0, \ldots, y_0^{(n-1)} \right)$, *where* $a, b_0, b_1, \ldots, b_{n-1}$ *are the positive numbers. Let the partial derivatives of f with respect to* $y, y', \ldots, y^{(n-1)}$ *be finite in R. Then, the unique solution of the Cauchy problem (3.1) and (3.2) exists in a vicinity of the initial point.*

Note that the condition of finiteness of partial derivatives of f can be replaced by Lipschitz conditions similarly to the one formulated in Chap. 2.

Example 3.1   An object that has a time-dependent mass $m(t)$ moves under the action of a force $F(t)$. It is assumed that the functions $m(t)$ and $F(t)$ are continuous. The motion is governed by the equation:

$$m(t)x''(t) = F(t). \tag{3.3}$$

Find the solution of Eq. (3.3) satisfying the initial conditions:

$$x(t_0) = x_0, \quad x'(t_0) = v_0. \tag{3.4}$$

Solution   Let us rewrite Eq. (3.3) as

$$x''(t) = \frac{F(t)}{m(t)}. \tag{3.5}$$

Obviously, the conditions of the Existence and Uniqueness Theorem are satisfied. We integrate Eq. (3.5) twice, each time starting the integration at the point $t = t_0$:

$$x'(t) = \int_{t_0}^{t} dt_1 \frac{F(t_1)}{m(t_1)} + C_1, \tag{3.6}$$

$$x(t) = \int_{t_0}^{t} dt_2 \int_{t_0}^{t_2} dt_1 \frac{F(t_1)}{m(t_1)} + C_1(t-t_0) + C_2. \tag{3.7}$$

Notice that t_1 and t_2 are dummy variables. Substituting (3.6) and (3.7) into (3.4), we find that $C_1 = v_0$ and $C_2 = x_0$.

3.2   Second-Order Differential Equations

In this section we consider second-order differential equations in more detail. This class of ODEs is especially important for applications.

For the second-order equation

$$y'' = f(x, y, y'), \tag{3.8}$$

the Cauchy problem has two initial conditions:

$$y(x_0) = y_0, \quad y'(x_0) = y'_0, \tag{3.9}$$

where y_0 and y'_0 are the given values of $y(x)$ and $y'(x)$ at x_0. y'_0 gives the angle of the tangent line (the slope) to the solution curve (the integral curve) at the point (x_0, y_0): $\alpha = \arctan y'_0$ (see Fig. 3.1).

According to the Existence and Uniqueness Theorem, two initial conditions (3.9) are necessary and sufficient to construct a particular solution on some interval $[a, b]$ containing x_0, inside a region R of the xy-space, where $f(x,y,y')$ is continuous and

Fig. 3.1 Solution to the Cauchy problem (3.8) and (3.9)

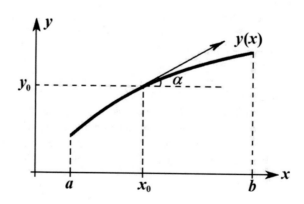

$\partial f / \partial y, \partial f / \partial y'$ are finite. These conditions guarantee that values of y and y' at a point $x_0 + h$ (where h is small) do not differ drastically from y_0 and y_0', respectively.

A general solution of (3.8) contains two arbitrary constants, which usually are denoted as C_1 and C_2. The number of initial conditions (3.9) matches the number of arbitrary constants, and a substitution of the general solution into the initial conditions results in C_1 and C_2 values. Substituting these values into a general solution gives a solution of IVP (3.8) and (3.9). The arbitrary constants are often termed parameters, and the general solution is termed *the two-parameter family of solutions*; a particular solution satisfying the initial conditions is just one member of this family.

Example 3.2 Find the solution of IVP $y'' = 1$, $y(1) = 0$, $y'(1) = 1$.

Solution Integrating the equation twice, we obtain a general solution: $y(x) = x^2/2 + C_1 x + C_2$. Substituting this solution into the initial conditions gives $y(1) = 1/2 + C_1 + C_2 = 0$, $y'(1) = 1 + C_1 = 1$; hence, $C_1 = 0$, $C_2 = -1/2$. Thus, $y(x) = x^2/2 + x$ is the particular solution of the equation for the given initial conditions, i.e., the solution of the IVP.

Example 3.3 Find the particular solution of $y'' = 1/(x+2)^2$ for the initial conditions $y(-1) = 1$ and $y'(-1) = 2$.

Solution A general solution is $y(x) = -\ln |x + 2| + C_1 x + C_2$. Notice that at $x = -2$, this solution diverges since the function $f = 1/(x+2)^2$ is not finite at $x = -2$ and violates the conditions of the Existence and Uniqueness Theorem.

On the other hand, $x = -1$ is a regular point for the function f, and a particular solution in the vicinity of this point does exist. Substitution of the general solution into the initial conditions gives $y(-1) = -C_1 + C_2 = 1$, $y'(-1) = -1/(-1+2) + C_1 = 2$. Thus, $C_1 = 3$, $C_2 = 4$, and the particular solution satisfying the initial conditions is $y(x) = -\ln |x + 2| + 3x + 4$.

Example 3.4 Find the particular solutions of $yy'' - 2y'^2 = 0$ for two sets of initial conditions: (a) $y(0) = 0$, $y'(0) = 2$, and (b) $y(0) = 1$, $y'(0) = 0$. Notice that the equation is given not in the form resolved with respect to the derivative, $y'' = f(x, y, y')$.

Solution
(a) Notice that the original equation has a special solution $y(x) = 0$, which does not exist for the equation resolved with respect to the derivative, $y'' = 2y'^2/y$.

The general solution, $y(x) = 1/(C_1 x + C_2)$, will be obtained in Example 3.8 – here its correctness can be checked by substituting this function into the equation. The initial condition $y(0) = 0$ contradicts this general solution; hence, the solution of the IVP (a) does not exist.

To understand how that result is related with the Existence and Uniqueness Theorem, let us rewrite the equation in the form $y'' = f(x, y, y')$, where $f(x, y, y') = 2y'^2/y$; we see that the function f cannot be finite on the line $y = 0$ if $y' \neq 0$. Therefore, the conditions of Theorem are not satisfied.

(b) For this set of the initial conditions, the function f and its derivatives $f'_y = -2y'^2/y^2$ and $f'_{y'} = 4y'/y$ are finite in the vicinity of the initial point, and we obtain $C_1 = 0$, $C_2 = 1$, thus $y(x) = 1$.

3.3   Reduction of Order

In this section we consider equations of arbitrary order n.

In some cases, the order of a differential equation can be reduced – usually this makes the solution easier. In this section we discuss several simple situations that are often encountered in practice.

1. Left side of equation

$$F\left(x, y, y', \ldots, y^{(n)}\right) = 0 \tag{3.10}$$

is a total derivative of some function G, i.e., Eq. (3.10) can be written as

$$\frac{d}{dx} G\left(x, y, y', \ldots, y^{(n-1)}\right) = 0.$$

This means that function G is a constant:

$$G\left(x, y, y', \ldots, y^{(n-1)}\right) = C_1. \tag{3.11}$$

This is the *first integral* of Eq. (3.10).

Example 3.5 Solve equation $yy'' + (y')^2 = 0$ using the reduction of order.

Solution This equation can be written as $\frac{d}{dx}(yy') = 0$, which gives $yy' = C_1$, or $ydy = C_1 dx$. Thus, $y^2 = C_1 x + C_2$ is the general solution.

2. Equation does not contain the unknown function y and its first $k-1$ derivatives

$$F\left(x, y^{(k)}, y^{(k+1)}, \ldots, y^{(n)}\right) = 0. \tag{3.12}$$

Substitution

$$y^{(k)} = z(x) \tag{3.13}$$

reduces the order of the equation to $n - k$.

Consider a second-order equation:

$$F(x, y', y'') = 0. \tag{3.14}$$

Note that y is absent from F, but y' and y'' are present there. This is the case $k = 1$. Substitution (3.13) reads

$$y' = z(x). \tag{3.15}$$

Then $y'' = z'$, and (3.14) reduces to the first-order equation $F(x, z, z') = 0$. Its general solution contains one arbitrary constant, and we can write this solution as $z(x, C_1) = y'$. Integrating this first-order equation, we obtain the general solution of (3.5), $y(x) = \int z(x, C_1)dx + C_2$.

Example 3.6 Solve $y^{(5)} - y^{(4)}/x = 0$ using the reduction of order.

Solution Substitution $y^{(4)} = z(x)$ gives $z' - z/x = 0$, thus, $z = Cx$, i.e., $y^{(4)} = Cx$, where C is arbitrary constant. Integrating four times gives

$$y = C_1 x^5 + C_2 x^3 + C_3 x^2 + C_4 x + C_5,$$

where all constants C_i, $i = 1, \ldots, 5$ are arbitrary.

Example 3.7 Solve $y'' - y'/x = x$ using the reduction of order.

Solution Note that equation above is a linear second-order nonhomogeneous equation. With $y' = z(x)$, we have $y'' = z'(x)$, and the equation transforms into $z' - z/x = x$, which is the linear nonhomogeneous first-order equation. Recall that its solution is the sum of the general solution of the homogeneous equation and the particular solution of the nonhomogeneous equation.

The homogeneous equation is $z' - \frac{z}{x} = 0$, or $\frac{dz}{z} = \frac{dx}{x}$. Its general solution is $z = C_1 x$.

The solution of the nonhomogeneous equation can be found using the method of variation of a parameter: $z = u_1(x)x$. Substituting this expression into $z' - z/x = x$ gives $u'_1 = 1$; hence, $u_1(x) = x + C_1$. Thus, $z(x) = x^2 + xC_1$. Finally

$$y(x) = \int z(x)dx = \int (x^2 + C_1 x)dx = \frac{x^3}{3} + C_1 \frac{x^2}{2} + C_2.$$

We can see that the general solution of the nonhomogeneous second-order equation is indeed the sum of the general solution of the homogeneous equation, $y_h(x) = C_1 \frac{x^2}{2} + C_2$, and the particular solution of the nonhomogeneous equation, $y_{nh}(x) = \frac{x^3}{3}$. In Sect. 3.4, we will consider this property of linear equations in detail.

3. Equation does not contain the independent variable x. For example, the second-order equation is

$$F(y, y', y'') = 0. \tag{3.16}$$

To reduce the order, we make the substitution:

$$y' = z(y). \tag{3.17}$$

Notice that here z is a function of y, not x. Differentiation using the chain rule gives $y'' = \frac{dz}{dy}\frac{dy}{dx} = \frac{dz}{dy}z(y)$. Equation (3.16) becomes $F(y, z'(y), z(y')) = 0$, which has a general solution $z(y, C_1)$. Therefore, we arrive to the first-order equation $y'(x) = z(y, C_1)$, where variables can be separated. Then, $\int \frac{dy}{z(y, C_1)} = x + C_2$, which gives the general solution.

Example 3.8 Solve $yy'' - 2y'^2 = 0$ using the reduction of order.

Solution Differentiation of $z(y) = y'$ gives $y'' = \frac{dz}{dy}\frac{dy}{dx}$, and replacing here y' by $z(y)$ gives $y'' = z(y)\frac{dz}{dy}$. Equation now reads $yz\frac{dz}{dy} - 2z^2 = 0$. Its solutions are as follows:
(a) $z(y) = 0$, or $y' = 0$, thus $y = C$; (b) $\frac{dz}{z} = 2\frac{dy}{y}$; then, $z = C_1y^2$, which gives $y' = C_1y^2$, $dy/y^2 = C_1dx$, $-1/y = C_1x + C_2$, $y(x) = -\frac{1}{C_1x+C_2}$ (the minus sign can be dropped after redefining the arbitrary constants C_1 and C_2).

Example 3.9 Solve the IVP $yy'' - 2y'^2 = 0$, $y(x_0) = y_0$, $y'(x_0) = y_0'$.

Solution The general solution (a), $y = C$, of the previous example is a constant continuous function. This solution corresponds to the initial conditions $y(x_0) = C$ (including the value $C = 0$) and $y'(x_0) = 0$.

General solution (b), $y(x) = -1/(C_1x + C_2)$ with $C_1 \neq 0$, describes hyperbolas with two asymptotes: a vertical one, $x = -C_2/C_1$, and a horizontal one, x. After rewriting equation in the form $y'' = 2y'^2/y = f$ and calculating the partial derivatives $f_y = -2y'^2/y^2$, $f_{y'} = 4y'/y$, we notice that $f, f_y, f_{y'}$ are discontinuous at $y = 0$. At this point the conditions of the Existence and Uniqueness Theorem are not satisfied.

Treating C_1 and C_2 as parameters, we can think of the solution as a two-parameter set of integral curves, $y(x, C_1, C_2)$. The sets separated by the asymptotes are unrelated, and two initial conditions select a particular curve from one of these four sets. If $x_0 < -C_2/C_1$ and $y(x_0) = y_0 > 0$, then we have the solution of IVP in the domain $-\infty < x < -C_2/C_1$, $y > 0$. There are also three other possibilities.

To illustrate this example, we plot four integral curves, $y(x, C_1, C_2)$, for the case of the vertical asymptote at $x = 1.5$ (i.e., values of C_1 and C_2 in $y(x) = -1/(C_1x + C_2)$ are chosen such that $-C_2/C_1 = 1.5$). In Fig. 3.2, the four particular solutions are shown.

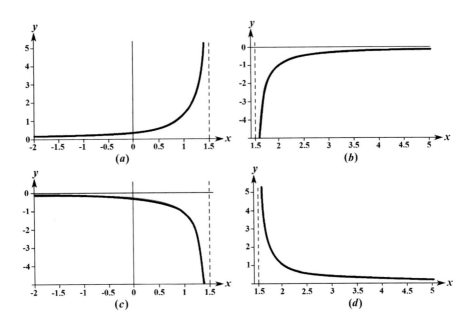

Fig. 3.2 Four particular solutions for different initial conditions. Dashed line is the vertical asymptote $x = -C_2/C_1 = 1.5$

1. Using the initial conditions $x_0 = 1$, $y_0 = 1$, $y'_0 = 2$, we obtain $C_1 = 2$, $C_2 = -3$; thus, the particular solution is $y = -\frac{1}{2x-3}$. This solution exists on the interval $x \in (-\infty, 1.5)$.

   Similarly:

2. $x_0 = 3, y_0 = -1/3, y'_0 = 2/9, C_1 = 2, C_2 = -3, y = -\frac{1}{2x-3}, x \in (1.5, \infty)$;
3. $x_0 = 1, \ y_0 = -1, \ y'_0 = -2, C_1 = -2, C_2 = 3, y = \frac{1}{2x-3}, x \in (-\infty, 1.5)$;
4. $x_0 = 3, \ y_0 = 1/3, \ y'_0 = -2/9, C_1 = -2, C_2 = 3, y = \frac{1}{2x-3}, x \in (1.5, \infty)$.

It is obvious from this example that knowing analytical solution is very important: without analytical solution a numerical procedure may not "detect" the vertical asymptote at $x = -C_2/C_1$, the numerical solution may "slip" to an adjacent integral curve, and thus the obtained solution may be completely unrelated to the initial conditions.

Problems

1. $(x - 3)y'' + y' = 0$.
2. $y^3 y'' = 1$.
3. $y'^2 + 2yy'' = 0$.
4. $y''(e^x + 1) + y' = 0$.
5. $yy'' = y'^2 - y'^3$.
6. $x(y'' + 1) + y' = 0$.
7. $y'' = \sqrt{1 - (y')^2}$.
8. $y''' + 2xy'' = 0$.
9. $y'' + 2y' = e^x y'^2$.
10. $yy'' = (y')_3$.
11. $(1 - x^2)y'' - 2y' + 2 = 0$.
12. $xy''' - y'' = 0$.

3.4 Linear Second-Order Differential Equations

Linear differential equations arise frequently in applications. We start with exploring the properties of a second-order linear differential equations:

$$y'' + p(x)y' + q(x)y = f(x), \tag{3.18}$$

where $p(x)$, $q(x)$ and $f(x)$ are given functions.

(Notice that in (3.18) and in (3.8), $f(x)$ have different meaning.)

Equation (3.18) is *nonhomogeneous*; if $f(x) = 0$ for any x, it is *homogeneous*:

$$y'' + p(x)y' + q(x)y = 0. \tag{3.19}$$

If the functions $p(x)$, $q(x)$ and $f(x)$ are continuous in the vicinity of $x_0 \in [a, b]$, then the function $F(x, y, y', y'') = y'' + p(x)y' + q(x)y - f(x)$ and its derivatives $F_y = q(x)$ and $F_{y'} = p(x)$ are also continuous and bounded on $[a, b]$. Then, the requirements of the Existence and Uniqueness Theorem are satisfied, and a particular solution of (3.18) that corresponds to the initial conditions $y(x_0) = y_0, y'(x_0) = y'_0$ (where y_0 and y'_0 are any values) exists and is unique. It can be shown that the boundness of $F_{y'}$ and F_y for any y and y' guarantees the existence and uniqueness of the solution on the whole interval $[a, b]$.

3.4.1 Homogeneous Equations

Here, we consider the homogeneous Eq. (3.19) and prove several basic theorems. As usual, any function that is free of arbitrary parameters and that satisfies Eq. (3.19) is called the particular solution.

Theorem 3.1

Let $y_1(x)$ and $y_2(x)$ be any two particular solutions of the homogeneous Eq. (3.19). Then

$$y(x) = C_1 y_1(x) + C_2 y_2(x), \tag{3.20}$$

where C_1 and C_2 are two arbitrary constants, is also a solution of this equation.

Proof Substitute $y(x)$ and its derivatives, $y' = C_1 y'_1 + C_2 y'_2$ and $y'' = C_1 y''_1 + C_2 y''_2$, into Eq. (3.19) and combine separately the terms with C_1 and C_2. Then, the left-hand side becomes $C_1 \left(y''_1 + p y'_1 + q y_1 \right) + C_2 \left(y''_2 + p y'_2 + q y_2 \right)$. Since $y_1(x)$ and $y_2(x)$ are solutions of (3.19), the expressions in both parentheses are zero. Thus, $y(x) = C_1 y_1(x) + C_2 y_2(x)$ is the solution of (3.19).

The expression $y(x) = C_1 y_1(x) + C_2 y_2(x)$ is called the *linear combination* of functions $y_1(x)$ and $y_2(x)$. More generally,

$$y(x) = C_1 y_1(x) + C_2 y_2(x) + \ldots + C_n y_n(x) \qquad (3.21)$$

is called the linear combination of n functions. Thus, Theorem 3.1 states that a linear combination of the particular solutions of a homogeneous Eq. (4.2) is also its solution. This property of the solutions of linear homogeneous equations is referred to as the *superposition principle*.

Functions $y_1(x)$, $y_2(x)$, $\ldots$, $y_n(x)$ are called *linearly dependent* on the interval $[a, b]$, if $C_1 y_1(x) + C_2 y_2(x) + \ldots + C_n y_n(x) = 0$ for all $x \in [a, b]$, and not all constants C_i are zero. Otherwise, functions are called *linearly independent*. Linear dependence of two functions, $C_1 y_1(x) + C_2 y_2(x) = 0$, simply means that functions are proportional: $y_2(x) = -C_1 y_1(x)/C_2$, if $C_2 \neq 0$, or $y_1(x) = -C_2 y_2(x)/C_1$, if $C_1 \neq 0$.

Next, we introduce a determinant, which is called a Wronskian determinant (it was introduced by Jozef Wronski (1812)), or simply a Wronskian:

$$W(y_1(x), y_2(x)) = \begin{vmatrix} y_1(x) & y_2(x) \\ y_1'(x) & y_2'(x) \end{vmatrix}. \qquad (3.22)$$

This determinant can be defined for any pair of differentiable functions.

Obviously, if $y_1(x)$ and $y_2(x)$ are linearly dependent, then the Wronskian is identically equal to zero. Indeed, substituting $y_2(x) = -C_1 y_1(x)/C_2$ and $y_2'(x) = -C_1 y_1'(x)/C_2$ into (4.5) gives immediately $W(y_1, y_2) = 0$. Note that for arbitrary differentiable functions, the inverse assertion is not true. As a counterexample, assume that on a certain subinterval $[a, c]$, $y_1(x) = 0$, $y_1'(x) = 0$, while $y_2(x)$ is not identically zero, while on a subinterval $[b, c]$, $y_2(x) = 0$, $y_2'(x) = 0$, while $y_1(x)$ is not identically zero. Though the Wronskian (4.5) is equal to zero everywhere in $[a, b]$, the condition $C_1 y_1(x) + C_2 y_2(x) = 0$ cannot be satisfied with the same constants C_1 and C_2 in both subintervals.

In the case where $y_1(x)$ and $y_2(x)$ are particular solutions of a homogeneous linear Eq. (4.2), their Wronskian has some special properties that are described by the following theorem.

Theorem 3.2

Let $y_1(x)$ and $y_2(x)$ be any two solutions of (3.19) on the interval $[a, b]$, where $p(x)$ and $q(x)$ are continuous functions. Then:

(a) *The Wronskian (3.22) either vanishes identically or is never zero on the interval $[a, b]$.*

(b) *The Wronskian (3.22) is different from zero if and only if $y_1(x)$ and $y_2(x)$ are linearly independent.*

Proof

(a) Because $y_1(x)$ and $y_2(x)$ are solutions of (3.19),

$$y_1'' + py_1' + qy_1 = 0, \quad y_2'' + py_2' + qy_2 = 0. \tag{3.23}$$

Multiplying the first equation by $-y_2$, the second equation by y_1, and adding both equations, we find

$$\left(y_1 y_2'' - y_2 y_1''\right) + p\left(y_1 y_2' - y_2 y_1'\right) = 0,$$

or

$$W'(x) + p(x)W(x) = 0. \tag{3.24}$$

Denote the value of $W(x)$ at a certain point $x = x_0$ inside the interval $[a, b]$ as $W(x_0) = W_0$. Then, solution of Eq. (3.24) is

$$W(x) = W(x_0)e^{\displaystyle -\int_{x_0}^{x} p(x_1)dx_1}, \tag{3.25}$$

where x_1 is a dummy variable. The exponential function is always positive. Thus, there are the following alternatives: (i) if $W(x_0) = 0$, then $W(x) = 0$ on the whole interval; (ii) if $W(x_0) \neq 0$, then $W(x) \neq 0$ on the whole interval.

Relation (3.25) is called *Abel's formula (Niels Henrik Abel (1802–1829), a Norwegian mathematician).*

(b) First, let us prove that if $y_1(x)$ and $y_2(x)$ are linearly independent, then $W(y_1, y_2) \neq 0$ on the whole interval.

Assume that $W(y_1, y_2) = 0$ at some point $x = x_0$. Consider the following homogeneous system of algebraic equations for the unknown quantities α and β:

$$\alpha y_1(x_0) + \beta y_2(x_0) = 0,$$
$$\alpha y_1'(x_0) + \beta y_2'(x_0) = 0.$$

Because the determinant of this system $W(y_1(x_0), y_2(x_0)) = 0$, it has nontrivial solutions such that α and β are not equal to zero simultaneously. With such choice of α and β, a linear combination $y = \alpha y_1(x) + \beta y_2(x)$ is the solution of IVP comprised of Eq. (4.2) and zero initial conditions $y(x_0) = 0$, $y'(x_0) = 0$. Clearly, with such initial conditions, Eq. (4.2) has a unique trivial solution: $y(x) = 0$. Thus, $\alpha y_1(x) + \beta y_2(x) \equiv 0$, which proves the linear dependence of functions $y_1(x)$ and $y_2(x)$ in the whole interval $[a, b]$. The obtained contradiction shows that if $y_1(x)$ and $y_2(x)$ are linearly independent, the Wronskian cannot vanish at any point on the interval $[a, b]$.

Let us prove now that if $W(y_1, y_2) \neq 0$ in $[a, b]$, then solutions $y_1(x)$ and $y_2(x)$ are linearly independent. Assume that they are linearly dependent. We have shown already that if two differentiable functions $y_1(x)$ and $y_2(x)$ (not necessarily solutions of the equation) are linearly dependent, then $W(y_1, y_2) = 0$. Thus, we get a contradiction that finishes the proof of the theorem.

Part (b) of Theorem 3.2 can be also formulated as follows: solutions $y_1(x)$ and $y_2(x)$ are linearly independent if and only if their Wronskian is nonzero at least at one point of the interval $[a, b]$.

Let us consider some examples.

1. $y_1(x) = e^x$, $y_2(x) = e^{-x}$. The Wronskian $W(y_1, y_2) = -2$; hence it is nonzero everywhere. Therefore, functions y_1 and y_2 are linearly independent, and they can be solutions of a certain linear second-order homogeneous Eq. (3.19) with continuous functions $p(x)$ and $q(x)$. Indeed, substituting y_1 and y_2 into (3.23), we find:

$$(1 + p + q)e^x = 0, (1 - p + q)e^{-x} = 0;$$

hence $p(x) = 0$, $q(x) = -1$. Thus, y_1 and y_2 are linearly independent solutions of the equation $y'' - y = 0$.

2. $y_1(x) = \cos \alpha x$, $y_2(x) = \sin \alpha x$, and $\alpha \neq 0$. The Wronskian $W(y_1, y_2) = \alpha \neq 0$. Thus, $y_1(x)$ and $y_2(x)$ are linearly independent. Substituting y_1 and y_2 into (3.23), we obtain two equations:

$$(q - \alpha^2) \cos \alpha x - \alpha p \sin \alpha x = 0,$$
$$\alpha p \cos \alpha x + (q - \alpha^2) \sin \alpha x = 0.$$

The equations are compatible if $(q - \alpha^2)^2 + \alpha^2 p^2 = 0$; hence, $p = 0$, $q = \alpha^2$. This means that y_1 and y_2 are the linearly independent solutions of the equation $y'' + \alpha^2 y = 0$.

3. $y_1(x) = x$, $y_2(x) = e^x$. The functions y_1 and y_2 are linearly independent. Indeed, assuming $C_1 x + C_2 e^x = 0$, i.e., $xe^{-x} = C_2/C_1$, we obtain a contradiction: a nonconstant function on the left-hand side equals a constant at the right-hand side. However, the Wronskian $W(y_1, y_2) = (x - 1)e^x$ vanishes in the point $x = 1$. This means that y_1 and y_2 cannot be the solutions of (3.19) with $p(x)$ and $q(x)$ continuous at any x. Indeed, substituting y_1 and y_2 into (4.6), we find:

$$p + qx = 0, (1 + p + q)e^x = 0;$$

thus $p = -x/(x - 1)$, $q = 1/(x - 1)$. Functions $p(x)$ and $q(x)$ are not continuous at $x = 1$.

4. $y_1(x) = e^x$, $y_2(x) = 5e^x$. The Wronskian $W(y_1, y_2) = 0$ everywhere, and the functions are linearly dependent: $5y_1 - y_2 = 0$.

Theorem 3.3

Let $y_1(x)$ and $y_2(x)$ be two linearly independent particular solutions of the homogeneous Eq. (3.19) on $[a,b]$. Then, its general solution on $[a,b]$ is

$$Y(x) = C_1 y_1(x) + C_2 y_2(x), \tag{3.26}$$

where C_1 and C_2 are two arbitrary constants (we will use capital $Y(x)$ for a general solution of a homogeneous equation).

Proof Let $Y(x)$ be an arbitrary solution of the homogeneous linear Eq. (4.2). Let us take an arbitrary point x_0 in the interval $[a, b]$, and define $Y(x_0) = Y_0$ and $Y'(x_0) = Y'_0$. Then, $Y(x)$ can be considered a particular solution of (4.2) satisfying boundary conditions $Y(x_0) = Y_0$, $Y'(x_0) = Y'_0$.

The idea of the proof is the construction of the solution of Eq. (3.19) which satisfies the boundary conditions:

$$y(x_0) = Y_0, \quad y'(x_0) = Y'_0 \tag{3.27}$$

in the form

$$y(x) = C_1 y_1(x) + C_2 y_2(x). \tag{3.28}$$

From Theorem 3.1 it follows that $y(x)$ is a solution of Eq. (3.19). Initial conditions (3.27) give:

$$\begin{aligned} C_1 y_1(x_0) + C_2 y_2(x_0) &= Y_0, \\ C_1 y'_1(x_0) + C_2 y'_2(x_0) &= Y'_0. \end{aligned} \tag{3.29}$$

The determinant of this linear algebraic system of equations for C_1 and C_2 is $W(y_1(x_0), y_2(x_0)) \neq 0$; thus, these coefficients can be unambiguously determined from the system (3.29). Thus, the function $y(x)$ determined by formula (3.28) is a solution of Eq. (3.19) with boundary conditions (3.27), i.e., $y(x)$ is the solution of the same initial value problem as $Y(x)$. Because the solution of the initial value problem is unique, $Y(x) = y(x)$. That means that relation (3.26) holds.

Corollary The maximum number of linearly independent solutions of the linear second-order homogeneous Eq. (4.2) is two.

Any two linearly independent solutions of a second-order linear homogeneous equation are called its *fundamental systems* of solutions.

Example 3.10 Find the general solution of equation $y'' - y = 0$.

Solution One can easily check that $y_1(x) = e^x$ and $y_2(x) = e^{-x}$ are the particular solutions of this equation. The Wronskian $W = \begin{vmatrix} e^x & e^{-x} \\ e^x & -e^{-x} \end{vmatrix} = -2 \neq 0$; thus $y_1(x) = e^x$ and $y_2(x) = e^{-x}$ are linearly independent and their linear combination, $Y(x) = C_1 e^x + C_2 e^{-x}$, is the general solution.

Problems

Check if the following functions are linearly independent:

1. e^x, e^{x-1}.   2. $\cos x, \sin x$.
3. $1, x, x^2$.   4. $4 - x, 2x + 3, 6x + 8$.
5. e^x, e^{2x}, e^{3x}.   6. x, e^x, xe^x.

3.4.2   Reduction of Order for a Linear Homogeneous Equation

If one particular solution of Eq. (3.19), $y_1(x)$, is known, one can find the second, linearly independent, solution using the approach described below.

Let us define a new variable $v(x)$ by the relation

$$y(x) = v(x)y_1(x). \tag{3.30}$$

Substituting (3.30) into (3.19), we obtain equation

$$y_1 v'' + (2y_1' + py_1)v' = 0$$

that does not contain a term with v. That is not surprising: according to (3.30), $v_1 = 1$ is a particular solution of the equation; therefore, a term with v is not possible. That circumstance allows to reduce the order of the equation taking $w = v'$:

$$y_1 w' + (2y_1' + py_1)w = 0. \tag{3.31}$$

This is a separable equation. Dividing (3.31) by $y_1 w$, we get:

$$\frac{w'}{w} = -\frac{2y_1'}{y_1} - p;$$

hence

$$\ln|w(x)| = -2\ln|y_1(x)| - \int_{x_0}^{x} p(x_1)dx_1 + C,$$

where x_1 is a dummy variable, x_0 is an arbitrary point in the region of the definition of equation, and C is an arbitrary constant.

Let us present the obtained solution as

$$v'(x) = w(x) = C_2 \left(\frac{1}{y_1^2(x)} \right) e^{-\int_{x_0}^{x} p(x_1)dx_1},$$

where C_2 is an arbitrary constant. Integrating the obtained equation, we find:

$$v(x) = C_1 + C_2 \int_{x_0}^{x} \frac{dx_2}{y_1(x_2)} e^{-\int_{x_0}^{x_2} p(x_1)dx_1}.$$

Finally, we obtain:

$$y(x) = C_1 y_1(x) + C_2 y_2(x),$$

where

$$y_2(x) = y_1(x) \int_{x_0}^{x} \frac{dx_2}{y_1(x_2)} e^{-\int_{x_0}^{x_2} p(x_1)dx_1}.$$

Example 3.11 Find the general solution of equation

$$x^2 y'' + 4xy' - 4y = 0$$

using the reduction of order.

Solution One can guess that $y_1(x) = x$ is a particular solution of the equation. Substituting $y = vx$, $y' = v + v'x$, and $y'' = 2v' + v''x$ into the equation, we obtain:

$$v''x + 6v' = 0$$

that does not contain a term with v. Denoting $w = v'$, we obtain a separable equation $w'x + 6w = 0$ which solution is $w = C_2 x^{-6}$. Thus, $v' = C_2 x^{-6}$ and $v = -\frac{C_2}{5} x^{-5} + C_1$; hence, the general solution of the equation can be written as $y = C_1 x + C_2/x^4$.

3.4.3 Nonhomogeneous Equations

Next, we consider the nonhomogeneous Eq. (3.18) and prove several useful theorems.

Theorem 3.4
Let $Y(x) = C_1y_1(x) + C_2y_2(x)$ be the general solution of (3.19) on $[a, b]$, and let $\bar{y}(x)$ be the particular solution of the nonhomogeneous Eq. (3.18). Then

$$y(x) = Y(x) + \bar{y}(x) \tag{3.32}$$

is the general solution of (3.18) on $[a, b]$.

Proof First, let us prove that the difference between two arbitrary solutions of the nonhomogeneous Eq. (3.18) is a solution of the homogeneous Eq. (3.19). Assume that $y_1(x)$ and $y_2(x)$ are solutions of (3.19):

$$y_1'' + p(x)y_1' + q(x)y_1 = f(x),$$
$$y_2'' + p(x)y_2' + q(x)y_2 = f(x).$$

Subtracting the first equation from the second equation, we find:

$$\frac{d^2(y_2 - y_1)}{d^2x} + p(x)\frac{d(y_2 - y_1)}{dx} + q(x)(y_2 - y_1) = 0.$$

Thus, $y_2 - y_1$ is a solution of the homogeneous equation.

Assume now that $\bar{y}(x)$ is a particular solution of the nonhomogeneous equation. Then for any other solution $y(x)$ of the nonhomogeneous equation, the difference $y(x) - \bar{y}(x)$ is a certain solution of the homogeneous equation, i.e., for any solution $y(x)$:

$$y(x) - \bar{y}(x) = C_1y_1(x) + C_2y_2(x)$$

with some C_1 and C_2. Thus, all the solutions of the nonhomogeneous equation can be presented in the form:

$$y(x) = \bar{y}(x) + C_1y_1(x) + C_2y_2(x),$$

i.e., the general solution of the nonhomogeneous equation is

$$y(x) = \bar{y}(x) + Y(x).$$

Theorem 3.5 (*Superposition principle for nonhomogeneous equations*).
   Let $y_1(x)$ be a solution of the linear nonhomogeneous equation:

$$y_1'' + p(x)y_1' + q(x)y_1 = f(x) \tag{3.33}$$

and $y_2(x)$ be a solution of the linear nonhomogeneous equation

$$y_2'' + p(x)y_2' + q(x)y_2 = \varphi(x). \tag{3.34}$$

Then, the sum

$$y(x) = y_1(x) + y_2(x) \tag{3.35}$$

is the solution of

$$y'' + p(x)y' + q(x)y = f(x) + \varphi(x). \tag{3.36}$$

Proof Adding Eqs. (3.33) and (3.34) immediately gives (3.36).
   Notice that the superposition principle is valid only for linear equations.

Example 3.12 Solve equation $y'' - y = 1 - 2\sin x$.

Solution It is easy to guess that a particular solution of equation $y_1'' - y_1 = 1$ is
$y_1 = -1$, and a particular solution of equation $y_2'' - y_2 = -2\sin x$ is $y_2 = \sin x$.
Using Theorem 3.5, we get that the particular solution of the equation is
$\bar{y}(x) = -1 + \sin x$, and using the result of Example 3.10, we have
$y(x) = -1 + \sin x + C_1 e^x + C_2 e^{-x}$.
   If a general solution $Y(x)$ of the homogeneous Eq. (3.19) has been determined,
then a particular solution $\bar{y}(x)$ of the nonhomogeneous Eq. (3.18) can be found by the
method of *variation of the parameters*.
   This method implies that we look for the particular solution of the
nonhomogeneous Eq. (3.18) in the form:

$$\bar{y}(x) = u_1(x)y_1(x) + u_2(x)y_2(x). \tag{3.37}$$

   Because we have introduced two new unknown functions, u_1 and u_2, to replace
one unknown function, $\bar{y}(x)$, we can add one more relation, to define u_1 and u_2
uniquely. That will be done below.
   To find functions $u_1(x)$ and $u_2(x)$, we substitute $\bar{y}(x)$, $\bar{y}'(x)$, and $\bar{y}''(x)$ into (3.18).
The first derivative is

$$\bar{y}' = u_1'y_1 + u_1y_1' + u_2'y_2 + u_2y_2'.$$

   Let us impose the additional relation:

$$u_1' y_1 + u_2' y_2 = 0. \tag{3.38}$$

Then

$$\bar{y}' = u_1 y_1' + u_2 y_2'. \tag{3.39}$$

Differentiating that relation gives

$$\bar{y}'' = u_1 y_1'' + u_1' y_1' + u_2 y_2'' + u_2' y_2'. \tag{3.40}$$

Substitution of (3.37), (3.39), and (3.40) into (3.18) gives

$$u_1 \left[y_1'' + p(x)y_1' + q(x)y_1 \right] + u_2 \left[y_2'' + p(x)y_2' + q(x)y_2 \right] + u_1' y_1' + u_2' y_2' = f(x).$$

Since $y_1(x)$ and $y_2(x)$ are the particular solutions of the homogeneous Eq. (3.19), $y_1'' + p(x)y_1' + q(x)y_1 = 0$ and $y_2'' + p(x)y_2' + q(x)y_2 = 0$. Thus, we arrive to the system of algebraic equations for u_1' and u_2':

$$\begin{aligned} y_1 u_1' + y_2 u_2' &= 0, \\ y_1' u_1' + y_2' u_2' &= f(x). \end{aligned} \tag{3.41}$$

The determinant of this system is the Wronskian $W(y_1, y_2)$. Functions $y_1(x)$ and $y_2(x)$ are linearly independent since they are two particular solutions of the homogeneous Eq. (3.19); thus, $W \neq 0$ and the system (3.41) for u_1' and u_2' is uniquely solvable. After u_1' and u_2' have been found, the functions $u_1(x)$ and $u_2(x)$ can be determined by integration. (Notice, because we are searching for a particular solution, the constants of integrations must not be added.) This completes the construction of $\bar{y}(x)$.

Example 3.13 Solve the nonhomogeneous equation $y'' - y = 0$. Then, find the solution of IVP with initial conditions $y(0) = 0$, $y'(0) = 1$.

Solution We guessed the particular solution of that equation in Example 3.12. Now we shall solve the problem using the universal method of the variation of parameters.

The general solution of the homogeneous equation $\bar{y}$ is $Y(x) = C_1 y_1(x) + C_2 y_2(x) = C_1 e^x + C_2 e^{-x}$. Thus, the system (3.41) is $e^x u_1' + e^{-x} u_2' = 0$, $e^x u_1' - e^{-x} u_2' = 1$.

Subtracting the second equation from the first gives $u u_2' = - e^x/2$, and after integration, $u_2(x) = - e^x/2$. Next, adding equations results in $u_1' = e^{-x}/2$, and after integration, $u_1(x) = - e^{-x}/2$. Thus, the particular solution of nonhomogeneous equation is $\bar{y} = u_1(x)y_1(x) + u_2(x)y_2(x) = - 1$ (substitute $\bar{y}(x) = - 1$ in $y'' - y = 1$ to verify that). Finally, $y(x) = Y(x) + \bar{y}(x) = C_1 e^x + C_2 e^{-x} - 1$ is the general solution of the nonhomogeneous equation $y'' - y = 1$. To find the solution of IVP, substitute this $y(x)$ and $y'(x) = Y'(x) + \bar{y}'(x) = C_1 e^x - C_2 e^{-x}$ in the initial conditions

$y(0) = 0$, $y'(0) = 1$, which gives $C_1 + C_2 - 1 = 0$, $C_1 - C_2 = 1$. Solution of this algebraic system is $C_1 = 1$, $C_2 = 0$, and the solution of IVP is $y(x) = e^x - 1$ (to verify, substitute this in the equation and the initial conditions).

3.5   Linear Second-Order Equations with Constant Coefficients

In this section we discuss Eq. (3.18), where the coefficients $p(x)$ and $q(x)$ are constants:

$$y'' + py' + qy = f(x). \tag{3.42}$$

Such constant-coefficient linear equation always can be solved analytically.

3.5.1   Homogeneous Equations

First, consider the corresponding homogeneous equation:

$$y'' + py' + qy = 0. \tag{3.43}$$

Let us search for the particular solutions of (3.43) in the form $y = e^{kx}$. Substitution of this exponent in (3.43) gives $(k^2 + pk + q)e^{kx} = 0$, and then division by $e^{kx} \neq 0$ gives the *characteristic equation*:

$$k^2 + pk + q = 0. \tag{3.44}$$

This quadratic equation has two roots:

$$k_{1,2} = -\frac{p}{2} \pm \sqrt{\frac{p^2}{4} - q}, \tag{3.45}$$

which can be real or complex.

Theorem 3.6

1. *If the roots of the characteristic equation are real and distinct, then $y_1(x) = e^{k_1 x}$ and $y_2(x) = e^{k_2 x}$ are two linearly independent particular solutions of (3.43), and thus*

$$y(x) = C_1 e^{k_1 x} + C_2 e^{k_2 x} \tag{3.46}$$

is the general solution of Eq. (3.43).

2. *If the roots of the characteristic equation are real and repeated, $k_1 = k_2 \equiv k$, then $y_1(x) = e^{kx}$ and $y_2(x) = xe^{kx}$ are two linearly independent particular solutions of (3.43), and thus*

$$y(x) = C_1 e^{kx} + C_2 x e^{kx} \qquad\qquad (3.47)$$

is the general solution of Eq. (3.43).

3. *If the roots of the characteristic equation are complex, $k_{1,\,2} = \alpha \pm i\beta$, then the complex form of two linearly independent particular solutions of (3.43) is $y_1(x) = e^{k_1 x}$ and $y_2(x) = e^{k_2 x}$; a pair of real linearly independent particular solutions is $y_1(x) = e^{\alpha x} \sin \beta x$ and $y_2(x) = e^{\alpha x} \cos \beta x$. Thus, the general solution of (3.43) can be written in two ways:*

complex-valued:

$$y(x) = C_1 e^{(\alpha + i\beta)x} + C_2 e^{(\alpha - i\beta)x}, \qquad\qquad (3.48)$$

or real-valued:

$$y(x) = e^{\alpha x}(C_1 \cos \beta x + C_2 \sin \beta x). \qquad\qquad (3.49)$$

In (3.48) and (3.49), the coefficients C_1 and C_2 are real.

Proof

1. When real $k_1 \neq k_2$ ($q < p^2/4$), the functions $y_1(x) = e^{k_1 x}$ and $y_2(x) = e^{k_2 x}$ are the particular solutions of (3.43) by construction. They are linearly independent because

$$W(y_1, y_2) = \begin{vmatrix} y_1 & y_2 \\ y_1' & y_2' \end{vmatrix} = \begin{vmatrix} e^{k_1 x} & e^{k_2 x} \\ k_1 e^{k_1 x} & k_2 e^{k_2 x} \end{vmatrix} = e^{(k_1 + k_2)x}(k_2 - k_1) \neq 0$$

for all x. Thus $y(x) = C_1 e^{k_1 x} + C_2 e^{k_2 x}$ is the general solution of (3.43).

2. In the case of equal real roots, $q = p^2/4$, $k_1 = k_2 = k$, then our guess, $y_1(x) = e^{kx}$, allows us to find only one particular solution of (3.43); note that $k = -p/2$. Therefore, we have to find another particular solution that is linearly independent of y_1.

For this goal, let us apply the method of the order reduction. Following (3.30), introduce a new variable $v(x)$ via $y(x) = v(x)e^{kx}$. Equation (3.43) becomes: $(v'' + 2kv' + k^2 v) + p(v' + kv) + qv = 0$. Substituting $k = -p/2$, $q = p^2/4$, we find $v'' = 0$. We know that the latter equation has two linearly independent solutions, $v_1(x) = 1$ and $v_2(x) = x$, which correspond to $y_1(x) = e^{kx}$ and $y_2(x) = xe^{kx}$. The linear independence of $y_1 = e^{kx}$ and $y_2 = xe^{kx}$ can be easily checked by calculating Wronskian:

$$W = \begin{vmatrix} y_1 & y_2 \\ y_1' & y_2' \end{vmatrix} = \begin{vmatrix} e^{kx} & xe^{kx} \\ ke^{kx} & e^{kx}(1+xk) \end{vmatrix} = e^{2kx} \neq 0 \text{ for all } x.$$

Thus, we proved that $y(x) = C_1 e^{kx} + C_2 x e^{kx}$ is the general solution of (5.2) when $k = k_1 = k_2$.

3. If $q > p^2/4$, then $k_{1,2} = \alpha \pm i\beta$, where $\alpha = -\frac{p}{2}, \beta = \sqrt{q - p^2/4}$. Thus, we obtain two complex-valued solutions, $y_1 = e^{(\alpha + i\beta)x}$ and $y_2 = e^{(\alpha - i\beta)x}$. These functions are linearly independent, since

$$W = \begin{vmatrix} y_1 & y_2 \\ y_1' & y_2' \end{vmatrix} = \begin{vmatrix} e^{(\alpha+i\beta)x} & e^{(\alpha-i\beta)x} \\ (\alpha + i\beta)e^{(\alpha+i\beta)x} & (\alpha - i\beta)e^{(\alpha-i\beta)x} \end{vmatrix}$$
$$= (\alpha - i\beta)e^{2\alpha x} - (\alpha + i\beta)e^{2\alpha x} = -2i\beta e^{2\alpha x} \neq 0$$

for all x. Therefore, we can take

$$y(x) = C_1 y_1(x) + C_2 y_2(x) = C_1 e^{(\alpha+i\beta)x} + C_2 e^{(\alpha - i\beta)x}$$

as the general solution of (3.43).

To avoid dealing with complex functions when solving equations, where the unknown function $y(x)$ is real by definition, one can use construct real particular solutions using the superposition principle (Theorem 3.1). Let us define two functions:

$$\tilde{y}_1(x) = (y_1 + y_2)/2 = \operatorname{Re} y_1 = e^{\alpha x} \cos \beta x$$

and

$$\tilde{y}_2(x) = (y_1 - y_2)/2i = \operatorname{Im} y_1 = e^{\alpha x} \sin \beta x.$$

The linear independence of these functions is checked by calculating the Wronskian:

$$W = \begin{vmatrix} \tilde{y}_1 & \tilde{y}_2 \\ \tilde{y}_1' & \tilde{y}_2' \end{vmatrix}$$
$$= \begin{vmatrix} e^{\alpha x} \sin \beta x & e^{\alpha x} \cos \beta x \\ \alpha e^{\alpha x} \sin \beta x + \beta e^{\alpha x} \cos \beta x & \alpha e^{\alpha x} \cos \beta x - \beta e^{\alpha x} \sin \beta x \end{vmatrix} = \beta e^{2\alpha x} \neq 0$$

for all x. Thus,

$$y(x) = e^{\alpha x}(C_1 \cos \beta x + C_2 \sin \beta x)$$

is the general solution of (3.43).

Obviously, the combination $C_1 \cos \beta x + C_2 \sin \beta x$ can be presented in more "physical" form:

$$A \cos(\beta x + \varphi) \quad \text{or} \quad A \sin(\beta x + \varphi),$$

where constants A (amplitude) and φ (phase) play the role of two arbitrary constants (instead of C_1 and C_2). Thus, the solution in the case of complex roots, $k_{1,2} = \alpha \pm i\beta$, of a characteristic equation can be presented in the form:

$$y(x) = A e^{\alpha x} \cos(\beta x + \varphi),$$

or

$$y(x) = A e^{\alpha x} \sin(\beta x + \varphi).$$

Below we show several examples based on Theorem 3.6.

Example 3.14 Solve IVP $y'' + y' - 2y = 0$, $y(0) = 1$, $y'(0) = 2$.

Solution The characteristic equation $k^2 + k - 2 = 0$ has two real and distinct roots, $k_1 = 1$, $k_2 = -2$. Thus, the general solution is $y(x) = C_1 e^x + C_2 e^{-2x}$. In order to solve IVP, we first calculate $y'(x) = C_1 e^x - 2C_2 e^{-2x}$, then substitute $y(x)$ and $y'(x)$ in the initial conditions, which gives:

$$\begin{cases} C_1 + C_2 = 1, \\ C_1 - 2C_2 = 2. \end{cases}$$

The solution of this system is $C_1 = 4/3$ and $C_2 = -1/3$. Thus, the solution of IVP is $y(x) = (4e^x - e^{-2x})/3$.

Example 3.15 Solve IVP $y'' - 4y' + 4y = 0$, $y(1) = 1$, $y'(1) = 1$.

Solution The characteristic equation $k^2 - 4k + 4 = 0$ has a repeated real root, $k_{1,2} = 2$; thus, the general solution is $y(x) = C_1 e^{2x} + C_2 x e^{2x}$. In order to solve IVP, substitute $y(x)$ and $y'(x) = 2C_1 e^{2x} + C_2 e^{2x} + 2C_2 x e^{2x}$ in the initial conditions, which gives:

$$\begin{cases} C_1 e^2 + C_2 e^2 = 1, \\ 2C_1 e^2 + 3C_2 e^2 = 1. \end{cases}$$

The solution of this system is $C_1 = 2e^{-2}$ and $C_2 = -e^{-2}$. Thus, the solution of IVP is $y(x) = (2 - x)e^{2(x-1)}$.

Example 3.16 Solve IVP $y'' - 4y' + 13y = 0$, $y(0) = 1$, $y'(0) = 0$.

Solution The characteristic equation $k^2 - 4k + 13 = 0$ has complex conjugated roots $k_{1,2} = 2 \pm 3i$.

The general real-valued solution is $y(x) = e^{2x}(C_1 \cos 3x + C_2 \sin 3x)$. The initial conditions give $C_1 = 1$, $C_2 = -2/3$; thus, the solution of IVP is $y(x) = e^{2x}\left(\cos 3x - \frac{2}{3} \sin 3x\right)$.

Problems

Solve equations:

1. $y'' - 5y' - 6y = 0$.
2. $y''' - 6y'' + 13y' = 0$.
3. $y^{(4)} - y = 0$.
4. $y^{(4)} + 13y^{(2)} + 36y = 0$.
5. $y'' - 5y' + 6y = 0$.
6. $y''' - 4y'' + 3y' = 0$.
7. $y''' + 6y'' + 25y' = 0$.
8. $y''' + 5y'' = 0$.
9. $y''' - 3y'' + 3y' - y = 0$.
10. $y^{(4)} - 8y^{(2)} - 9y = 0$.
11. $y'' + 4y' + 3y = 0$.
12. $y'' - 2y' = 0$.
13. $2y'' - 5y' + 2y = 0$.
14. $y'' - 4y' + 5y = 0$.
15. $y''' - 8y = 0$.
16. $y^{(4)} + 4y = 0$.
17. $y'' - 2y' + y = 0$.
18. $4y'' + 4y' + y = 0$.
19. $y^{(5)} - 6y^{(4)} + 9y''' = 0$.
20. $y''' - 5y'' + 6y' = 0$.
21. $y'' + 2y = 0$, $y(0) = 0$, $y'(0) = 1$.
22. $y''' - y' = 0$, $y(2) = 1$, $y'(2) = 0$, $y''(2) = 0$.
23. $y'' + 4y' + 5y = 0$, $y(0) = -3$, $y'(0) = 0$.

3.5.2 Nonhomogeneous Equations: Method of Undetermined Coefficients

A general solution of the nonhomogeneous Eq. (3.42), $y(x)$, is the sum of a general solution of the corresponding homogeneous Eq. (3.43), which we denote as $Y(x)$, and a particular solution of the nonhomogeneous Eq. (3.42), denoted as $\bar{y}(x)$; thus $y(x) = Y(x) + \bar{y}(x)$. A particular solution $\bar{y}(x)$ can be found using the method of variation of the parameters discussed in Sect. 3.4. This method requires integrations, which is often complicated. For special cases of the right-hand side of Eq. (3.42), which are very common in applications, this equation can be solved using simpler method, called the *method of undetermined coefficients*.

The idea of the method is to choose the particular solution $\bar{y}(x)$ of the nonhomogeneous Eq. (3.42) in the form similar to the term in the right-hand side, $f(x)$, and then determine the coefficients in $\bar{y}(x)$ by substituting $\bar{y}(x)$ in (3.42). Below we consider functions $f(x)$ for which the method is applicable, and in each case give the form of a particular solution $\bar{y}(x)$. We will not prove the method rigorously and simply provide the solution scheme and examples.

1. *Function f(x) is a polynomial.*

Let $f(x)$ be a polynomial of degree n:

$$f(x) = P_n(x). \tag{3.50}$$

Function $\bar{y}(x)$ has to be taken in the form:

$$\bar{y}(x) = x^r Q_n(x), \tag{3.51}$$

where $Q_n(x)$ is a polynomial that has same order as $P_n(x)$ (with the coefficients to be determined), and the value of r equals the number of zero roots of the characteristic equation.

Example 3.17 Solve $y'' - 2y' + y = x$.

Solution First, solve the homogeneous equation $y'' - 2y' + y = 0$. The characteristic equation $k^2 - 2k + 1 = 0$ has the roots $k_1 = k_2 = 1$. Thus, the general solution of the homogeneous equation is $Y(x) = C_1 e^x + C_2 x e^x$.

The right side of the given equation is the first-degree polynomial; thus, we search the particular solution as $\bar{y} = x^r (Ax + B)$ with $r = 0$ (since there are no roots $k = 0$ of the characteristic equation) and A, B will be determined by substituting this $\bar{y}$ and its derivatives, $\bar{y}' = A$, $\bar{y}'' = 0$ into the nonhomogeneous equation. This gives $-2A + Ax + B = x$. Next, equating the coefficients of the like terms (the zero and first power of x) at both sides of the latter algebraic equation gives:

$$\begin{cases} x^0 : & -2A + B = 0 \\ x^1 : & A = 1. \end{cases}$$

From this system, $A = 1$, $B = 2$; thus $\bar{y} = x + 2$. Finally,

$$y(x) = Y(x) + \bar{y}(x) = e^x(C_1 + C_2 x) + x + 2$$

is the general solution of the given equation.

Example 3.18 Solve $y'' - y' = 1$.

Solution The characteristic equation $k^2 - k = k(k - 1) = 0$ has the roots $k_1 = 0$, $k_2 = 1$; thus, the general solution of the homogeneous equation is $Y(x) = C_1 + C_2 e^x$. The right side of the equation is zero-degree polynomial, also $r = 1$ (there is one zero root, $k = 0$, of the characteristic equation); thus, we search for the particular solution in the form $\bar{y} = Ax$. Substituting this $\bar{y}$ and its derivatives, $\bar{y}' = A$, $\bar{y}'' = 0$ into equation results in $A = -1$, thus $\bar{y} = -x$. Finally,

$$y(x) = Y(x) + \bar{y}(x) = C_1 + C_2 e^x - x$$

is the general solution of the given equation.

2. *Function f(x) is a product of a polynomial and an exponent.*

Let $f(x)$ be a product of a polynomial of degree n and $e^{\gamma x}$:

$$f(x) = e^{\gamma x} P_n(x). \tag{3.52}$$

The previous situation (3.50) obviously is a particular case of (3.52) with $\gamma = 0$. Take $\bar{y}(x)$ in the form:

$$\bar{y}(x) = x^r e^{\gamma x} Q_n(x), \tag{3.53}$$

where $Q_n(x)$ is a polynomial of the same degree as $P_n(x)$ (with the coefficients to be determined), and the value of r equals the number of roots of the characteristic equation that are equal to γ.

Example 3.19 Solve $y'' - 4y' + 3y = xe^x$.

Solution Solve the homogeneous equation first: $y'' - 4y' + 3y = 0$, $k^2 - 4k + 3 = 0$, $k_1 = 3$, $k_2 = 1$; thus, $Y(x) = C_1 e^{3x} + C_2 e^x$.

As can be seen, $\gamma = 1$ and since $k_2 = 1 = \gamma$ is a simple root of the characteristic equation, we must take $r = 1$; therefore, $\bar{y} = xe^x(Ax + B)$. Differentiation of $\bar{y}$ gives:

$$\bar{y}' = e^x(Ax^2 + Bx) + e^x(2Ax + B) = e^x(Ax^2 + 2Ax + Bx + B),$$
$$\bar{y}'' = e^x(Ax^2 + 4Ax + Bx + 2A + 2B).$$

Next, substitution of $\bar{y}$, $\bar{y}'$ and $\bar{y}''$ in the nonhomogeneous equation results in

$$e^x\left[(Ax^2 + 4Ax + Bx + 2A + 2B) - 4(Ax^2 + 2Ax + Bx + B) + 3(Ax^2 + Bx)\right] = xe^x.$$

Cancelling $e^x \neq 0$ and equating coefficients of the same powers of x at both sides gives the system:

$$\begin{cases} x^0 : 2A - 2B = 0, \\ x^1 : -4A = 1, \end{cases}$$

which has the solution $A = B = -1/4$. Thus, the particular solution is $\bar{y} = -xe^x(x + 1)/4$. Finally, the general solution is

$$y(x) = Y(x) + \bar{y}(x) = C_1 e^{3x} + C_2 e^x - x(x+1)e^x/4.$$

For comparison, we next solve this problem using the method of variation of the parameters. Varying parameters in the general solution $Y(x) = C_1 e^{3x} + C_2 e^x$ gives $y(x) = u_1(x)e^{3x} + u_2(x)e^x$, and then the system of algebraic equations to find u_1' and u_2' is (see Sect. 3.4)

$$e^{3x}u_1' + e^x u_2' = 0, \quad 3e^{3x}u_1' + e^x u_2' = xe^x.$$

Solving the system gives $u_1' = xe^{-2x}/2$, $u_2' = -x/2$. Then, $u_2 = -x^2/4 + C_2$, and integration by parts for u_1' gives $u_1 = -xe^{-2x}/4 - e^{-2x}/8 + C_2$. Finally, substitution of $u_1(x)$ and $u_2(x)$ into the expression for $y(x)$ gives $y(x) = C_1 e^{3x} + C_2 e^x - x(x+1)e^x/4$. This is the same result that was obtained using the method of undetermined coefficients.

3. *Function $f(x)$ contains sine and/or cosine functions.*

Let

$$f(x) = a \cos \delta x + b \sin \delta x. \tag{3.54}$$

Take $\bar{y}(x)$ in the form:

$$\bar{y} = x^r (A \cos \delta x + B \sin \delta x), \tag{3.55}$$

where A and B are the coefficients to be determined, and the value of r equals the number of roots of the characteristic equation equal to $i\delta$.

Note that in (3.55) for $\bar{y}(x)$, we have to keep both terms even if either a or b is zero in $f(x)$.

Example 3.20 Solve $y'' - 3y' + 2y = \sin x$.

Solution The characteristic equation $k^2 - 3k + 2 = 0$ has the roots $k_1 = 1$, $k_2 = 2$. Thus, $Y(x) = C_1 e^x + C_2 e^{2x}$. Next, we find the particular solution of the nonhomogeneous equation, $\bar{y}$. Since $\delta = 1$ and there is no root equal to $i\delta = i$, we have $r = 0$ and then we search for $\bar{y}$ in the form $\bar{y} = A \cos x + B \sin x$. Substituting $\bar{y}, \bar{y}', \bar{y}''$ in the given equation and equating the coefficients of the like terms at both sides give $A = 0.3$, $B = 0.1$; thus, $y(x) = Y(x) + \bar{y} = C_1 e^x + C_2 e^{2x} + 0.3 \cos x + 0.1 \sin x$.

Example 3.21 Solve $y'' + y = \sin x$.

Solution The characteristic equation $k^2 + 1 = 0$ has the roots $k_1 = i$, $k_2 = -i$. Thus, $Y = C_1 \cos x + C_2 \sin x$. Next, we find the particular solution, $\bar{y}$, of the nonhomogeneous equation. Since $\delta = 1$ and $k_1 = i = i\delta$, we have $r = 1$; thus $\bar{y} = x(A \cos x + B \sin x)$. Substituting $\bar{y}, \bar{y}'$ and $\bar{y}''$

$$\bar{y}' = A\cos x + B\sin x + x(-A\sin x + B\cos x),$$
$$\bar{y}'' = -2A\sin x + 2B\cos x + x(-A\cos x - B\sin x)$$

in equation and equating the coefficients of the like terms at both sides give

$$\begin{cases} \sin x: \ -2A = 1 \\ \cos x: \ 2B = 0. \end{cases}$$

Thus, $\bar{y} = -\frac{x}{2}\cos x$ and $y(x) = C_1\cos x + C_2\sin x - \frac{x}{2}\cos x$ are the general solutions of the given equation.

Notice that often a function $f(x)$ has a form

$$f(x) = c\sin(\delta x + \varphi), \text{ or } f(x) = c\cos(\delta x + \varphi).$$

The second form can be obtained from the first one by adding $\pi/2$ to φ. These forms clearly are equivalent to (3.54) with constants c and φ instead of a and b. In this case a particular solution $\bar{y}(x)$ of the nonhomogeneous equation can be found as

$$\bar{y} = x^r[A\cos(\delta x + \varphi) + B\sin(\delta x + \varphi)],$$

where A and B are the coefficients to be determined and the value of r equals the number of roots of the characteristic equation equal to $i\delta$. Another form of $\bar{y}(x)$ can be

$$\bar{y} = Ax^r\cos(\delta x + \gamma),$$

where the coefficients to be determined are A and γ.

4. *Function $f(x)$ is a product of polynomial, exponent, sine, and/or cosine functions.*

Let $f(x)$ be a form

$$f(x) = e^{\gamma x}(P_m(x)\cos\delta x + Q_n(x)\sin\delta x), \tag{3.56}$$

where $P_m(x)$ and $Q_n(x)$ are polynomials of degrees m and n, respectively.
Take $\bar{y}(x)$ in the form:

$$\bar{y} = e^{\gamma x}(M_p(x)\cos\delta x + N_p(x)\sin\delta x)x^r, \tag{3.57}$$

where M_p and N_p are polynomials of degree p, with p being the largest of m and n, and the value of r equals the number of roots of the characteristic equation that are equal to $\gamma + i\delta$. Note that in (3.57) for $\bar{y}(x)$, we have to keep both terms even if either $P_m(x)$ or $Q_n(x)$ is zero in $f(x)$.
Obviously (3.56) includes the cases (3.50), (3.52), and (3.55).

Example 3.22 Solve $y'' - 2y' + 5y = e^x(\cos 2x - 3 \sin 2x)$.

Solution The characteristic equation $k^2 - 2k + 5 = 0$ has the roots $k_{1,\,2} = 1 \pm 2i$; thus, the general solution of the homogeneous equation is $Y(x) = e^x(C_1 \cos 2x + C_2 \sin 2x)$.

Next, we find the particular solution $\bar{y}(x)$. Since $\gamma + i\delta = 1 + i2 = k_1$, we have $r = 1$ and $\bar{y} = xe^x(A\cos 2x + B\sin 2x)$. Substituting $\bar{y}, \bar{y}', \bar{y}''$ in the nonhomogeneous equation and equating the coefficients of the like terms, $xe^x \cos 2x$ and $xe^x \sin 2x$, at the both sides give $A = 3/4$, $B = 1/4$. Thus, $\bar{y} = xe^x\left(\frac{3}{4} \cos 2x + \frac{1}{4} \sin 2x\right)$, and the general solution of the equation is

$$y(x) = e^x(C_1 \cos 2x + C_2 \sin 2x) + xe^x\left(\frac{3}{4} \cos 2x + \frac{1}{4} \sin 2x\right).$$

The principle of superposition helps to handle the situations where $f(x)$ is a sum of two or more functions considered above.

Example 3.23 Solve IVP $y'' + y = \sin x + x$. $y(0) = 1$, $y'(0) = 0$.

Solution Here $f(x) = f_1(x) + f_2(x) = \sin x + x$. The general solution of the homogeneous equation is $Y(x) = C_1 \cos x + C_2 \sin x$. The particular solution of equation $y'' + y = \sin x$ was obtained in Example 3.21: $\bar{y}_1 = -\frac{x}{2} \cos x$. It remains to find the particular solution of $y'' + y = x$. According to (3.51), we take the particular solution in the form $\bar{y}_2 = Ax + B$. Substitution of $\bar{y}_2$ and $\bar{y}_2'' = 0$ in $y'' + y = x$ gives $A = 1$ and $B = 0$. Thus, the general solution of the equation $y'' + y = x + \sin x$ is the sum, $y(x) = Y(x) + \bar{y}_1(x) + \bar{y}_2(x) = C_1 \cos x + C_2 \sin x - \frac{x}{2} \cos x + x$.

The initial conditions give $C_1 = 1$, $C_2 = -1/2$; thus, the solution of IVP is $y(x) = \cos x - \frac{1}{2} \sin x - \frac{x}{2} \cos x + x$.

3.6 Linear Second-Order Equations with Periodic Coefficients

3.6.1 Hill Equation

Let us consider now the linear homogeneous equation:

$$y'' + p(x)y' + q(x)y = 0, \quad -\infty < x < \infty, \tag{3.58}$$

where $p(x)$ and $q(x)$ are periodic functions

$$p(x + L) = p(x), \quad q(x + L) = q(x). \tag{3.59}$$

Before studying the properties of the solutions of that kind of equations, let us note that it can be transformed to the *Hill equation* (George Hill (1838–1914), an American astronomer and mathematician):

$$u'' + Q(x)u = 0 \tag{3.60}$$

by means of the change of the variable

$$y(x) = a(x)u(x), \tag{3.61}$$

where $a(x)$ is the function that we find below. Substituting (3.61) into (3.58), we obtain:

$$au'' + (2a' + pa)u' + (a'' + pa' + qa)u = 0. \tag{3.62}$$

Thus, the term with the first derivative of u is eliminated, if $a(x)$ satisfies the equation:

$$2a' + pa = 0, \tag{3.63}$$

i.e.,

$$a(x) = \exp\left[-\frac{1}{2}\int_0^x p(x_1)dx_1\right]. \tag{3.64}$$

Substituting (3.64) into (3.62), we obtain (3.60) with

$$Q(x) = -\frac{p'(x)}{2} - \frac{p^2(x)}{4} + q(x).$$

Obviously, $Q(x)$ is periodic:

$$Q(x + L) = Q(x). \tag{3.65}$$

Later on, we consider Eqs. (3.60) and (3.65). Let $u_1(x)$ and $u_2(x)$ be the linearly independent solutions of (3.60), e.g., they are solutions satisfying initial conditions $u_1(0) = 1$, $u_1'(0) = 0$, and $u_2(0) = 0$, $u_2'(1) = 0$, so that the Wronskian

$$W(u_1(x), u_2(x)) = \begin{vmatrix} u_1(x) & u_2(x) \\ u_1'(x) & u_2'(x) \end{vmatrix}$$

is equal to 1 in the point $x = 0$. Due to Abel's formula (3.25), the Wronskian is equal to 1 at any x.

Due to the periodicity of the coefficient $Q(x)$, solutions of Eq. (3.60) have some specific features described below. Let us consider Eq. (3.60) in the point $x + L$. Due to the property (8), we find that

$$u''(x + L) + Q(x)u(x + L) = 0.$$

Thus, $u(x + L)$ itself is a solution of Eq. (3.60). Therefore, $u_1(x + L)$ and $u_2(x + L)$ are linear combinations of $u_1(x)$ and $u_2(x)$:

$$u_1(x + L) = T_{11}u_1(x) + T_{12}u_2(x), \quad u_2(x + L) = T_{21}u_1(x) + T_{22}u_2(x), \quad (3.66)$$

where $\mathbf{T} = \{T_{ij}\}$, $i, j = 1, 2$ is a real matrix called *translation matrix*.

Let us calculate the Wronskian in the point $x + L$ using relation (3.66). We find:

$$W(u_1(x + L), u_2(x + L)) = u_1(x + L)u_2'(x + L) - u_1'(x + L)u_2(x + L)$$
$$= (T_{11}T_{22} - T_{12}T_{21})W(u_1(x), u_2(x)) = \det \mathbf{T} W(u_1(x), u_2(x)).$$

Because $W(u_1(x + L), u_2(x + L)) = W(u_1(x), u_2(x))$, we find that

$$\det \mathbf{T} = 1. \tag{3.67}$$

The eigenvalues of matrix $\mathbf{T}$, *multipliers* $\lambda_{\pm}$, correspond to linear combinations $u_{\pm}(x)$ of functions $u_1(x)$ and $u_2(x)$ satisfying the relations:

$$u_{\pm}(x + L) = \lambda_{\pm} u_{\pm}(x). \tag{3.68}$$

Those eigenvalues are obtained from the equation:

$$\begin{vmatrix} T_{11} - \lambda & T_{12} \\ T_{21} & T_{22} - \lambda \end{vmatrix} = \lambda^2 - \lambda \operatorname{tr} \mathbf{T} + \det \mathbf{T} = 0, \tag{3.69}$$

where $\operatorname{tr} \mathbf{T} = T_{11} + T_{22}$. Solving Eq. (3.69) and taking into account (3.67), we find that

$$\lambda_{\pm} = \frac{1}{2} \operatorname{tr} \mathbf{T} \pm \sqrt{\frac{1}{4} \operatorname{tr} \mathbf{T}^2 - 1}, \tag{3.70}$$

and the product of multipliers

$$\lambda_+ \lambda_- = 1.$$

If $|\operatorname{tr} \mathbf{T}| > 2$, $\lambda_{\pm}$ are real

$$\lambda_{\pm} = \frac{1}{2} \pm i\sqrt{1 - \frac{1}{4}\,\mathrm{tr}\,\mathbf{T}^2}$$

are complex conjugate numbers with $|\lambda_{\pm}| = 1$. We can define real numbers $k_{\pm}$, $k_{+} = -k_{-}$, such that

$$\lambda_{\pm} = \exp(ik_{\pm}L).$$

In that case, functions $u_{\pm}(x)$ can be written as *Floquet function* (Achille Floquet (1847–1920), a French mathematician):

$$u_{\pm}(x) = \exp(ik_{\pm}x)v_{\pm}(x), \tag{3.71}$$

where $v_{\pm}(x)$ are periodic function

$$v_{\pm}(x + L) = v_{\pm}(x). \tag{3.72}$$

Therefore, solutions $u_{\pm}(x)$ are bounded, and $|u_{\pm}(x)|$ are periodic. If $|\mathrm{tr}\,\mathbf{T}| > 2$, $\lambda_{\pm}$ are real:

$$\lambda_{\pm} = \exp(q_{\pm}L), \quad q_{+} = -q_{-},$$

where $q_{\pm}$ are real. In the latter case

$$u_{\pm}(x) = \exp(q_{\pm}x)v_{\pm}(x), \quad v_{\pm}(x + L) = v_{\pm}(x). \tag{3.73}$$

Thus, one of the functions $u_{\pm}(x)$ is unbounded, as $x \to +\infty$, and another one is unbounded, as $x \to -\infty$.

3.6.2   Mathieu Equation

As a useful example of the Hill equation, let us consider the Mathieu equation:

$$u'' + (\omega^2 - h\cos\Omega x)u = 0, \quad -\infty < x < \infty. \tag{3.74}$$

Comparing Eq. (3.74) with Eq. (3.60) governing the linear oscillator, we understand that the physical meaning of the former equation is as follows: the "internal" frequency of the oscillator is not constant, but it is changed with the "external" frequency Ω.

A comprehensive description of solutions of (3.74), known as *Mathieu functions* (Emile Mathieu (1835–1890), a French mathematician), can be found in the books on special functions (see, e.g., [4]). Here we consider only the behavior of solutions in the case where the ratio ω/Ω is close to $1/2$.

Let us take $\Omega = 2$ and $\omega = 1 + \varepsilon$, $|\varepsilon| \gg 1$. Assume that h is also small, $h = O(\varepsilon)$. In the previous subsection, we have seen that either solutions of a Hill equation are bounded or grow exponentially at $|x| \to \infty$. Our goal is to find the conditions for those qualitatively different kinds of behavior.

According to the theory described above, the general solution of the Mathieu equation is the linear superposition of solutions of the type (3.71–14) or (3.73–16). First, let us consider the case $\varepsilon = 0$, $h = 0$. In that case, the general solution of (17) is just

$$u(x) = A \cos x + B \sin x, \tag{3.75}$$

where A and B are constant. It is reasonable to assume that small corrections to the equation only slightly distort the solutions. Actually, for small ε and $h = O(\varepsilon)$, the solution of (3.74) has the structure

$$u(x) = A(x) \cos x + B(x) \sin x + O(\varepsilon), \tag{3.76}$$

where $A(x)$ and $B(x)$ are *slow* functions of x, i.e., A', B' are $O(\varepsilon)$ while A'' and B'' are $O(\varepsilon^2)$. The last term $O(\varepsilon)$ in (3.76) denotes terms with higher harmonics $\cos 3x$ and $\sin 3x$. The expression (3.76) can be obtained by means of the singular perturbation theory that is beyond the scope of the present book (see, e.g., [5]).

Let us substitute (3.76) into (3.74). The terms of order $O(1)$ are cancelled. Leaving only the terms of order $O(\varepsilon)$, we get:

$$- 2A'(x) \sin x + 2B'(x) \cos x + 2\varepsilon(A(x) \cos x + B(x) \sin x)$$
$$- h \cos 2x(A(x) \cos x + B(x) \sin x) = O(\varepsilon),$$

where terms $O(\varepsilon)$ in the right-hand side contain higher harmonics. Taking into account that

$$\cos 2x \cos x = \frac{1}{2}(\cos x + \cos 3x), \qquad \cos 2x \sin x = \frac{1}{2}(- \sin x + \sin 3x)$$

and omitting the higher harmonics, we find:

$$\sin x\left[- 2A'(x) + \left(2\varepsilon + \frac{1}{2}h\right)B(x)\right] + \cos x\left[2B'(x) + \left(2\varepsilon - \frac{1}{2}h\right)A(x)\right] = 0.$$

Thus,

$$A'(x) = \frac{1}{4}(h + 4\varepsilon)B(x), \tag{3.77}$$

$$B'(x) = \frac{1}{4}(h - 4\varepsilon)A(x). \tag{3.78}$$

Differentiating (3.77) with respect to x and using (3.78), we obtain:

$$A''(x) + \frac{1}{16}\left(16\varepsilon^2 - h^2\right)A(x) = 0. \tag{3.79}$$

The function $B(x)$ satisfies a similar equation.

Solving (3.79), we find that if $|\varepsilon| > |h|/4$, the functions $A(x)$ and $B(x)$ are bounded; hence, the solution (3.76) is bounded as well. However, if $|\varepsilon| < |h|/4$, i.e., the ratio of the "internal" frequency of the oscillator ω to the "external" frequency Ω is close to 1/2, then solution (3.79) has exponentially growing solutions; hence, $u(x)$ is also unbounded. The latter phenomenon is called *the parametric resonance.*

In a similar way, one can show that the unbounded growth of solutions is possible also when the ratio $r = \omega/\Omega$ is close to $n/2$, where n is a natural number. The interval of r, where the parametric resonance takes place, is $O(h^n)$ at small h.

3.7 Linear Equations of Order $n > 2$

Everything that has been said above about linear second-order equations extends straightforwardly to linear equations of higher orders. The linear equation of order $\bar{y}(x)$ has the form:

$$y^{(n)} + p_{n-1}(x)y^{(n-1)} + \ldots + p_1(x)y' + p_0(x)y = f(x). \tag{3.80}$$

Its general solution is

$$y(x) = Y(x) + \bar{y}(x), \tag{3.81}$$

where $Y(x)$ is a general solution of the *homogeneous equation*

$$Y^{(n)} + p_{n-1}(x)Y^{(n-1)} + \ldots + p_1(x)Y' + p_0(x)Y = 0 \tag{3.82}$$

and $\bar{y}(x)$ is a particular solution of the *nonhomogeneous Eq. (3.80).*

Function $Y(x)$ is a linear combination of n linearly independent particular solutions (fundamental solutions) of the linear homogeneous Eq. (3.82). The maximum number of such solutions is equal to the order of the equation, n:

$$Y(x) = C_1 y_1(x) + C_2 y_2(x) + \ldots + C_n y_n(x). \tag{3.83}$$

Functions $y_i(x)$ are linearly independent on some interval if and only if the Wronskian, which is the $n \times n$ determinant, is not zero on this interval:

$$W(y_1(x), \ldots, y_n(x)) = \begin{vmatrix} y_1(x) & y_2(x) & \cdots\cdots & y_n(x) \\ y_1'(x) & y_2'(x) & \cdots\cdots & y_n'(x) \\ \cdots & \cdots & \cdots & \cdots \\ y_1^{(n-1)}(x) & y_2^{(n-1)}(x) & \cdots\cdots & y_n^{(n-1)}(x) \end{vmatrix} \neq 0. \tag{3.84}$$

Any n linearly independent solutions of a n-th order linear homogeneous equation are called its *fundamental system of solutions*.

The method of variation of the parameters implies that we search for a particular solution of the nonhomogeneous equation in the form of a function $Y(x)$, where the constants C_i are replaced by the functions $u_i(x)$:

$$\bar{y}(x) = u_1(x)y_1(x) + u_2(x)y_2(x) + \ldots + u_n(x)y_n(x). \tag{3.85}$$

The system of equations for the functions $u_i'(x)$ is a generalization of the system (3.41), and it can be derived by the Reader as an *Exercise*:

$$\begin{aligned} y_1 u_1' + y_2 u_2' + \ldots + y_n u_n' &= 0, \\ y_1' u_1' + y_2' u_2' + \ldots + y_n' u_n' &= 0, \\ \cdots\cdots\cdots\cdots\cdots\cdots\cdots\cdots\cdots\cdots & \\ y_1^{(n-1)} u_1' + y_2^{(n-1)} u_2' + \ldots + y_n^{(n-1)} u_n' &= f(x). \end{aligned} \tag{3.86}$$

Hint Differentiate $\bar{y}(x)$ in Eq. (3.85) and impose the first relation of the system (3.86). Then, differentiate the rest of $\bar{y}'(x)$ and impose the second relation of (3.86), etc., until $\bar{y}^{(n)}$ has been found. Substitute $\bar{y}'(x)$, $\bar{y}''(x)$, $\ldots$, $\bar{y}^{(n)}$ into (3.80), combine the terms with $u_1(x)$, with $u_2(x)$, etc., take into account that the functions $y_i(x)$ obey the homogeneous equation, and you will arrive to the last equation in (3.86).

The determinant of this system is $W(y_1, y_2 \ldots, y_n) \neq 0$; thus, the solution of the system (3.86) exists and is unique. After the functions $u_i'(x)$ have been found from this system, the functions $u_i(x)$ can be determined by integration. When we are searching for a particular solution $y_p(x)$, the constants of integrations should not be added when we obtain $u_i(x)$ from $u_i'(x)$. The constants of integrations enter the general solution of the homogeneous equation, $Y(x)$, which appears in the expression for the general solution of the nonhomogeneous equation, $y(x) = y_p(x) + Y(x)$.

Equation (3.80) is often conveniently written in the *operator* form:

$$\widehat{L}(y) = f(x), \tag{3.87}$$

where

$$\widehat{L}(y) = y^{(n)} + p_{n-1}(x)y^{(n-1)} + \ldots + p_1(x)y' + p_0(x)y \tag{3.88}$$

is called a *linear differential operator*. Its linearity means that

$$\widehat{L}(Cy) = C\widehat{L}(y), \tag{3.89}$$

and for several functions

$$\widehat{L}(C_1y_1 + C_2y_2 + \ldots + C_my_m) = C_1\widehat{L}(y_1) + C_2\widehat{L}(y_2) + \ldots + C_m\widehat{L}(y_m). \tag{3.90}$$

When the right side of a linear nonhomogeneous equation is the sum of several functions

$$f(x) = f_1(x) + f_2(x) + \ldots + f_k(x).$$

Then, the particular solution of a nonhomogeneous equation

$$\widehat{L}(y) = \sum_{i=1}^{k} f_i(x) \tag{3.91}$$

is

$$y(x) = \sum_{i=1}^{k} y_i(x), \tag{3.92}$$

where $y_i(x)$ is a particular solution of equation

$$\widehat{L}(y_i) = f_i(x). \tag{3.93}$$

Using this *principle of superposition*, the solution of a linear differential equation with a complicated right side can be reduced to a solution of several simpler equations.

Theorem 3.7
If $y = U + iV$ is the complex solution of a linear homogeneous Eq. (3.82) with the real coefficients $p_i(x)$, then the imaginary and the real parts of this complex solution (i.e., the real functions $U(x)$ and $V(x)$) are the solutions of this equation.

Proof The proof uses a linear operator $\widehat{L}(y)$. From $\widehat{L}(U + iV) = 0$, it follows that $\widehat{L}(U) + i\widehat{L}(V) = 0$. Thus, the real and imaginary parts of the complex expression at the left-hand side must be zero: $\widehat{L}(U) = \widehat{L}(V) = 0$. Thus, U and V are the solutions of (3.82).

This theorem allows to present a complex form of solution of a linear equation in a real form, like we already did in Sect. 3.5.

Next, consider an IVP for the nth order linear equation. If the functions $p_i(x)$ ($i = 0, \ldots, n - 1$) and $f(x)$ are continuous on the interval $[a, b]$, then a solution of (3.80) or (3.82) with n initial conditions at $x_0 \in [a, b]$

$$y(x_0) = y_0, \quad y'(x_0) = y'_0, \quad y''(x_0) = y''_0, \quad \ldots, \quad y^{(n-1)}(x_0) = y_0^{(n-1)} \tag{3.94}$$

exists and is unique. Here $y_0, y'_0, y''_0, \ldots, y_0^{(n-1)}$ are n real arbitrary numbers – the values of the function $y'' + y = \sin x + x$. and its first $n - 1$ derivatives at x_0. To find the solution of IVP, substitute

$$y(x) = Y(x) + \bar{y}(x) = C_1 y_1(x) + C_2 y_2(x) + \ldots + C_n y_n(x) + \bar{y}(x)$$

(take here $\bar{y}(x) = 0$ in the case of a homogeneous equation) into the initial conditions (3.94). The coefficients C_i can be unambiguously determined from the system of algebraic linear equations:

$$C_1 y_1(x_0) + C_2 y_2(x_0) + \ldots + C_n y_n(x_0) = y_0 - \bar{y}(x_0)$$
$$C_1 y'_1(x_0) + C_2 y'_2(x_0) + \ldots + C_n y'_n(x_0) = y'_0 - \bar{y}'(x_0)$$

$$\cdots\cdots\cdots\cdots\cdots\cdots\cdots\cdots\cdots\cdots\cdots\cdots\cdots\cdots\cdots\cdots\cdots\cdots\cdots\cdots\cdots\cdots$$

$$C_1 y_1^{(n-1)}(x_0) + C_2 y_2^{(n-1)}(x_0) + \ldots + C_n y_n^{(n-1)}(x_0) = y_0^{(n-1)} - \bar{y}^{(n-1)}(x_0)$$

$$\tag{3.95}$$

with a nonzero determinant $W(y_1(x_0), y_2(x_0)\ldots, y_n(x_0)) \neq 0$.

Example 3.24 Find the general solution of the linear nonhomogeneous equation $y''' - y' = 3e^{2x}$. Then, find the solution of IVP with the initial conditions $y(0) = 0$, $y'(0) = 1$, and $y''(0) = 2$.

Solution First, notice that $y_1(x) = e^x$, $y_2(x) = e^{-x}$, and $y_3(x) = 1$ (or any constant) are the particular solutions of the homogeneous equation $y''' - y' = 0$. These solutions are linearly independent, since the Wronskian

$$W(y_1(x), y_2(x), y_3(x)) = \begin{vmatrix} e^x & e^{-x} & 1 \\ e^x & -e^{-x} & 0 \\ e^x & e^{-x} & 0 \end{vmatrix} = 1 \cdot \begin{vmatrix} e^x & e^{-x} \\ e^x & -e^{-x} \end{vmatrix} = -2 \neq 0.$$

Thus, the general solution of the homogeneous equation is

$$Y(x) = C_1 e^x + C_2 e^{-x} + C_3.$$

Next, using variation of the parameters, we look for the particular solution of the nonhomogeneous equation in the form $\bar{y}(x) = C_1(x)e^x + C_2(x)e^{-x} + C_3(x)$. The system of algebraic Eq. (3.86) for C_1', C_2', and C_3' (with the determinant $W(y_1, y_2, y_3) = -2$) is

$$e^x C_1' + e^{-x} C_2' + C_3' = 0,$$
$$e^x C_1' - e^{-x} C_2' = 0,$$
$$e^x C_1' + e^{-x} C_2' = 3e^{2x}.$$

Solving the system (add and subtract the second and the third equations to find C_1' and C_2'; then C_3' is found from the first equation), and then integrating C_1', C_2', and C_3', gives $C_1(x) = 3e^x/2$, $C_2(x) = e^{3x}/2$, and $C_3(x) = -3e^{2x}/2$. This gives $\bar{y}(x) = e^{2x}/2$. Finally, the general solution of the nonhomogeneous equation is

$$y(x) = Y(x) + \bar{y}(x) = C_1 e^x + C_2 e^{-x} + C_3 + e^{2x}/2.$$

Next, substitute this general solution into the initial conditions, which gives $C_1 = C_2 = 0, C_3 = -1/2$. Thus, the solution of IVP is $y(x) = -1/2 + e^{2x}/2$.

Example 3.25 Find the general solution of equation $y''' - y' = 3e^{2x} + 5x$.

Solution The right side of this equation is the sum of two functions, $f_1(x) = 3e^{2x}$ and $f_2(x) = 5x$. The particular solution of the nonhomogeneous equation $y''' - y' = 3e^{2x}$ was found in the previous example ($\bar{y}_1(x) = e^{2x}/2$); the particular solution of the nonhomogeneous equation $y''' - y' = 5x$ is easy to guess: $\bar{y}_2(x) = -5x^2/2$. Using the principle of superposition, we conclude that the particular solution of the nonhomogeneous equation is $\bar{y}(x) = \bar{y}_1(x) + \bar{y}_2(x) = e^{2x}/2 - 5x^2/2$. Thus, the general solution of this equation is

$$y(x) = Y(x) + \bar{y}(x) = C_1 e^x + C_2 e^{-x} + C_3 + e^{2x}/2 - 5x^2/2.$$

Reduction of the Order

In Sect. 3.4, it was shown that the order of a linear homogeneous Eq. (3.19) can be reduced if one particular solution of that equation, $y_1(x)$, is known. For that goal, it is sufficient to carry out the following transformation of the unknown function:

$$y(x) = v(x)y_1(x). \tag{3.96}$$

The known solution of (3.19), $y = y_1(x)$, corresponds to the solution $v = v_1 = 1$ of the linear homogeneous equation for $v(x)$, which is obtained when (3.96) is substituted into (3.19). Therefore, the equation for $v(x)$ does not contain the term proportional to $v(x)$; it contains only the derivatives of $v(x)$. That allows to define $z(x) = v'(x)$ and arrive at the equation of a lower order for $z(x)$.

The method described above works also in the case of Eq. (3.82) of arbitrary order. If a function $y_1(x)$ is known such that $\hat{L}[y_1(x)] = 0$, then, substituting

$$Y(x) = v(x)y_1(x) \tag{3.97}$$

into (3.82), we get a linear homogeneous equation of the nth order for $v(x) = Y(x)/y_1(x)$ that does not contain a term with $v(x)$. That allows to introduce $z(x) = v'(x) = (Y(x)/y_1(x))'$ and obtain an equation of the $(n-1)$-th order for $z(x)$.

If several independent particular solutions, $y_1(x)$, $y_2(x)$, ..., $y_m(x)$, are known, the approach described above can be repeated m times. At the first step, we obtain equation of the $(n-1)$-th order for $z(x) = (Y(x)/y_1(x))'$. We know $m-1$ particular solutions of that equation:

$$z_1 = (y_2/y_1)', z_2 = (y_3/y_1)', \ldots, z_{m-1} = (y_m/y_1)'.$$

It is essential that if functions $y_1(x)$, ..., $y_m(x)$ are linearly independent, then functions $z_1(x)$, ..., $z_{m-1}(x)$ are also linear independent. Indeed, assume that they are linearly dependent, i.e., there are constants C_1, ..., C_{m-1}, not equal to zero simultaneously, that

$$C_1 z_1 + \ldots + C_{m-1} z_{m-1} = 0,$$

i.e.,

$$\frac{d}{dx}\left[C_1 \frac{y_2}{y_1} + \ldots + C_{m-1} \frac{y_m}{y_1} \right] = 0.$$

Thus,

$$C_1 \frac{y_2}{y_1} + \ldots + C_{m-1} \frac{y_m}{y_1} = C_m, \tag{3.98}$$

where C_m is a certain constant. Relation (3.98) can be written as

$$C_1 y_2 + \ldots + C_{m-1} y_m + (-C_m)y_1 = 0. \tag{3.99}$$

Because functions $y_1(x)$, ..., $y_m(x)$ are linearly independent, relation (3.99) holds only if $C_1 = \ldots = C_m = 0$. The obtained contradiction proves the linear independence of $z_1(x)$, ..., $z_{m-1}(x)$.

Applying the transformation, say, $z(x) = u(x)z_1(x)$, $u'(x) = w(x)$, we obtain equation for $w(x)$ which is of order $(n-2)$, etc. Repeating the approach m times, we arrive at the equation of the $(n-m)$th order.

Example 3.26 Solve the following differential equation:

$$x^3 y''' - 3x^2 y'' + 6xy' - 6y = 0, \quad x > 0. \tag{3.100}$$

Solution One can guess that $y_1(x) = x$ is the solution of equation. Hence, it is reasonable to apply the transformation $y(x) = xv(x)$. Substituting

$$y = xv, \quad y' = xv' + v, \quad y'' = xv'' + 2v', \quad y''' = xv''' + 3v''$$

into (3.100), we obtain (for $x \neq 0$):

$$v''' = 0.$$

The general solution of the obtained equation is $v = C_1 + C_2 x + C_3 x^2$, which corresponds to $y = C_1 x + C_2 x^2 + C_3 x^3$.

3.8   Linear Equations of Order $n > 2$ with Constant Coefficients

Everything said in Sect. 3.5 about linear second-order equations with constant coefficients extends in the obvious manner to equations of higher orders. The linear equation of order n with constant coefficients has the form

$$y^{(n)} + a_{n-1} y^{(n-1)} + \ldots + a_1 y' + a_0 y = f(x). \tag{3.101}$$

The *homogeneous* equation is

$$y^{(n)} + a_{n-1} y^{(n-1)} + \ldots + a_1 y' + a_0 y = 0. \tag{3.102}$$

Solution of (3.102) is taken in the form $y = e^{kx}$, and after substitution in (3.102) and cancellation of $e^{kx} \neq 0$, we obtain the *characteristic equation*:

$$k^n + a_{n-1} k^{n-1} + \ldots + a_1 k + a_0 = 0. \tag{3.103}$$

This algebraic equation of n-th order has n roots.

1. For each *simple* real root k of (3.103), the corresponding particular solution is

$$y = e^{kx}. \tag{3.104}$$

2. For real root k of *multiplicity* $r \geq 2$, there are r linearly independent particular solutions:

$$y_1 = e^{kx}, \quad y_2 = xe^{kx}, \quad y_3 = x^2 e^{kx}, \quad \ldots, \quad y_r = x^{r-1} e^{kx}. \tag{3.105}$$

3. For a *simple pair* of complex-conjugate solutions $k_{1,\,2} = \alpha \pm i\beta$, there are two linearly independent, real particular solutions:

$$y_1(x) = e^{\alpha x} \cos \beta x, \tag{3.106}$$
$$y_2(x) = e^{\alpha x} \sin \beta x.$$

They are obtained by taking real and imaginary parts of complex solutions $y_+ = e^{k_+ x}$ and $y_- = e^{k_- x}$ according to Theorem 3.7.

4. For a *pair of complex-conjugate* solutions $k_{1,\,2} = \alpha \pm i\beta$ of *multiplicity* $r \geq 2$, there are $2r$ linearly independent, real particular solutions:

$$y_1(x) = e^{\alpha x} \cos \beta x, \quad y_2(x) = xe^{\alpha x} \cos \beta x, \quad \ldots, \quad y_r(x) = x^{r-1} e^{\alpha x} \cos \beta x,$$
$$y_{r+1}(x) = e^{\alpha x} \sin \beta x, \quad y_{r+2}(x) = xe^{\alpha x} \sin \beta x, \quad \ldots, \quad y_{2r}(x) = x^{r-1} e^{\alpha x} \sin \beta x. \tag{3.107}$$

A general solution of a homogeneous Eq. (3.102) is a linear combination of linearly independent particular (fundamental) solutions $y_i(x)$:

$$Y(x) = \sum_{i=1}^{n} C_i y_1(x). \tag{3.108}$$

Example 3.27 Solve IVP $y''' + 25y' = 0$, $y(\pi) = 0$, $y'(\pi) = 1$, $y''(\pi) = 0$.

Solution The characteristic equation $k^3 + 25k = 0$ has the roots $k_1 = 0$, $k_{2,3} = \pm 5i$. These roots are simple (not repeating); thus, the general solution is $Y(x) = C_1 + C_2 \cos 5x + C_3 \sin 5x$. Substituting $Y(x)$, $Y'(x)$, $Y''(x)$ in initial conditions gives $C_1 = C_2 = 0$, $C_3 = -1/5$; thus, the solution of the IVP is $y(x) = -\frac{1}{5} \sin 5x$.

Example 3.28 Solve IVP $y^{(4)} = 8y'' - 16y$, $y(0) = 1$, $y'(0) = 0$, $y''(0) = 2$, $y'''(0) = 1$.

Solution The characteristic equation corresponding to this homogeneous equation is $k^4 - 8k^2 + 16 = 0$, or $(k^2 - 4)^2 = (k - 2)^2(k + 2)^2 = 0$. Its solutions, $k_{1,2} = 2$, $k_{3,4} = -2$, are real roots, each having multiplicity two; thus, the general solution of the equation is $Y(x) = C_1 e^{2x} + C_2 xe^{2x} + C_3 e^{-2x} + C_4 xe^{-2x}$. Calculating the derivatives and substituting initial conditions gives the system:

$$C_1 + C_3 = 1,$$
$$2C_1 + C_2 - 2C_3 + C_4 = 0,$$
$$4C_1 + 4C_2 + 4C_3 - 4C_4 = 2,$$
$$8C_1 + 12C_2 - 8C_3 + 12C_4 = 1.$$

The solution of the system is $C_1 = 17/12$, $C_2 = -49/12$, $C_3 = -5/12$, $C_4 = 5/12$. Thus, $y(x) = \frac{17}{12}e^{2x} - \frac{49}{12}xe^{2x} - \frac{5}{12}e^{-2x} + \frac{5}{12}xe^{-2x}$ is the solution of IVP.

Now consider *nonhomogeneous* equations. The method of variation of the parameters discussed in Sect. 3.5 gives a particular solution of a nonhomogeneous Eq. (3.101). This method is universal, but often it needs rather difficult integrations to solve the system of equations for functions $C_i'(x)$, and then more integrations to find functions $C_i(x)$.

In some cases, the method of *undetermined coefficients* allows to obtain a particular solution of a nonhomogeneous equation substantially easier.

Example 3.29 Solve $y^{(4)} - 16y = \cos 3x$.

Solution The characteristic equation is $k^4 - 16 = 0$, or $(k^2 - 4)(k^2 + 4) = 0$, and its roots are $k_{1,2} = \pm 2$, $k_{2,3} = \pm 2i$. Thus, the solution of the homogeneous equation is $Y = C_1 e^{2x} + C_2 e^{-2x} + C_3 \cos 2x + C_4 \sin 2x$. Since there is no root $k = 3i$, we have $r = 0$ and then we take the particular solution in the form $\bar{y} = A \cos 3x + B \sin 3x$. Substituting $\bar{y}$ and $\bar{y}^{(4)} = 81A \cos 3x + 81B \sin 3x$ in the given equation and equating the coefficients of sine and cosine terms at both sides give $A = 1/65$, $B = 0$. Thus, the general solution of the nonhomogeneous equation is

$$y = C_1 e^{2x} + C_2 e^{-2x} + C_3 \cos 2x + C_4 \sin 2x + \frac{1}{65} \cos 3x.$$

3.9 Euler Equation

Finally, consider the *Euler equation*:

$$a_0 x^n y^{(n)} + a_1 x^{n-1} y^{(n-1)} + \ldots + a_{n-1} x y' + a_n y = 0. \tag{3.109}$$

Its solution can be searched as $y = x^k$. Substituting in Eq. (3.109) and canceling x^k, we obtain the equation for k:

$$a_0 k(k-1) \ldots (k-n+1) + a_1 k(k-1) \ldots (k-n+2) + \ldots + a_n = 0. \tag{3.110}$$

It can be checked that (for $x > 0$), a real root k_i of a multiplicity α_i of this equation gives the following particular solutions of Eq. (3.109):

$$x^{k_i}, \quad x^{k_i} \ln x, \quad x^{k_i} \ln^2 x, \quad \ldots, \quad x^{k_i} \ln^{\alpha_i - 1} x, \tag{3.111}$$

complex roots $p \pm qi$ of a multiplicity α give the particular solutions:

$$
\begin{aligned}
&x^p \cos(q\ln x), \quad x^p \ln x \cos(q\ln x), \quad \ldots, \quad x^p \ln^{\alpha-1} x \cos(q\ln x), \\
&x^p \sin(q\ln x), \quad x^p \ln x \sin(q\ln x), \quad \ldots, \quad x^p \ln^{\alpha-1} x \sin(q\ln x).
\end{aligned}
\tag{3.112}
$$

Example 3.30 Solve $x^2 y'' + \frac{5}{2} x y' - y = 0$.

Solution Search solution in the form $y = x^k$. This gives equation $k(k-1) + \frac{5}{2} k - 1 = 0$, which has two roots: $k_1 = 1/2$, $k_2 = -2$. Thus, the general solution (for $x > 0$) is

$$y = C_1 x^{\frac{1}{2}} + C_2 x^{-2}.$$

Example 3.31 Solve $x^3 y''' - x^2 y'' + 2xy' - 2y = 0$.

Solution Equation for k is $k(k-1)(k-2) - k(k-1) + 2k - 2 = 0$, or $(k-1)(k^2 - 3k + 2) = 0$. The roots are $k_1 = k_2 = 1$, $k_3 = 2$; thus, a general solution (for $x > 0$) is

$$y = (C_1 + C_2 \ln x)x + C_2 x^2.$$

Example 3.32 Solve $x^2 y'' + xy' + y = 0$.

Solution Equation for k is $k(k-1) + k + 1 = 0$, its roots are $k_{1,2} = \pm i$ and the general solution for $x > 0$ is

$$y = C_1 \cos \ln x + C_2 \sin \ln x.$$

To obtain a solution for $x < 0$, simply replace x by $-x$ in the solution on the interval $x > 0$. Notice that a solution at $x = 0$ cannot be found by the described method.

Problems

1. $y'' + 6y' + 5y = 25x^2 - 2$.
2. $y^{(4)} + 3y'' = 9x^2$.
3. $y'' + 6y = 5e^x$.
4. $y'' + 6y' + 9y = 10 \sin x$.
5. $\frac{d^2 x}{dt^2} - 2x = te^{-t}$.
6. $y'' - 2y' = 4(x + 1)$.
7. $y'' - 2y' - 3y = e^{4x}$.
8. $y'' + y = 4xe^x$.
9. $y'' - y = 2e^x - x^2$.
10. $y'' + y' - 2y = 3xe^x$.
11. $y'' - 3y' + 2y = \sin x$.
12. $y'' + y = 4 \sin x$.
13. $y'' - 3y' + 2y = x \cos x$.
14. $y'' + 3y' - 4y = e^{-4x} + xe^{-x}$.
15. $y'' - 4y' + 8y = e^{2x} + \sin 2x$.
16. $y'' + y = x \sin x$.
17. $y'' + 4y' + 4y = xe^{2x}$.
18. $y'' - 5y' = 3x^2 + \sin 5x$.

19. $y'' + y = 4e^x$, $y(0) = 0$, $y'(0) = -3$.
21. $y'' + 2y' + 2y = xe^{-x}$,
    $y(0) = y'(0) = 0$.
23. $y'' + y = e^x + \cos x$.
25. $y^{(4)} + 2y'' + y = \cos x$.
27. $y'' - y = x$, $y(0) = 1$, $y'(0) = -1$.

29. $x^2 y'' - xy' + y = 0$.
31. $x^2 y'' - xy' - 3y = 0$.

20. $y'' - 2y' = 2e^x$, $y(1) = -1$, $y'(1) = 0$.
22. $y'' - y = 2 \sin x - 4 \cos x$.

24. $y'' + y = \cos x + \cos 2x$.
26. $y'' - 4y = \sin 2x$, $y(0) = 0$, $y'(0) = 0$.
28. $y'' + 4y' + 4y = 3e^{-2x}$,
    $y(0) = 0$, $y'(0) = 0$.
30. $x^2 y'' - 4xy' + 6y = 0$.
32. $y = x(C_1 + C_2 \ln |x| + C_3 \ln^2 |x|)$.

A homogeneous cable of length L with a linear mass density ρ was placed on a horizontal table in such a way that its part of length l_0 is hanging from the table. At time $t = 0$ the cable is released. Find the time when the whole cable will slide off the table for two situations: (a) friction is neglected; (b) friction is taken into account.

Hint The force pulling the cable down is $m(t)g = l(t)\rho g$, where $l(t)$ is the cable's length hanging from the table at time t. The force of friction is $f_{fr} = \mu N$, where μ is the coefficient of friction, and N equals the weight of the part of the cable remaining on the table at time t.

Newton's law of motion for part (a) gives $\rho L l'' = \rho g l$, or $l'' = g l / L$. The initial condition is $l(0) = l_0$. What is the second initial condition? To check your solution, the numerical answer for $L = 2.5m$ and $l_0 = 0.1m$ is time $t \approx 2s$. Next, do part (b).

3.10 Applications

The linear second-order differential equation

$$L(x) \equiv x'' + 2\lambda x' + \omega_0^2 x = f_{max} \sin \omega t \tag{3.113}$$

describes oscillations generated by a source $f(t) = f_{max} \sin \omega t$ with frequency ω. Here ω_0 and λ are constants: the first one is the so-called natural frequency (of a system); the second is the *damping coefficient*.

The solution to this problem can be expressed as the sum of functions:

$$x(t) = \bar{x}(t) + C_1 x_1(t) + C_2 x_2(t),$$

where $\bar{x}(t)$ is a particular solution to the nonhomogeneous problem and $x_1(t)$, $x_2(t)$ are the fundamental solutions (two linearly independent solutions) for the respective homogeneous equation, $L(x) = 0$. First, let us consider the case $\lambda > 0$ (damped oscillator). It is easy to check that:

(a) If $\lambda < \omega_0$, then $x_1(t) = e^{-\lambda t}\cos\tilde{\omega}t$, $x_2(t) = e^{-\lambda t}\sin\tilde{\omega}t$.

(b) If $\lambda = \omega_0$, then $x_1(t) = e^{-\lambda t}$, $x_2(t) = te^{-\lambda t}$.

(c) If $\lambda > \omega_0$, then $x_1(t) = e^{k_1 t}$, $x_2(t) = e^{k_2 t}$.

Here

$$\tilde{\omega} = \sqrt{\omega_0^2 - \lambda^2},\ k_1 = -\lambda - \sqrt{\lambda^2 - \omega_0^2},\ k_2 = -\lambda + \sqrt{\lambda^2 - \omega_0^2}\ \text{(clearly } k_{1,\,2} < 0).$$

Let us seek the particular solution of the nonhomogeneous equation as

$$\bar{x}(t) = a\cos\omega t + b\sin\omega t.$$

Comparing the coefficients of $\cos\omega t$ and $\sin\omega t$ in both sides of equation $\bar{x}'' + 2\lambda\bar{x}' + \omega_0^2\bar{x} = f(t)$, we find that

$$\begin{cases} -b\omega^2 - 2\lambda a\omega + b\omega_0^2 = f_{\max} \\ -a\omega^2 + 2\lambda b\omega + a\omega_0^2 = 0. \end{cases}$$

Thus,

$$a = -\frac{2f_{\max}\lambda\omega}{\left(\omega_0^2 - \omega^2\right)^2 + (2\lambda\omega)^2},\qquad b = \frac{f_{\max}\left(\omega_0^2 - \omega^2\right)}{\left(\omega_0^2 - \omega^2\right)^2 + (2\lambda\omega)^2}.$$

The particular solution $\bar{x}(t)$ can also be written as

$$\bar{x} = A\cos(\omega t + \varphi).$$

With $\cos(\omega t + \varphi) = \cos\omega t\cos\varphi - \sin\omega t\sin\varphi$, we see that $a = A\cos\varphi$, $b = -A\sin\varphi$; thus, the amplitude A and the phase shift φ are

$$A = \sqrt{a^2 + b^2} = \frac{f_{\max}}{\left(\omega_0^2 - \omega^2\right)^2 + (2\lambda\omega)^2},\ \varphi = \arctan\left(\frac{2\lambda\omega}{\omega^2 - \omega_0^2}\right).$$

The initial conditions $x(0)$ and $x'(0)$ determine the constants C_1 and C_2. Obviously, for big t a general solution $C_1 x_1(t) + C_2 x_2(t)$ vanishes because of dissipation and only particular solution $\bar{x}(t)$ describing the forced oscillations survives. For ω close to ω_0, we have resonance with the maximum amplitude $A = f_{\max}/(2\lambda\omega)^2$.

Physical examples for the linear oscillator might include mechanical systems such as a body attached to a string, electrical circuits driven by an oscillating voltage, etc.

If the oscillator gets energy from a certain source rather than dissipates it, the coefficient λ can be negative. In that case, solution (a) is obtained for $\lambda > -\omega_0$, solution (b) for $\lambda = -\omega_0$, and solution (c) for $\lambda < -\omega_0$.

3.10.1 Mechanical Oscillations

Consider a motion of a body of mass m under the action of a periodic force, $F(t)$, when the body is attached to a horizontal spring, like shown in Fig. 3.3. Other forces acting on the body are the force of friction for which we assume a proportionality to body's velocity and the force from the spring, $-kx$, that is, Hooke's law (Robert Hooke (1635–1703), a British scientist), k is the spring constant); $x(t)$ is a displacement of the body from the equilibrium position, $x = 0$.
   The equation of motion is

$$mx'' = F_{tot} = F - kx + F_{fr}.$$

   With $F_{fr}/m = -2\lambda x'$ (here $2\lambda \equiv k/m$) and $F/m = f_{max} \sin \omega t$, this is the Eq. (3.113). In the simplest case, when there is no driving force and force of friction, the solution of the reduced equation, $y'' + \omega_0^2 y = 0$, is $y = A \cos(\omega_0 t + \varphi)$ – thus, oscillations are purely harmonic with the period $T = 2\pi/\omega_0$. For instance, if at $t = 0$, the spring is stretched (or compressed) by $x = A$ and then released with zero initial velocity, the phase $\varphi = 0$. The energy is conserved, and the initial potential energy of the spring, $kA^2/2$, transforms into the kinetic energy of the body, $mv^2/2$ at $x = 0$ in a quarter of the period, $t = T/4$.

3.10.2 RLC Circuit

Consider current oscillations driven by a periodic source (emf), $V(t) = V_{max} \sin \omega t$. The circuit resistance is R, an inductance is L, and a capacitance is C. The current starts at $t = 0$ when the switch is getting closed. The corresponding diagram is show in Fig. 3.4.

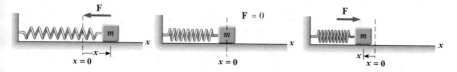

Fig. 3.3 Mechanical oscillations

Fig. 3.4 RLC circuit

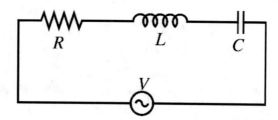

Let us start with the simplest case of the LC circuit when the circuit has only (initially charged) capacitor and an inductor. The *emf* is also absent. When at $t = 0$ the switch is closed, the capacitor discharging causes the current $i(t)$ in the circuit, creating a magnetic field in L, which, in turn, creates an induced voltage in the inductor, $V_L = L\frac{di}{dt}$, acting against the voltage across the capacitor, $V_C = \frac{q}{C}$; $q(t)$ is the charge on the capacitor. The total voltage in the circuit is zero; thus,

$$L\frac{di}{dt} + \frac{q}{C} = 0,$$

which, with $i = \frac{dq}{dt}$, gives the equation

$$q'' + \frac{1}{LC}q = 0,$$

or $q'' + \omega_0^2 q = 0$, where $\omega_0 = 1/\sqrt{LC}$. The solution of this equation describes the harmonic oscillations, $q(t) = q_{max} \cos(\omega_0 t + \varphi)$. If the capacitor initially is charged, then $\varphi = 0$. Physically, these oscillations with the period $T = 2\pi\sqrt{LC}$ occur because the energy of an electric field $\frac{q^2}{2C}$ in C every quarter of the period transforms into the energy of a magnetic field $\frac{Li^2}{2}$ in L.

The voltage across the capacitor is $V(t) = q(t)/C = V_{max} \cos(\omega_0 t + \varphi)$ with $V_{max} = q_{max}/C$. The current in the circuit is

$$i(t) = dq/dt = -\omega_0 q_{max} \sin(\omega_0 t + \varphi) = i_{max} \cos(\omega_0 t + \varphi + \pi/2);$$

thus, when $i = i_{max}$, the charge is $q = 0$ (and $V = 0$), and when $q = q_{max}$, the current and the magnetic field in L are zero.

Now it is easier to discuss a general case of RLC circuit driven by the *emf*. The voltage across the resistance is $Ri(t)$, and because the total voltage across R, L, and C equals the *emf* of the source, we have

$$L\frac{di}{dt} + Ri + \frac{q}{C} = V_{max} \sin \omega t,$$

or

$$q'' + 2\lambda q' + \omega_0^2 q = \frac{V_{max}}{L} \sin \omega t.$$

As we already discussed, the particular solution of this equation is

$$q = q_{max} \cos(\omega t + \varphi),$$

where $q_{max} = \dfrac{V_{max}/L}{\sqrt{\left(\omega_0^2 - \omega^2\right)^2 + (2\lambda\omega)^2}}$, $\varphi = \arctan\left(\dfrac{2\lambda\omega}{\omega^2 - \omega_0^2}\right)$.

Fig. 3.5 Floating body oscillations

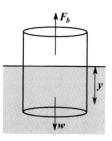

As an interesting result, notice that at $\omega = \omega_0$ the voltage across the capacitor, $V_C = q/C$, is $V_C = \frac{V_{max}}{R}\sqrt{\frac{L}{C}} = V_{max}Q$ (Q is called the quality factor) and can be many times the *emf* amplitude.

3.10.3 Floating Body Oscillations

Consider a body of mass m and uniform cross section A, partially submerged and floating in a fluid. The equilibrium position of the body is determined by the balance of the buoyant force and the force of gravity:

$$F_b = \text{weight of replaced fluid} = \rho V_{sub}g = \rho A y_0 g = mg;$$

thus, $y_0 = m/(\rho A)$. In Fig. 3.5 y is the distance between the bottom of the cylinder and the fluid surface.

Assume that the body initially is moved from the equilibrium, for instance, by pushing down (but still remaining not fully below the fluid surface). The equation of motion of the body (neglecting viscosity) is

$$my''(t) = F_b - w = -\rho A g y(t) + mg;$$

thus,

$$y'' + \frac{\rho A g}{m}y = g.$$

This equation is a particular case of Eq. (3.91), and its solution

$$y = y_0 + A\sin \omega_0 t, \qquad \omega_0 = \rho A g/m,$$

describes the floating body oscillations.

Chapter 4
Systems of Differential Equations

4.1 General Considerations

A system of ordinary differential equations consists of several equations containing derivatives of unknown functions of one variable. The general form of such system of n equations is

$$F_j\left(t, x_1, x_1', \ldots, x_1^{(m_1)}, x_2, x_2', \ldots, x_2^{(m_2)}, \ldots, x_n, x_n', \ldots, x_n^{(m_n)}\right) = 0,$$
$$j = 1, \ldots, n,$$

(4.1)

where $x_j(t)$ are unknown functions, t is the independent variable and F_j are given functions. A general solution of system (4.1) is comprised of all particular solutions, i.e., the sets of functions $x_j(t)$, $j = 1, \ldots, n$, such that when they are substituted into the system, each equation becomes the identity $(0 = 0)$.

As an example, let us consider Newton's equation describing the motion of a mass m under the influence of a force $\vec{f}\left(t, \vec{r}, \vec{r}'\right)$:

$$m\frac{d^2\vec{r}}{dt^2} = \vec{f}\left(t, \vec{r}, \vec{r}'\right),$$

(4.2)

where $\vec{r}(t) = (x(t), y(t), z(t))$ is a vector pointing from the origin of the Cartesian coordinate system to the current position of the mass. Equation (4.2) is equivalent to the system of three second-order differential equations:

$$\begin{cases} m\dfrac{d^2x}{dt^2} = f_x(t,x,y,z,x',y',z'), \\[2ex] m\dfrac{d^2y}{dt^2} = f_y(t,x,y,z,x',y',z'), \\[2ex] m\dfrac{d^2z}{dt^2} = f_z(t,x,y,z,x',y',z'). \end{cases} \tag{4.3}$$

To find the unknown functions $x(t)$, $y(t)$, $z(t)$, one has to specify (at $t = t_0$) the initial position of the mass, $(x(t_0), y(t_0), z(t_0))$, and the three projections of the velocity on the coordinate axes, $(x'(t_0), y'(t_0), z'(t_0))$. This is the total of six initial conditions, i.e., two conditions (position and velocity) for each of the three unknown functions:

$$\begin{cases} x(t_0) = x_0, \\ y(t_0) = y_0, \\ z(t_0) = z_0, \\ x'(t_0) = x'_0, \\ y'(t_0) = y'_0, \\ z'(t_0) = z'_0. \end{cases} \tag{4.4}$$

Equations (4.3) together with the initial conditions (4.4) form the Cauchy problem, or the IVP. If a force $\vec{f}\left(t, \vec{r}, \vec{r}'\right)$ is a sufficiently smooth function, this IVP has a unique solution that describes a specific trajectory of the mass.

It is important that this problem can be reduced to a system of six *first-order* equations by introducing new functions. This is done as follows. Let us rename the functions x, y, z as x_1, x_2, x_3 and introduce three new functions x_4, x_5, x_6 defined as

$$\begin{cases} x_4 = x', \\ x_5 = y', \\ x_6 = z'. \end{cases}$$

System (4.3) now is rewritten as

$$\begin{cases} mx_4' = f_x, \\ mx_5' = f_y, \\ mx_6' = f_z, \\ x_1' = x_4, \\ x_2' = x_5, \\ x_3' = x_6. \end{cases}$$

Thus, we arrive at a system of six first-order differential equations for the functions $x_1, x_2, x_3, x_4, x_5, x_6$. Initial conditions for this system have the form:

$$\begin{cases} x_1(t_0) = x_0, \\ x_2(t_0) = y_0, \\ x_3(t_0) = z_0, \\ x_4(t_0) = x_0', \\ x_5(t_0) = y_0', \\ x_6(t_0) = z_0'. \end{cases}$$

The first three conditions specify the initial coordinates of the mass and the last three ones the projections of the initial velocity on the coordinate axes.

As a particular case, any equation of order $n > 1$ can be rewritten as a system of n first-order equations.

Example 4.1 Rewrite equation $x'' + x = 0$ as a system of two first-order equations.

Solution Denote $x_1 = x$ and $x_2 = x'$. Then, the given equation can be written as the system $x_1' = x_2$, $x_2' = -x_1$.

Example 4.2 Rewrite the system of two second-order equations:

$$\begin{cases} x'' = a_1 x + b_1 y + c_1 y' + f_1(t), \\ y'' = a_2 x + b_2 y + c_2 x' + f_2(t) \end{cases}$$

as a system of four first-order equations.

Solution Let us introduce the notations:

$$x_1 = x, x_2 = y, x_3 = x', x_4 = y'.$$

Then, the system takes the form:

$$x_3' = a_1 x_1 + b_1 x_2 + c_1 x_4 + f_1(t),$$
$$x_4' = a_2 x_1 + b_2 x_2 + c_2 x_3 + f_2(t).$$

Note that these equations are the first-order equations. Two additional first-order equations are (see the notations):

$$x_3 = x_1',$$
$$x_4 = x_2'.$$

Finally, the first-order system that is equivalent to the original second-order system is

$$\begin{cases} x_1' = x_3, \\ x_2' = x_4, \\ x_3' = a_1 x_1 + b_1 x_2 + c_1 x_4 + f_1(t), \\ x_4' = a_2 x_1 + b_2 x_2 + c_2 x_3 + f_2(t). \end{cases}$$

These examples illustrate the general principle: any system of differential equations can be reduced to a larger system of a first-order equation. Thus, it is sufficient to study only the systems of a first-order differential equation. In the next section, we introduce such systems.

4.2 Systems of First-Order Differential Equations

A general system of n first-order differential equations can be written as

$$\begin{cases} \dfrac{dx_1}{dt} = f_1(t, x_1, x_2, \ldots, x_n), \\ \dfrac{dx_2}{dt} = f_2(t, x_1, x_2, \ldots, x_n), \\ \quad \cdot \ \cdot \ \cdot \ \cdot \ \cdot \ \cdot \ \cdot \ \cdot \ \cdot \ \cdot \ \cdot \\ \dfrac{dx_n}{dt} = f_n(t, x_1, x_2, \ldots, x_n), \end{cases} \tag{4.5}$$

where $x_i(t)$ are the unknown functions and f_i are the given functions. A *general solution* of a system (4.5) contains n arbitrary constants, and it has the form:

$$x_i = x_i(t, C_1, C_2, \ldots, C_n), \quad i = \overline{1, n}.$$

When solving a real-world problem using a system (4.5), a solution of IVP (a *particular solution*) is determined from a general solution by specifying a certain *initial condition* for each function $x_i(t)$:

$$x_i(t_0) = x_{i0} \quad i = \overline{1, n}. \tag{4.6}$$

Solution $x_i(t)$ determines a solution curve (called an *integral curve*) in the space $(t, x_1, x_2, \ldots, x_n)$.

Next, we state the existence and uniqueness theorem for the solution of a system (4.5).

Theorem (*Existence and Uniqueness of a Solution*)

System (4.5) together with the initial conditions (4.6) has a unique solution (which is determined by a single set of constants C_i), if the functions f_i and their partial derivatives with respect to all arguments, $\frac{\partial f_i}{\partial x_k}$ $(i, k = \overline{1, n})$, are continuous in the vicinity of the initial condition point (that is in some domain D in the space $(t, x_1, x_2, \ldots, x_n)$ that contains the initial point $(t_0, x_{10}, x_{20}, \ldots, x_{n0})$).

Solution $x_i(t)$ $(i = 1, \ldots, n)$ of a system (4.5) can be represented by a vector function X, whose components are $x_i(t)$. Similarly, the set of functions f_i can be represented by a vector function F (we use capital letters for vector functions):

$$X = \begin{pmatrix} x_1 \\ x_2 \\ \vdots \\ x_n \end{pmatrix}, \quad F = \begin{pmatrix} f_1 \\ f_2 \\ \vdots \\ f_n \end{pmatrix}.$$

Using this notation, we can compactly rewrite the original system (4.5) and the initial conditions (4.6) in the *vector form*:

$$\frac{dX}{dt} = F(t), \tag{4.7}$$

$$X(t_0) = X_0. \tag{4.8}$$

In some cases, one can solve a system (4.5) by reducing it to a single high-order equation. Using Eq. (4.5), as well as equations that are obtained by differentiating (4.5), we can get, and then solve, a single n-th order equation for one of the unknown functions $x_i(t)$. The remaining unknown functions are found from the equations of the original system (4.5) and from the intermediate equations that have been obtained by differentiation. Examples 4.3 and 4.4 illustrate this method.

Example 4.3 Find the general solution of the system:

$$\begin{cases} \dfrac{dx}{dt} = y, \\ \dfrac{dy}{dt} = -x, \end{cases}$$

by reducing it to a single second-order equation.

Solution Let us differentiate the first equation:

$$\frac{d^2x}{dt^2} = \frac{dy}{dt}.$$

Using the second equation, we arrive at a second-order equation for x:

$$x'' + x = 0.$$

Solution of this equation is $x(t) = C_1 \cos t + C_2 \sin t$. Substitution of this solution into $y = \frac{dx}{dt}$ and differentiation give $y(t)$. Thus, we finally have the general solution:

$$\begin{cases} x(t) = C_1 \cos t + C_2 \sin t, \\ y(t) = -C_1 \sin t + C_2 \cos t. \end{cases}$$

Remark We found $y(t)$ from the equation $y = \frac{dx}{dt}$. It seems at first glance that the same solution can be obtained by substituting $x(t) = C_1 \cos t + C_2 \sin t$ into the second equation of the system:

$$\frac{dy}{dt} = -x = -C_1 \cos t - C_2 \sin t,$$

and then integrating. However, upon integration there appears an additional constant:

$$y(t) = -C_1 \sin t + C_2 \cos t + C_3.$$

This function $y(t)$ cannot be the solution of a first-order system, since the number of constants exceeds the number of equations in the system. Indeed, when this $y(t)$ (and $x(t)$ that we had found) is substituted in the system, it becomes clear that $y(t)$ is a solution only when $C_3 = 0$ (check). Thus, the second function should preferably be determined without integration.

To visualize our solution, let us square $x(t)$ and $y(t)$, and add:

$$x^2(t) + y^2(t) = C_1^2 + C_2^2.$$

This equation describes a family of circles in the (x, y)-plane, centered at the origin (see Fig. 4.1). The curves described parametrically by equations $x(t) = C_1 \cos t + C_2 \sin t$ and $y(t) = -C_1 \sin t + C_2 \cos t$ are called the *phase curves*, and the plane in which they are located is the *phase plane*. Substituting initial conditions into the general solution, we can obtain values of the integration constants C_1, C_2, and thus obtain a circle of a certain radius $\sqrt{C_1^2 + C_2^2}$ in the phase plane. Thus, to every set of initial conditions, there corresponds a certain particular phase

Fig. 4.1 Phase plane in
Example 4.3

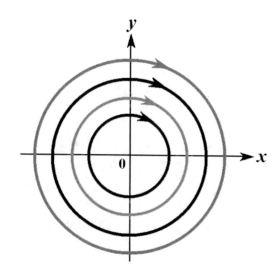

curve. Take, for instance, the initial conditions $x(0) = 0$, $y(0) = 1$. Substitution into
the general solution results in $C_1 = 0$, $C_2 = 1$, and thus the IVP solution has the form
$x(t) = \sin t$, $y(t) = \cos t$. When t increases from zero on the interval $[0, 2\pi]$, the
circular phase curve is traced clockwise: point $(0, 1)$ on the y-axis corresponds to
value $t = 0$, point $(1, 0)$ on the x-axis corresponds to $t = \pi/2$, point $(0, -1)$
corresponds to $t = \pi$, $(-1, 0)$ corresponds to $t = 3\pi/2$, and finally, at $t = 2\pi$ the
phase curve returns to the initial point $(0, 1)$.

Note that the phase plane is an efficient way to "visualize" the qualitative
behavior of solutions. We shall use it extensively in Chap. 5.

Example 4.4

(a) Find the general solution of the system of two first-order equations:

$$\begin{cases} \dfrac{dx}{dt} = -x - 5y, \\ \dfrac{dy}{dt} = x + y \end{cases}$$

by reducing the system to a single second-order equation; (b) Solve IVP, given
the initial conditions $x(\pi/2) = 0$, $y(\pi/2) = 1$.

Solution Let us differentiate the second equation:

$$\frac{d^2y}{dt^2} = \frac{dx}{dt} + \frac{dy}{dt}.$$

Substitute the derivative $\frac{dx}{dt}$ from the first equation:

$$\frac{d^2y}{dt^2} = -x - 5y + \frac{dy}{dt}.$$

The variable x can be taken from the second equation:

$$x = \frac{dy}{dt} - y,$$

and then the second-order equation for $y(t)$ is

$$\frac{d^2y}{dt^2} + 4y = 0.$$

We obtained a homogeneous second-order linear differential equation with constant coefficients. Its characteristic equation is

$$\lambda^2 + 4 = 0,$$

which has solutions $\lambda_{1,2} = \pm 2i$. Then, the first of two functions in the general solution is

$$y(t) = C_1 \cos 2t + C_2 \sin 2t.$$

Next, using $x = \frac{dy}{dt} - y$, we find $x(t)$, the second function in the general solution:

$$x(t) = (2C_2 - C_1) \cos 2t - (2C_1 + C_2) \sin 2t.$$

To solve the IVP, we substitute the initial conditions in the general solution:

$$\begin{cases} 0 = C_1 - 2C_2, \\ 1 = -C_1 \end{cases}$$

and obtain the integration constants: $C_1 = -1$, $C_2 = -1/2$.
Finally, the solution of the IVP is

$$\begin{cases} x(t) = \dfrac{5}{2} \sin 2t, \\ y(t) = -\cos 2t - \dfrac{1}{2} \sin 2t. \end{cases}$$

Functions $x(t)$ and $y(t)$ are shown in Fig. 4.2.
Note that not every system of n first-order equations can be written as an nth order equation (see Example 4.5). Otherwise, there would be little need in the present chapter. The methods for solving systems of first-order equations presented in this

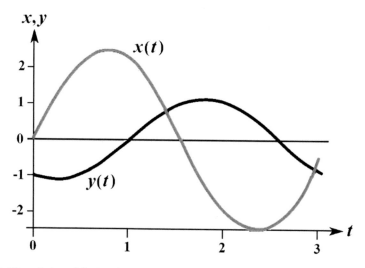

Fig. 4.2 The solution of the IVP in Example 4.4 (on the interval [0,3])

chapter can be applied to problems considered in the previous chapter, but their scope is wider.

Example 4.5 Even when it is impossible to transform a system to a single n-th order equation, the elimination of a subset of variables can be helpful for finding a general solution. As the example, consider the following system of equations:

$$\begin{cases} \dfrac{dx}{dt} = -2x + y - 2z, \\[2mm] \dfrac{dy}{dt} = x - 2y + 2z, \\[2mm] \dfrac{dz}{dt} = 3x - 3y + 5z. \end{cases} \qquad (4.9)$$

Eliminating z from the first equation

$$z = -\frac{1}{2}\frac{dx}{dt} - x + y/2, \qquad (4.10)$$

and substituting (4.10) into the other two equations in (4.9), we find:

$$\frac{dy}{dt} = -\frac{dx}{dt} - x - y \qquad (4.11)$$

and

$$\frac{d^2x}{dt^2} - 3\frac{dx}{dt} - \frac{dy}{dt} - 4x - y = 0. \tag{4.12}$$

Substituting the expression for dy/dt given by (4.11) into (4.12), we find that the variable y cancels, and we obtain a closed second-order equation for x:

$$\frac{d^2x}{dt^2} - 2\frac{dx}{dt} - 3x = 0. \tag{4.13}$$

Thus, the construction of a single third-order ODE that is equivalent to (4.9) is impossible here.

Solving (4.13) by the method described in Chap. 3, we find:

$$x = C_1 e^{3t} + C_2 e^{-t} \tag{4.14}$$

where C_1 and C_2 are arbitrary constants. Substituting (4.14) into (4.11), we obtain the inhomogeneous equation:

$$\frac{dy}{dt} + y = -4C_1 e^{3t}.$$

Solving that equation, we find:

$$y = -C_1 e^{3t} + C_3 e^{-t}, \tag{4.15}$$

where C_3 is another arbitrary constant. Finally, substituting (4.14) and (4.15) into (4.10), we find:

$$z = -3C_1 e^{3t} - \frac{1}{2}C_2 e^{-t} + \frac{1}{2}C_3 e^{-t}. \tag{4.16}$$

Collecting (4.14), (4.15), and (4.16), we obtain the general solution of the system (4.9):

$$(x, y, z) = C_1(1, -1, -3)e^{3t} + C_2(1, 0, -1/2)e^{-t} + C_3(0, 1, 1/2)e^{-t}.$$

Notice that we got two terms with different arbitrary constants C_2 and C_3, and the same exponential function e^{-t}. That never happens in the case of higher-order ODEs, as explained in Sect. 4.6.

Sometimes a system of differential equations can be easily solved by finding *integrating combinations* of the unknown functions. Next, we consider the example of using this solution method.

Example 4.6 Solve the system:

$$\begin{cases} \dfrac{dx}{dt} = 3x + y, \\ \dfrac{dy}{dt} = x + 3y, \end{cases}$$

by finding an integrating combination.

Solution Addition of two equations (a first integrating combination) gives a separable equation for $x + y$:

$$\frac{d(x+y)}{dt} = 4(x+y).$$

Separating variables and integrating, we obtain:

$$\ln|x+y| = 4t + \ln C_1, \quad \text{or} \quad x+y = C_1 e^{4t}.$$

Next, subtract the second equation from the first one (a second integrating combination). This results in the separable equation for $x - y$:

$$\frac{d(x-y)}{dt} = 2(x-y).$$

Separation of variables and integration gives $x - y = C_2 e^{2t}$.

From the two linear algebraic equations for x and y, one can easily obtain the general solution of the system:

$$\begin{cases} x(t) = C_1 e^{4t} + C_2 e^{2t}, \\ y(t) = C_1 e^{4t} - C_2 e^{2t}. \end{cases}$$

Problems

Find general solutions of systems of differential equations by increasing the order. In the problems 1–6, convert a system to a second-order equation for $y(t)$; in the problems 7–10, convert a system to a second-order equation for $x(t)$. Then, solve the IVP. In the problems 1–5, use the initial conditions $x(0) = 1$, $y(0) = 0$. In the problems 6–10, use the initial conditions $x(0) = 0$, $y(0) = 1$. Plot the graph of the obtained particular solution with the program *ODE first order*.

1. $\begin{cases} x' = -2x + 4y, \\ y' = -x + 3y. \end{cases}$

   6. $\begin{cases} x' = 5x + 2y, \\ y' = -4x - y. \end{cases}$

2. $\begin{cases} x' - 2y + 3x = 0, \\ y' - y + 2x = 0. \end{cases}$
7. $\begin{cases} x' = 3x + y, \\ y' = -x + y. \end{cases}$

3. $\begin{cases} x' + 2x + 3y = 0, \\ y' + x = 1. \end{cases}$
8. $\begin{cases} x' = 3x + 2y, \\ y' = -x - y. \end{cases}$

4. $\begin{cases} x' = -3x - y - 1, \\ y' = x - y. \end{cases}$
9. $\begin{cases} x' = -x - 2y, \\ y' = 3x + 4y. \end{cases}$

5. $\begin{cases} x' = x - 3y, \\ y' = 3x + y. \end{cases}$
10. $\begin{cases} x' = 2x - y, \\ y' = x + 2y. \end{cases}$

4.3   Systems of First-Order Linear Differential Equations

A system of ordinary differential equations is *linear* if the right-hand sides of equations are linear with respect to all unknown functions. Consider a system of n linear differential equations:

$$\begin{cases} \dfrac{dx_1}{dt} = a_{11}(t)x_1 + a_{12}(t)x_2 + \cdots + a_{1n}(t)x_n + f_1(t), \\[2mm] \dfrac{dx_2}{dt} = a_{21}(t)x_1 + a_{22}(t)x_2 + \cdots + a_{2n}(t)x_n + f_2(t), \\[2mm] \cdot\;\cdot\;\cdot\;\cdot\;\cdot\;\cdot\;\cdot\;\cdot\;\cdot\;\cdot\;\cdot\;\cdot\;\cdot\;\cdot\;\cdot\;\cdot\;\cdot\;\cdot\;\cdot\;\cdot \\[2mm] \dfrac{dx_n}{dt} = a_{n1}(t)x_1 + a_{n2}(t)x_2 + \cdots + a_{nn}(t)x_n + f_n(t), \end{cases} \tag{4.17}$$

where $a_{ij}(t)$ $(i,j = \overline{1,n})$ and $f_i(t)$ $(i = \overline{1,n})$ are given functions. (Notice that in (4.17) and in (4.5), $f_i(t)$ have different meaning.)

System (4.17) also can be written in a vector form:

$$\frac{dX}{dt} = AX + F(t), \tag{4.18}$$

where $A = \left[a_{ij}(t) \right]$ $(i,j = \overline{1,n})$ is the coefficient matrix, $F(t)$ is the n-dimensional vector with the components $f_1(t), f_2(t), \ldots, f_n(t)$:

$$A(t) = \begin{pmatrix} a_{11}(t) & \cdots & a_{1n}(t) \\ \vdots & \ddots & \vdots \\ a_{n1}(t) & \cdots & a_{nn}(t) \end{pmatrix}, \quad F(t) = \begin{pmatrix} f_1(t) \\ \vdots \\ f_n(t) \end{pmatrix}, \tag{4.19}$$

and $X(t)$ is a vector with the components $x_1, x_2, \ldots, x_n$. We assume that functions $a_{ij}(t)$ and $f_i(t)$ are continuous on the interval $a < t < b$; thus, by the existence and uniqueness theorem system (4.17) has a unique solution for any given initial conditions at a point $t = t_0$ in that interval.

The properties of first-order linear differential systems are generally analogous to the properties of higher-order linear differential equations. Below we briefly state some of these properties.

Consider a *homogeneous system* $(F = 0)$:

$$\frac{dX}{dt} = AX, \qquad (4.20)$$

and let us assume that we know n solutions of that system:

$$X^{(1)} = \begin{pmatrix} x_1^{(1)}(t) \\ x_2^{(1)}(t) \\ \vdots \\ x_n^{(1)}(t) \end{pmatrix}, \quad X^{(2)} = \begin{pmatrix} x_1^{(2)}(t) \\ x_2^{(2)}(t) \\ \vdots \\ x_n^{(2)}(t) \end{pmatrix}, \quad \ldots, \quad X^{(n)} = \begin{pmatrix} x_1^{(n)}(t) \\ x_2^{(n)}(t) \\ \vdots \\ x_n^{(n)}(t) \end{pmatrix}. \qquad (4.21)$$

The combination of these vectors

$$\sum_{k=1}^{n} C_k X^{(k)}, \qquad (4.22)$$

is called a *linear combination* of the solutions $X^{(k)}$. Here $C_1, C_2, \ldots, C_n$ are constants. This vector function has the following properties.

Property 4.1 (*Principle of Superposition*). A linear combination of the solutions of a homogeneous system is also the solution of a system.

This property is easily verified by a substitution of $\sum_{k=1}^{n} C_k X^{(k)}$ into the homogeneous system (4.20).

We say that particular solutions (4.21) are linearly dependent on the interval (a, b), if there exists a nonzero set of n coefficients C_k, such that at every point in (a, b)

$$\sum_{k=1}^{n} C_k X^{(k)} = 0. \qquad (4.23)$$

If no such set of coefficients exists, then the vector functions (4.21) are linearly independent. Set of n linearly independent solutions (4.21) of a system (4.20) is called *the fundamental solution set* of this system. Note that for linearly independent solutions, the determinant of the matrix

$$Y(t) = \begin{pmatrix} x_1^{(1)} & x_1^{(2)} & \cdots & x_1^{(n)} \\ x_2^{(1)} & x_2^{(2)} & \cdots & x_2^{(n)} \\ \cdots & \cdots & \cdots & \cdots \\ x_n^{(1)} & x_n^{(2)} & \cdots & x_n^{(n)} \end{pmatrix},$$

which is called the fundamental matrix, is nonzero at any t.

Property 4.2 For a linear system of a differential equations (4.20), a fundamental solution set always exists, and arbitrary linear combinations of the solutions in a fundamental set form a *general solution* of a linear homogeneous system:

$$X(t) = \sum_{k=1}^{n} C_k X^{(k)}(t). \tag{4.24}$$

For an initial value problem that includes given n initial conditions

$$x_i(t_0) = x_{i0}, \quad i = \overline{1, n}, \tag{4.25}$$

the corresponding values of the coefficients C_1, C_2, ..., C_n are calculated by a substitution of (4.24) into (4.25) and the solution of the obtained system of algebraic equations with respect to the unknowns C_1, C_2, ..., C_n. Solution of that system is unique, because a determinant of a fundamental matrix $Y(t_0)$ is nonzero. Substitution of the coefficients' values into the general solution gives IVP solution for a system (4.20).

Property 4.3 A general solution of a nonhomogeneous system is a sum of a general solution of a homogeneous system and a particular solution of a nonhomogeneous system:

$$x_i(t) = X_i(t) + x_i(t), \quad (i = \overline{1, n}).$$

The proof of this property is similar to the one in the case of a single linear equation presented in Chaps. 2 and 3.

A self-consistent theory describing solution of first-order systems of linear differential equations is presented in Sect. 4.6.

4.4 Systems of Linear Homogeneous Differential Equations with Constant Coefficients

Discussion in Sect. 4.3 provides the solution scheme for a system of linear equations with constant coefficients a_{ij}. For such (nonhomogeneous) system

$$\begin{cases} \dfrac{dx_1}{dt} = a_{11}x_1 + a_{12}x_2 + \cdots + a_{1n}x_n + f_1(t), \\[2mm] \dfrac{dx_2}{dt} = a_{21}x_1 + a_{22}x_2 + \cdots + a_{2n}x_n + f_2(t), \\[1mm] \cdots \cdots \cdots \cdots \cdots \cdots \cdots \cdots \cdots \cdots \cdots \\[1mm] \dfrac{dx_n}{dt} = a_{n1}x_1 + a_{n2}x_2 + \cdots + a_{nn}x_n + f_n(t), \end{cases} \tag{4.26}$$

an analytical solution is always possible.

Solution of system (4.26) starts with a determination of a general solution of a corresponding homogeneous system:

$$\begin{cases} \dfrac{dx_1}{dt} = a_{11}x_1 + a_{12}x_2 + \cdots + a_{1n}x_n, \\[2mm] \dfrac{dx_2}{dt} = a_{21}x_1 + a_{22}x_2 + \cdots + a_{2n}x_n, \\[1mm] \cdots \cdots \cdots \cdots \cdots \cdots \cdots \cdots \cdots \cdots \cdots \\[1mm] \dfrac{dx_n}{dt} = a_{n1}x_1 + a_{n2}x_2 + \cdots + a_{nn}x_n. \end{cases} \tag{4.27}$$

The vector form of the system (4.27) is

$$\frac{dX}{dt} = AX, \tag{4.28}$$

where A is the constant coefficient matrix. Next, we describe a standard solution method for (4.27) which does not involve raising a system's order. The method is based on the determination of n linearly independent particular solutions (a fundamental set) and then forming a linear combination of these solutions – which is a general solution of a system by Property 4.2.

We look for solutions of a system (4.27) in the form:

$$X = V e^{\lambda t}, \tag{4.29}$$

where $V = \begin{pmatrix} v_1 \\ \cdots \\ v_n \end{pmatrix}$ is a column vector of the coefficients that must be determined

(below we write this vector in the form $(v_1, \ldots, v_n)$).

Let us substitute (4.29) in (4.27), move all terms to the right-hand side, and cancel the exponent $e^{\lambda t}$. This results in a system of n linear homogeneous algebraic equations:

$$\begin{cases} (a_{11} - \lambda)v_1 + a_{12}v_2 + \cdots + a_{1n}v_n = 0, \\ a_{21}v_1 + (a_{22} - \lambda)v_2 + \cdots + a_{2n}v_n = 0, \\ \cdots \cdots \cdots \cdots \cdots \cdots \cdots \cdots \cdots \\ a_{n1}v_1 + a_{n2}v_2 + \cdots + (a_{nn} - \lambda)v_n = 0. \end{cases} \tag{4.30}$$

This system is compactly written as follows (I is the identity $n \times n$ matrix):

$$(A - \lambda I)V = 0. \tag{4.31}$$

As it is well known from linear algebra, the system (4.30) has a nontrivial solution only when its determinant is zero:

$$\Delta = \begin{vmatrix} a_{11} - \lambda & a_{12} & \cdots & a_{1n} \\ a_{21} & a_{22} - \lambda & \cdots & a_{2n} \\ \cdots & \cdots & \cdots & \cdots \\ a_{n1} & a_{n2} & \cdots & a_{nn} - \lambda \end{vmatrix} = 0, \tag{4.32}$$

or, in a vector notation:

$$\det|A - \lambda I| = 0. \tag{4.33}$$

Equation (4.32) (or 4.33) is called a *characteristic equation*, and it is an algebraic equation of order n; thus, it has n roots. For each root determined from the characteristic equation, one finds the set of coefficients v_i from the system (4.30). Then, after a substitution in (4.29), the solution is complete. The particulars depend on whether the roots of a characteristic equation, λ, are (i) real and distinct, (ii) complex and distinct, or (iii) repeated, i.e., the number of different roots is less than n.

We consider these three cases separately.

1. Roots of Eq. (4.32) are real and distinct

System (4.30) is solved for $V = (v_1, \ldots, v_n)$ with each root k_i. Since we are looking for the solution of a system (4.27) in the form (4.29), then for k_1 we write the particular solution:

$$x_1^{(1)} = v_1^{(1)} e^{\lambda_1 t}, \quad x_2^{(1)} = v_2^{(1)} e^{\lambda_1 t}, \quad \ldots, \quad x_n^{(1)} = v_n^{(1)} e^{\lambda_1 t}.$$

Superscript indicates that the corresponding particular solution and the set of coefficients v_i are associated with the root λ_1. Similarly, for other roots of the characteristic equation, we write:

$$x_1^{(2)} = v_1^{(2)} e^{\lambda_2 t}, \quad x_2^{(2)} = v_2^{(2)} e^{\lambda_2 t}, \quad \ldots, \quad x_n^{(2)} = v_n^{(2)} e^{\lambda_2 t};$$

$$\cdots\cdots\cdots\cdots\cdots\cdots\cdots\cdots\cdots\cdots\cdots\cdots\cdots\cdots\cdots ;$$

$$x_1^{(n)} = v_1^{(n)} e^{\lambda_n t}, \quad x_2^{(n)} = v_2^{(n)} e^{\lambda_n t}, \quad \ldots, \quad x_n^{(n)} = v_n^{(n)} e^{\lambda_n t}.$$

Note that the coefficients $v_i^{(j)}$ found from the system (4.30) are not unique, since by construction that system is homogeneous and has zero determinant. This means that one of the coefficients $v_i^{(j)}$ for each j can be set equal to an arbitrary constant.

Recall that exponents with different arguments are linearly independent functions. A linear combination of vector functions $(x_1^{(1)}, x_2^{(1)}, \ldots, x_n^{(1)})$, $(x_1^{(2)}, x_2^{(2)}, \ldots, x_n^{(2)})$, $\ldots$, $(x_1^{(n)}, x_2^{(n)}, \ldots, x_n^{(n)})$ with arbitrary coefficients is the general solution of a linear system of equations.

Thus, the general solution of a homogeneous system (4.27) has the form:

$$\begin{cases} x_1(t) = C_1 v_1^{(1)} e^{\lambda_1 t} + C_2 v_1^{(2)} e^{\lambda_2 t} + \cdots + C_n v_1^{(n)} e^{\lambda_n t}, \\ x_2(t) = C_1 v_2^{(1)} e^{\lambda_1 t} + C_2 v_2^{(2)} e^{\lambda_2 t} + \cdots + C_n v_2^{(n)} e^{\lambda_n t}, \\ \cdots\cdots\cdots\cdots\cdots\cdots\cdots\cdots\cdots\cdots\cdots \\ x_n(t) = C_1 v_n^{(1)} e^{\lambda_1 t} + C_2 v_n^{(2)} e^{\lambda_2 t} + \cdots + C_n v_n^{(n)} e^{\lambda_n t}. \end{cases} \quad (4.34)$$

Example 4.7 Solve the system:

$$\begin{cases} \dfrac{dx}{dt} = -x + 8y, \\ \dfrac{dy}{dt} = x + y. \end{cases}$$

Solution We look for solutions in the form:

$$x = v_1 e^{\lambda t},$$
$$y = v_2 e^{\lambda t}.$$

Next, substitute these expressions into the system, move all terms to the right side, cancel the exponent $e^{\lambda t}$, and equate the determinant of the resulting system of homogeneous algebraic equations to zero:

$$\begin{vmatrix} -1-\lambda & 8 \\ 1 & 1-\lambda \end{vmatrix} = 0, \quad \text{or} \quad \lambda^2 - 9 = 0.$$

This characteristic equation has the roots $\lambda_1 = 3$; $\lambda_2 = -3$. Particular solutions are

$$\begin{cases} x^{(1)} = v_1^{(1)} e^{3t}, \\ y^{(1)} = v_2^{(1)} e^{3t} \end{cases} \quad \text{and} \quad \begin{cases} x^{(2)} = v_1^{(2)} e^{-3t}, \\ y^{(2)} = v_2^{(2)} e^{-3t}. \end{cases}$$

Next, we find coefficients $v_i^{(j)}$. System (4.30) for $\lambda = 3$ is

$$\begin{cases} -4v_1^{(1)} + 8v_2^{(1)} = 0, \\ v_1^{(1)} - 2v_2^{(1)} = 0. \end{cases}$$

As can be seen, these equations are redundant (the first one is four times the second one); thus, one of the unknown coefficients can be chosen arbitrarily. Since $v_1^{(1)} = 2v_2^{(1)}$, it is convenient to set $v_2^{(1)}$ equal to one. Then, $v_1^{(1)} = 2$.

Similarly, we find the second pair of coefficients (for $\lambda = -3$). System (4.30) is

$$\begin{cases} 2v_1^{(2)} + 8v_2^{(2)} = 0, \\ v_1^{(2)} + 4v_2^{(2)} = 0. \end{cases}$$

One of its solutions is $v_1^{(2)} = -4$, $v_2^{(2)} = 1$.
The particular solutions are

$$\begin{cases} x^{(1)} = 2e^{3t}, \\ y^{(1)} = e^{3t}, \end{cases} \quad \begin{cases} x^{(2)} = -4e^{-3t}, \\ y^{(2)} = e^{-3t} \end{cases}$$

and they form a fundamental set. The general solution of the original system of differential equations is the linear combination of the solutions in the fundamental set:

$$\begin{cases} x(t) = 2C_1 e^{3t} - 4C_2 e^{-3t}, \\ y(t) = C_1 e^{3t} + C_2 e^{-3t}. \end{cases}$$

Let the initial conditions for the IVP be

$$\begin{cases} x(0) = 1, \\ y(0) = 2. \end{cases}$$

Substitution into the general solution gives:

$$\begin{cases} 1 = 2C_1 - 4C_2, \\ 2 = C_1 + C_2. \end{cases}$$

The solution of this system is $C_1 = 3/2$, $C_2 = 1/2$. Thus, the solution of the IVP is

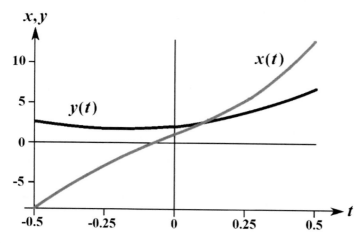

Fig. 4.3 Solution of IVP in Example 4.7

$$\begin{cases} x(t) = 3e^{3t} - 2e^{-3t}, \\ y(t) = \dfrac{3}{2}e^{3t} + \dfrac{1}{2}e^{-3t}. \end{cases}$$

In Fig. 4.3 the functions $x(t)$ and $y(t)$ are shown on the interval $[-0.5, 0.5]$.

Example 4.8 Find solution of the system:

$$\begin{cases} \dfrac{dx}{dt} = 3x - y + z, \\ \dfrac{dy}{dt} = x + y + z, \\ \dfrac{dz}{dt} = 4x - y + 4z \end{cases}$$

subject to the initial conditions

$$\begin{cases} x(0) = 3, \\ y(0) = 0, \\ z(0) = -5. \end{cases}$$

Solution Consider solution of the system in the form:

$$x = v_1 e^{\lambda t},$$
$$y = v_2 e^{\lambda t},$$
$$z = v_3 e^{\lambda t}.$$

Substitution of these expressions into the system gives the characteristic equation:

$$\begin{vmatrix} 3 - \lambda & -1 & 1 \\ 1 & 1 - \lambda & 1 \\ 4 & -1 & 4 - \lambda \end{vmatrix} = 0.$$

By calculating the determinant, we obtain the equation $(2 - \lambda)(\lambda^2 - 6\lambda + 5) = 0$. Its roots are $\lambda_1 = 1$, $\lambda_2 = 2$, $\lambda_3 = 5$.

For $\lambda_1 = 1$ the system (4.30) is

$$\begin{cases} 2v_1^{(1)} - v_2^{(1)} + v_3^{(1)} = 0, \\ v_1^{(1)} + v_3^{(1)} = 0, \\ 4v_1^{(1)} - v_2^{(1)} + 3v_3^{(1)} = 0. \end{cases}$$

This is the homogeneous algebraic system with zero determinant; thus, it has infinitely many solutions. We find one of these solutions. Let $v_1^{(1)} = 1$; then, from any two equations of this system, we obtain $v_2^{(1)} = 1$, $v_3^{(1)} = -1$.

For the second root $\lambda_2 = 2$, the system (4.30) is

$$\begin{cases} v_1^{(2)} - v_2^{(2)} + v_3^{(2)} = 0, \\ v_1^{(2)} - v_2^{(2)} + v_3^{(2)} = 0, \\ 4v_1^{(2)} - v_2^{(2)} + 2v_3^{(2)} = 0. \end{cases}$$

Choosing again $v_1^{(2)} = 1$ gives $v_2^{(2)} = -2$, $v_3^{(2)} = -3$.

Finally, for the third root $\lambda_3 = 3$, we have:

$$\begin{cases} -2v_1^{(3)} - v_2^{(3)} + v_3^{(3)} = 0, \\ v_1^{(3)} - 4v_2^{(3)} + v_3^{(3)} = 0, \\ 4v_1^{(3)} - v_2^{(3)} - v_3^{(3)} = 0, \end{cases}$$

and $v_1^{(3)} = 1, v_2^{(3)} = 1, v_3^{(3)} = 3$.

Now we can write the general solution of the system:

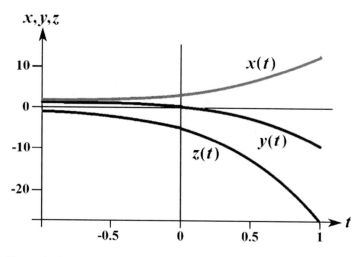

Fig. 4.4 The particular solution of the system in Example 4.8

$$\begin{cases} x(t) = C_1 e^t + C_2 e^{2t} + C_3 e^{5t}, \\ y(t) = C_1 e^t - 2C_2 e^{2t} + C_3 e^{5t}, \\ z(t) = -C_1 e^t - 3C_2 e^{2t} + 3C_3 e^{5t}. \end{cases}$$

Finally, substitution into the initial conditions gives the algebraic system:

$$\begin{cases} 3 = C_1 + C_2 + C_3, \\ 0 = C_1 - 2C_2 + C_3, \\ -5 = -C_1 - 3C_2 + 3C_3. \end{cases}$$

Its solution is $C_1 = 2$, $C_2 = 1$, $C_3 = 0$. Thus, the solution of IVP is

$$\begin{cases} x(t) = 2e^t + e^{2t}, \\ y(t) = 2e^t - 2e^{2t}, \\ z(t) = -2e^t - 3e^{2t}. \end{cases}$$

Graphs of $x(t)$, $y(t)$ and $z(t)$ are shown in Fig. 4.4.

2. Roots of Eq. (4.32) are complex and distinct

Note first that for a system of differential equations with a real-valued coefficient, the complex roots occur only in conjugate pairs. To a pair of a complex conjugate roots $\lambda_{1,\,2} = p \pm iq$, there correspond two particular solutions:

$$x_1^{(1)} = v_1^{(1)} e^{(p+iq)t}, \quad x_2^{(1)} = v_2^{(1)} e^{(p+iq)t}, \quad \ldots, \quad x_n^{(1)} = v_n^{(1)} e^{(p+iq)t},$$
$$x_1^{(2)} = v_1^{(2)} e^{(p-iq)t}, \quad x_2^{(2)} = v_2^{(2)} e^{(p-iq)t}, \quad \ldots, \quad x_n^{(2)} = v_n^{(2)} e^{(p-iq)t}.$$

As in the case of real roots, in order to determine coefficients $v_i^{(j)}$, one has to solve a system (4.30) with λ_1 and λ_2. Recall that exponents with different arguments, or their real and imaginary parts, are linearly independent functions that form a fundamental solution set. Thus, instead of complex functions $x_n^{(1)}$ and $x_n^{(2)}$, their real and imaginary parts can be used.

Example 4.9 Solve the system:

$$\begin{cases} \dfrac{dx}{dt} = -7x + y, \\[2mm] \dfrac{dy}{dt} = -2x - 5y. \end{cases}$$

Solution We look for solution in the form:

$$x = v_1 e^{\lambda t},$$
$$y = v_2 e^{\lambda t}.$$

Substitution of this ansatz into the system gives the characteristic equation:

$$\begin{vmatrix} -7 - \lambda & 1 \\ -2 & -5 - \lambda \end{vmatrix} = 0.$$

The roots are $\lambda_{1,2} = -6 \pm i$. System (4.30) for $\lambda_1 = -6 + i$ is

$$\begin{cases} (-1 - i)v_1^{(1)} + v_2^{(1)} = 0, \\ -2v_1^{(1)} + (1 - i)v_2^{(1)} = 0. \end{cases}$$

The equations in the latter system are redundant: we see that a multiplication of the first equation by $1 - i$ gives the second equation. Then, from either equation, we obtain the following relation between $v_1^{(1)}$ and $v_2^{(1)}$:

$$v_2^{(1)} = (1 + i)v_1^{(1)}.$$

If we let $v_1^{(1)} = 1$, then $v_2^{(1)} = 1 + i$. Next, we repeat the steps for $\lambda_2 = -6 - i$. System (4.30) in this case is

$$\begin{cases} (-1+i)v_1^{(2)} + v_2^{(2)} = 0, \\ -2v_1^{(2)} + (1+i)v_2^{(2)} = 0. \end{cases}$$

Omitting the second equation (due to redundancy) and letting $v_1^{(2)} = 1$ in the first equation result in $v_2^{(2)} = 1 - i$.

Now we can write two particular solutions of the original differential system:

$$x^{(1)} = e^{(-6+i)t}, \qquad x^{(2)} = e^{(-6-i)t},$$
$$y^{(1)} = (1+i)e^{(-6+i)t}, \quad y^{(2)} = (1-i)e^{(-6-i)t}.$$

Obviously, $x^{(2)} = (x^{(1)})^*$ and $y^{(2)} = (y^{(1)})^*$, i.e., the obtained particular solutions $(x^{(1)}, y^{(1)})$ and $(x^{(2)}, y^{(2)})$ are mutually complex conjugate. Their linear combination is the general solution:

$$\begin{cases} x(t) = C_1 e^{(-6+i)t} + C_2 e^{(-6-i)t}, \\ y(t) = C_1(1+i)e^{(-6+i)t} + C_2(1-i)e^{(-6-i)t}. \end{cases}$$

For arbitrary coefficients C_1 and C_2, the obtained solution is complex, though our goal is to find a real solution. The solution is real, if $C_2 = C_1^*$.

Note that a general solution containing complex exponents is often convenient for further calculations (if these are following), but for an IVP solution (where initial conditions are real numbers), it must be transformed to a real-valued form. It is a fact that for differential equations with real coefficients, the real and the imaginary parts of a complex solutions are also solutions, because they are linear combinations of the obtained complex solutions:

$$\text{Re}\left(x^{(1)}, y^{(1)}\right) = \text{Re}\left(x^{(2)}, y^{(2)}\right) = \left[\left(x^{(1)}, y^{(1)}\right) + \left(x^{(2)}, y^{(2)}\right)\right]/2,$$
$$\text{Im}\left(x^{(1)}, y^{(1)}\right) = -\text{Im}\left(x^{(2)}, y^{(2)}\right) = \left[\left(x^{(1)}, y^{(1)}\right) - \left(x^{(2)}, y^{(2)}\right)\right]/2i.$$

Let us apply *Euler's formula* to the complex solutions $x^{(1)}$ and $x^{(2)}$:

$$x^{(1)} = e^{-6t}(\cos t + i\sin t),$$
$$x^{(2)} = e^{-6t}(\cos t - i\sin t).$$

Then,

$$\text{Re}\,x^{(1)} = e^{-6t}\cos t, \qquad \text{Re}\,x^{(2)} = \text{Re}\,x^{(1)},$$
$$\text{Im}\,x^{(1)} = e^{-6t}\sin t. \qquad \text{Im}\,x^{(2)} = -\text{Im}\,x^{(1)}.$$

A linear combination of linearly independent functions $\mathrm{Re}\,x^{(1)}$ and $\mathrm{Im}\,x^{(1)}$ gives the first component, $x(t)$, of the general solution:

$$x(t) = C_1\,\mathrm{Re}\,x^{(1)} + C_2\mathrm{Im}\,x^{(1)}.$$

Obviously, if we use $\mathrm{Re}\,x^{(2)}$ and $\mathrm{Im}\,x^{(2)}$, we obtain the same result.
Then, take $y^{(1)}$ and also isolate the real and the imaginary parts:

$$\mathrm{Re}\,y^{(1)} = e^{-6t}(\cos t - \sin t), \qquad \mathrm{Re}\,y^{(2)} = \mathrm{Re}\,y^{(1)},$$
$$\mathrm{Im}\,y^{(1)} = e^{-6t}(\cos t + \sin t), \qquad \mathrm{Im}\,y^{(2)} = -\,\mathrm{Im}\,y^{(1)}.$$

Forming a linear combination of $\mathrm{Re}\,y^{(1)}$ and $\mathrm{Im}\,y^{(1)}$ with the same coefficients C_1 and C_2, we have:

$$y(t) = C_1\,\mathrm{Re}\,y^{(1)} + C_2\mathrm{Im}\,y^{(1)}.$$

Substitution of $\mathrm{Re}\,x^{(1)}$, $\mathrm{Im}\,x^{(1)}$, $\mathrm{Re}\,y^{(1)}$, and $\mathrm{Im}\,y^{(1)}$ gives:

$$\begin{cases} x(t) = e^{-6t}(C_1\cos t + C_2 \sin t), \\ y(t) = e^{-6t}(C_1 + C_2)\cos t + e^{-6t}(C_2 - C_1)\sin t. \end{cases}$$

Thus, it can be seen that in order to form a real-valued general solution, one can limit the calculation to the determination of the functions $x^{(1)}$ and $y^{(1)}$.
Let the following initial conditions be given:

$$\begin{cases} x(0) = 2, \\ y(0) = 0. \end{cases}$$

Substituting the general solution into the initial conditions, we find the integration constants: $C_1 = 2$, $C_2 = -2$. Thus, the solution of IVP is

$$\begin{cases} x(t) = 2e^{-6t}(\cos t - \sin t), \\ y(t) = -4e^{-6t}\sin t. \end{cases}$$

The graphs $x(t)$ and $y(t)$ are shown in Fig. 4.5.
Next, consider a situation when the characteristic equation has different kinds of roots.

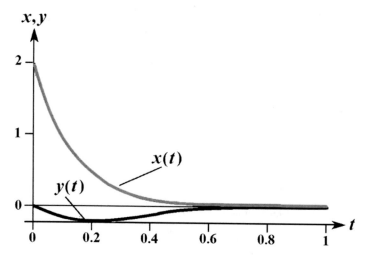

Fig. 4.5 The solution of the IVP in Example 4.9 (on the interval [0,1])

Example 4.10 Find a general solution of the system:

$$\begin{cases} \dfrac{dx}{dt} = 2x - y + 2z, \\[2mm] \dfrac{dy}{dt} = x + 2z, \\[2mm] \dfrac{dz}{dt} = -2x + y - z. \end{cases}$$

Solution Characteristic equation for this system is

$$\begin{vmatrix} 2-\lambda & -1 & 2 \\ 1 & -\lambda & 2 \\ -2 & 1 & -1-\lambda \end{vmatrix} = 0,$$

with roots $\lambda_1 = 1,\ \lambda_{2,\,3} = \pm i$.
  Consider system (4.30) for these roots.
  For $\lambda_1 = 1$:

$$\begin{cases} v_1^{(1)} - v_2^{(1)} + 2v_3^{(1)} = 0, \\ v_1^{(1)} - v_2^{(1)} + 2v_3^{(1)} = 0, \\ -2v_1^{(1)} + v_2^{(1)} - 2v_3^{(1)} = 0. \end{cases}$$

Choosing $v_3^{(1)} = 1$ gives $v_1^{(1)} = 0,\ v_2^{(1)} = 2$.

For $\lambda_2 = i$:

$$\begin{cases} (2-i)v_1^{(2)} - v_2^{(2)} + 2v_3^{(2)} = 0, \\ v_1^{(2)} - iv_2^{(2)} + 2v_3^{(2)} = 0, \\ -2v_1^{(2)} + v_2^{(2)} - (1+i)v_3^{(2)} = 0. \end{cases}$$

Choosing $v_1^{(2)} = 1$ gives $v_2^{(2)} = 1$, $v_3^{(2)} = \frac{1}{2}(i-1)$; thus,

$$x^{(2)} = e^{it}, \quad y^{(2)} = e^{it}, \quad z^{(2)} = \frac{1}{2}(i-1)e^{it}.$$

Instead of these complex functions, one can use their real and imaginary parts (below the indices 2 and 3 denote these linearly independent functions):

$$\begin{aligned} x^{(2)} &= \cos t, & x^{(3)} &= \sin t, \\ y^{(2)} &= \cos t, & y^{(3)} &= \sin t, \\ z^{(2)} &= -\frac{1}{2}(\cos t + \sin t), & z^{(3)} &= \frac{1}{2}(\cos t - \sin t). \end{aligned}$$

In this approach the solution $(x^{(2)*}, y^{(2)*}, z^{(2)*})$ for the complex-conjugate root $\lambda_3 = -i$ is used for formation of real solutions:

$$\mathrm{Re}\, x^{(2)} = \left(x^{(2)} + x^{(2)*}\right)/2, \quad \mathrm{Im}\, x^{(2)} = \left(x^{(2)} - x^{(2)*}\right)/2i, \quad \text{etc.}$$

One can directly check that the triplet $x^{(1)}, y^{(1)}, z^{(1)}$, as well as the other two triplets (these three triplets are the components of three particular vector solutions), satisfy the original differential system.

Linear combinations of these functions give the general solution:

$$\begin{cases} x(t) = C_2 \cos t + C_3 \sin t, \\ y(t) = 2C_1 e^t + C_2 \cos t + C_3 \sin t, \\ z(t) = C_1 e^t - \frac{C_2}{2}(\cos t + \sin t) + \frac{C_3}{2}(\cos t - \sin t). \end{cases}$$

Using the initial conditions

$$\begin{cases} x(0) = 0, \\ y(0) = 1, \\ z(0) = -1, \end{cases}$$

results in the constants $C_1 = 0.5$, $C_2 = 0$, $C_3 = -3$, and then the solution of the IVP is

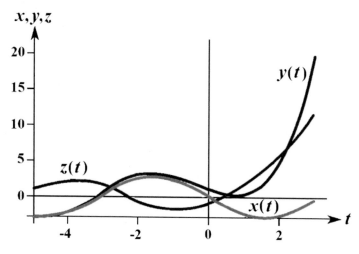

Fig. 4.6 The solution of the IVP in Example 4.10 (on the interval $[-5,3]$)

$$\begin{cases} x(t) = -3\sin t, \\ y(t) = e^t - 3\sin t, \\ z(t) = \frac{1}{2}e^t - \frac{3}{2}(\cos t - \sin t). \end{cases}$$

The graphs $x(t)$, $y(t)$ and $z(t)$ are shown in Fig. 4.6.

3. Roots of the characteristic equation are repeated (the number of different roots is less than n)

A particular solution of a system corresponding to a root λ of multiplicity r should be constructed as

$$X(t) = \begin{pmatrix} \alpha_1^{(1)} + \alpha_1^{(2)}t + \cdots + \alpha_1^{(r)}t^{r-1} \\ \alpha_2^{(1)} + \alpha_2^{(2)}t + \cdots + \alpha_2^{(r)}t^{r-1} \\ \cdots\cdots\cdots\cdots\cdots\cdots\cdots \\ \alpha_n^{(1)} + \alpha_n^{(2)}t + \cdots + \alpha_n^{(r)}t^{r-1} \end{pmatrix} e^{\lambda t}. \tag{4.35}$$

Substituting this ansatz $X(t)$ into the system (4.27) and equating the coefficients at the same powers of t, we can find coefficients $\alpha_i^{(j)}$ ($i=1, \ldots, n$, $j=1, \ldots, r$).

Example 4.11 Solve the system:

$$\begin{cases} \dfrac{dx}{dt} = 5x + 3y, \\ \dfrac{dy}{dt} = -3x - y. \end{cases}$$

Solution The characteristic equation is

$$\begin{vmatrix} 5 - \lambda & 3 \\ -3 & -1 - \lambda \end{vmatrix} = 0.$$

The roots are $\lambda_{1,\,2} = 2$. We look for solution in the form:

$$\begin{cases} x(t) = (\alpha_1 + \beta_1 t)e^{2t}, \\ y(t) = (\alpha_2 + \beta_2 t)e^{2t}, \end{cases}$$

where $\alpha_1, \alpha_2, \beta_1, \beta_2$ are the coefficients to be determined. These expressions must be substituted in one of the system's equations, for instance, into the first one (substitution in another equation gives the same result):

$$2\alpha_1 + \beta_1 + 2\beta_1 t = 5\alpha_1 + 5\beta_1 t + 3\alpha_2 + 3\beta_2 t,$$

from which

$$\alpha_2 = \frac{\beta_1}{3} - \alpha_1,$$
$$\beta_2 = -\beta_1.$$

Coefficients α_1 and β_1 are still arbitrary. We introduce the notation C_1 and C_2 for them, respectively, and then write the general solution:

$$\begin{cases} x(t) = (C_1 + C_2 t)e^{2t}, \\ y(t) = (C_2/3 - C_1 - C_2 t)e^{2t}. \end{cases}$$

Let the initial conditions be

$$\begin{cases} x(0) = 1, \\ y(0) = 1. \end{cases}$$

Substituting these initial conditions gives $C_1 = 1$, $C_2 = 6$; thus, the solution of the IVP is

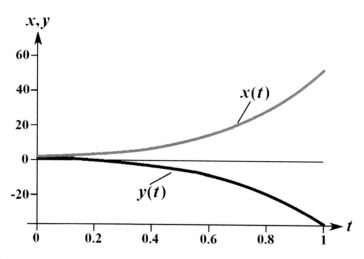

Fig. 4.7 The solution of the IVP in Example 4.11 (on the interval [0,1])

$$\begin{cases} x(t) = (1 + 6t)e^{2t}, \\ y(t) = (1 - 6t)e^{2t}. \end{cases}$$

The graphs $x(t)$ and $y(t)$ are shown in Fig. 4.7.

Example 4.12 Find the general solution of the system:

$$\begin{cases} \dfrac{dx}{dt} = -2x + y - 2z, \\[2mm] \dfrac{dy}{dt} = x - 2y + 2z, \\[2mm] \dfrac{dz}{dt} = 3x - 3y + 5z. \end{cases}$$

This system is exactly the system (4.9) that was solved in Sect. 4.2 by elimination of variables. Now we shall solve it using the general algorithm described above.

Solution Characteristic equation for this system is

$$\begin{vmatrix} -2-\lambda & 1 & -2 \\ 1 & -2-\lambda & 2 \\ 3 & -3 & 5-\lambda \end{vmatrix} = 0.$$

Roots are $\lambda_1 = 3$, $\lambda_2 = \lambda_3 = -1$. For $\lambda_1 = 3$ system (4.30) becomes:

$$\begin{cases} -5v_1^{(1)} + v_2^{(1)} - 2v_3^{(1)} = 0, \\ v_1^{(1)} - 5v_2^{(1)} + 2v_3^{(1)} = 0, \\ 3v_1^{(1)} - 3v_2^{(1)} + 2v_3^{(1)} = 0. \end{cases}$$

As always, this system of homogeneous linear algebraic equations with zero determinant has infinitely many solutions. Choosing $v_1^{(1)} = 1$, we get $v_2^{(1)} = -1, v_3^{(1)} = -3$. This gives the first particular solution:

$$x^{(1)} = e^{3t}, \quad y^{(1)} = -e^{3t}, \quad z^{(1)} = -3e^{3t}.$$

That solution will enter the general solution multiplied by the coefficient C_1. The solution for the repeated root $\lambda_{2,3} = -1$ has the form:

$$\begin{cases} x(t) = (\alpha_1 + \beta_1 t)e^{-t}, \\ y(t) = (\alpha_2 + \beta_2 t)e^{-t}, \\ z(t) = (\alpha_3 + \beta_3 t)e^{-t}. \end{cases}$$

To determine the coefficients, these expressions must be substituted in the equations of the differential system.

Substitution into the first equation gives:

$$-\alpha_1 + \beta_1 - \beta_1 t = -2\alpha_1 - 2\beta_1 t + \alpha_2 + \beta_2 t - 2\alpha_3 - 2\beta_3 t.$$

Equating coefficients of the equal powers of t, we find:

$$\alpha_3 = \frac{1}{2}(\alpha_2 - \alpha_1 - \beta_1),$$
$$\beta_3 = \frac{1}{2}(\beta_2 - \beta_1).$$

Substitution into the second equation gives:

$$\alpha_3 = (\alpha_2 - \alpha_1 + \beta_2)/2,$$
$$\beta_3 = (\beta_2 - \beta_1)/2.$$

Substitution into the third equation gives:

$$\alpha_3 = (\alpha_2 - \alpha_1 + \beta_3/3)/2,$$
$$\beta_3 = (\beta_2 - \beta_1)/2.$$

Thus, we obtained four independent equations for six constants α_i, β_i. Comparing expressions for α_3, it becomes clear that they are satisfied for $\beta_1 = \beta_2 = \beta_3 = 0$.

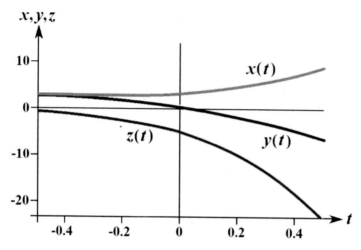

Fig. 4.8 The solution of the IVP in Example 4.12 (on the interval $[-0.5, 0.5]$)

Note that the general form of the solution containing terms with a product of t and exponential function turned out to be needless. The explanation of that circumstance will be given in Sect. 4.6. With $\alpha_3 = (\alpha_2 - \alpha_1)/2$, we obtain $\alpha_2 = \alpha_1 + 2\alpha_3 = C_1 + 2C_3$. The general solution can be written as

$$\begin{cases} x(t) = C_1 e^{3t} + C_2 e^{-t}, \\ y(t) = -C_1 e^{3t} + e^{-t}(C_2 + 2C_3), \\ z(t) = -3C_1 e^{3t} + C_3 e^{-t}. \end{cases} \qquad (4.36)$$

The initial conditions

$$\begin{cases} x(0) = 3, \\ y(0) = 0, \\ z(0) = -5 \end{cases}$$

give $C_1 = 1.75$, $C_2 = 1.25$, $C_3 = 0.25$, and the solution of IVP is

$$\begin{cases} x(t) = 1.75 e^{3t} + 2.25 e^{-t}, \\ y(t) = -1.75 e^{3t} + 1.75 e^{-t}, \\ z(t) = -5.25 e^{3t} + 0.25 e^{-t}. \end{cases}$$

Graphs of this solution are shown in Fig. 4.8.

A choice of coefficients α_1 and α_3 for C_1 and C_3, respectively, is only one possibility. Choosing instead $\alpha_2 \equiv C_2$, $\alpha_3 \equiv C_3$ gives $\alpha_1 = C_2 - 2C_3$. In this case the general solution looks differently:

$$\begin{cases} x(t) = C_1 e^{3t} + (C_2 - 2C_3)e^{-t}, \\ y(t) = -C_1 e^{3t} + C_2 e^{-t}, \\ z(t) = -3C_1 e^{3t} + C_3 e^{-t}. \end{cases}$$

But obviously, the solution of the IVP coincides with the one obtained above.

Example 4.13 Find the general solution of the system:

$$\begin{cases} \dfrac{dx}{dt} = 2x - y - z, \\ \dfrac{dy}{dt} = 2x - y - 2z, \\ \dfrac{dz}{dt} = -x + y + 2z. \end{cases}$$

Solution The characteristic equation for this system

$$\begin{vmatrix} 2 - \lambda & -1 & -1 \\ 2 & -1 - \lambda & -2 \\ -1 & 1 & 2 - \lambda \end{vmatrix} = 0$$

has three coincident roots: $\lambda_1 = \lambda_2 = \lambda_3 = 1$. In general, a solution of a system of differential equations when a root of a characteristic equation is real and three times repeating is sought in the form:

$$\begin{cases} x(t) = (\alpha_1 + \beta_1 t + \gamma_1 t^2)e^t, \\ y(t) = (\alpha_2 + \beta_2 t + \gamma_2 t^2)e^t, \\ z(t) = (\alpha_3 + \beta_3 t + \gamma_3 t^2)e^t. \end{cases}$$

Substituting these expressions into the first equation of the differential system and dividing the equation by e^t, we have:

$$\beta_1 + 2\gamma_1 t + \alpha_1 + \beta_1 t + \gamma_1 t^2 = 2\alpha_1 + 2\beta_1 t + 2\gamma_1 t^2 - \alpha_2 - \beta_2 - \gamma_2 t^2 - \alpha_3 - \beta_3 t - \gamma_3 t^2.$$

Now let us compare the coefficients of the same powers of t:

$$\begin{aligned} \beta_1 &= \alpha_1 - \alpha_2 - \alpha_3, \\ 2\gamma_1 &= \beta_1 - \beta_2 - \beta_3, \\ 0 &= \gamma_1 - \gamma_2 - \gamma_3. \end{aligned}$$

Similarly, substituting $x(t)$, $y(t)$, $z(t)$ in the second equation of the differential system, we have:

$$\beta_2 = 2(\alpha_1 - \alpha_2 - \alpha_3),$$
$$\gamma_2 = \beta_1 - \beta_2 - \beta_3,$$
$$0 = \gamma_1 - \gamma_2 - \gamma_3.$$

Substitution into the third equation gives:

$$\beta_3 = -(\alpha_1 - \alpha_2 - \alpha_3), \quad 2\gamma_3 = -(\beta_1 - \beta_2 - \beta_3), \quad 0 = -(\gamma_1 - \gamma_2 - \gamma_3).$$

Let us denote $\alpha = \alpha_1 - \alpha_2 - \alpha_3$; then $\beta_1 = \alpha$, $\beta_2 = 2\alpha$ and $\beta_3 = -\alpha$. We find that $\beta_1 - \beta_2 - \beta_3 = 0$; hence, $\gamma_1 = \gamma_2 = \gamma_3 = 0$.
Denote $\alpha_1 = C_1$, $\alpha_2 = C_2$, and $\alpha = C_3$. Then, $\alpha_3 = C_1 - C_2 - C_3$, $\beta_1 = C_3$, $\beta_2 = 2C_3$, $\beta_3 = -C_3$.
Thus, the general solution of the system of differential equations is

$$\begin{cases} x = (C_1 + C_3 t)e^t, \\ y = (C_2 + 2C_3 t)e^t, \\ z = (C_1 - C_2 - C_3 - C_3 t)e^t. \end{cases}$$

Problems

Solve differential systems without resorting to the increase of the system's order. Then, solve the IVP. In the problems 1–5, use the initial conditions $x(0) = 1$, $y(0) = 1$, in the problems 6–12, the initial conditions $x(0) = 1$, $y(0) = 2$, and in the problems 13–17 the initial conditions $x(0) = 1$, $y(0) = 2$, $z(0) = 3$. Plot the graph of the IVP solution using the program *ODE 1ˢᵗ order*.
To simplify the task, the roots of the characteristic equation are stated for systems of three equations:

1. $\begin{cases} x' + 2x + 3y = 0, \\ y' + x = 0. \end{cases}$ 2. $\begin{cases} x' = -3x - y, \\ y' = x - y. \end{cases}$

3. $\begin{cases} x' = x - 3y, \\ y' = 3x + y. \end{cases}$ 4. $\begin{cases} x' = 3x - 2y, \\ y' = 4x - y. \end{cases}$

5. $\begin{cases} x' = -x - 2y, \\ y' = 3x + 4y. \end{cases}$ 6. $\begin{cases} x' = x + y, \\ y' = 2x. \end{cases}$

7. $\begin{cases} x' = x - 2y, \\ y' = 2x - 3y. \end{cases}$ 8. $\begin{cases} x' = 4x - y, \\ y' = x + 2y. \end{cases}$

9. $\begin{cases} x' = 2x - y, \\ y' = x + 2y. \end{cases}$ 10. $\begin{cases} x' = x - 2y, \\ y' = x - y. \end{cases}$

11. $\begin{cases} x' = x - y, \\ y' = -4x + 4y. \end{cases}$ 12. $\begin{cases} x' = 2x - 3y, \\ y' = 3x + 2y. \end{cases}$

13. $\begin{cases} x' = 4y - 2z - 3x, \\ y' = z + x, \\ z' = 6x - 6y + 5z. \end{cases}$ 14. $\begin{cases} x' = x - y - z, \\ y' = x + y, \\ z' = 3x + z. \end{cases}$

$\lambda_1 = 1, \lambda_2 = 2, \lambda_3 = -1.$ $\lambda_1 = 1, \lambda_{2,3} = 1 \pm 2i.$

$$15. \begin{cases} x' = 2x - y - z, \\ y' = 3x - 2y - 3z, \\ z' = -x + y + 2z. \end{cases}$$

$\lambda_1 = 0, \lambda_2 = \lambda_3 = 1.$

$$16. \begin{cases} x' = 2x + y, \\ y' = x + 3y - z, \\ z' = -x + 2y + 3z. \end{cases}$$

$\lambda_1 = 2, \lambda_{2,\,3} = 3 \pm i.$

$$17. \begin{cases} x' = 3x + 12y - 4z, \\ y' = -x - 3y + z, \\ z' = -x - 12y + 6z. \end{cases}$$

$\lambda_1 = 1, \lambda_2 = 2, \lambda_3 = 3.$

4.5   Systems of Linear Nonhomogeneous Differential Equations with Constant Coefficients

When solving nonhomogeneous systems of the form

$$\frac{dX(t)}{dt} = AX(t) + F(t), \tag{4.37}$$

we first determine a general solution of a corresponding homogeneous system (with $F = 0$).

Assume that the solution of a homogeneous system is known: $x_i(t) = \sum_{k=1}^{n} C_k x_i^{(k)}$, $(i = \overline{1, n})$. Then, we can find the solution of a nonhomogeneous system, using, for instance, *variation of parameters*. In this method the solution is sought in the form:

$$x_i(t) = \sum_{k=1}^{n} u_k(t) x_i^{(k)}, \quad (i = \overline{1, n}), \tag{4.38}$$

where $u_k(t)$ are unknown functions. Substitution of this form in the original nonhomogeneous system gives a system of equations for $u'_k(t)$. Integration gives functions $u_k(t)$. Then, from (4.38) one finds the solution of a nonhomogeneous system (4.37). Its structure is the sum of a general solution of a homogeneous system and a particular solution of a nonhomogeneous system.

The method can be used also in the general case when the coefficients are not constant.

Example 4.14 Solve nonhomogeneous differential system:

$$\begin{cases} \dfrac{dx}{dt} = -x + 8y + 1, \\ \dfrac{dy}{dt} = x + y + t. \end{cases}$$

Solution The corresponding homogeneous system is

$$\begin{cases} \dfrac{dx_h}{dt} = -x + 8y, \\ \dfrac{dy_h}{dt} = x + y. \end{cases}$$

Its general solution was obtained in Example 4.5:

$$\begin{cases} x_h(t) = 2C_1 e^{3t} - 4C_2 e^{-3t}, \\ y_h(t) = C_1 e^{3t} + C_2 e^{-3t}. \end{cases}$$

Next, we find the solution of the nonhomogeneous differential system. Varying the parameters

$$\begin{cases} x(t) = 2u_1(t)e^{3t} - 4u_2(t)e^{-3t}, \\ y(t) = u_1(t)e^{3t} + u_2(t)e^{-3t}. \end{cases}$$

Substituting these forms into the first equation of the nonhomogeneous system gives:

$$2u_1' e^{3t} + 6u_1 e^{3t} - 4u_2' e^{-3t} + 12u_2 e^{-3t} = -2u_1 e^{3t} + 4u_2 e^{-3t} + 8u_1 e^{3t} + u_2 e^{-3t} + 1.$$

The terms that contain functions $u_1(t)$ and $u_2(t)$ cancel (this always happens by the design of the method), and we get:

$$2u_1' e^{3t} - 4u_2' e^{-3t} = 1.$$

Repeating for the second equation of the system gives:

$$u_1' e^{3t} + u_2' e^{-3t} = t.$$

Thus, we arrive to the algebraic system of two equations for two unknowns u_1' and u_2':

$$\begin{cases} 2u_1' e^{3t} - 4u_2' e^{-3t} = 1, \\ u_1' e^{3t} + u_2' e^{-3t} = t. \end{cases}$$

Its solution is

$$u_1' = \frac{1}{6}(1 + 4t)e^{-3t},$$

$$u_2' = \frac{1}{6}(2t - 1)e^{3t}.$$

Integration gives u_1' и u_2':

$$u_1(t) = -\frac{1}{18}e^{-3t}\left(\frac{7}{3} + 4t\right) + C_1,$$

$$u_2(t) = \frac{1}{18}e^{3t}\left(2t - \frac{5}{3}\right) + C_2.$$

Substituting these functions into the expressions for $x(t)$ and $y(t)$ gives the general solution of the inhomogeneous differential system:

$$\begin{cases} x(t) = 2C_1 e^{3t} - 4C_2 e^{-3t} + \frac{1}{9}(1 - 8t), \\ y(t) = C_1 e^{3t} + C_2 e^{-3t} - \frac{1}{9}(2 + t), \end{cases}$$

which is the sum of the general solution of the homogeneous system and the particular solution of the nonhomogeneous one.

Let the initial conditions be

$$\begin{cases} x(0) = 1, \\ y(0) = -1. \end{cases}$$

Substitution in the general solution gives:

$$\begin{cases} 1 = 2C_1 - 4C_2 + \frac{1}{9}, \\ -1 = C_1 + C_2 - \frac{2}{9}. \end{cases}$$

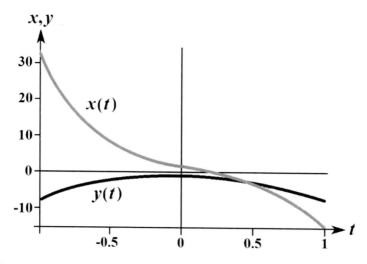

Fig. 4.9 The solution of the IVP in Example 4.14 (on the interval $[-1,1]$)

Solving gives $C_1 = -\frac{10}{27}$, $C_2 = -\frac{11}{27}$ and then the solution of the IVP is

$$\begin{cases} x(t) = -\dfrac{20}{27}e^{3t} + \dfrac{44}{27}e^{-3t} + \dfrac{1}{9}(1-8t), \\[2mm] y(t) = -\dfrac{10}{27}e^{3t} - \dfrac{11}{27}e^{-3t} - \dfrac{1}{9}(2+t). \end{cases}$$

Graphs of this solution are shown in Fig. 4.9.

Problems

Solve nonhomogeneous systems of differential equations. Then, solve the IVP for the initial conditions $x(0) = 1$, $y(0) = 0$. The graph of the obtained solution can be plotted with the program ODE first order.

1. $\begin{cases} x' = 2x - y, \\ y' = 2y - x - 5e^t \sin t. \end{cases}$

2. $\begin{cases} x' + 4y = \cos 2t, \\ y' + 4x = \sin 2t. \end{cases}$

3. $\begin{cases} x' = 5x - 3y + 2e^{3t}, \\ y' = x + y + 5e^{-t}. \end{cases}$

4. $\begin{cases} x' = -2x + 4y, \\ y' = -x + 3y + 3t^2. \end{cases}$

5. $\begin{cases} x' = 2x + y + 2e^t, \\ y' = x + 2y - 3e^{4t}. \end{cases}$

6. $\begin{cases} x' = y - 5\cos t, \\ y' = 2x + y. \end{cases}$

7. $\begin{cases} x' = 2x + y - 7te^{-t} - 3, \\ y' = -x + 2y - 1. \end{cases}$

4.6   Matrix Approach

In the present section, we describe a general approach for solving systems of first-order linear ODEs based on matrix representation of a fundamental solution.

4.6.1   Homogeneous Systems of Equations

4.6.1.1   Matrix Equation

As explained in Sect. 4.3, the set of n linearly independent vector solutions

$$X^{(k)}(t) = \left\{ x_1^{(k)}(t), x_2^{(k)}(t), \ldots, x_n^{(k)}(t) \right\}, \quad k = 1, 2, \ldots, n \tag{4.39}$$

of a vector equation

$$\frac{dX^{(k)}(t)}{dt} = A(t)X^{(k)}(t) \tag{4.40}$$

determines a general solution of (4.20), which can be presented as

$$X(t) = \sum_{k=1}^{n} C_k X^{(k)}(t), \tag{4.41}$$

where C_k, $k = 1, 2, \ldots, n$ are arbitrary constant coefficients.

Let us construct a matrix $Y(t)$ (called the *fundamental matrix*), such that its columns are the vectors $X(t)$, i.e.,

$$Y(t) = \left(X^{(1)}(t)\ X^{(2)}(t)\ \ldots\ X^{(n)}(t) \right) = \begin{pmatrix} x_1^{(1)} & x_1^{(2)} & \ldots & x_1^{(n)} \\ x_2^{(1)} & x_2^{(2)} & \ldots & x_2^{(n)} \\ \ldots & \ldots & \ldots & \ldots \\ x_n^{(1)} & x_n^{(2)} & \ldots & x_n^{(n)} \end{pmatrix}. \tag{4.42}$$

Differentiating matrix (4.42) and using (4.40), we obtain the matrix equation:

$$\begin{aligned} \frac{dY(t)}{dt} &= \left(\frac{dX^{(1)}(t)}{dt}\ \frac{dX^{(2)}(t)}{dt} \ldots \frac{dX^{(n)}(t)}{dt} \right) \\ &= \left(A(t)X^{(1)}(t)\,A(t)X^{(2)}(t) \ldots A(t)X^{(n)}(t) \right) \\ &= A(t)\left(X^{(1)}(t)\,X^{(2)}(t) \ldots X^{(n)}(t) \right) = A(t)Y(t). \end{aligned} \tag{4.43}$$

Because the solutions $X^{(k)}$, $k = 1, 2, \ldots, n$, are linearly independent, the determinant of matrix Y

$$\det Y(t) = \det\left(X^{(1)}(t)\ X^{(2)}(t)\ \ldots\ X^{(n)}(t)\right) \equiv W(t) \neq 0. \tag{4.44}$$

Finding solutions $X^{(k)}(t))$, $k = 1, 2, \ldots, n$, is equivalent to solving the matrix equation (4.43), subject to an initial condition:

$$Y(t)|_{t=t_0} = Y(t_0), \tag{4.45}$$

where $Y(t_0)$ is an arbitrary matrix satisfying the single condition

$$\det Y(t_0) = W(t)|_{t=t_0} \neq 0. \tag{4.46}$$

For instance, one can choose the identity matrix:

$$I = \begin{pmatrix} 1 & 0 & \ldots & 0 \\ 0 & 1 & \ldots & 0 \\ \ldots & \ldots & \ldots & \ldots \\ 0 & 0 & \ldots & 1 \end{pmatrix}$$

as $Y(t_0)$.

When a matrix $Y(t)$ is found, the general solution (4.41) can be constructed as

$$X(t) = \sum_{k=1}^{n} X^{(k)}(t)C_k = \left(X^{(1)}\ \ldots\ X^{(n)}\right)\begin{pmatrix} C_1 \\ \ldots \\ C_n \end{pmatrix} = Y(t)C, \tag{4.47}$$

where C is an arbitrary constant vector.

Let us consider now the initial value problem that consists of Eq. (4.40) and the initial condition

$$X(t_0) = X_0. \tag{4.48}$$

According to (4.47)

$$X(t_0)C = X_0,$$

hence

$$C = Y^{-1}(t_0)X_0. \tag{4.49}$$

Note that the inverse matrix always exists because of (4.44). Substituting (4.49) into (4.47), we find:

$$X(t) = Y(t)Y^{-1}(t_0)X_0. \tag{4.50}$$

4.6.1.2   Series Solution for a Constant Matrix A

If A is a constant matrix, one can easily obtain a general solution of Eq. (4.40) in the form of a series. Let us consider the problem:

$$Y'(t) = AY(t), \tag{4.51}$$
$$Y(0) = I. \tag{4.52}$$

One can easily check that the series

$$Y(t) = I + tA + \frac{t^2}{2!}A^2 + \cdots = \sum_{k=0}^{\infty} \frac{t^k}{k!}A^k \tag{4.53}$$

satisfies Eq. (4.51):

$$Y'(t) = \frac{d}{dt}\sum_{k=0}^{\infty} \frac{t^k}{k!}A^k = \sum_{k=1}^{\infty} \frac{t^{k-1}}{(k-1)!}A^k = A\sum_{k=1}^{\infty} \frac{t^{k-1}}{(k-1)!}A^{k-1} = AY,$$

and the initial condition (4.52). Here we omit the proof of convergence of the series (4.53) and the possibility of its term-by-term differentiation. We can use the natural notation:

$$\sum_{k=0}^{\infty} \frac{t^k}{k!}A^k \equiv \exp(tA). \tag{4.54}$$

($\exp(tA)$ is called *a matrix exponential*.) Thus, the general solution of the system (4.40) of the system of ODEs is

$$X(t) = \exp(tA)C,$$

where C is an arbitrary constant vector.

Example 4.15 Solve the system:

$$\frac{dx_1}{dt} = x_2, \quad \frac{dx_2}{dt} = -x_1. \tag{4.55}$$

Solution The matrix of coefficients

$$A = \begin{pmatrix} 0 & 1 \\ -1 & 0 \end{pmatrix}$$

is constant. Therefore, the fundamental solution of (4.55) is

$$Y = \exp(tA).$$

Let us calculate the matrix A^2:

$$A^2 = \begin{pmatrix} 0 & 1 \\ -1 & 0 \end{pmatrix}\begin{pmatrix} 0 & 1 \\ -1 & 0 \end{pmatrix} = \begin{pmatrix} -1 & 0 \\ 0 & -1 \end{pmatrix} = -I.$$

Thus, $A^3 = A \cdot A^2 = -A$, $A^4 = A(-A) = I$, etc. Hence,

$$Y = \exp(tA) = I + tA + \frac{t^2}{2!}(-I) + \frac{t^3}{3!}(-A) + \frac{t^4}{4!}I + \cdots$$

$$= I\left(1 - \frac{t^2}{2!} + \frac{t^4}{4!} + \cdots\right) + A\left(t - \frac{t^3}{3!} + \frac{t^5}{5!} + \cdots\right).$$

Using the Taylor series expansion of the functions $\cos t$ and $\sin t$, we find that the fundamental matrix

$$Y = \exp(tA) = I\cos t + A\sin t = \begin{pmatrix} \cos t & \sin t \\ -\sin t & \cos t \end{pmatrix}.$$

The columns of Y are the linearly independent solutions of the system (4.55). According to (4.47), the general solution is

$$\begin{pmatrix} y_1 \\ y_2 \end{pmatrix} = \begin{pmatrix} \cos t & \sin t \\ -\sin t & \cos t \end{pmatrix}\begin{pmatrix} C_1 \\ C_2 \end{pmatrix} = C_1\begin{pmatrix} \cos t \\ -\sin t \end{pmatrix} + C_2\begin{pmatrix} \sin t \\ \cos t \end{pmatrix}.$$

This result coincides with that of Example 4.3.

Generally, a direct calculation of a series (4.53) is not the best way of calculating a matrix exponential. Before introducing a more practical way, let us list some properties of the matrix $\exp(tA)$ that will be used below:

1. $\exp(t_1 A) \exp(t_2 A) = \exp[(t_1 + t_2)A]$.
2. $[\exp(tA)]^{-1} = \exp(-tA)$.
3. If $AB = BA$, then $\exp(tA) \exp(tB) = \exp(t(A + B))$.
4. For any invertible matrix P,

$$\exp[tP^{-1}AP] = P^{-1} \exp(tA)P.$$

Indeed, for any k,

$$(P^{-1}AP)^k = P^{-1}APP^{-1}AP \ldots PP^{-1}AP = P^{-1}A^k P.$$

Therefore,

$$\exp[tP^{-1}AP] = \sum_{k=0}^{\infty} \frac{t^k}{k!} (P^{-1}AP)^k = \sum_{k=0}^{\infty} \frac{t^k}{k!} P^{-1}A^k P = P^{-1} \sum_{k=0}^{\infty} \frac{t^k}{k!} A^k P = P^{-1} \exp(tA)P.$$

The last property can be written also as

$$\exp(tA) = P \exp[tP^{-1}AP]P^{-1}. \qquad (4.56)$$

4.6.1.3 The Case of a Diagonalizable Constant Matrix A

From the course of a linear algebra, it is known that some matrices can be *diagonalized*, i.e., for a given matrix A, there exists a matrix:

$$P = \left(V^{(1)} \ldots V^{(n)}\right), \qquad (4.57)$$

such that

$$P^{-1}AP \equiv \Lambda = \begin{pmatrix} \lambda_1 & 0 & \ldots & 0 \\ 0 & \lambda_2 & \ldots & 0 \\ \ldots & \ldots & \ldots & \ldots \\ 0 & 0 & \ldots & \lambda_n \end{pmatrix}.$$

Vectors $V^{(k)}$, $k = 1, \ldots, n$, which are the columns of that matrix, are the eigenvectors of a matrix A corresponding to eigenvalues λ_k. All vectors $V^{(k)}$ are linearly independent; otherwise, det P would be zero and an inverse matrix P^{-1} would not exist. Thus, we can construct a matrix P, if a matrix A has n linearly independent eigenvectors. In that case

$$\exp\left(tP^{-1}AP\right) = \exp(t\Lambda) = \sum_{k=0}^{\infty}\frac{t^k}{k!}\Lambda^k$$

$$= \sum_{k=0}^{\infty}\frac{t^k}{k!}\begin{pmatrix} \lambda_1^k & 0 & \cdots & 0 \\ 0 & \lambda_2^k & \cdots & 0 \\ \cdots & \cdots & \cdots & \cdots \\ 0 & 0 & \cdots & \lambda_n^k \end{pmatrix} = \begin{pmatrix} e^{\lambda_1 t} & 0 & \cdots & 0 \\ 0 & e^{\lambda_2 t} & \cdots & 0 \\ \cdots & \cdots & \cdots & \cdots \\ 0 & 0 & \cdots & e^{\lambda_n t} \end{pmatrix}.$$

$$(4.58)$$

Recall that according to (4.56)

$$Y(t) = \exp(tA) = P\exp\left(tP^{-1}AP\right)P^{-1} = P\exp(t\Lambda)P^{-1},$$

and the general solution of the system (see Eq. (4.47)) is

$$X(t) = Y(t)\,C = P\exp(t\Lambda)P^{-1}C, \tag{4.59}$$

where C is an arbitrary constant vector. Introducing another arbitrary constant vector

$$c = P^{-1}C, \tag{4.60}$$

and using (4.57), (4.58) and (4.60), we rewrite (4.59) as

$$X = \left(V^{(1)} \ldots V^{(n)}\right)\begin{pmatrix} e^{\lambda_1 t} & \cdots & 0 \\ \cdots & \cdots & \cdots \\ 0 & \cdots & e^{\lambda_n t} \end{pmatrix}\begin{pmatrix} c_1 \\ \cdots \\ c_n \end{pmatrix}$$

$$= \left(e^{\lambda_1 t}V^{(1)} \ldots e^{\lambda_n t}V^{(n)}\right) = \sum_{i=1}^{n} c_i e^{\lambda_i t}V^{(i)}, \tag{4.61}$$

which coincides with expression (4.34) in Sect. 4.4.

Note that if a matrix A is diagonalizable (i.e., if it has n linearly independent eigenvectors), then expression (4.61) is valid even when some of λ_i are equal. In that case, no terms containing powers of x are needed in the ansatz (4.11).

We considered such example in Sect. 4.4 (Example 4.11). In that example, though there are equal eigenvalues, the matrix A has $n = 3$ linearly independent eigenvectors; therefore, the terms with powers do not appear in the solution.

Generally, for any root λ_i of a (polynomial) characteristic equation, one has to distinguish between its *algebraic multiplicity*, which is a root multiplicity, and its *geometric multiplicity* (a number of a linearly independent eigenvectors corresponding to the eigenvalue λ_i). A matrix is diagonalizable if it has n linearly independent eigenvectors, i.e., for each eigenvalue a geometric multiplicity equals an algebraic multiplicity. In that case, a system has n solutions in the exponential form. Otherwise, an additional analysis is needed, which is described in the next subsection.

4.6.1.4   The Case of a Non-diagonalizable Constant Matrix A

The number of a linearly independent eigenvectors (a geometric multiplicity of an eigenvalue) can be smaller than its algebraic multiplicity. The simple example is the matrix

$$\begin{pmatrix} 1 & 1 \\ 0 & 1 \end{pmatrix}$$

with the characteristic polynomial $(\lambda - 1)^2$ (the algebraic multiplicity of the eigenvalue $\lambda = 1$ is two) and only one eigenvector

$$\begin{pmatrix} 1 \\ 0 \end{pmatrix},$$

i.e., the geometric multiplicity is one. The matrix cannot be diagonalized, and the number of exponential solutions is less than two (the size of the matrix), that is, there is only one exponential solution.

A non-diagonalizable matrix can be transformed to the *Jordan form*. That means that one can find a matrix $P = (V^{(1)} \; V^{(2)} \ldots)$, such that all nonzero elements of the matrix $P^{-1}AP$ form the *Jordan blocks* $J_1, \ldots, J_s$ having the dimensions ("algebraic multiplicities") $r_1, \ldots, r_s$:

$$P^{-1}AP = \begin{pmatrix} J_1 & 0 & \cdots & 0 \\ 0 & J_2 & \cdots & 0 \\ \cdots & \cdots & \cdots & \cdots \\ 0 & 0 & \cdots & J_s \end{pmatrix}.$$

The Jordan blocks have the structure $J_i = \lambda_i I + Z_i$, where I is a $r_i \times r_i$ unit matrix and Z_i is a matrix of the dimension $r_i \times r_i$:

$$Z_i = \begin{pmatrix} 0 & 1 & 0 & \cdots & 0 & 0 \\ 0 & 0 & 1 & \cdots & 0 & 0 \\ \cdots & \cdots & \cdots & \cdots & \cdots & \cdots \\ 0 & 0 & 0 & \cdots & 0 & 1 \\ 0 & 0 & 0 & \cdots & 0 & 0 \end{pmatrix},$$

with nonzero elements $(Z_i)_{12} = (Z_i)_{23} = \cdots = (Z_i)_{r-1,\,r} = 1$ populating the first super-diagonal. Calculating the powers of the matrix Z_i, we arrive to a conclusion that each additional multiplication by Z_i shifts the line of nonzero elements to the right, i.e.,

$$Z_i^2 = \begin{pmatrix} 0 & 0 & 1 & 0 & \ldots & 0 \\ 0 & 0 & 0 & 1 & \ldots & 0 \\ \ldots & \ldots & \ldots & \ldots & \ldots & \ldots \\ 0 & 0 & 0 & 0 & \ldots & 1 \\ 0 & 0 & 0 & 0 & \ldots & 0 \\ 0 & 0 & 0 & 0 & \ldots & 0 \end{pmatrix}$$

has the nonzero elements

$$\left(Z_i^2\right)_{13} = \left(Z_i^2\right)_{24} = \cdots = \left(Z_i^2\right)_{r_i-2,r_i} = 1,$$
$$\cdots,$$
$$Z_i^{r_i-1} = \begin{pmatrix} 0 & 0 & \ldots & 0 & 1 \\ 0 & 0 & \ldots & 0 & 0 \\ \ldots & \ldots & \ldots & \ldots & \ldots \\ 0 & 0 & \ldots & 0 & 0 \end{pmatrix}$$

has the only nonzero element $(Z_i)_{1,r_i} = 1$, and $Z_i^{r_i} = 0$. Therefore,

$$\exp(tZ_i) = I + tZ_i + \frac{t^2}{2!}Z_i^2 + \cdots + \frac{t^{r_i-1}}{(r_i-1)!}Z_i^{r_i-1}$$
$$= \begin{pmatrix} 1 & t & t^2/2! & \ldots & t^{r_i-1}/(r_i-1)! \\ 0 & 1 & t & \ldots & t^{r_i-2}/(r_i-2)! \\ \ldots & \ldots & \ldots & \ldots & \ldots \\ 0 & 0 & 0 & \ldots & 1 \end{pmatrix}.$$

Let us calculate the matrix:

$$\exp(tJ_i) = \exp(t\lambda_i I + tZ_i).$$

Because the matrices $\lambda_i I$ and Z_i commute:

$$\exp(tJ_i) = \exp(t\lambda_i I)\exp(tZ_i),$$

where

$$\exp(t\lambda_i I) = \sum_{k=0}^{\infty} \frac{(\lambda_i t)^k}{k!} I^k = e^{\lambda_i t} I.$$

Thus,

$$\exp(t J_i) = e^{\lambda_i t} \exp(t Z_i).$$

For the sake of simplicity, consider below the case where we have only one Jordan block J of the dimension r with the eigenvalue λ_1, and calculate solutions (see Eq. (4.59)):

$$U^{(k)} = P \exp\left(t \cdot P^{-1} A P\right) c^{(k)}, \quad k = 1, \ldots, r,$$

with

$$c^{(1)} = \begin{pmatrix} 1 \\ 0 \\ \ldots \\ 0 \end{pmatrix}, \quad c^{(2)} = \begin{pmatrix} 0 \\ 1 \\ \ldots \\ 0 \end{pmatrix}, \quad \ldots.$$

We obtain:

$$X^{(1)} = \begin{pmatrix} V^{(1)} & V^{(2)} & \ldots \end{pmatrix} \begin{pmatrix} e^{\lambda_1 t} & t e^{\lambda_1 t} & t^2 e^{\lambda_1 t}/2 & \ldots & \ldots \\ 0 & e^{\lambda_1 t} & t e^{\lambda_1 t} & \ldots & \ldots \\ \ldots & \ldots & \ldots & \ldots & \ldots \\ 0 & 0 & 0 & \ldots & \ldots \end{pmatrix} \begin{pmatrix} 1 \\ 0 \\ \ldots \\ 0 \end{pmatrix}$$

$$= \begin{pmatrix} V^{(1)} & V^{(2)} & \ldots \end{pmatrix} \begin{pmatrix} e^{\lambda_1 t} \\ 0 \\ \ldots \\ 0 \end{pmatrix} = V^{(1)} e^{\lambda_1 t}, \tag{4.62}$$

$$X^{(2)} = \begin{pmatrix} V^{(1)} & V^{(2)} \ldots \end{pmatrix} \begin{pmatrix} e^{\lambda_1 t} & t e^{\lambda_1 t} & t^2 e^{\lambda_1 t}/2 & \ldots & \ldots \\ 0 & e^{\lambda_1 t} & t e^{\lambda_1 t} & \ldots & \ldots \\ \ldots & \ldots & \ldots & \ldots & \ldots \\ 0 & 0 & 0 & \ldots & e^{\lambda_1 t} \end{pmatrix} \begin{pmatrix} 0 \\ 1 \\ \ldots \\ 0 \end{pmatrix} \tag{4.63}$$

$$= V^{(1)} t e^{\lambda_1 t} + V^{(2)} e^{\lambda_1 t},$$

$$X^{(3)} = V^{(1)} \frac{t^2}{2} e^{\lambda_1 t} + V^{(2)} t e^{\lambda_1 t} + V^{(3)} e^{\lambda_1 t}, \tag{4.64}$$

$$X^{(r)} = V^{(1)} \frac{t^{r-1}}{(r-1)!} e^{\lambda_1 t} + \cdots + V^{(r)} e^{\lambda_1 t}. \tag{4.65}$$

Thus, we arrive to the conclusion formulated in Sect. 4.4: in the case where the eigenvalue λ_i has an algebraic multiplicity r_i, there are r_i solutions in the form:

$$e^{\lambda_i t} \begin{pmatrix} p_1(t) \\ \ldots \\ p_n(t), \end{pmatrix},$$

where $p_k(t)$, $k = 1, \ldots, n$ are the polynomials of the degree not higher than $r_i - 1$.

Now we have to calculate the vectors $V^{(1)}$, $V^{(2)}, \ldots, V^{(r)}$ that appear in expressions (4.62)–(4.65).

All solutions $X^{(k)}$ satisfy Eq. (4.40):

$$\frac{dX^{(k)}}{dt} = AX^{(k)}. \tag{4.66}$$

Substituting (4.62) into (4.66), we find:

$$\lambda_1 V^{(1)} = AV^{(1)},$$

or

$$(A - \lambda_1 I)V^{(1)} = 0. \tag{4.67}$$

Thus, $V^{(1)}$ is the eigenvector of a matrix A, corresponding to the eigenvalue λ_1.

Let us substitute now (4.63) into (4.66):

$$V^{(1)} + \lambda_1 \left(V^{(1)}t + V^{(2)} \right) = A \left(V^{(1)}t + V^{(2)} \right).$$

Taking into account (4.67), we find:

$$(A - \lambda_1 I)V^{(2)} = V^{(1)}. \tag{4.68}$$

Similarly, substituting (4.64) into (4.66) and cancelling several terms due to relations (4.67) and (4.68), we find:

$$(A - \lambda_1 I)V^{(3)} = V^{(2)}, \tag{4.69}$$

etc. The last equation obtained by substitution of (4.65) into (4.66) is

$$(A - \lambda_1 I)V^{(r)} = V^{(r-1)}. \tag{4.70}$$

Thus, we arrived to the following algorithm for finding linearly independent solutions in the case of a multiple eigenvalue:

1. Solve the sequence of Eqs. (4.67), (4.68), . . ., (4.70) *until that is possible*.
2. Find the corresponding solutions using formulas (4.62)–(4.65).

After finding all n linearly independent solutions corresponding to all eigenvalues, the general solution of the system is found as a linear combination of the solutions with arbitrary coefficients.

Example 4.16 Consider the system of equations:

$$\frac{dx}{dt} = 2x - y - z, \quad \frac{dy}{dt} = 2x - y - 2z, \quad \frac{dz}{dt} = -x + y + 2z, \tag{4.71}$$

that we have solved already by another approach (see Example 4.13). Following the algorithm described above, we formulate the eigenvalue problem (4.67):

$$\begin{pmatrix} 2 - \lambda & -1 & -1 \\ 2 & -1 - \lambda & -2 \\ -1 & 1 & 2 - \lambda \end{pmatrix} \begin{pmatrix} v_1 \\ v_2 \\ v_3 \end{pmatrix} = \begin{pmatrix} 0 \\ 0 \\ 0 \end{pmatrix}.$$

As shown in Example 4.13, there is a single eigenvalue of algebraic multiplicity three, $\lambda = 1$. Thus, we obtain the problem:

$$v_1 - v_2 - v_3 = 0, \quad 2v_1 - 2v_2 - 2v_3 = 0, \quad -v_1 + v_2 + v_3 = 0.$$

These equations are equivalent; therefore, v_1 and v_2 are arbitrary, and $v_3 = v_1 - v_2$. Choosing (i) $v_1 = 1$, $v_2 = 0$ or (ii) $v_1 = 0$, $v_2 = 1$, we obtain two linearly independent eigenvectors:

$$\begin{pmatrix} v_1 \\ v_2 \\ v_3 \end{pmatrix} = \begin{pmatrix} 1 \\ 0 \\ 1 \end{pmatrix} \quad \text{and} \quad \begin{pmatrix} v_1 \\ v_2 \\ v_3 \end{pmatrix} = \begin{pmatrix} 0 \\ 1 \\ -1 \end{pmatrix}, \tag{4.72}$$

which form a two-dimensional space of vectors in the form

$$\begin{pmatrix} v_1^{(1)} \\ v_2^{(1)} \\ v_3^{(1)} \end{pmatrix} = \begin{pmatrix} C_1 \\ C_2 \\ C_1 - C_2 \end{pmatrix}, \tag{4.73}$$

where C_1 and C_2 are arbitrary real numbers.

The corresponding linearly independent particular solutions of (4.71) are

$$\begin{pmatrix} x \\ y \\ z \end{pmatrix} = \begin{pmatrix} 1 \\ 0 \\ 1 \end{pmatrix} e^t \tag{4.74}$$

and

$$\begin{pmatrix} x \\ y \\ z \end{pmatrix} = \begin{pmatrix} 0 \\ 1 \\ -1 \end{pmatrix} e^t. \tag{4.75}$$

We have to find one more solution that will be related to Eq. (4.68):

$$v_1^{(2)} - v_2^{(2)} - v_3^{(2)} = C_1, \quad 2v_1^{(2)} - 2v_2^{(2)} - 2v_3^{(2)} = C_2,$$
$$-v_1^{(2)} + v_2^{(2)} + v_3^{(2)} = C_1 - C_2. \tag{4.76}$$

Comparing the first two equations, we find that the system is solvable only if $C_2 = 2C_1$, where C_1 is arbitrary. For instance, we can choose $C_1 = 1$; then, (4.73) becomes:

$$V^{(1)} = \begin{pmatrix} 1 \\ 2 \\ -1 \end{pmatrix}. \tag{4.77}$$

All three Eq. (4.76) give:

$$V_1^{(2)} - V_2^{(2)} - V_3^{(2)} = 1.$$

We can choose, e.g., $V_1^{(2)} = V_2^{(2)} = 0$; then $v_3^{(2)} = -1$ and

$$V^{(2)} = \begin{pmatrix} 0 \\ 0 \\ -1 \end{pmatrix}. \tag{4.78}$$

Using expression (4.63), we obtain the additional linearly independent solution of the problem:

$$\begin{pmatrix} x \\ y \\ z \end{pmatrix} = \begin{pmatrix} t \\ 2t \\ -t-1 \end{pmatrix} e^t. \tag{4.79}$$

Combining (4.74), (4.75), and (4.79), we obtain the general solution of the problem:

$$\begin{pmatrix} x \\ y \\ z \end{pmatrix} = \begin{pmatrix} C_1 + C_3 t \\ C_2 + 2C_3 t \\ C_1 - C_2 - C_3(t+1) \end{pmatrix} e^t, \tag{4.80}$$

which coincides with that obtained in Example 4.13. The existence of *two eigenvectors* and two solutions (4.74) and (4.75) explains the absence of the terms proportional to $t^2 e^t$ in the solution (4.80).

4.6.2   Nonhomogeneous Systems of Equations

4.6.2.1   The General Case

Let us consider now a nonhomogeneous system:

$$\frac{dX(t)}{dt} = A(t)X(t) + F(t). \tag{4.81}$$

Denote the corresponding homogeneous system as

$$\frac{dX_h(t)}{dt} = A(t)X_h(t) \tag{4.82}$$

and its fundamental matrix as $Y(t)$. According to (4.47), an arbitrary solution of a homogeneous system (4.82) is

$$X_h(t) = Y(t)C, \tag{4.83}$$

where C is an arbitrary constant vector.

As shown in Sect. 4.5, a solution of a nonhomogeneous system can be found in the framework of the *method of variation of parameters* by replacing constant vector C in a general solution of a corresponding homogeneous system with unknown functions. Following that idea, we search the solution of (4.81) in the form:

$$X(t) = Y(t)U(t), \tag{4.84}$$

where $U(t)$ is an unknown vector function. Substituting (4.84) into (4.82) and taking into account that

$$\frac{dY(t)}{dt} = A(t)Y(t)$$

(see Eq. (4.43)), we find:

$$Y(t)\frac{dU(t)}{dt} = F(t);$$

hence,

$$\frac{dU(t)}{dt} = Y^{-1}(t)F(t). \tag{4.85}$$

Integrating (4.85), we find:

$$U(t) = \int_{t_0}^{t} Y^{-1}(s)F(s)ds + C, \tag{4.86}$$

where t_0 is a point in the region where the system is defined and C is an arbitrary constant vector. Substituting (4.86) into (4.84), we obtain the general solution of Eq. (4.81):

$$X(t) = Y(t)U(t) = Y(t)\int_{t_0}^{t} Y^{-1}(s)F(s)ds + Y(t)C. \tag{4.87}$$

The first term in the right-hand side of (4.87) is a particular solution $X_p(t)$ of a nonhomogeneous system, while the second term, according to (4.83), is a general solution of a homogeneous system. Thus, formula (4.87) can be rewritten as

$$X(t) = \int\limits_{t_0}^{t} Y(t)Y^{-1}(s)F(s)ds + X_u(t). \qquad (4.88)$$

A matrix $Y(t)Y^{-1}(s)$, which "converts" the right-hand side of an equation into its particular solution, is called *the Cauchy kernel*. Note that according to (4.50), the same matrix governs the evolution of the solution of a homogeneous equation.

If Eq. (4.81) is supplemented with an initial condition

$$X(t_0) = X_0, \qquad (4.89)$$

then a solution of a Cauchy problem (4.81), (4.89) can be written as

$$X(t) = \int\limits_{X_0}^{X} Y(t)Y^{-1}(s)F(s)ds + X_0.$$

4.6.2.2   The Case of a Constant Matrix A

If a matrix A is constant, then $Y(t) = \exp(tA)$ (see Sect. 4.6.1.2), $Y^{-1}(s) = \exp(-sA)$, and $Y(t)Y^{-1}(s) = \exp[(t - s)A]$. Thus, (4.88) becomes:

$$X(t) = X_h(t) + \int\limits_{t_0}^{t} \exp[(t - s)A]F(s)ds.$$

If

$$F(t) = e^{\alpha t}(a \cos \beta t + b \sin \beta t) \begin{pmatrix} p_1(t) \\ \cdots \\ p_n(t) \end{pmatrix},$$

where $p_1(t), \ldots, p_n(t)$ are polynomials of degree not higher than m, one can apply *the method of undetermined coefficients* (obviously, $\alpha = 0$ and $\beta = 0$ are just particular cases). One can search a particular solution of the nonhomogeneous system in the form:

$$X_p(t) = e^{\alpha t} \cos \beta t \begin{pmatrix} Q_1(t) \\ \cdots \\ Q_n(t) \end{pmatrix} + e^{\alpha t} \sin \beta t \begin{pmatrix} R_1(t) \\ \cdots \\ R_n(t) \end{pmatrix}, \qquad (4.90)$$

where $Q_j, R_j, j = 1, \ldots, n$ are polynomials. If $\alpha \pm i\beta$ are not the eigenvalues of A, then the degree of those polynomials is not higher than m. If $\alpha \pm i\beta$ are the eigenvalues of A with the algebraic multiplicity r, then the degree of the polynomials is not higher than $m + r$. Expression (4.90) is substituted into the system, and the polynomials are found by comparing the expressions at both sides of equations.

Example 4.17 Find the general solution of the system:

$$\frac{dX}{dt} = \begin{pmatrix} 1 & 1 \\ 4 & 1 \end{pmatrix} X + \begin{pmatrix} e^{5t} \\ e^{7t} \end{pmatrix}. \tag{4.91}$$

First, let us consider the homogeneous problem:

$$\frac{dX_h}{dt} = \begin{pmatrix} 1 & 1 \\ 4 & 1 \end{pmatrix} X.$$

The eigenvalues are determined from the characteristic equation:

$$\begin{vmatrix} 1-\lambda & 1 \\ 4 & 1-\lambda \end{vmatrix} = (\lambda - 1)^2 - 4 = 0;$$

thus, $\lambda_1 = 3$ and $\lambda_2 = -1$. Let us find the eigenvectors corresponding to each eigenvalue. For $\lambda_1 = 3$, we obtain:

$$\begin{pmatrix} -2 & 1 \\ 4 & -2 \end{pmatrix} \begin{pmatrix} v_1^{(1)} \\ v_2^{(1)} \end{pmatrix} = 0;$$

hence,

$$V^{(1)} = \begin{pmatrix} 1 \\ 2 \end{pmatrix}; \tag{4.92}$$

for $\lambda_1 = -1$, we get

$$\begin{pmatrix} 2 & 1 \\ 4 & 2 \end{pmatrix} \begin{pmatrix} v_1^{(1)} \\ v_2^{(2)} \end{pmatrix} = 0;$$

hence,

$$V^{(2)} = \begin{pmatrix} 1 \\ -2 \end{pmatrix}. \tag{4.93}$$

Now we can compare two approaches.

Cauchy Kernel

Using (4.92) and (4.93), we construct the fundamental matrix of the homogeneous system:

$$Y(t) = \begin{pmatrix} e^{3t} & e^{-t} \\ 2e^{3t} & -2e^{-t} \end{pmatrix}.$$

Using the formula

$$\begin{pmatrix} a & b \\ c & d \end{pmatrix}^{-1} = \frac{1}{ad - bc} \begin{pmatrix} d & -b \\ -c & a \end{pmatrix},$$

we find:

$$Y^{-1}(t) = \begin{pmatrix} \dfrac{1}{2}e^{-3t} & \dfrac{1}{4}e^{-3t} \\ \dfrac{1}{2}e^{t} & -\dfrac{1}{4}e^{t} \end{pmatrix}.$$

It is convenient to take the low bound of the integral in (4.88) equal to $-\infty$. Then, we get:

$$
\begin{aligned}
X_p(t) &= \begin{pmatrix} e^{3t} & e^{-t} \\ 2e^{3t} & -2e^{-t} \end{pmatrix} \int_{-\infty}^{t} \begin{pmatrix} \dfrac{1}{2}e^{-3s} & \dfrac{1}{4}e^{-3s} \\ \dfrac{1}{2}e^{s} & -\dfrac{1}{4}e^{s} \end{pmatrix} \begin{pmatrix} e^{5s} \\ e^{7s} \end{pmatrix} ds \\
&= \begin{pmatrix} e^{3t} & e^{-t} \\ 2e^{3t} & -2e^{-t} \end{pmatrix} \int_{-\infty}^{t} \begin{pmatrix} \dfrac{1}{2}e^{2s} + \dfrac{1}{4}e^{4s} \\ \dfrac{1}{2}e^{6s} - \dfrac{1}{4}e^{8s} \end{pmatrix} ds \\
&= \begin{pmatrix} e^{3t} & e^{-t} \\ 2e^{3t} & -2e^{-t} \end{pmatrix} \begin{pmatrix} \dfrac{1}{4}e^{2t} + \dfrac{1}{16}e^{4t} \\ \dfrac{1}{12}e^{6t} - \dfrac{1}{32}e^{8t} \end{pmatrix} = \begin{pmatrix} \dfrac{1}{3}e^{5t} + \dfrac{1}{32}e^{7t} \\ \dfrac{1}{3}e^{5t} + \dfrac{3}{16}e^{7t} \end{pmatrix};
\end{aligned}
\tag{4.94}
$$

thus, the general solution of the nonhomogeneous system (4.91) is

$$X(t) = \begin{pmatrix} \dfrac{1}{3}e^{5t} + \dfrac{1}{32}e^{7t} + C_1 e^{3t} + C_2 e^{-t} \\ \dfrac{1}{3}e^{5t} + \dfrac{3}{16}e^{7t} + 2C_1 e^{3t} - 2C_2 e^{-t} \end{pmatrix}, \tag{4.95}$$

where C_1 and C_2 are arbitrary real numbers.

Method of Undetermined Coefficients

Because 5 and 7 in the exponents forming a vector $\begin{pmatrix} e^{5t} \\ e^{7t} \end{pmatrix}$ are not the eigenvalues of the matrix, we search the particular solution of the nonhomogeneous problem as

$$X_p(t) = e^{5t} \begin{pmatrix} Q_1 \\ Q_2 \end{pmatrix} + e^{7t} \begin{pmatrix} R_1 \\ R_2 \end{pmatrix}, \tag{4.96}$$

where Q_1, Q_2, R_1, and R_2 are numbers. Substituting (4.96) into Eq. (4.91), we find:

$$e^{5t} \begin{pmatrix} 5Q_1 \\ 5Q_2 \end{pmatrix} + e^{7t} \begin{pmatrix} 7R_1 \\ 7R_2 \end{pmatrix} = e^{5t} \begin{pmatrix} Q_1 + Q_2 + 1 \\ 4Q_1 + Q_2 \end{pmatrix} + e^{7t} \begin{pmatrix} R_1 + R_2 \\ 4R_1 + R_2 + 1 \end{pmatrix}.$$

Equating the coefficients, we obtain two uncoupled systems of linear algebraic equations:

$$5Q_1 = Q_1 + Q_2 + 1, \quad 5Q_2 = 4Q_1 + Q_2$$

and

$$7R_1 = R_1 + R_2, \quad 7R_2 = 4R_1 + R_2 + 1.$$

Solving these equations, we find:

$$Q_1 = Q_2 = \frac{1}{3}, \quad R_1 = \frac{1}{32}, \quad R_2 = \frac{3}{16}.$$

We obtained exactly the same particular solution of the nonhomogeneous equation (4.94). Combining (4.92), (4.93), and (4.94), we find the same general solution (4.95).

Problems

1. Solve the problems 1–17 of Sect. 4.5 using the algorithm described in Sect. 4.6.1.
2. Solve the problems 1–7 of Sect. 4.5 using the methods described in Sect. 4.6.2.

After an analytical solution of an IVP has been obtained, you can use the program *ODE 1st order* to compare your result with the numerical one.

4.7 Applications

In the following section, we present some physical examples that illustrate the applications of systems of linear differential equations.

4.7.1 Charged Particle in a Magnetic Field

A particle with the charge q in a constant magnetic field $\vec{B}$ is subject to the *Lorentz force* $\vec{F}$ (Hendrik Lorentz (1853–1928), a Dutch physicist):

$$\vec{F} = q\vec{v} \times \vec{B},$$

where v is the particle velocity and $\times$ denotes vector product. Let the axis z is directed along the direction of $\vec{B}$. The Newton's second law for the particle motion reads:

$$m\frac{d^2x}{dt^2} = qB\frac{dy}{dt}, \quad m\frac{d^2y}{dt^2} = -qB\frac{dx}{dt}, \quad m\frac{d^2z}{dt^2} = 0. \tag{4.97}$$

System (4.97) has to be solved subject to initial conditions:

$$\begin{aligned} x(0) &= x_0, \quad y(0) = y_0, \quad z(0) = z_0, \\ x'(0) &= u_0, \quad y'(0) = v_0, \quad z'(0) = w_0. \end{aligned} \tag{4.98}$$

Equation for $z(t)$ is easily integrated:

$$z = z_0 + w_0 t. \tag{4.99}$$

Let us define $u(t) = dx(t)/dt$ and $v(t) = dy(t)/dt$, and write the initial value problem (4.97), (4.98) as

$$\frac{du}{dt} = \frac{qB}{m}v, \quad \frac{dv}{dt} = -\frac{qB}{m}u, \quad \frac{dx}{dt} = u, \quad \frac{dy}{dt} = v, \tag{4.100}$$

$$u(0) = u_0, \quad v(0) = v_0, \quad x(0) = x_0, \quad y(0) = y_0. \tag{4.101}$$

Finding v from the first equation of (4.100), $v = (m/qB)du/dt$, and substituting it into the second equation, we obtain equation:

$$\frac{d^2u}{dt^2} + \left(\frac{qB}{m}\right)^2 u = 0 \tag{4.102}$$

with the initial conditions

$$u(0) = u_0, \quad u'(0) = \frac{qB}{m}v_0. \tag{4.103}$$

The general solution of (4.102) is

$$u(t) = C_1 \cos \omega t + C_2 \sin \omega t,$$

where

$$\omega = \frac{qB}{m}$$

is *the cyclotron frequency*. The solution satisfying the initial conditions is

$$u(t) = u_0 \cos \omega t + v_0 \sin \omega t; \tag{4.104}$$

hence,

$$u(t) = u_0 \cos \omega t + v_0 \sin \omega t. \tag{4.105}$$

The trajectory of the particle determined by Eqs. (4.99), (4.104), and (4.105) is a cylindrical helical line.

4.7.2 Precession of a Magnetic Moment in a Magnetic Field

Consider a particle with the angular momentum, $\vec{J}$, and the magnetic moment, $\vec{\mu}$, related via

$$\vec{\mu} = \gamma \vec{J},$$

where γ is called *the gyromagnetic ratio*. When this particle is placed in a magnetic field $\vec{B}$, the field exerts a torque $\vec{\tau} = \vec{\mu} \times \vec{B}$ on the particle, and from the rotation equation, $d\vec{J}/dt = \vec{\tau}$, we have:

$$\frac{d\vec{J}}{dt} = \vec{\mu} \times \vec{B}.$$

It follows that the magnetic moment of the particle is subject to the rotation governed by the equation:

$$\frac{d\vec{\mu}}{dt} = \gamma \vec{\mu} \times \vec{B}. \tag{4.106}$$

The same equation describes the angular momentum $\vec{J}$.

If the z-axis is chosen along the direction of vector $\vec{B}$, the system (4.106) can be written as

$$\frac{d\mu_x}{dt} = \gamma B\mu_y, \quad \frac{d\mu_y}{dt} = -\gamma B\mu_x, \quad \frac{d\mu_z}{dt} = 0. \tag{4.107}$$

Solving this system with the initial conditions

$$\mu_x(0) = \mu_x^0, \quad \mu_y(0) = \mu_y^0, \quad \mu_z(0) = \mu_z^0, \tag{4.108}$$

we find:

$$\mu_x(t) = \mu_x^0 \cos \omega t + \mu_y^0 \sin \omega t, \quad \mu_y(t) = -\mu_x \sin \omega t + \mu_y \cos \omega t,$$

where $\omega = \gamma B$ is *the Larmor frequency* (Joseph Larmor (1857–1942), a British physicist). Thus, the vector $\vec{\mu}(t)$ (as well as the vector $\vec{J}(t)$) is rotating about the direction of vector $\vec{B}$ – this phenomenon is called *Larmor precession*. For instance, for the hydrogen nucleus (its magnetic moment is related to the nucleus spin, $\vec{\mu} = \gamma \vec{S}$, rather than to its mechanical momentum) in the field $B = 1$ T, the Larmor frequency $\omega = 45.57$ MHz.

4.7.3   Spring-Mass System

Let us consider the system of two masses m_1 and m_2 and three springs with stiffness coefficients k_1, k_2, and k_3 connected as shown in Fig. 4.10.

Applying the second Newton's law, we obtain the following system of equations (for the sake of simplicity, below we disregard the friction):

$$m_1 \frac{d^2 x_1}{dt^2} = -k_1 x_1 + k_2(x_2 - x_1), \quad m_2 \frac{d^2 x_2}{dt^2} = -k_2(x_2 - x_1) - k_3 x_2. \tag{4.109}$$

Here x_1 and x_2 are displacements of masses m_1 and m_2 from their equilibrium positions. Define $x_3 = dx_1/dt$ and $x_4 = dx_2/dt$, and rewrite system (4.109) as

$$\frac{dx_1}{dt} = x_3, \quad \frac{dx_2}{dt} = x_4,$$

$$\frac{dx_3}{dt} = -\frac{k_1 + k_2}{m_1} x_1 + \frac{k_2}{m_1} x_2, \quad \frac{dx_4}{dt} = \frac{k_2}{m_2} x_1 - \frac{k_2 + k_3}{m_2} x_2.$$

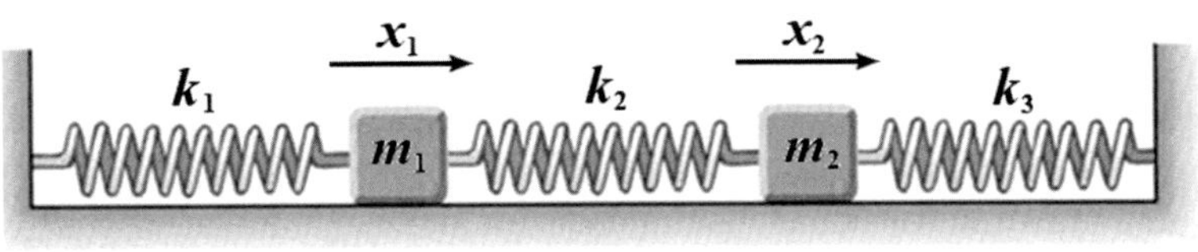

Fig. 4.10 Spring-mass system

In the vector form, the system is written as

$$\frac{dX}{dt} = AX,$$

where

$$X = \begin{pmatrix} x_1(t) \\ x_2(t) \\ x_3(t) \\ x_4(t) \end{pmatrix}, \quad A = \begin{pmatrix} 0 & 0 & 1 & 0 \\ 0 & 0 & 0 & 1 \\ -\dfrac{k_1 + k_2}{m_1} & \dfrac{k_2}{m_1} & 0 & 0 \\ \dfrac{k_2}{m_2} & -\dfrac{k_2 + k_3}{m_2} & 0 & 0 \end{pmatrix}.$$

The determinant of the matrix $A - \lambda I$ is

$$\det(A - \lambda I) = \begin{vmatrix} -\lambda & 0 & 1 & 0 \\ 0 & -\lambda & 0 & 1 \\ -\dfrac{k_1 + k_2}{m_1} & \dfrac{k_2}{m_1} & -\lambda & 0 \\ \dfrac{k_2}{m_2} & -\dfrac{k_2 + k_3}{m_2} & 0 & -\lambda \end{vmatrix}$$

$$= \lambda^4 + \left(\frac{k_1 + k_2}{m_1} + \frac{k_2 + k_3}{m_2} \right) \lambda^2 + \frac{(k_1 + k_2)(k_2 + k_3) - k_2^2}{m_1 m_2} = 0.$$

Substituting $\lambda = i\omega$, we obtain the equation that determines the *eigenfrequencies* of the mass-spring systems:

$$\omega^4 - \left(\frac{k_1 + k_2}{m_1} + \frac{k_2 + k_3}{m_2} \right) \omega^2 + \frac{(k_1 + k_2)(k_2 + k_3) - k_2^2}{m_1 m_2} = 0. \qquad (4.110)$$

Equation (4.110) can be written as

$$\left(\omega^2 - \omega_+^2 \right) \left(\omega^2 - \omega_-^2 \right) = 0,$$

where

$$\omega_\pm^2 = \frac{1}{2} \left\{ \left(\frac{k_1 + k_2}{m_1} + \frac{k_2 + k_3}{m_2} \right) \pm \left[\left(\frac{k_1 + k_2}{m_1} - \frac{k_2 + k_3}{m_2} \right)^2 + \frac{4 k_2^2}{m_1 m_2} \right]^{1/2} \right\}.$$

According to (4.110), the sum and the product of the quantities ω_+^2 and ω_-^2 are positive; therefore, both ω_+^2 and ω_-^2 are positive. The general real solution of the system is

$$X(t) = c_+ X_+ e^{i\omega_+ t} + c_+^* X_+^* e^{-i\omega_+ t} + c_- X_- e^{i\omega_- t} + c_-^* e^{-i\omega_- t},$$

where

$$X_\pm = \begin{pmatrix} x_1^\pm \\ x_2^\pm \\ x_3^\pm \\ x_4^\pm \end{pmatrix}$$

are eigenvectors corresponding to eigenvalues $i\omega_\pm$ and $c_\pm$ are arbitrary complex numbers. The calculations give the following expressions for the components of eigenvectors:

$$
\begin{aligned}
x_1^\pm &= 1, \\
x_2^\pm &= \frac{k_1 + k_2 + i\omega_\pm}{k_2}, \\
x_3^\pm &= i\omega_\pm, \\
x_4^\pm &= i\omega_\pm x_2^\pm.
\end{aligned}
$$

More complex mass-spring systems can be considered in a similar way.

4.7.4 Mutual Inductance

It is known that the change of electric current in one inductor induces a voltage in another nearby inductor. This phenomenon is called a *mutual inductance* and it has numerous applications. A mutual inductance is also a measure of the coupling between two inductors.

Consider two RL circuits (see Example 2.40) with a mutual inductance M, and let in the first circuit be a sinusoidal alternating *emf*, $E_0 \sin \omega t$. Then, we want to find the *induced current* $i_2(t)$ in the second circuit.

Denoting as i_1, L_1, R_1 current, inductance, and resistance in the first circuit, i_2, L_2, and R_2 in the second, and taking into account that the voltages across the resistors equal $R_1 i_1$ and $R_2 i_2$, Kirchhoff's equations for the circuits are

$$
\begin{aligned}
L_1 i_1' + M i_2' + R_1 i_1 &= E_0 \sin \omega t, \\
L_2 i_2' + M i_1' + R_2 i_2 &= 0.
\end{aligned}
\tag{4.111}
$$

Notice that this system of equations does not have a standard form (4.17) with i_1' and i_2' standing in the left side of equations. But any system of linear DE can be converted to the form (4.17).

As a *Reading Exercise*, check that the system (4.111) can be written as

$$i_1' = (-R_1 L_2 i_1 + M R_2 i_2 + L_2 E_0 \sin \omega t)/\Delta,$$
$$i_2' = (M R_1 i_1 - R_2 L_1 i_2 - M E_0 \sin \omega t)/\Delta,$$

(4.112)

where $\Delta = \begin{vmatrix} L_1 & M \\ M & L_2 \end{vmatrix} = L_1 L_2 - M^2.$

The general solution of this system of nonhomogeneous linear equations is the sum of the solution of the associated homogeneous equations, which exponentially decay in time, and the particular solution of the nonhomogeneous equations which can be found by the method of variation of parameters.

As an example, find the induced current i_2 for the following set of parameters: $E_0 = 50$, $\omega = 40$, $L_1 = 2.7$, $R_1 = 0$ (an "ideal" first circuit), $L_2 = 5.1$, $R_2 = 200$, $M = 3.7$ (L_1, L_2, M are in Henry, E_0 is in volts, $R_{1,2}$ are in ohms, ω is in hertz).

Answer:

$$i_2 \approx 0.002 \cos 40t - 0.343 \sin 40t + C e^{-6750t},$$

where time is in seconds and current is in amperes.

The steady-state regime is reached almost instantaneously, and the steady-state induced current is

$$i_2 \approx 0.002 \cos 40t - 0.343 \sin 40t.$$

Chapter 5
Qualitative Methods and Stability of ODE Solutions

5.1 Phase Plane Approach

The phase plane method is a useful qualitative approach to study some important properties of solutions to systems of two *autonomous* ordinary differential equations. (Autonomous means that the right-hand sides of the system's equations do not depend explicitly on the independent variable.) This method has all advantages (and disadvantages) over the numerical and analytical solution methods for systems that the phase line method presented in Sect. 2.7 has over the corresponding methods for a single first-order equation. A phase plane method can be also applied to a second-order autonomous equation, because such equation can be reduced to a system of two autonomous first-order equations. That is, with a designation $x_1(t) = y(t)$ and $x_2(t) = y'(t)$, a second-order autonomous equation

$$y'' = f(y, y'),\qquad(5.1)$$

is equivalent to the first-order autonomous system

$$\begin{aligned} x_1' &= x_2, \\ x_2' &= f(x_1, x_2). \end{aligned}\qquad(5.2)$$

(In what follows, we omit "autonomous.")

In vector form, a general system of two first-order equations is written as

$$X' = F(X),\qquad(5.3)$$

where

© The Author(s), under exclusive license to Springer Nature Switzerland AG 2023
V. Henner et al., *Ordinary Differential Equations*,
https://doi.org/10.1007/978-3-031-25130-6_5

$$X = \begin{pmatrix} x_1(t) \\ x_2(t) \end{pmatrix}, \quad F(X) = \begin{pmatrix} f_1(x_1, x_2) \\ f_2(x_1, x_2) \end{pmatrix} \tag{5.4}$$

are the column vectors whose components are the unknown functions $x_1(t)$, $x_2(t)$ and the known functions f_1 and f_2, respectively.

How can we visualize solutions of system (5.3)? Once the pair of solutions $(x_1(t)$, $x_2(t))$ is known, it can be thought of as a point that sweeps out a solution curve in the $x_1 x_2$-plane as time varies. The $x_1 x_2$-plane in this context is called a phase plane, and the solution curve is called a *phase trajectory*, or *orbit*. (In other words, a phase trajectory is a parametric curve, where the parameter is t.) The shape of a phase trajectory is determined by the functions f_1 and f_2 and the initial conditions for $x_1(t)$ and $x_2(t)$. The set of trajectories that correspond to different initial conditions form a *phase portrait* of a system, from which one can see the behavior of all solutions. Isolated points in the phase portrait mark equilibrium (constant) solutions, also called the critical points (more on this terminology below). Since the derivative of a constant solution is zero, it follows from system (5.3) that equilibrium points are the solutions of an algebraic system:

$$\begin{aligned} f_1(x_1, x_2) &= 0, \\ f_2(x_1, x_2) &= 0. \end{aligned} \tag{5.5}$$

A trajectory that is a closed curve corresponds to periodic solutions, since the solution that starts at any point on a curve returns to this point after some time (this time is, of course, the period of the solution $(x_1(t), x_2(t))$).

A vector function $F(X)$ defines a *vector field* (also called a *direction plot*) in the phase plane. That is, every point (x_1, x_2) in the plane could be a base of a vector with the components:

$$f_1(x_1, x_2) / \sqrt{f_1(x_1, x_2)^2 + f_2(x_1, x_2)^2} \text{ and } f_2(x_1, x_2) / \sqrt{f_1(x_1, x_2)^2 + f_2(x_1, x_2)^2};$$

the radical ensures that vectors at different points have a unit length.

Choosing enough points at random (or according to a smart rule) and plotting a vector at each point is a way of obtaining the information on solutions, *since a vector at a point is tangent to a trajectory there*. Thus, trajectories are easy to visualize from a vector field. (Of course, the trajectories always can be found numerically if high solution accuracy is needed, but a single computation results in a single trajectory only, one that originates at a chosen initial condition, whereas a direction plot shows multiple trajectories for the initial conditions in, say, a rectangular region of the $x_1 x_2$-plane.)

Construction of a direction plot is rarely done by hand, since this would be very tedious for even a few tens of vectors – the number that usually is not sufficient for plotting all representative trajectories.

To understand why equilibrium points are also called critical points, note that an equilibrium point is equivalently defined as such point, where the tangent to the solution curve does not exist, that is, $dx_2/dx_1 = f_2(x_1, x_2)/f_1(x_1, x_2) = 0/0$ (see (5.2) and (5.5)). For a *linear* system, $f_1(x_1, x_2) = ax_1 + bx_2$, and $f_2(x_1, x_2) = cx_1 + dx_2$ (where a, b, c, d are constants) and when the determinant $\begin{vmatrix} a & b \\ c & d \end{vmatrix} \neq 0$, the indeterminate form $dx_2/dx_1 = 0/0$ is possible only at the origin $(x_1, x_2) = (0, 0)$. *Thus, for linear system the origin is the only equilibrium point when the coefficient matrix is non-singular.* This is also stated in Theorem 5.1 below.

If a system (linear or nonlinear) is autonomous, then it can be proved that there is a unique trajectory through each point in the phase plane – in other words, trajectories do not cross or merge, except for equilibrium points.

Example 5.1 (*Mass-Spring*)

Consider a mass-spring system – a one-dimensional harmonic oscillator with zero damping. The equation of motion for the mass m is $y'' + \omega^2 y = 0$. Here $y(t)$ is the position of a mass as a function of time, $\omega = \sqrt{k/m}$ is the frequency of the oscillation, and k is a spring constant. The corresponding system of two first-order equations is

$$\begin{aligned} x_1' &= f_1(x_2) = x_2 \\ x_2' &= f_2(x_1, x_2) = -\omega^2 x_1. \end{aligned} \tag{5.6}$$

The only equilibrium point is the origin, $(x_1, x_2) = (0, 0)$. Let, for simplicity, $\omega = 1$. Solving the homogeneous equation $y'' + y = 0$ gives the general solution $y = C_1 \cos t + C_2 \sin t$. Next, suppose that the initial condition is $(y, y') = (1, 0)$, which translates into the initial condition $(x_1, x_2) = (1, 0)$ in the phase plane. The solution of the IVP is $y = \cos t$. Its derivative is $y' = -\sin t$. Since $y^2 + (y')^2 = 1$, the corresponding trajectory in the phase plane is $x_1^2 + x_2^2 = 1$, which is the circle of radius one centered at the origin. As t increases from zero, the point moves clockwise along the trajectory, and returns to $(1, 0)$ at $t = 2\pi$ (the period of the solution $y = \cos t$). Clearly, the trajectories corresponding to other initial conditions are also circles centered at the origin, and the period of motion along all trajectories is constant and equal to 2π when $\omega = 1$ (the period is $2\pi/\omega$). Saying this a little differently, in the phase plot, the signature of a periodic oscillation of the solutions can be seen in that x_1 and x_2 change the sign periodically and for as long as the trajectory is traced (i.e., forever), also the trajectory returns to where it started after a fixed and constant time elapsed (the period) (Fig. 5.1).

Notice that the family of a circular trajectory $x_1^2 + x_2^2 = const.$ can be obtained directly from the system (5.6). Multiplying first equation by $\omega^2 x_1$, second equation by x_2 and adding equations gives $\omega^2 x_1' x_1 + x_2' x_2 = 0$, or $\frac{d}{dt}(\omega^2 x_1^2 + x_2^2) = 0$. Thus, $\omega^2 x_1^2 + x_2^2 = const.$, which at $\omega = 1$ yields $x_1^2 + x_2^2 = const.$

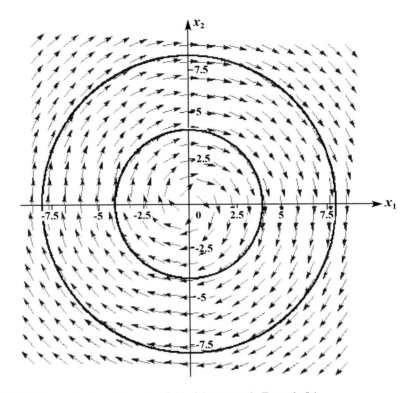

Fig. 5.1 Phase portrait and the vector field of the system in Example 5.1

A phase portrait also allows to visualize the properties of a solution set in a situation when the exact analytical general solution is not available, such as for the nonlinear equation in the next example.

Example 5.2
Consider a nonlinear second-order equation $y'' = y^2 - (y')^2$. The corresponding system of two first-order equations is

$$x_1' = f_1(x_2) = x_2$$
$$x_2' = f_2(x_1, x_2) = x_1^2 - x_2^2.$$

Notice that the second equation of the system is nonlinear. Equilibrium points are found from the system:

$$x_2 = 0$$
$$x_1^2 - x_2^2 = 0.$$

Fig. 5.2 Monotonicity
regions

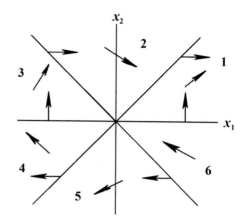

The only solution of this system is $(x_1, x_2) = (0, 0)$, which is the origin of a phase
plane. The vector field is $F\begin{pmatrix} x_1 \\ x_2 \end{pmatrix} = \begin{pmatrix} x_2 \\ x_1^2 - x_2^2 \end{pmatrix}$.

First, let us determine the monotonicity regions for functions $x_1(t)$ and $x_2(t)$ in the
phase plane. Function $x_1(t)$ grows when $x_2 > 0$ (i.e., in the upper half-plane) and
decreases when $x_2 < 0$ (i.e., in the lower half-plane). Function $x_2(t)$ grows as $|x_1| > |x_2|$ and decreases as $|x_1| < |x_2|$. Thus, the phase plane can be divided into six regions
(see Fig. 5.2):

1.	$0 < x_1 < x_2 : \quad x_1' > 0, \ x_2' > 0.$
2.	$x_2 > 0, \ -x_2 < x_1 < x_2 : \quad x_1' > 0, \ x_2' < 0.$
3.	$x_2 > 0, \ x_1 < -x_2 : \quad x_1' > 0, \ x_2' > 0.$
4.	$x_2 < 0, \ x_1 < x_2 : \quad x_1' < 0, \ x_2' > 0.$
5.	$x_2 < 0, \ x_2 < x_1 < -x_2 : \quad x_1' < 0, \ x_2' < 0.$
6.	$x_1 < x_2 < 0 : \quad x_1' < 0, \ x_2' > 0.$

Let us now clarify what happens on the boundaries between the regions. On the
boundary between regions 1 and 2, $x_1' > 0$ and $x_2 = 0$. That means that all trajectories
cross that boundary from region 2 to region 1, and they cannot return.

On the boundary between 2 and 3, also $x_1' > 0$ and $x_2 = 0$, i.e., all trajectories go
from 3 to 2.

On the boundary between 3 and 4, $x_1' = 0$ and $x_2' > 0$; that corresponds to crossing
the boundary from 4 to 3.

On the boundary between 4 and 5, $x_1' < 0$ and $x_2' = 0$; the trajectories cross the
boundary from 5 to 4.

On the boundary between 5 and 6, $x_1' < 0$ and $x_2' = 0$; hence, the trajectories go
from 6 to 5.

Finally, on the boundary between regions 6 and 1, $x_1' = 0$ and $x_2' > 0$; therefore, all trajectories cross that boundary from 6 to 1.

Thus, we obtain the following sequences of crossings:

$$6 \rightarrow 5 \rightarrow 4 \rightarrow 3 \rightarrow 2 \rightarrow 1 \text{ and } 6 \rightarrow 1.$$

We can see that for trajectories that start in region 4, the only way is to pass through regions 3 and 2 and reach region 1. No trajectory can leave region 1, where $x_1(t)$ and $x_2(t)$ are monotonically growing functions. Their growth is unbounded. At large t, $x_1 \sim x_2 \sim e^t$, while $x_1 - x_2 = O(1)$.

For trajectories that start in region 6, there are two options: they can either go to region 1 (where x_1 and x_2 go to $+\infty$) or pass into region 5. Thus, region 6 consists of two subregions separated by a line called a separatrix. On that line, $x_1 \rightarrow +\infty$ and $x_2 \rightarrow -\infty$ as $t \rightarrow -\infty$. When $t \rightarrow +\infty$, that line tends to the equilibrium point $(0,0)$.

Concerning region 5, we know that some trajectories come from region 6 and some of them pass to region 4. However, there is also another option: $x_1(t)$ and $x_2(t)$ can tend to $-\infty$ as $t \rightarrow -\infty$. In order to find the asymptotic behavior of the corresponding solutions, let us assume that $|x_2| \gg |x_1| \gg 1$ at large t. Then, equation for x_2 is simplified as $x_2' = -x^2$; hence, x_2 tends to $-\infty$ during a finite time: $x_2(t) \sim 1/(t - t_*)$, $t < t_*$, where t_* is a constant. From the equation for x_1, we find $x_1(t) \sim \ln|t - t_*|$, which confirms the assumption $|x_2| \gg |x_1|$.

Thus, by means of the qualitative analysis, we obtained the predictions of the possible solution behaviors for arbitrary initial conditions, almost without any calculation. The direct numerical simulation of the system confirms the predictions of the qualitative analysis (see Fig. 5.3).

Examples 5.1 and 5.2 show that the analysis of the phase portrait allows to obtain the qualitative properties of a solution without calculating solutions analytically or numerically.

Generally, the construction of a phase portrait starts with finding the equilibrium points and the analysis of trajectories in their vicinity. To better understand the behavior of trajectories near equilibrium points, we shall start with the analytically tractable case of linear autonomous systems.

5.2   Phase Portraits and Stability of Solutions in the Case of Linear Autonomous Systems

5.2.1   Equilibrium Points

Consider first the important case of a linear, homogeneous second-order equation with constant coefficients:

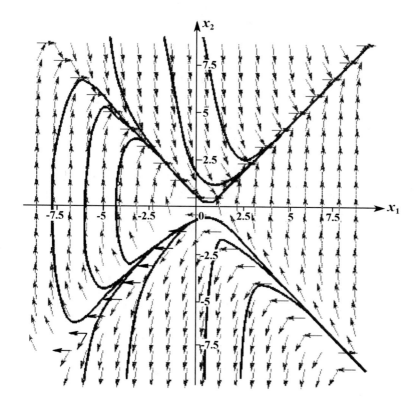

Fig. 5.3 Phase portrait and the vector field of the system in Example 5.2

$$y'' + py' + qy = 0. \tag{5.7}$$

The corresponding planar system is

$$\begin{aligned} x_1' &= f_1(x_2) = x_2 \\ x_2' &= f_2(x_1, x_2) = -qx_1 - px_2. \end{aligned} \tag{5.8}$$

Clearly, the only equilibrium point of this system is $(0, 0)$. The vector form of the system is

$$X' = F(X) = AX, \tag{5.9}$$

where $X = \begin{pmatrix} x_1 \\ x_2 \end{pmatrix} = \begin{pmatrix} y \\ y' \end{pmatrix}$ is the column vector of unknowns, and

$A = \begin{pmatrix} 0 & 1 \\ -q & -p \end{pmatrix}$ is the two-by-two matrix of coefficients. Equations (5.8) and (5.9) are the particular case of a more general system:

$$x_1' = f_1(x_1, x_2) = ax_1 + bx_2$$
$$x_2' = f_2(x_1, x_2) = cx_1 + dx_2, \tag{5.10}$$

where a, b, c, d are constants; it is assumed that at least one of these constants is nonzero. In a vector form

$$X' = AX, A = \begin{pmatrix} a & b \\ c & d \end{pmatrix}. \tag{5.11}$$

Next, we focus the attention on the system (5.10) (or (5.11)). The analysis that we present applies to the system (5.8) (or (5.9)) as well, if in the formulas one replaces (a, b, c, d) by $(0, 1, -q, -p)$.

We state the following theorem without proving it:

Theorem 5.1 *The planar linear system* $X' = AX$, $A \neq 0$ *has a unique equilibrium point if* $\det(A) \neq 0$, *and a straight line of (infinitely many) equilibrium points if* $\det(A) = 0$.

We will further consider only the first case, i.e., we assume that matrix A is non-singular.

The topology of a phase portrait and thus the behavior of solutions to system (5.10) depend at large on *eigenvalues* of A.

Eigenvalues λ are the solutions of the equation:

$$\det(A - \lambda I) = 0, \tag{5.12}$$

where I is the identity matrix, $I = \begin{pmatrix} 1 & 0 \\ 0 & 1 \end{pmatrix}$. Equation (5.12) is called the *characteristic equation* for matrix A. In expanded form, it reads:

$$(a - \lambda)(d - \lambda) - bc = \lambda^2 - \lambda(a + d) + (ad - bc) =$$
$$= \lambda^2 - T(A)\lambda + \det(A) = 0, \tag{5.13}$$

where $T(A)$ is the trace of A. We can immediately notice that the assumption we made above, $\det(A) \neq 0$, is equivalent to $\lambda \neq 0$.

Thus, the system has the unique equilibrium point $(0, 0)$ only if there is no zero eigenvalue(s). The quadratic characteristic Equation, (5.13), has two solutions. These solutions can be real, or two complex-conjugate solutions may occur. Since our interest in this section is in determining the topology of a phase portrait, we briefly consider all possible cases for eigenvalues (excluding a zero eigenvalue). The reader is referred to Chap. 4 for further details.

5.2.2 Stability: Basic Definitions

In this section we take a close look at the stability of solutions. In order to be able to introduce all necessary concepts, we omit the discussion of the simplest case, that is, of a single first-order equation, and start with the higher-order equations, or equivalently by virtue of a reduction method, with systems.

The following Definition 5.1 is for *linear*, autonomous systems of n first-order equations with a non-singular $n \times n$ matrix A. In particular, when $n = 2$, Definition 5.1 is valid for Eq. (5.11).

Definition 5.1 (*Lyapunov stability of the equilibrium solution of a linear autonomous system*):

Let $\overline{X} = 0$ be the equilibrium solution of $X' = AX$. Also let X be the perturbation, at $t = t_0$ (*where t_0 is any time*), of the equilibrium solution. Let $\|X(t)\| = \sqrt{x_1(t)^2 + x_2(t)^2 + \ldots + x_n(t)^2}$.

Then, the equilibrium solution is said to be:

(i) *Stable*, if for all $\varepsilon > 0$ there exists $\delta(\varepsilon)$, such that $\|X(t_0)\| < \delta$ implies that $\|X(t)\| < \varepsilon$ for all $t > t_0$.

This means that solution that is inside the "ball" of radius δ (with the center at the origin, $\overline{X} = 0$) at time t_0, stays within the ball of radius ε (which depends on δ and t_0) for all $t > t_0$.

Note that $\varepsilon(\delta)$ can be larger than δ (see Example 5.4).

(ii) *Asymptotically stable*, if stable and there exists such $\delta > 0$ that $\|X(t_0)\| < \delta$ implies that $\|X(t)\| \to 0$ as $t \to \infty$.

This means that if the solution is inside the disc of radius δ (with the center at the origin, $\overline{X} = 0$) at time t_0, then it approaches the origin as $t \to \infty$.

Clearly, asymptotic stability is a stronger property than stability, since the former implies the latter.

If (i) does not hold, the solution $\overline{X} = 0$ is said to be *unstable*.

(Alexander Lyapunov (1857–1918) was the prominent Russian mathematician, one of the founders of the theory of dynamical systems.)

The definitions presented above can be generalized for time-dependent solutions $\overline{X}(t)$, of a linear, not necessarily homogeneous, system. The stability of an arbitrary solution is defined in a similar way.

Definition 5.1a (*Lyapunov stability of an arbitrary solution of a linear autonomous system*)

The solution $\overline{X}(t)$ of $X' = AX + b$ is said to be:

(i) *Stable*, if for all $\varepsilon > 0$ there exists $\delta(\varepsilon, t_0)$, such that for any other solution $X(t)$ of the system, $\|X(t_0) - \overline{X}(t_0)\| < \delta$ implies that $\|X(t) - \overline{X}(t)\| < \varepsilon$ for all $t > t_0$.

(ii) *Asymptotically stable*, if stable and $\|X(t_0) - \overline{X}(t_0)\| < \delta$ implies that $\|X(t) - \overline{X}(t)\| \to 0$ as $t \to \infty$.

For such solutions we will now demonstrate that it is sufficient to analyze the stability of the equilibrium point of a homogeneous system: any other solution of the linear system has the same stability property as the equilibrium solution.

Indeed, consider another solution, $X(t)$, and let $Y(t) = X(t) - \overline{X}(t)$, that is, $Y(t)$ is thought of as perturbation (not necessarily a small one), of the solution whose stability is in question, $\overline{X}(t)$. Differentiation gives:

$$Y'(t) = X'(t) - \overline{X}'(t) = AX(t) + b - \left(A\overline{X}(t) + b\right) = A\left(X(t) - \overline{X}(t)\right) = AY(t).$$

This shows that the perturbation is governed by the same homogeneous ODE system, independently of the choice of solution $\overline{X}(t)$. Thus, the behavior of $Y(t)$ is the same for any $\overline{X}(t)$. If the equilibrium point $\overline{X}(t) = 0$ of the homogeneous system is stable ($\|Y(t)\|$ does not grow) or asymptotically stable ($\|Y(t)\| \to 0$), then any solution $\overline{X}(t)$ of the inhomogeneous system is also stable; if the equilibrium point is unstable, then the perturbation grows and $\overline{X}(t)$ is unstable.

Below we consider the phase portraits and stability of trajectories in different cases.

5.2.3 Real and Distinct Eigenvalues

Obviously, there are three cases to consider:

1	$\lambda_1 < \lambda_2 < 0$; the (stable) equilibrium point $(0,0)$ is called *an attractor, or a sink, or a stable node.*
2	$\lambda_1 < 0 < \lambda_2$; the (unstable) equilibrium point $(0,0)$ is called *a saddle.*
3	$0 < \lambda_1 < \lambda_2$; the (unstable) equilibrium point $(0,0)$ is called *a repeller, or a source, or an unstable node.*

We discuss one example for each case.

Example 5.3 (*Sink*)

Consider (5.11), where $A = \begin{pmatrix} -2 & -2 \\ -1 & -3 \end{pmatrix}$. The characteristic Eq. (5.13) is $\lambda^2 + 5\lambda + 4 = 0$, which has solutions $\lambda_1 = -1$, $\lambda_2 = -4$. The corresponding eigenvectors are $X_1 = \begin{pmatrix} -2 \\ 1 \end{pmatrix}$, $X_2 = \begin{pmatrix} 1 \\ 1 \end{pmatrix}$. The general solution of the system is

$$X = C_1 X_1 e^{\lambda_1 t} + C_2 X_2 e^{\lambda_2 t} = C_1 \begin{pmatrix} -2 \\ 1 \end{pmatrix} e^{-t} + C_2 \begin{pmatrix} 1 \\ 1 \end{pmatrix} e^{-4t}.$$

Linearly independent eigenvectors X_1 and X_2 define two distinct lines through the origin, given by equations $x_2 = x_1$, $x_2 = -x_1/2$. In Fig. 5.4 the vector field and

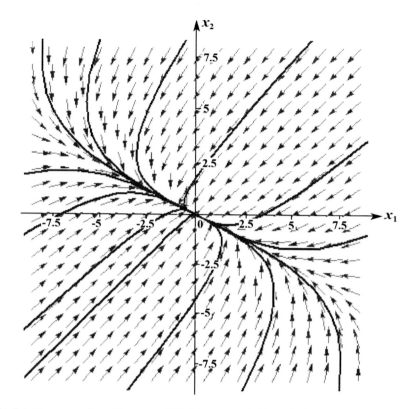

Fig. 5.4 Phase portrait and the vector field of the system in Example 5.3

several phase trajectories are plotted. Solution curves come from infinity along the line $x_2 = x_1$, corresponding to the largest (in magnitude) negative eigenvalue, bend near the equilibrium point so that they become tangent to the line $x_2 = -x_1/2$, and then "sink into" the equilibrium.

According to the definitions given in Sect. 5.2.2, the sink (and all the trajectories tending to the sink) is asymptotically stable.

Let us emphasize that generally the eigenvectors of matrix A are not orthogonal to each other, because matrix A is generally non-normal. Therefore, in the case where $\lambda_1 < 0$, $\lambda_2 < 0$, though the norms of each vector $C_1 X_1 e^{\lambda_1 t}$ and $C_2 X_2 e^{\lambda_2 t}$ decrease monotonically, the norm of their sum, the vector X, can change nonmonotonically. Let us consider the following example.

Example 5.4

Let us consider the system:

$$\frac{dx_1}{dt} = -x_1 + 10x_2, \qquad \frac{dx_2}{dt} = -2x_2$$

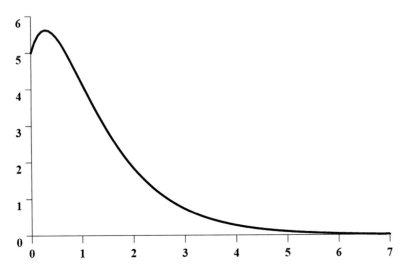

Fig. 5.5 The norm $\|X(t)\|$ of the solution vector in Example 5.4

with initial conditions $x_1(0) = 5$, $x_2(0) = 1$. The solution is

$$x_1(t) = 15e^{-t} - 10e^{-2t}, \quad x_2(t) = e^{-2t}.$$

Thus,

$$\|X(t)\|^2 = x_1^2(t) + x_2^2(t) = 225e^{-2t} - 300e^{-3t} + 101e^{-4t}.$$

One can see in Fig. 5.5 that in the beginning, the norm of the solution grows, but for any initial conditions, $\|X(t)\| \to 0$ when $t \to \infty$. In some cases, that growth can be significant. The investigation of the evolution of the solution norm $\|X(t)\|$ at finite t is the subject of the non-modal (or non-normal) stability analysis, which is not considered in this book.

Example 5.5 (Saddle)

Let in (5.11), $A = \begin{pmatrix} 0 & 1 \\ 4 & 3 \end{pmatrix}$. Comparing this matrix to the matrix in (5.9), we see that they coincide with $q = 3$ and $p = 4$. Thus, system (5.11) corresponds to the second-order equation $y'' - 3y' - 4y = 0$, where $y = x_1$, $y' = x_2$. The characteristic equation for this homogeneous differential equation is $\lambda^2 - 3\lambda - 4 = 0$, which has solutions $\lambda_1 = -1$, $\lambda_2 = 4$.

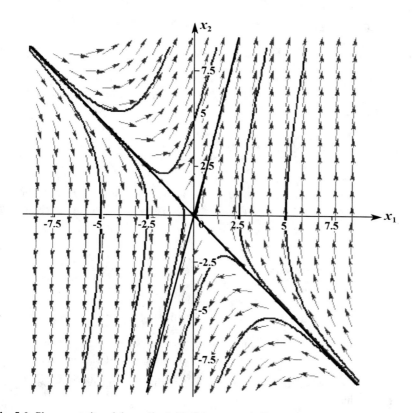

Fig. 5.6 Phase portrait and the vector field of the system in Example 5.5

Thus, the general solution is $y = C_1 e^{-t} + C_2 e^{4t}$. Now we can write the vector form of the general solution using $y' = -C_1 e^{-t} + 4C_2 e^{4t}$ as the second component:

$$X = \begin{pmatrix} y \\ y' \end{pmatrix} = \begin{pmatrix} x_1 \\ x_2 \end{pmatrix} = C_1 X_1 e^{\lambda_1 t} + C_2 X_2 e^{\lambda_2 t} = C_1 \begin{pmatrix} 1 \\ -1 \end{pmatrix} e^{-t} + C_2 \begin{pmatrix} 1 \\ 4 \end{pmatrix} e^{4t}.$$

This is the same general solution that we would have obtained if we decided instead to solve (5.13) directly. Indeed, the characteristic equation for A is $\lambda^2 - 3\lambda - 4 = 0$, which is the same equation as the characteristic equation we just solved. Thus, the eigenvalues are the same two numbers, $\lambda_1 = -1$, $\lambda_2 = 4$. The eigenvectors are $\begin{pmatrix} 1 \\ -1 \end{pmatrix}$ and $\begin{pmatrix} 1 \\ 4 \end{pmatrix}$, i.e., they coincide with X_1 and X_2 found above. The phase portrait is shown in Fig. 5.6. Note that trajectories approach the origin along the line of eigenvectors $x_2 = -x_1$, corresponding to the negative eigenvalue (the stable axis), and tend to infinity along the line of eigenvectors $x_2 = 4x_1$, corresponding to the positive eigenvalue (the unstable axis).

Example 5.6 (*Source*)

Let in (5.11), $A = \begin{pmatrix} 2 & 0 \\ 0 & 3 \end{pmatrix}$. The characteristic equation for A is $(2 - \lambda)(3 - \lambda) = 0$.

Its solutions are $\lambda_1 = 2$, $\lambda_2 = 3$. As in Example 5.4, eigenvectors can be easily determined. We notice that $a = 2$, $d = 3$, $b = c = 0$, and thus equations in system (5.10) are completely decoupled:

$$x_1' = 2x_1$$
$$x_2' = 3x_2.$$

Thus, each equation can be solved independently by using, say, separation of variables. The solutions are $x_1 = C_1 e^{2t}$, $x_2 = C_2 e^{3t}$. The vector form of the solution to the system is

$$X = \begin{pmatrix} x_1 \\ x_2 \end{pmatrix} = C_1 X_1 e^{2t} + C_2 X_2 e^{3t} = C_1 \begin{pmatrix} 1 \\ 0 \end{pmatrix} e^{2t} + C_2 \begin{pmatrix} 0 \\ 1 \end{pmatrix} e^{3t},$$

where $\begin{pmatrix} 1 \\ 0 \end{pmatrix}$ and $\begin{pmatrix} 0 \\ 1 \end{pmatrix}$ are the eigenvectors. The lines of eigenvectors are $x_2 = 0$ and $x_1 = 0$. The solutions leave the equilibrium tangent to the second line (the smallest eigenvalue) and tend to infinity along the first line (the largest eigenvalue).

The phase portrait is shown in Fig. 5.7. Notice that the eigenvectors are aligned with the coordinate axes.

5.2.4   Complex Eigenvalues

There are three cases to consider:

1.	$\lambda_1 = \lambda_r + i\lambda_i$, $\lambda_2 = \lambda_r - i\lambda_i$, $\lambda_r < 0$; the (stable) equilibrium point $(0,0)$ is called *a spiral attractor, or a spiral sink, or a stable focus.*
2.	$\lambda_1 = \lambda_r + i\lambda_i$, $\lambda_2 = \lambda_r - i\lambda_i$, $\lambda_r > 0$; the (unstable) equilibrium point $(0,0)$ is called *a spiral repeller, or a spiral source, or an unstable focus.*
3.	$\lambda_1 = i\lambda_i$, $\lambda_2 = -i\lambda_i$, $\lambda_r = 0$; the (stable) equilibrium point $(0,0)$ is called *a center.*

We discuss one example for each case.

Example 5.7 (*Spiral Sink*)

Let in (5.11), $A = \begin{pmatrix} -2 & -3 \\ 3 & -2 \end{pmatrix}$.   The   characteristic   equation   for A is $\lambda^2 + 4\lambda + 13 = 0$. Its solutions are $\lambda_1 = -2 + 3i$, $\lambda_2 = -2 - 3i$. The direction field and the phase portrait are shown in Fig. 5.8. The trajectories spiral toward the origin. The period of rotation about the origin is $2\pi/|\lambda_i| = 2\pi/3$. Note that the

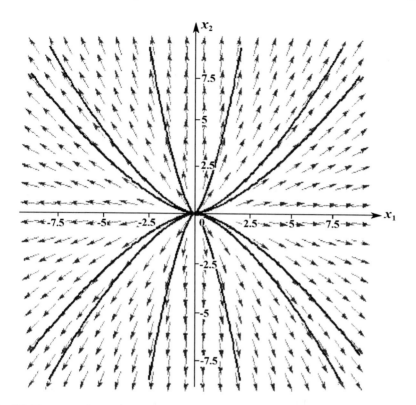

Fig. 5.7 Phase portrait and the vector field of the system in Example 5.6

eigenvectors are complex-valued and thus they can't be plotted. Clearly, the phase plane points to both the oscillatory and decaying behaviors of solutions. Indeed (see Chap. 4), the general solution of the system is

$$X = \begin{pmatrix} x_1 \\ x_2 \end{pmatrix} = C_1 \begin{pmatrix} -\sin 3t \\ \cos 3t \end{pmatrix} e^{-2t} + C_2 (\cos 3t \sin 3t) e^{-2t},$$

from which it is obvious that solutions $x_1(t)$ and $x_2(t)$ oscillate with the period $2\pi/3$ and their amplitude is exponentially decaying.

Obviously, the stable sink and all other trajectories are asymptotically stable.

Example 5.8 (Spiral Source)

Let in (5.11), $A = \begin{pmatrix} 0 & 1 \\ -5 & 2 \end{pmatrix}$. The characteristic equation is $\lambda^2 - 2\lambda + 5 = 0$, and the eigenvalues are $\lambda_1 = 1 + 2i$, $\lambda_2 = 1 - 2i$. The direction field and the phase portrait are shown in Fig. 5.9. The trajectories spiral away from the origin. The period of the rotation about the origin is $2\pi/|\lambda_i| = 2\pi/2 = \pi$. Comparing matrix A to the matrix in (5.9), we see that $q = 5, p = -2$, and therefore the system (5.11) corresponds to the

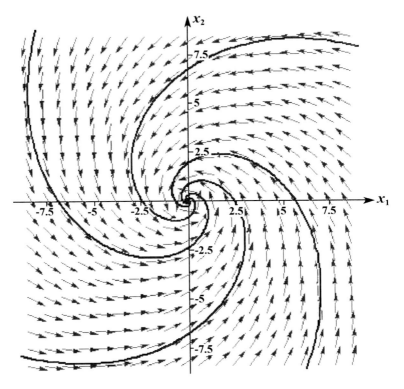

Fig. 5.8 Phase portrait and the vector field of the system in Example 5.7

second-order equation $y'' - 2y' + 5y = 0$, where $y = x_1$, $y' = x_2$. This is the equation of a harmonic oscillator with a negative damping. Thus, from the form of the equation only, we expect oscillatory solutions with increasing amplitude, and the phase portrait shows precisely that. Solving the oscillator gives $r_1 = \lambda_1$, $r_2 = \lambda_2$, and the general solution is $y = C_1 e^t \cos 2t + C_2 e^t \sin 2t$, which has the period π.

Example 5.9 (*Center*)

Let in (5.11), $A = \begin{pmatrix} 0 & 1 \\ -2 & 0 \end{pmatrix}$. The characteristic equation is $\lambda^2 + 2 = 0$, and the eigenvalues are $\lambda_1 = i\sqrt{2}, \lambda_2 = -i\sqrt{2}$. The direction field and the phase portrait are shown in Fig. 5.10. The trajectories are the circles centered at the origin. The period of the rotation about the origin is $2\pi/|\lambda_i| = 2\pi/\sqrt{2} = \pi\sqrt{2}$. Comparing matrix A to the matrix in (5.9), we see that $q = 2$, $p = 0$, and therefore the system (5.10) corresponds to the second-order equation $y'' + 2y = 0$. This is the equation of an undamped harmonic oscillator. Solving the oscillator gives $r_1 = \lambda_1$, $r_2 = \lambda_2$, and the general solution is $y = C_1 \cos\sqrt{2}t + C_2 \sin\sqrt{2}t$. A careful reader may notice that this example coincides with Example 5.2, if one takes there $\omega = \sqrt{2}$. Then, the

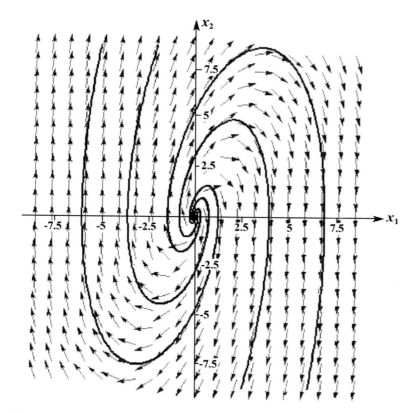

Fig. 5.9 Phase portrait and the vector field of the system in Example 5.8

system's matrix in Example 5.2 is precisely the matrix A in Example 5.9, and the phase portraits are identical.

In general, all solution curves in a system with purely imaginary eigenvalues lie on ellipses that enclose the origin.

The center is a stable equilibrium point: the trajectory never leaves an arbitrary small vicinity of the center if its starting point is sufficiently close to the center. But it is not asymptotically stable, because the trajectories do not tend to it when $t \to \infty$. The same is true for any other trajectory.

5.2.5   Repeated Real Eigenvalues

Example 5.10

Let in (5.11), $A = \begin{pmatrix} 0 & 1 \\ -4 & -4 \end{pmatrix}$. The characteristic equation is $\lambda^2 + 4\lambda + 4 = 0$, and the eigenvalues are $\lambda_1 = \lambda_2 = \lambda = -2$. Thus, we expect the origin to be a sink.

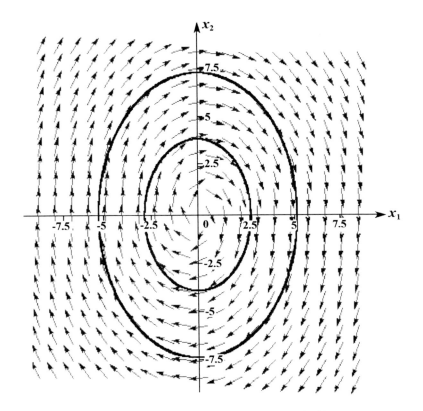

Fig. 5.10 Phase portrait and the vector field of the system in Example 5.9

Comparing matrix A to the matrix in (5.9), we see that $q = p = 4$, and therefore the system (5.11) corresponds to the second-order equation $y'' + 4y' + 4y = 0$, where $y = x_1$, $y' = x_2$. This is the equation of the critically damped oscillator. Solution gives $r_1 = r_2 = \lambda = -2$, and the general solution is $y = C_1 e^{-2t} + C_2 t e^{-2t}$. The vector form of the solution is obtained by differentiating this and writing the two formulas as a single vector formula. This gives $X = C_1 \begin{pmatrix} 1 \\ -2 \end{pmatrix} e^{-2t} +$

$C_2 \left[\begin{pmatrix} 1 \\ -2 \end{pmatrix} t + \begin{pmatrix} 0 \\ 1 \end{pmatrix} \right] e^{-2t}$. The direction plot and the phase portrait are shown in Fig. 5.11. Notice how the phase trajectories try to spiral around the origin, but the line of eigenvectors $x_2 = -2x_1$ terminates the spiraling, indicating attempted but not completed oscillation.

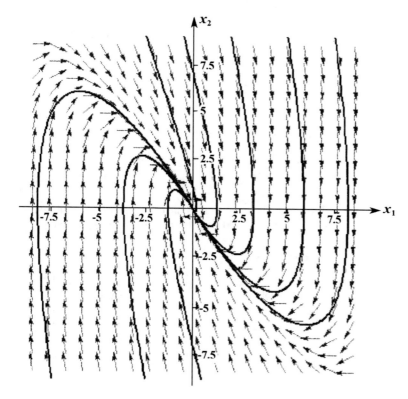

Fig. 5.11 Phase portrait and the vector field of the system in Example 5.10

If the eigenvalue is repeated ($\lambda_1 = \lambda_2$) and its geometric multiplicity is two (which means that the number of linearly independent eigenvectors is two), then the equilibrium point (sink, if $\lambda_1 = \lambda_2 < 0$, or source if $\lambda_1 = \lambda_2 > 0$) is called a *proper node*. In this case the general solution is $X = C_1 X_1 e^{\lambda t} + C_2 X_2 e^{\lambda t}$, where $\lambda = \lambda_1 = \lambda_2$, and X_1, X_2 are two linearly independent eigenvectors corresponding to eigenvalue λ. If there is only one linearly independent eigenvector that corresponds to the repeated eigenvalue (geometric multiplicity is one), then the equilibrium point is called a *degenerate*, or *improper*, *node*.

Degenerate nodes do not occur if A is symmetric ($a_{ij} = a_{ji}$), or skew-symmetric ($a_{ij} = -a_{ji}$; thus, $a_{ii} = 0$).

Phase portraits of all planar autonomous systems retain the typical topology demonstrated in the Examples. That is, depending on eigenvectors, all saddles differ only in the angle of rotation of the axes of a saddle with respect to the coordinate axes x_1 and x_2, and also in the angle that the saddle's axes make. The same is true for phase portraits near other equilibrium points.

Problems

In problems 1–3, write a system of first-order equations, corresponding to the given equation or the IVP:

1. $2y'' - 3y' + 11y = 2$.
2. $y'' + (t + 1)y' - 5t = 0$, $\quad y(0) = 0, y'(0) = 1$.
3. $y^{(4)} = y + y''$.

In problems 4 and 5, replace the planar ODE system by a second-order equation.

4.	$X' = \begin{pmatrix} 0 & 1 \\ -2 & 3 \end{pmatrix} X.$	5.	$X' = \begin{pmatrix} 0 & 1 \\ t-4 & 1/t \end{pmatrix} X.$

In problems 6–17, solve the linear system, state the equilibrium point type, and, if the system is planar, use CAS to plot the vector field and phase portrait.

6.	System in Exercise 5.4.	7.	$X' = \begin{pmatrix} 8 & -3 \\ 16 & -8 \end{pmatrix} X.$
8.	$X' = \begin{pmatrix} 12 & -15 \\ 4 & -4 \end{pmatrix} X.$	9.	$X' = \begin{pmatrix} -1 & 0 \\ 0 & -4 \end{pmatrix} X.$
10.	$X' = \begin{pmatrix} 1 & -1 & -1 \\ 0 & 1 & 3 \\ 0 & 3 & 1 \end{pmatrix} X,$ $\lambda_{1,2,3} = 1, 4, -2.$	11.	$X' = \begin{pmatrix} 2 & -5 \\ 2 & -4 \end{pmatrix} X.$
12.	$X' = \begin{pmatrix} 4 & 1 \\ -8 & 8 \end{pmatrix} X.$	13.	$X' = \begin{pmatrix} 4 & 5 \\ -4 & -4 \end{pmatrix} X.$
14.	$X' = \begin{pmatrix} 4 & -13 \\ 2 & -6 \end{pmatrix} X.$	15.	$X' = \begin{pmatrix} 1 & 2 & -1 \\ 0 & 1 & 1 \\ 0 & -1 & 1 \end{pmatrix} X,$ $\lambda_{1,2,3} = 1, 1+i, 1-i.$
16.	$X' = \begin{pmatrix} 8 & -1 \\ 4 & 12 \end{pmatrix} X.$	17.	$X' = \begin{pmatrix} 4 & 1 \\ -1 & 2 \end{pmatrix} X.$

5.2.6 Summary

Summarizing the examples presented above, we formulate the theorem that relates the eigenvalues and stability. The theorem will be formulated for a linear autonomous system of an arbitrary order n.

In the case of a multiple root λ_i of the characteristic equation, it is necessary to distinguish between its *algebraic multiplicity*, which is the root multiplicity, and its *geometric multiplicity*, i.e., the number of linearly independent eigenvectors corresponding to the eigenvalue λ_i. For instance, for the matrix $\begin{pmatrix} 1 & 0 \\ 0 & 1 \end{pmatrix}$, there are two eigenvectors $\begin{pmatrix} 1 \\ 0 \end{pmatrix}$ and $\begin{pmatrix} 0 \\ 1 \end{pmatrix}$ corresponding to the double root λ_i; hence, both the algebraic multiplicity and the geometric multiplicity are equal to 2. For the matrix $\begin{pmatrix} 1 & 1 \\ 0 & 1 \end{pmatrix}$, there is only one eigenvector $\begin{pmatrix} 1 \\ 0 \end{pmatrix}$ for the double root λ_i; therefore, its algebraic multiplicity is 2 and its geometric multiplicity is 1.

Theorem 5.2 *Suppose that* $\lambda_1, \ldots, \lambda_n$ *are the eigenvalues of matrix A:*

(i) *If* $\mathrm{Re}(\lambda_i) < 0$ *for all* $i = 1, \ldots, n$, *then all solutions of* $X' = AX$ *are asymptotically stable.*

(ii) *If* $\mathrm{Re}(\lambda_i) \leq 0$ *for all* $i = 1, \ldots, n$ *and the algebraic multiplicity equals the geometric multiplicity whenever* $\mathrm{Re}(\lambda_j) = 0$ *for any j, then all solutions are stable.*

(iii) *If* $\mathrm{Re}(\lambda_j) > 0$ *for at least one j, or the algebraic multiplicity is greater than the geometric multiplicity should* $\mathrm{Re}(\lambda_j) = 0$, *then all solutions are unstable.*

Rigorous proof of Theorem 5.2 is difficult, but it is easy once again to get the feeling that this theorem is correct. Consider, for instance, the simplest case of a system that has n distinct eigenvalues and the corresponding n different, linearly independent eigenvectors X_i. Then, the general solution is the linear combination $X = X_1 e^{\lambda_1 t} + X_2 e^{\lambda_2 t} + \ldots + X_n e^{\lambda_n t}$. If all eigenvalues have negative real parts, then X approaches zero as $t \to \infty$; indeed, by Euler's formula $e^{\lambda_i t} = e^{(r_i \pm i\theta_i)t} = e^{r_i t} e^{\pm i\theta_i t} \to 0$ as $t \to \infty$. Such solution is asymptotically stable. If at least one eigenvalue has a positive real part, then the corresponding term in the general solution grows exponentially, while the terms that correspond to eigenvalues with positive real parts decay to zero. Thus, such solution is unstable, etc.

Applying Theorem 5.2 to the case $n = 2$ (system (5.10) with $\det(A) \neq 0$), we arrive to the following summary of stability.

Eigenvalues	Type of equilibrium point	Stability
$\lambda_1 > \lambda_2 > 0$	Source (unstable node)	Unstable
$\lambda_1 < \lambda_2 < 0$	Sink (stable node)	Asymptotically stable
$\lambda_2 < 0 < \lambda_1$	Saddle	Unstable
$\lambda_1 = \lambda_2 > 0$	Source (proper or improper (degenerate) node)	Unstable
$\lambda_1 = \lambda_2 < 0$	Sink (proper or improper (degenerate) node)	Asymptotically stable
$\lambda_1, \lambda_2 = r \pm i\theta$	Spiral	
$r > 0$	Spiral source (unstable focus)	Unstable
$r < 0$	Spiral sink (stable focus)	Asymptotically stable
$r_1 = i\theta, r_2 = -i\theta$	Center	Stable

In the case of a linear autonomous system, sink and spiral sink are not only asymptotically stable equilibrium points, but they are also *global attractors*, i.e., all nonzero initial conditions give rise to phase trajectories that asymptotically approach the equilibrium.

5.2.7 Stability Diagram in the Trace-Determinant Plane

There is another, graphical, way of summarizing stability results for planar linear systems. Referring to Eq. (5.13), we observe that eigenvalues λ_1 and λ_2 can be determined in terms of a trace and a determinant of A. One obtains $\lambda_1\lambda_2 = \det(A)$, $\lambda_1 + \lambda_2 = T(A)$, and $\lambda_{1,2} = \frac{T(A) \pm \sqrt{T(A)^2 - 4\det(A)}}{2}$. Using, say, the last equation, one may plot all equilibrium point types in the trace-determinant plane. Such plot is called the stability diagram. The curve $\Delta(A) \equiv T(A)^2 - 4\det(A) = 0$ (equivalently, the curve $\det(A) = T(A)^2/4$) is the *repeated roots parabola*, since whenever $\Delta(A) = 0$, $\lambda_1 = \lambda_2 = T(A)/2$. For all points above, the repeated root parabola $\Delta(A) < 0$, and thus λ_1 and λ_2 are complex (and conjugate). Thus, in this region only the spiral equilibrium is possible. Additionally, if $T(A) < 0$ (which is the region to the left of the vertical axis), $\mathrm{Re}(\lambda_{1,2}) < 0$ (spiral sink), if $T(A) > 0$ (which is the region to the right of the vertical axis), $\mathrm{Re}(\lambda_{1,2}) > 0$ (spiral source), and if $T(A) = 0$ (on the vertical axis), $\mathrm{Re}(\lambda_{1,2}) = 0$ (center). Similar analysis demarcates other regions of the trace-determinant plane; see Fig. 5.12.

Example 5.11

In Example 5.3, $A = \begin{pmatrix} -2 & -2 \\ -1 & -3 \end{pmatrix}$. Thus, $T(A) = -5$, $\det(A) = 4$. The point $(-5, 4)$ is below the repeated root parabola (since the point on the repeated root parabola that correspond to $T = -5$ is $(-5, 25/4)$), and it is in the second quadrant. The stability diagram indicates that the equilibrium point is a sink (a stable node). Indeed, it is, since $\lambda_1 = -1$, $\lambda_2 = -4$ – we determined this in Example 5.3.

Example 5.12

In Example 5.10, $A = \begin{pmatrix} 0 & 1 \\ -4 & -4 \end{pmatrix}$. Thus, $T(A) = -4$, $\det(A) = 4$. The point $(-4, 4)$ is on the repeated root parabola. The stability diagram indicates that the equilibrium point is a stable proper or improper node. Using only the trace and determinant, it is not possible to further conclude whether it is proper or improper, but indeed it is stable (a sink), since $\lambda_1 = \lambda_2 = -2$.

5.3 Stability of Solutions in the Case of Nonlinear Systems

5.3.1 Definition of Lyapunov Stability

Next, consider *general (nonautonomous and nonlinear) system of n first-order ODEs* $X' = F(X)$.

Let all f_i and $\partial f_i/\partial x_k$ be continuous for all $t \geq t_0$. Lyapunov stability of any solution of this system is established by Definition 5.2, which is similar to Definition 5.1 for a linear system.

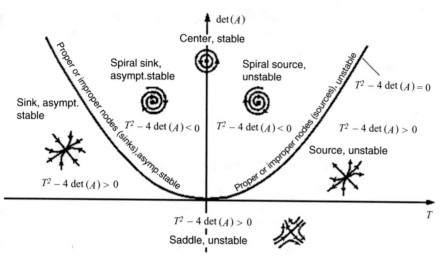

Fig. 5.12 Stability diagram

Definition 5.2 (*Lyapunov stability of a solution of a general first-order system*):
Let $\Phi(t)$ be a solution of (4.5). Then, the solution $\Phi(t)$ is:

(i) *Stable*, if for all $\varepsilon > 0$ and any t_0 there exists $\delta(\varepsilon, t_0) > 0$, such that for any solution $X(t)$ of (4.5) that satisfies $\|X(t_0) - \Phi(t_0)\| < \delta$ it is true that $\|X(t) - \Phi(t)\| < \varepsilon$ for all $t > t_0$.

(ii) *Asymptotically stable*, if stable and $\|X(t_0) - \Phi(t_0)\| < \delta$ implies that $\|X(t) - \Phi(t)\| \to 0$ as $t \to \infty$.

If (i) does not hold, the solution $\Phi(t)$ is said to be *unstable*. Of course, when (4.5) defines a linear system and $\Phi(t) = 0$ (equilibrium solution), Definition 5.2 is equivalent to Definition 5.1a.

5.3.2   Stability Analysis of Equilibria in Nonlinear Autonomous Systems

As application, we now consider a frequently encountered issue of determining stability of the critical points of *a planar* ($n = 2$) *nonlinear autonomous system* (5.3). All critical points are determined by solving system (5.5). Let $f_1(\alpha, \beta) = f_2(\alpha, \beta) = 0$, that is, let the critical point be $x_1(t) = \alpha$, $x_2(t) = \beta$, where α and β are any real values (including zero). We assume that $f_1, f_2, \partial f_i / \partial x_j$, $i, j = 1..2$ are continuous in a neighborhood of the critical point, and also that the critical point is isolated, that is, there are no other critical points in the immediate vicinity (in fact, in the infinitesimally small disc centered at (α, β)). To determine stability, introduce new variables $u(t)$ and $v(t)$ by the translation $u = x_1 - \alpha$, $v = x_2 - \beta$. Substituting in (5.3) gives:

$$
\begin{aligned}
u' &= f_1(u + \alpha, v + \beta), \\
v' &= f_2(u + \alpha, v + \beta).
\end{aligned}
\tag{5.14}
$$

Clearly, system (5.14) has the critical point $u(t) = v(t) = 0$; thus, the translation moved the critical point (α, β) to the origin. Next, we expand f_1 and f_2 in a double Taylor series about (α, β):

$$
\begin{aligned}
u' &= f_1(\alpha, \beta) + \frac{\partial f_1}{\partial u} u + \frac{\partial f_1}{\partial v} v + h_1(u, v) = \frac{\partial f_1}{\partial x_1} u + \frac{\partial f_1}{\partial x_2} v + h_1(u, v), \\
v' &= f_2(\alpha, \beta) + \frac{\partial f_2}{\partial u} u + \frac{\partial f_2}{\partial v} v + h_2(u, v) = \frac{\partial f_2}{\partial x_1} u + \frac{\partial f_2}{\partial x_2} v + h_2(u, v),
\end{aligned}
\tag{5.15}
$$

where the partial derivatives are evaluated at $(x_1, x_2) = (\alpha, \beta)$ and nonlinear functions $h_1(u, v)$, $h_2(u, v)$ are the remainders of the expansions. The nonlinear system (5.15) can be compactly written in a vector form:

$$
\Psi' = J\Psi + H(\Psi),
\tag{5.16}
$$

where

$$
\Psi = \begin{pmatrix} u(t) \\ v(t) \end{pmatrix}, \quad J = \begin{pmatrix} a_{11} & a_{12} \\ a_{21} & a_{22} \end{pmatrix}, \quad H(\Psi) = \begin{pmatrix} h_1(u, v) \\ h_2(u, v) \end{pmatrix},
\tag{5.17}
$$

and $a_{ij} = \partial f_i / \partial x_j$, $i, j = 1..2$ (at (α, β)). Matrix J is called the *Jacobian* (named after the German mathematician Carl Gustav Jacob Jacobi (1804–1851)) of the system (5.3) at (α, β). We now assume that (i) $\|H(\Psi)\|/\|\Psi\| \to 0$ as $\|\Psi\| \to 0$, that is, near the origin $\|H(\Psi)\|$ is small in comparison with $\|\Psi\|$. The norms here are $\|\Psi\|^2 = u^2 + v^2$, $\|H(\Psi)\|^2 = h_1^2 + h_2^2$.

The linearized system is obtained by dropping $H(\Psi)$:

$$
\Psi' = J\Psi.
\tag{5.18}
$$

Next, we also assume, as usual, that (ii) $\det(J) \neq 0$, that is, the origin is the only critical point of the linear system (5.18). When the conditions (i) and (ii) are satisfied, the system (5.16) is called an *almost linear system* in the neighborhood of the critical point (α, β). (It can be shown that the condition (i) is equivalent to the requirement that not only f_1, f_2, $\partial f_i / \partial x_j$, $i, j = 1..2$ are continuous in the neighborhood of (α, β), but also the second partial derivatives are continuous there. This makes the condition (i) very easy to check.) Finally, we can state the following Theorem.

Theorem 5.3 *If* (5.16) *is an almost linear system in the neighborhood of the critical point* (α, β), *then the linearized system* (5.16) *provides the type and stability of the critical point* (α, β) *if:*

(i) *All eigenvalues of J are negative or have a negative real part* (*stability*).
(ii) *One of real eigenvalues or a real part of one of complex eigenvalues is positive* (*instability*).

In the case where the largest real eigenvalue of J is zero, or all eigenvalues are pure imaginary, the linear analysis is not sufficient, and nonlinear analysis is needed (see Sect. 5.4).

Also, unless the eigenvalues are pure imaginary or the largest real eigenvalue of J is zero, the trajectories of the linear system (5.18) near the origin are a good approximation to the trajectories of the nonlinear system (5.3) near (α, β).

This Theorem shows that the center equilibrium point and the degenerate case $\lambda_1 = \lambda_2$ are special. They are special due to their sensitivity to a small perturbation $H(\Psi)$. For all other combinations of eigenvalues from the table, the corresponding critical point is insensitive to a small perturbation $H(\Psi)$. Next, if a critical point of (5.3) is asymptotically stable, then it is only *locally* asymptotically stable, i.e., it is asymptotically stable in some finite domain of the phase space, called a basin of attraction that includes the critical point. And finally, the above ideas and analysis carry over, with only some minor modifications, to higher dimensional autonomous nonlinear systems ($n > 2$). Of course, the phase portraits of such systems are not easy to visualize.

5.3.3 Orbital Stability

There is also another notion of stability, called orbital (trajectorial), or Poincare, stability. (French mathematician Henri Poincare (1854–1912) made many original fundamental contributions to pure and applied mathematics, mathematical physics, and celestial mechanics.) Roughly speaking, a solution that has an orbit C is called orbitally stable if a small perturbation to C remains small. In other words, if a solution which is close to C at some $t = t_0$ remains close to C for any $t > t_0$, then the orbit is stable. It can be shown that Lyapunov stability implies orbital stability, but the converse is not true – an orbit may be stable simultaneously with the solution, defining the orbit, being unstable in the Lyapunov sense. The following definition of orbital stability is for a periodic solution, i.e., $X(t + T) = X(t)$ for all t, defining a closed orbit C in the phase plane, enclosing the equilibrium point at the origin (a center; see for instance Fig. 5.10).

Definition 5.3 (*Orbital stability*):

(i) Suppose that for any solution $Y(t)$ of (4.5), the following property holds: for all $\varepsilon > 0$ and any t_0, there is $\delta(\varepsilon) > 0$ such that $d(Y(t_0), C) < \delta$ implies that $d(Y(t), C) < \varepsilon$ for all $t > t_0$. Then, C is said to be *orbitally, or Poincare, stable*.

(ii) If C is orbitally stable, and $d(Y(t), C) \to 0$ as $t \to \infty$, then C is said to be *asymptotically orbitally, or asymptotically Poincare, stable*.

Here $d(Y(t), C) = \min\limits_{X \in C} \|X - Y(t)\|$ is the distance of $Y(t)$ from the orbit C.

Example 5.13 Let us consider the system:

$$x_1' = x_2\left(1 + x_1^2 + x_2^2\right), \quad x_2' = -x_1\left(\left(1 + x_1^2 + x_2^2\right)\right). \tag{5.19}$$

Multiplying the first equation with x_1, the second equation with x_2, and adding equations, we find:

$$x_1 \cdot x_1' + x_2 \cdot x_2' = 0;$$

hence,

$$\frac{d}{dt}\left(x_1^2 + x_2^2\right) = 0.$$

Thus, each trajectory is a circle:

$$x_1^2 + x_2^2 = r^2. \tag{5.20}$$

Substituting (5.20) into the right-hand side of (5.19), we obtain:

$$x_1' = x_2\left(1 + r^2\right), \quad x_2' = -x_1\left(1 + r^2\right).$$

Thus, for each orbit described by Eq. (5.20), we obtain the family of solutions:

$$x_1(t) = r \sin\left[\left(1 + r^2\right) + \varphi\right], \quad x_2(t) = r \cos\left[\left(1 + r^2\right) + \varphi\right], \tag{5.21}$$

where φ is arbitrary number. Solution (5.21) describes harmonic oscillations with an amplitude r and a frequency $\omega = 1 + r^2$.

Let us consider the orbit C_1 with the radius $r = r_1$ and another orbit C_2 with the radius $r = r_2$. The distance between two circles is a constant. Therefore, if the distance $|r_2 - r_1| = \varepsilon$ at $t = 0$, it is the same forever. Thus, all solutions are orbitally stable.

At the same time, because the frequencies of two solutions do not coincide, each of the differences $\left|r_1 \sin\left[\left(1 + r_1^2\right)t + \varphi_1\right] - r_2 \sin\left[\left(1 + r_2^2\right)t + \varphi_2\right]\right|$ and $\left|r_1 \cos\left[\left(1 + r_1^2\right)t + \varphi_1\right] - r_2 \cos\left[\left(1 + r_2^2\right)t + \varphi_2\right]\right|$ reaches the value $r_1 + r_2$ at a certain time instant, even if we choose the initial $|r_2 - r_1| = \varepsilon$ arbitrary small and also the initial $\varphi_2 - \varphi_1 = 0$, so that at $t = 0$ the distance between the solutions is equal to epsilon. Thus, the solutions are not stable in the sense of Lyapunov.

Problems

1. Using the trace and determinant of matrix A, confirm the equilibrium point type in problems 6–9, 11–14, 16, and 17 from Sect. 5.2

In problems 3–10, show that the system is almost linear and find the location and type of all critical points by linearization. Show the details of your work. Use CAS to plot phase portraits of nonlinear and linearized systems near the equilibrium point(s) and study differences and similarities.

2. Consider the system in Example 2.1.

3.	$x_1' = -x_1 + x_2 - x_2^2$	4.	$x_1' = x_1^2 - x_2^2 - 4$
	$x_2' = -x_1 - x_2.$		$x_2' = x_2$
5.	$x_1' = x_1 + x_2 + x_2^2$	6.	$x_1' = 4 - 2x_2$
	$x_2' = 2x_1 + x_2$		$x_2' = 12 - 3x_1^2$
7.	$x_1' = -(x_1 - x_2)(1 - x_1 - x_2)$	8.	$x_1' = x_1 - x_1^2 - x_1 x_2$
	$x_2' = x_1(2 + x_2)$		$x_2' = 3x_2 - x_1 x_2 - 2x_2^2$
9.	$x_1' = -2x_2 - x_1 \sin(x_1 x_2)$	10.	$x_1' = (1 + x_1) \sin x_2$
	$x_2' = -2x_1 - x_2 \sin(x_1 x_2).$		$x_2' = 1 - x_1 - \cos x_2$

In problems 11–13 find and classify all critical points by first converting the ODE to a system and then linearizing it. Note that (i) the equation in 12 is one of free undamped pendulum, where $k = g/L > 0$, g is the acceleration of gravity, and L is the pendulum length, and (ii) the equation in 13 is one of damped pendulum, where a nonzero $\gamma > 0$ is the damping coefficient:

11. $y'' - 9y + y^3 = 0.$
12. $\theta'' + k \sin \theta = 0.$
13. $\theta'' + \gamma \theta' + k \sin \theta = 0.$
14. The predator-prey population model, also called the *Lotka-Volterra model*, is given by a nonlinear system:

$$x_1' = ax_1 - bx_1 x_2$$
$$x_2' = -cx_2 + dx_1 x_2,$$

where all parameters are positive. Find and classify all equilibrium points.

5.4　Bifurcations and Nonlinear Oscillations

5.4.1　Systems Depending on Parameters

In applications, physical, chemical, and biological systems governed by ordinary differential equations are characterized by some *parameters* (temperature, concentration of a reagent, rate of energy input, etc.). Usually, a desired stable regime of system's functioning is possible only in a certain interval of those parameters. With the change of parameters, the stability of an equilibrium point corresponding to a desired regime can be lost. For instance, a stable equilibrium point corresponding to a constant rate of a certain chemical reaction can become unstable, leading to an explosion. Thus, it is important to understand the possible consequences of the equilibrium point instability.

In the present section, we consider systems of two first-order equations depending on a parameter μ:

$$\frac{dX}{dt} = F(X; \mu), \tag{5.22}$$

where

$$X = \begin{pmatrix} x_1(t) \\ x_2(t) \end{pmatrix}, \quad F(x, \mu) = \begin{pmatrix} f_1(x_1, x_2; \mu) \\ f_2(x_1, x_2; \mu) \end{pmatrix}. \tag{5.23}$$

Assume that for values of μ in a certain region, there exists an equilibrium point $x_1 = \alpha(\mu)$, $x_2 = \beta(\mu)$, and rewrite a system (5.23) in the form (5.15), (5.16):

$$\frac{d\Psi}{dt} = J\Psi + H(\Psi), \tag{5.24}$$

where

$$\Psi = \begin{pmatrix} u \\ v \end{pmatrix}, \quad J(\mu) = \begin{pmatrix} a_{11}(\mu) & a_{12}(\mu) \\ a_{21}(\mu) & a_{22}(\mu) \end{pmatrix}, \quad H(\Psi) = \begin{pmatrix} h_1(u, v, \mu) \\ h_2(u, v, \mu) \end{pmatrix}, \tag{5.25}$$

$$u = x_1 - \alpha(\mu), \quad v = x_2 - \beta(\mu). \tag{5.26}$$

As shown in the previous section, the local stability of an equilibrium point is typically determined by the eigenvalues of the Jacobian $J(\mu)$ obtained by the linearization of equations near the equilibrium point. Stability is guaranteed, if the matrix $J(\mu)$ satisfies two conditions:

$$\det(J(\mu)) = a_{11}(\mu)a_{22}(\mu) - a_{12}(\mu)a_{21}(\mu) > 0 \tag{5.27}$$

and

$$\operatorname{tr}(J(\mu)) = a_{11}(\mu) + a_{22}(\mu) < 0 \tag{5.28}$$

(see Fig. 5.12).

By changing μ, for a steady equilibrium point, there are two ways to leave the region determined by relations (5.27) and (5.28). It can cross the line $\det(J(\mu)) = 0$, where both eigenvalues $\lambda_{1,2}$ are real and one of them, say, λ_1 is equal to 0. After crossing, $\lambda_1 > 0$; therefore, the deviation from the equilibrium point along the corresponding instability grows monotonically as $\exp(\lambda_1 t)$. This type of instability is called *monotonic instability*.

Another way to lose stability is by crossing the line $\operatorname{tr}(J(\mu)) = 0$ while keeping det $(J(\mu)) > 0$. At that line, $\operatorname{Im}\lambda_1 = -\operatorname{Im}\lambda_2 = \sqrt{\det(J)}$, while $\operatorname{Re}\lambda_1 = \operatorname{Re}\lambda_2 = \operatorname{tr}(J)/2$ change their signs. In the region $\operatorname{tr}(J) > 0$, the deviations from the equilibrium point oscillate with a growing oscillation amplitude as $\exp(\operatorname{tr}(J)t/2)\cos(\sqrt{\det(J)}t + \varphi)$. This type of instability is called *oscillatory instability*.

The change of stability of an equilibrium point leads to a qualitative change of the system's phase portrait called *bifurcation*. Below we discuss each type of instability separately.

5.4.2   Bifurcations in the Case of Monotonic Instability

Let us consider some basic examples.

Example 5.14 Let us perform the qualitative analysis of trajectories of the following system of equations:

$$\frac{dx}{dt} = \mu - x^2, \quad \frac{dy}{dt} = -y. \tag{5.29}$$

First, let us consider the case $\mu > 0$. In that case, there exist two equilibrium points, $(\sqrt{\mu}, 0)$ and $(-\sqrt{\mu}, 0)$.

Define $u = x - \sqrt{\mu}$, $v = y$ and linearize equations for u, v. We obtain:

$$\frac{du}{dt} = -2\sqrt{\mu}u, \quad \frac{dv}{dt} = -v;$$

thus, $\lambda_1 = -2\sqrt{\mu} < 0$, $\lambda_2 = -1$; therefore, the point $(\sqrt{\mu}, 0)$ is a stable node.

Similarly, when we define $\bar{u} = x + \sqrt{\mu}$, $\bar{v} = y$, we obtain the following linearized system:

$$\frac{d\bar{u}}{dt} = 2\sqrt{\mu}\bar{u}, \quad \frac{d\bar{v}}{dt} = -\bar{v}$$

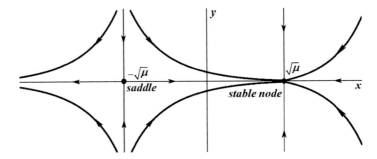

Fig. 5.13 Phase portrait of (5.29) for $\mu > 0$

Fig. 5.14 Phase portrait of
(5.29) for $\mu = 0$

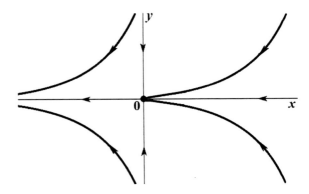

with the eigenvalues $\lambda_1 = 2\sqrt{\mu} > 0$, $\lambda_2 = -1 < 0$, i.e., the point $\left(-\sqrt{\mu}, 0\right)$ is a saddle.

Let us construct the full phase portrait of system (5.29) without solving it (though that system can be solved analytically). First, notice from the second equation that the quantity $|y|$ always decreases with time. The variable x decreases, if $x > \sqrt{\mu}$ or $x < -\sqrt{\mu}$, and grows, if $-\sqrt{\mu} < x < \sqrt{\mu}$; when $x = \pm\mu$, x does not change. Thus, the phase portrait looks as it is shown in Fig. 5.13.

With the decrease of μ, both equilibrium points approach each other, and they merge as $\mu = 0$. At $\mu = 0$, there is only one equilibrium point $(0,0)$ (see Fig. 5.14), and its eigenvalues are $\lambda_1 = 0$, $\lambda_2 = -1 < 0$ (hence, that is a nongeneric, "improper," point). That equilibrium point is unstable: the trajectories approach that point within the region $x \geq 0$, but they leave the vicinity of that point if $x < 0$. The left part of the phase plane ($x < 0$) looks like a vicinity of a saddle point, while its right part ($x > 0$) looks like a vicinity of a stable node. The improper equilibrium point of that kind is called *saddle-node point*.

At $\mu < 0$, there are no equilibrium points: $dx/dt < 0$ for any (x, y), and the phase portrait looks as in Fig. 5.15. Thus, merging of a saddle and a node leads to their *annihilation*.

Fig. 5.15 Phase portrait of
(5.29) for $\mu < 0$

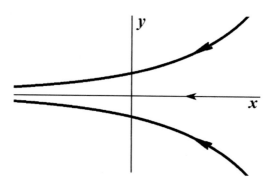

Fig. 5.16 Direct pitchfork
bifurcation

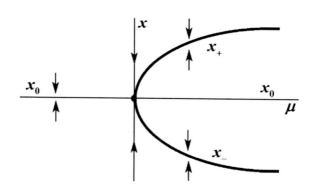

The transition "saddle point and node point $\rightarrow$ saddle-node point $\rightarrow$ absence of an equilibrium point" is called *saddle-node bifurcation*. The annihilation of equilibrium points is the most typical consequence of an eigenvalue equal to zero. However, in systems with symmetry, another scenario is possible, which is described by the next example.

Example 5.15 Let us consider the system:

$$\frac{dx}{dt} = x(\mu - x^2), \quad \frac{dy}{dt} = -y. \tag{5.30}$$

In a contradistinction to the previous example, the solution $x_0 = 0$, $y_0 = 0$ exists for any μ. Linearizing system (5.30) around that point, we find that $\lambda_1 = \mu$, $\lambda_2 = -1$. Thus, the equilibrium point $(0,0)$ is a stable node as $\mu < 0$ and a saddle as $\mu > 0$.

The transformation of a stable node into a saddle is accompanied by the birth of two new equilibrium points: $x_+ = \sqrt{\mu}$, $y_+ = 0$ and $x_- = -\sqrt{\mu}$, $y_- = 0$. This phenomenon is called a *direct pitchfork bifurcation*. The origin of the latter term is clear from Fig. 5.16 that displays x_0, x_+, and x_- as function of μ. The arrows in that figure show the direction of the change of x with time. From Fig. 5.16, one can conclude that solutions $(x_\pm, 0)$ are stable.

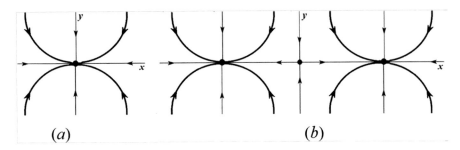

Fig. 5.17 Phase portrait for system (5.31): (a) $\mu \le 0$; (b) $\mu > 0$

Indeed, let us define $u_\pm = x \mp \sqrt{\mu}$, $v_\pm = y$ and linearize Eq. (5.30) with respect to $u_\pm$. We obtain:

$$\frac{du_\pm}{dt} = -2\mu u_\pm, \quad \frac{dv_\pm}{dt} = -v_\pm; \tag{5.31}$$

hence, the eigenvalues are $\lambda_1 = -2\mu < 0$, $\lambda_2 = -1 < 0$ for both "newborn" equilibrium points. The transformation of the phase portrait with the change of μ is shown in Fig. 5.17.

Example 5.16 Next, consider the system:

$$\frac{dx}{dt} = x(\mu + x^2), \quad \frac{dy}{dt} = -y. \tag{5.32}$$

Similar to Example 5.15, the equilibrium solution $x_0 = 0$, $y_0 = 0$ exists for any μ, and according to the analysis of the linearized system ($\lambda_1 = \mu, \lambda_2 = -1$), it is stable for $\mu < 0$ and unstable for $\mu > 0$. However, the additional equilibrium points $x_+ = \sqrt{-\mu}$, $y_+ = 0$ and $x_- = -\sqrt{-\mu}$, $y_- = 0$, exist in the region $\mu < 0$, where the equilibrium point $(0,0)$ is stable (see Fig. 5.18). The arrows in that figure, which indicate the change of x with time, show that the newborn stationary solutions are unstable. Indeed, when we define $u_\pm = x \mp \sqrt{-\mu}$, $v_\pm = y$ and linearize Eq. (5.32) about these solutions, we obtain Eq. (5.31), but this time $\lambda_1 = -2\mu > 0$.

The bifurcation of this kind is called *inverse pitchfork bifurcation*. The phase portraits in the regions $\mu < 0$ and $\mu \ge 0$ are shown in Fig. 5.19.

Note that in the case $\mu = 0$, where one of two eigenvalues is $\lambda_1 = 0$, the point $x_0 = 0$, $y_0 = 0$ is stable in the case (5.30) and unstable in the case (5.32), though the eigenvalues of the linearized systems are identical in both cases. Thus, for $\mu = 0$, the linear analysis is not sufficient for an equilibrium point in a nonlinear system.

Let us note that bifurcations that change the number of equilibrium points (annihilation/birth of the pair of stable and unstable equilibrium point, direct pitchfork bifurcation, and inverse pitchfork bifurcation) are possible also for single first-order equation (equation for y in the systems above can be dropped). In the next section, we consider the bifurcations that are possible only for system with $n \ge 2$.

Fig. 5.18 Inverse pitchfork
bifurcation

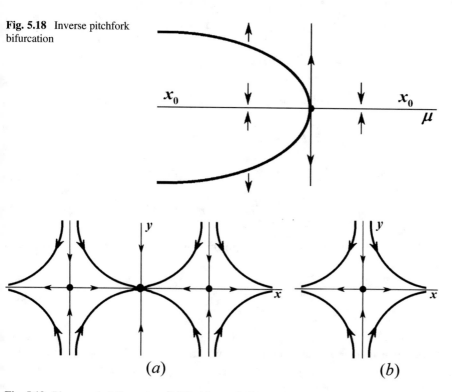

Fig. 5.19 Phase portrait for system (5.32): (a) $\mu < 0$; (b) $\mu \geq 0$

5.4.3   *Bifurcations in the Case of Oscillatory Instability*

Example 5.17 Let us next consider the following system:

$$\frac{dx}{dt} = -y + \mu x - x(x^2 + y^2), \quad \frac{dy}{dt} = -x + \mu y - y(x^2 + y^2). \quad (5.33)$$

The system of equations linearized about the only equilibrium point $x = 0$, $y = 0$, has
eigenvalues satisfying equation $(\mu - \lambda)^2 + 1 = 0$; thus,

$$\lambda_{1,2} = \mu \pm i.$$

The equilibrium point is a stable focus (spiral sink) at $\mu < 0$ and an unstable focus
(spiral source) at $\mu > 0$.

Let us discuss the change of the phase portrait of the system when the value $\mu = 0$
is crossed.

It is convenient to use polar coordinates (r, φ), which are related to Cartesian
coordinates (x, y) by the formulas:

$$x = r \cos \varphi, \quad y = r \sin \varphi. \tag{5.34}$$

Substituting (5.34) into (5.33), we obtain:

$$\cos \varphi \frac{dr}{dt} - r \sin \varphi \frac{d\varphi}{dt} = -r \sin \varphi + \mu r \cos \varphi - r^3 \cos \varphi, \tag{5.35}$$

$$\sin \varphi \frac{dr}{dt} + r \cos \varphi \frac{d\varphi}{dt} = r \cos \varphi + \mu r \sin \varphi - r^3 \sin \varphi. \tag{5.36}$$

Multiplying (5.35) by $\cos\varphi$, (5.36) by $\sin\varphi$ and adding both equations, we find:

$$\frac{dr}{dt} = \mu r - r^3. \tag{5.37}$$

Multiplying (5.35) by $(-\sin \varphi)$, (5.36) by $\cos\varphi$ and adding both equations, we find that at $r \neq 0$:

$$\frac{d\varphi}{dt} = 1. \tag{5.38}$$

Equation (5.38) describes a rotation of the point $(x(t), y(t))$ around the origin with the constant angular velocity: $\varphi(t) = t + C$, where $C = \varphi(t_0)$. The motion of that point in the radial direction is governed by Eq. (5.37) that was already analyzed in Sect. 5.4.2.

For $\mu \leq 0$, $dr/dt < 0$ for any $r \neq 0$, i.e., the trajectory is a spiral tending to the equilibrium point (see Fig. 5.20a).

If $\mu > 0$, the trajectories are attracted to the circle $r = \sqrt{\mu}$ from the inside and from the outside (see Fig. 5.20b). On the circle

$$x(t) = r \cos \varphi(t) = \sqrt{\mu} \cos(t + C), \quad y(t) = r \sin \varphi(t) = \sqrt{\mu} \sin(t + C). \tag{5.39}$$

As usual, the closed trajectory corresponds to a periodic solution. That attracting trajectory is called a *stable limit cycle*. The change of the phase portrait shown in Fig. 5.20, "stable focus $\rightarrow$ unstable focus and stable limit cycle," is called a *direct Hopf bifurcation* (discovered by Eberhard Hopf (1902–1983), the German-American mathematician, one of the founders of the theory of dynamical systems).

Note that stable limit cycle is orbitally asymptotically stable but not asymptotically stable in the sense of Lyapunov. Indeed, let us consider solutions with initial conditions in two different points on the limit cycle trajectory. Because both solutions oscillate periodically, the difference between those solutions also oscillates; hence, it does not tend to zero.

The meaning of the transition shown in Fig. 5.20 is as follows: for $\mu \leq 0$, the equilibrium point is globally stable, i.e., any disturbances imposed at $t = 0$ decay with time. For $\mu > 0$, the equilibrium point is unstable, and arbitrary small disturbances grow, leading to oscillations with a well-defined amplitude $\sqrt{\mu}$ at $t \rightarrow \infty$.

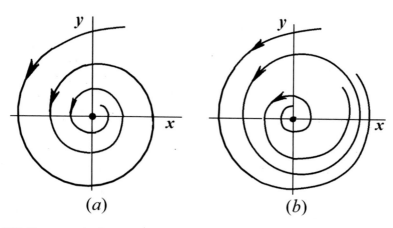

Fig. 5.20 Phase portrait of system (5.33): (a) $\mu \leq 0$; (b) $\mu > 0$

If the initial disturbance has the amplitude $\sqrt{x^2(0) + y^2(0)}$ that is larger than $\sqrt{\mu}$, the amplitude of the oscillations $\sqrt{x^2(t) + y^2(t)}$ decreases with time tending to the same value $\sqrt{\mu}$ at $t \to \infty$. That behavior of a nonlinear system is quite different from that of a linear system. We know that for a linear system with a complex eigenvalue $\lambda_{1,2} = \lambda_r \pm i\lambda_i$, the growth of disturbances is unbounded in the case $\lambda_r > 0$, while in the case $\lambda_r = 0$, the periodic oscillations may have an arbitrary amplitude (see Sect. 5.2).

Example 5.18 Let us next consider the system:

$$\frac{dx}{dt} = -y + \mu x + x(x^2 + y^2), \quad \frac{dy}{dt} = x + \mu y + y(x^2 + y^2), \quad (5.40)$$

which has the same linear terms as in the previous example; hence, it has the same eigenvalues $\lambda_{1,2} = \mu \pm i$ of the equilibrium point $x = y = 0$. However, the nonlinear terms at the right-hand side are of the opposite sign. Similar to the previous example, the system can be rewritten in polar coordinates as

$$\frac{dr}{dt} = \mu r + r^3, \quad \frac{d\varphi}{dt} = 1.$$

In the region $\mu < 0$, the equilibrium point is stable; however, $dr/dt < 0$ only for $r < \sqrt{-\mu}$, while for $r = \sqrt{-\mu}$, $dr/dt = 0$ (circular trajectory), and for $r > \sqrt{-\mu}$, $dr/dt > 0$, i.e., the spiral trajectory tends to infinity. Thus, though the equilibrium point $x = y = 0$ is *locally* stable, only disturbances with the amplitude $\sqrt{x^2(0) + y^2(0)} < \sqrt{-\mu}$ decay. The disturbances with amplitude $\sqrt{x^2(0) + y^2(0)} > \sqrt{-\mu}$ undergo oscillatory growth (there is no *global* stability). The phase portrait of the system is shown in Fig. 5.21a. The periodic solution

$$x(t) = \sqrt{-\mu}\cos(t + C), \quad y(t) = \sqrt{-\mu}\sin(t + C)$$

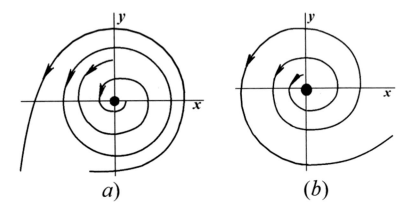

Fig. 5.21 Phase portrait of system (5.40): (**a**) $\mu < 0$; (**b**) $\mu \geq 0$

is obviously unstable. It corresponds to an *unstable limit cycle*, which separates the region of attraction to the equilibrium point from the region of unbounded growth of the oscillation amplitude.

In the case $\mu \geq 0$, we find that $dr/dt > 0$ for any $r \neq 0$, i.e., all the trajectories tend to infinity (see Fig. 5.21b).

The transition observed with the decrease of μ, "unstable equilibrium $\rightarrow$ stable equilibrium and unstable limit cycle," is called an *inverse Hopf bifurcation*.

The two scenarios of bifurcations described above are the most typical scenarios of a nonlinear development of an oscillatory instability.

5.4.4 Nonlinear Oscillations

Poincare-Bendixson Theorem

Solution (5.39) of Eq. (5.33) provides an example of *periodic nonlinear oscillations*. A qualitative analysis of the phase plane allows to find the location of closed curves, which correspond to periodic oscillations of definite shape and amplitude. The main tool for that goal is *Poincare-Bendixson theorem*. (Ivar Otto Bendixson (1861–1935) was a Swedish mathematician.)

If a trajectory $(x(t), y(t))$ is confined to a bounded region of the phase plane that does not contain equilibrium points, then that trajectory is attracted at $t \rightarrow \infty$ to a closed curve that corresponds to periodic solutions.

Let us return to Example 5.16 at $\mu > 0$:

$$\frac{dr}{dt} = \mu r - r^3, \quad \frac{d\varphi}{dt} = 1, \tag{5.41}$$

and consider a ring D: $r_1 < r < r_2$, where $r_1 < \sqrt{\mu}$ and $r_2 > \sqrt{\mu}$ (see Fig. 5.22).

Fig. 5.22 "Trapping ring"

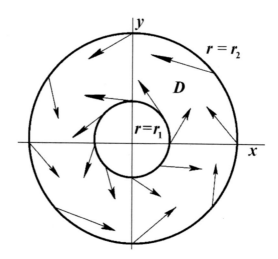

Because $(dr/dt)|_{r_1} > 0$, all trajectories starting at the circle $r = r_1$ enter D. Similarly, because $(dr/dt)|_{r_2} < 0$, all trajectories at the outer boundary of D, $r = r_2$, are directed inward. Thus, D is a "trapping region": the trajectory cannot leave D and it stays in that region forever. Because the only equilibrium point, $x = y = 0$, is outside D, according to Poincare-Bendixson theorem, there is (at least one) closed orbit in D. As we know, that is the circle $r = \sqrt{\mu}$.

The Poincare-Bendixson theorem allows to locate closed orbits also when the solutions cannot be found analytically.

Example 5.19 Consider the following modification of the system in Example 5.15:

$$\frac{dx}{dt} = -y + x(1 + ax - x^2 - y^2), \quad \frac{dy}{dt} = x + y(1 + ax - x^2 - y^2). \quad (5.42)$$

In polar coordinates, the equations are

$$\frac{dr}{dt} = r(1 - r^2 + ar\cos\varphi), \quad \frac{d\varphi}{dt} = 1.$$

The only equilibrium point is $x = y = 0$. At sufficiently small r, $dr/dt > 0$ for any φ; at sufficiently large r, $dr/dt < 0$. Therefore, one can construct a trapping region in the form of a ring between two circles and show the existence of a periodic solution.

Example 5.20 Let us consider the oscillator with a nonlinear friction:

$$\frac{d^2x}{dt^2} + \left[\left(\frac{dx}{dt}\right)^2 + x^2 - 1\right]\frac{dx}{dt} + x = 0. \quad (5.43)$$

We define $y = dx/dt$ and rewrite Eq. (5.43) as system:

$$\frac{dx}{dt} = y, \quad \frac{dy}{dt} = -x + \left(1 - x^2 - y^2\right)y. \tag{5.44}$$

In polar coordinates

$$\frac{dr}{dt} = r\left(1 - r^2\right)\sin^2\varphi, \quad \frac{d\varphi}{dt} = -1 + \left(1 - r^2\right)\sin\varphi\cos\varphi.$$

Let us consider the ring $r_1 < r < r_2$, $r_1 < 1$, $r_2 > 1$. At $r = r_1$, $dr/dt \geq 0$; at $r = r_2$, $dr/dt \neq 0$. Unlike the previous example, $dr/dt = 0$ at the points on the x-axis, $\varphi = 0$ and $\varphi = \pi$. In order to understand the behavior of the trajectories crossing the points $(r_j, 0)$, $j = 1, 2$, it is necessary to calculate higher derivatives:

$$\left.\frac{d^2 r}{dt^2}\right|_{(r_j,0)} = 0, \quad \left.\frac{d^3 r}{dt^3}\right|_{(r_j,0)} = 2r_j\left(1 - r_j^2\right).$$

Thus, if $r(0) = r_j$ and $\varphi(0) = 0$ or $\varphi(0) = \pi$,

$$r(t) = r_j + \frac{1}{3}r_j\left(1 - r_j^2\right)t^3 + O(t^3).$$

Thus, at $r_0 = r_1 < 1$, $r(t)$ grows with time, and at $r_0 = r_2 > 1$, $r(t)$ decreases with time, i.e., the trajectory enters the ring $r_1 < r < r_2$. That proves the existence of periodic nonlinear oscillations.

Example 5.21 (The Mathematical Pendulum)

We are ready now to solve Eq. (1.23) of Chap. 1:

$$\theta'' + \frac{g}{l}\sin\theta = 0 \tag{5.45}$$

that describes the mathematical pendulum swinging (see Fig. 1.4).

First, we transform Eq. (5.45) into a first-order equation by multiplying it by θ':

$$\theta'\theta'' + \frac{g}{l}\theta'\sin\theta = 0,$$

and rewriting:

$$\frac{d}{dt}\left(\frac{\theta'^2}{2} - \frac{g}{l}\cos\theta\right) = 0.$$

One can see that

$$\frac{\theta'^2}{2} + U(\theta) = E, \quad U(\theta) = -\frac{g}{l}\cos\theta, \tag{5.46}$$

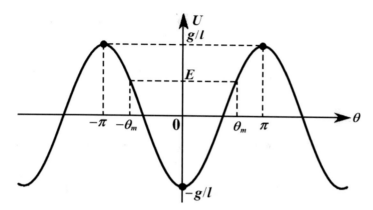

Fig. 5.23 The mathematical pendulum

where E is a constant that we shall call "energy" (actually, the mechanical energy of the particle is mE, where m is the mass).

Thus, the system governed by Eq. (5.46) is a *conservative* system: there exists a conserved combination of the function θ and its derivatives. Such systems have many specific features.

To visualize the behavior of the system, let us define $\omega(t) = \theta'(t)$ and plot the trajectories in the phase space (θ, ω). Note that θ is a cyclic variable: the points with values $\theta \pm 2\pi$ correspond to the same position of the pendulum as the point with the value θ. Thus, the phase space is a cylinder rather than a plane. Still, for the sake of simplicity, we shall call it "phase plane" below.

According to (5.46), all trajectories are described by the formula:

$$\theta'(t) = \pm \sqrt{2[E + (g\cos\theta)/l)]} \tag{5.47}$$

Thus, each trajectory is characterized by a certain value of energy E.

Let us discuss the shape of the trajectory at different E. To make the behavior of the system clearer, let us look at the plot of the function $U(\theta)$ shown in Fig. 5.23 and imagine another physical interpretation of the problem: a particle rolls over the surface that has the shape of $U(\theta)$. At the minimum point, the height of that surface is $U_{\min} = U(0) = -g/l$ and $U_{\max} = U(\pi) = g/l$.

There are several qualitatively different cases depending on the value of E.

1.	Expression (5.47) is meaningful if the expression under the sign of radical in (5.47) is non-negative; hence, the values given by (5.47) are real. Therefore, if $E < U_{\min}$, there are no solutions. Indeed, the "particle" cannot be below the surface $U(\theta)$
2.	If $E = -g/l$, there exists the unique physical solution, $\theta = 0$, which corresponds to the motionless pendulum in its lowest point (it is clear that other solutions, $\theta = 2\pi n$ with integer n, describe exactly the same state of the physical system)

(continued)

3. | Let E be slightly larger than $-g/l$. One can see from Fig. 5.23 that in that case, the amplitude of oscillations will be small, i.e., θ will be always small. That means that we can replace $\sin\theta$ with θ, i.e., linearize Eq. (5.45) reducing it to

$\theta'' + \frac{g}{l}\theta = 0.$ (5.48)

The general solution of that linear equation is

$\theta = C_1 e^{\lambda_1 t} + C_2 e^{\lambda_2 t},$

where

$\lambda_1 = i\omega_0, \lambda_2 = -i\omega_0,$

and $\omega_0 = \sqrt{g/l}$.

Note that $\lambda_2 = -\lambda_1$. The reason is that Eq. (5.48) (as well as the original Eq. (5.48)) is not changed by the replacement of t by $-t$. Therefore, if $e^{\lambda_1 t}$ is a solution, then $e^{-\lambda_1 t}$ is also a solution; hence, $\lambda_2 = -\lambda_1$.

Thus, the equilibrium solution is a center, rather than a focus. That is typical for conservative systems

The shape of the trajectories at small $E + g/l$ can be obtained by the replacement $\cos\theta \approx 1 - \theta^2/2$ in (5.46). We find that the trajectories are close to ellipses

$(\theta')^2 + g\theta^2/l = E + g/l.$

4. | If $-g/l \leq E \leq g/l$, the pendulum swings in the interval of angles $-\theta_m \leq \theta \leq \theta_m$, where the swing amplitude θ_m is determined by the relation:

$$-\frac{g}{l}\cos\theta_m = E.$$ (5.49)

All trajectories are closed lines, which is the sign that the solution is periodic (see Fig. 5.24).

Thus, instead of isolated limit cycles, the conservative systems possess a continuum of periodic solutions describing oscillations with different amplitudes. The solutions can be written as

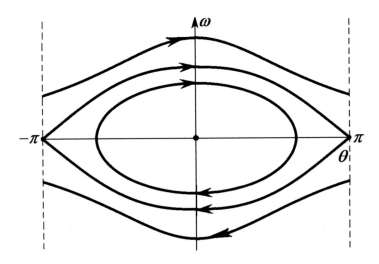

Fig. 5.24 Trajectories for the mathematical pendulum

$$t = \pm \int \frac{d\theta}{\sqrt{2(E + (g/l)\cos\theta}} + C, \qquad (5.50)$$

or

$$t = \pm \sqrt{\frac{l}{g}} \int \frac{d\theta}{\sqrt{2(\cos\theta - \cos\theta_m)}} + C. \qquad (5.51)$$

Using the formula

$$\cos\theta = 1 - 2\sin^2\frac{\theta}{2},$$

we can rewrite (5.51) as

$$t = \pm \frac{1}{2}\sqrt{\frac{l}{g}} \int \frac{d\theta}{\sqrt{\sin^2(\theta_m/2) - \sin^2(\theta/2)}} + C.$$

Let us introduce the new variable φ by the relation:

$$\sin\varphi = \frac{\sin(\theta/2)}{\sin(\theta_m/2)}.$$

Unlike the variable θ, which oscillates, the variable φ changes monotonically. We find:

$$t = \sqrt{\frac{l}{g}} \int \frac{d\phi}{\sqrt{1 - \sin^2(\theta_m/2)\sin^2\varphi}} + C \qquad (5.52)$$

(another sign of the integral does not yield a different solution).

The integral at the right-hand side of Eq. (5.52) cannot be expressed in terms of elementary functions. For its description, *special functions* called *elliptic functions* are introduced (for more details, see Sect. 10.4).

The *incomplete elliptic integral of the first kind* is defined as

$$u = F(\varphi, k) = \int\limits_0^\varphi \frac{d\varphi}{\sqrt{1 - k^2\sin^2\varphi}},$$

where the *elliptic modulus* k satisfies $0 < k^2 < 1$. The inverse function

$$\varphi = \mathrm{am}(u, k)$$

is called the *Jacobi amplitude*.

Using the definitions of elliptic functions, we can rewrite formula (5.52) as

$$t = \sqrt{\frac{l}{g}} F\left(\varphi,\ \sin\frac{\theta_m}{2}\right) + C;$$

hence,

$$\varphi = \operatorname{am}\left(\sqrt{\frac{g}{l}}(t - C),\ \sin\frac{\theta_m}{2}\right)$$

and

$$\theta = 2\arcsin\left[\sin\frac{\theta_m}{2} \cdot \sin\left(\operatorname{am}\left(\sqrt{\frac{g}{l}}(t - C),\ \sin\frac{\theta_m}{2}\right)\right)\right]. \tag{5.53}$$

Note that

$$\sin\frac{\theta_m}{2} = \sqrt{\frac{1}{2}(1 - \cos\theta_m)} = \sqrt{\frac{1}{2}\left(1 + \frac{lE}{g}\right)}.$$

The period of oscillations T can be expressed in terms of the *complete elliptic integral of the first kind*:

$$K(k) \equiv F\left(\frac{\pi}{2}, k\right) = \int_0^{\pi/2} 2\frac{d\varphi}{\sqrt{1 - k^2\sin^2\phi}},$$

$$T = \sqrt{\frac{l}{k}} \int_0^{2\pi} \frac{d\varphi}{\sqrt{1 - \sin^2(\theta_m/2)\sin^2\varphi}} = 4\sqrt{\frac{l}{k}}K\left(\sin\frac{\theta_m}{2}\right).$$

---

5. For $E = g/l$, there are two kinds of trajectories. One is the equilibrium point $\theta = \pi$: the pendulum is in its upright position. It is obvious that this solution must be unstable: a small disturbance will create a force that will move the pendulum downward. Let us check that by the linearization of the equation around that equilibrium point. Introduce $\tilde{\theta} = \theta - \pi$. Then, (5.45) becomes:
$\tilde{\theta}'' - (g/l)\sin\tilde{\theta} = 0$.
Linearizing, we obtain:
$\tilde{\theta}'' - (g/l)\tilde{\theta} = 0$;
hence,
$\tilde{\theta} = C_1 e^{\lambda_1 t} + C_2 e^{\lambda_2 t}$
with $\lambda_1 = \omega_0 > 0$, $\lambda_2 = -\omega_0 < 0$. Thus, the equilibrium point $\theta = \pi$ is a saddle. As explained above, the relation $\lambda_2 = -\lambda_1$ is caused by the invertibility of the equation

(continued)

6. Solutions of another kind are obtained from (5.53) in the limit $E \to g/l$, i.e., $\theta_m = \pi$. One can choose $\varphi = \theta/2$ or $\varphi = (\pi - \theta)/2$; then,

$$t = \sqrt{\frac{l}{g}} \int \frac{d\varphi}{\cos\varphi} + C.$$

Calculating the integral, we find:

$$t = \sqrt{\frac{l}{g}} \ln\left[\frac{1 + \tan(\varphi/2)}{1 - \tan(\varphi/2)}\right] + C;$$

hence,

$$\tan(\varphi/2) = \tanh\left[(t - C)\sqrt{g/l}/2\right].$$

We obtain two families of solutions:

$$\theta = 4\tan^{-1}\left[\tanh(t - C)\sqrt{g/l}/2\right]$$

and

$$\theta = -4\tan^{-1}\left[\tanh(t - C)\sqrt{g/l}/2\right].$$

The pendulum leaves the upper point and returns to that point in infinite time, moving counterclockwise or clockwise.

On the phase plane, these solutions correspond to the lines that connect the saddle points ($\theta = 0, \omega = \pm\pi$). Actually, those are same points corresponding to the upper equilibrium position of the pendulum, i.e., the trajectory leaves the saddle point along the direction of the eigenvector with a positive eigenvalue and returns to the same point along the direction of the eigenvector with a negative eigenvalue. The trajectories of that kind are called homoclinic trajectories (see Fig. 5.24)

5. Finally, let us consider the case $E > g/l$. In that case, there are no turning points, where $E = U(\theta)$; therefore, the pendulum *rotates* counterclockwise or clockwise, without changing the direction of the rotation. The corresponding trajectories on Fig. 5.24 are situated above and below homoclinic trajectories

Applying formula (5.46) and the relation $\cos\theta = 1 - 2\sin^2(\theta/2)$, we find:

$$t = \pm\int \frac{d\theta}{\sqrt{2[E + (g/l)\cos\theta]}} + C = \pm\sqrt{2}\int \frac{d(\theta/2)}{\sqrt{(E + g/l) - (2g/l)\sin^2(\theta/2)}} +$$

$$C = \pm\sqrt{\frac{2}{E + g/l}}\int \frac{d(\theta/2)}{\sqrt{1 - k^2\sin^2(\theta/2)}} + C = \pm\sqrt{\frac{2}{E + g/l}}F\left(\frac{\theta}{2}, k\right) + C, \quad \text{where}$$

$$k^2 = \frac{2g/l}{E + g/l}.$$

The revolution of the pendulum is described by the formula:

$$\theta = 2\ \mathrm{am}\left[\pm\sqrt{\frac{E + g/l}{2}}(t - C), k\right].$$

Example 5.22 (the mathematical pendulum with friction)

Let us consider now the following modification of Eq. (5.48):

$$\theta'' + 2\gamma\theta' + \omega_0^2 \sin\theta = 0; \quad \gamma > 0. \tag{5.54}$$

The term with θ' describes friction. Introducing $\omega = \theta'$, we can rewrite Eq. (5.54) as a system of two equations of the first order:

$$\theta' = 0, \quad \omega' = -2\gamma\omega - \omega_0^2 \sin\theta. \tag{5.55}$$

The equilibrium points $\theta = 0$ (the pendulum is in its lowest position) and $\theta = \pi$ (the pendulum is in its highest position) are not changed. Let us investigate the linear stability of those equilibria.

Linearizing (5.54) around the point $\theta = 0$, we obtain:

$$\theta'' + 2\gamma\theta' + \omega_0^2\theta = 0.$$

For the solution $\theta \sim \exp(\lambda t)$, we find:

$$\lambda^2 + 2\gamma\lambda + \omega_0^2 = 0;$$

hence,

$$\lambda_{1,2} = -\gamma \pm \sqrt{\gamma^2 - \omega_0^2}. \tag{5.56}$$

If $\gamma < \omega_0$

$$\lambda_{1,2} = -\gamma \pm i\sqrt{\omega_0^2 - \gamma^2};$$

hence, $\lambda_{1,2}$ are complex with $\mathrm{Re}\lambda_{1,2} = -\gamma < 0$. Thus, the equilibrium point $\theta = 0$ is a stable focus, and the trajectories near that point have the form of spirals. If $\gamma > \omega_0$, both roots (5.55) are real and negative; hence, that equilibrium point is a stable node.

Consider now the behavior of trajectories in the vicinity of the point $\theta = \pi$. Defining $\theta = \pi + \tilde{\theta}$ and linearizing Eq. (5.54) at small $\tilde{\theta}$, we find:

$$\tilde{\theta}'' + 2\gamma\tilde{\theta}' - \omega_0^2\tilde{\theta} = 0;$$

hence, the characteristic equation is

$$\lambda^2 + 2\gamma\lambda - \omega_0^2 = 0$$

and

$$\lambda_{1,2} = -\gamma \pm \sqrt{\gamma^2 + \omega_0^2}.$$

Obviously, $\lambda_1 > 0$ and $\lambda_2 < 0$; hence, for $\gamma > 0$, that point is a saddle, like in the case $\gamma = 0$.

In a contradistinction to Eq. (5.48), Eq. (5.54) is not subject to a conservation law. Multiplying (5.54) by θ', we find that the quantity

Fig. 5.25 Phase portrait for the mathematical pendulum with friction; $\gamma < \omega_0$

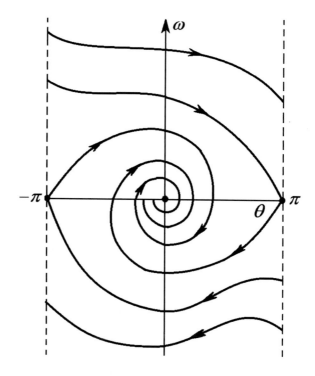

$$E(t) = \frac{(\theta')^2}{2} - \omega_0^2 \cos \theta \tag{5.57}$$

is not constant but satisfies the relation

$$\frac{dE}{dt} = -2\gamma(\theta')^2. \tag{5.58}$$

One can see that $E(t)$ decreases monotonically, except for constant solutions $\theta = 0$ and $\theta = \pi$ discussed above. Thus, unlike (5.48), the pendulum with friction is a dissipative system rather than a conservative one. The monotonically decreasing function $E(t)$ is called *Lyapunov function*. The existence of a Lyapunov function means that the system has no periodic solutions: would a periodic solution $\theta(t)$ exist, the corresponding function (5.57) would be periodic as well, i.e., nonmonotonic. That means that as $t \to \infty$, the trajectory has no choice except tending to an equilibrium point.

The phase portrait of system (5.55) for $\gamma < \omega_0$ is shown in Fig. 5.25. In the case $\gamma > \omega_0$, the trajectories approach the point $x = y = 0$ without rotations.

Chapter 6
Power Series Solutions of ODEs

In the previous chapters, we explored the general properties of ODEs and systems of ODEs. Typically, solutions of ODEs cannot be presented in an explicit form using elementary functions.

In the present chapter, we consider a universal approach for solving ODEs. It will be shown that for a wide class of equations, the solutions can be found in the form of a convergent power series. We shall concentrate on homogeneous linear second-order equations. As shown in previous sections, finding a general solution of a homogeneous equation opens the way to solving nonhomogeneous equations, e.g., by means of a variation of parameters. Higher-order equations can be solved using the approach developed for second-order equations.

Because the approach is based on using power series, we start with a review of some their properties.

6.1 Convergence of Power Series

Assume that a function $f(x)$ has infinitely many derivatives at the point $x = x_0$. From calculus we know that $f(x)$ can be approximated by a Taylor polynomial of degree N:

$$\sum_{n=0}^{N} a_n (x - x_0)^n,$$

where

$$a_0 = f(x_0), \quad a_1 = f'(x_0), \quad \ldots, \quad a_n = \frac{f^{(n)}}{n!}, \quad \ldots$$

© The Author(s), under exclusive license to Springer Nature Switzerland AG 2023
V. Henner et al., *Ordinary Differential Equations*,
https://doi.org/10.1007/978-3-031-25130-6_6

The limit

$$\lim_{N \to \infty} \sum_{n=0}^{N} a_n (x - x_0)^n \equiv \sum_{n=0}^{\infty} a_n (x - x_0)^n \tag{6.1}$$

is the *Taylor series* (named after Brook Taylor (1685–1731), an English mathematician) of $f(x)$ about the point $x = x_0$. There are three possibilities:

(i) A series (6.1) converges to $f(x)$.
(ii) A series (6.1) diverges, i.e., the limit (6.1) does not exist.
(iii) A series (6.1) converges to a function that is not $f(x)$.

For convergence analysis of a series, one can apply the following criteria.

d'Alembert criterion (named after Jean-Baptiste d'Alembert (1717–1783), a French mathematician and physicist), *also known as ratio test.*

Let $\{c_n\}$ be a sequence of *numbers* and $\lim_{n \to \infty} |c_{n+1}/c_n| = r$. Then, the series $\sum_{n=1}^{\infty} c_n$ converges *absolutely* (i.e., $\sum_{n=1}^{\infty} |c_n|$ also converges), if $r < 1$, and diverges if $r > 1$. If $r = 1$, other criteria are needed that we do not present here.

Weierstrass M-test (named after Karl Weierstrass (1815–1897), a German mathematician, one of the founders of mathematical analysis).

Let $\{f_n(x)\}$ be a sequence of *functions* defined in a certain common interval I and $\{M_n\}$ be a sequence of nonnegative numbers such that $|f_n(x)| < M_n$ for any n and $x \in I$. Then, if $\sum_{n=1}^{\infty} M_n$ converges, then the sequence $\{f_n(x)\}$ converges absolutely and *uniformly* on I (i.e., for any $\varepsilon > 0$, there exists $N(\varepsilon)$ independent of x, such that $\left| \sum_{n=N+1}^{\infty} f_n(x) \right| < \varepsilon$ for any $x \in I$).

Let us consider some examples.

Example 6.1 Let

$$f(x) = \frac{1}{1 - x}. \tag{6.2}$$

Since $f^{(n)}(x) = n \, ! \, (1 - x)^n$, we find $f^{(n)}(0) = n!$; hence $a_n = 1$ for any n, i.e., the Taylor (Maclaurin) series about $x = 0$ is

$$\sum_{n=0}^{\infty} x^n \equiv 1 + x + x^2 + \ldots + x^n + \ldots . \tag{6.3}$$

Obviously, (6.3) is the geometric series with a ratio x. This series converges to $f(x) = 1/(1 - x)$, if $|x| < 1$, and it does not converge if $|x| \geq 1$. It is not surprising that the Taylor series does not converge at $x = 1$, because $\lim_{x \to 1} f(x)$ does not exist. But we see that it does not converge also at $x = -1$, although this x value is in the domain of

$f(x)$ ($f(-1) = 1/2$). That is the general property of a power series about the point $x = x_0$: there exists a certain *radius of convergence R* (not necessarily finite), such that the series converges for all x, such that $|x - x_0| < R$ (and possibly converges also at one or both endpoints of this interval, $x = x_0 + R$ and $x = x_0 - R$) and diverges otherwise. For (6.3), where $x_0 = 0$, we just obtained $R = 1$. Alternatively, we can apply the d'Alembert criterion to find the radius of convergence:

$$\lim_{n \to \infty} \left| \frac{a_{n+1}(x - x_0)^{n+1}}{a_n(x - x_0)^n} \right| < 1$$

gives

$$|x - x_0| < \lim_{n \to \infty} \left| \frac{a_n}{a_{n+1}} \right|. \tag{6.4}$$

For (6.3), the limit on the right in (6.4) is equal to 1, since $a_n = 1$ for any n. Therefore, again $R = 1$.

Example 6.2 The radius of convergence is not necessarily determined by the location of the singularity at the *real axis*. For example, let

$$f(x) = \frac{1}{1 + x^2}.$$

By d'Alembert criterion, its Maclaurin series (named after Colin Maclaurin (1698–1746), a Scottish mathematician)

$$\sum_{n=0}^{\infty} (-1)^n x^{2n} = 1 - x^2 + x^4 + \dots$$

converges for $|x| < 1$, though $f(x)$ has no singularities at $x = \pm 1$. To understand this, consider f in the complex plane:

$$f(x + iy) = \frac{1}{1 + (x + iy)^2}. \tag{6.5}$$

Obviously, (6.5) has singularities at the points $x + iy = \pm i$. These singularities are at the distance 1 from $x_0 = 0$, which is the origin of the complex plane, $x = y = 0$. It follows that the radius of convergence, $R = 1$, is determined by the distance from x_0 to the nearest singularity in the complex plane.

Example 6.3 Consider the function:

$$f(x) = e^x.$$

Its Maclaurin series is

$$\sum_{n=0}^{\infty} \frac{x^n}{n!} = 1 + x + \frac{x^2}{2!} + \dots \tag{6.6}$$

Because

$$\lim_{n \to \infty} \left| \frac{a_{n+1}}{a_n} \right| = \lim_{n \to \infty} \frac{1}{n+1} = 0,$$

according to (6.4), the series converges for any x, i.e., the radius of convergence is infinite. Indeed, the function

$$e^{i(x+iy)} = e^{-y}(\cos x + i \sin x)$$

has no singularities in the entire complex plane.

Example 6.4 Let

$$f(x) = \exp\left(-\frac{1}{x^2}\right) \tag{6.7}$$

and consider that function and its derivatives

$$f' = \frac{2}{x^3} \exp\left(-\frac{1}{x^2}\right), \quad f'' = \left(-\frac{6}{x^4} + \frac{4}{x^6}\right) \exp\left(-\frac{1}{x^2}\right),$$

etc., in the limit $x \to 0$. Because

$$\lim_{x \to 0} x^{-m} \exp\left(-\frac{1}{x^2}\right) = 0$$

for any m, $f(0)$ and all the derivatives $f'(0)$, $f''(0)$ etc., can be taken equal to 0. The Maclaurin series is $0 + 0 + \dots = 0$ for any x. Thus, that series converges for any x, but not to $f(x)$.

In Examples 6.1, 6.2, and 6.3, the Taylor series converge to their generating function $f(x)$ within the intervals of their convergence, centered at x_0 ($x_0 = 0$ in those examples). Such $f(x)$ is said to be *analytic* at x_0. $f(x)$ also is analytic at all x in an open interval of convergence about x_0. Example 6.4 presents a function that is not analytic at $x_0 = 0$. If we consider this $f(x)$ in the complex plane, we get:

$$f(x+iy) = \exp\left(-\frac{1}{(x+iy)^2}\right) = \exp\left(\frac{y^2 - x^2 + 2ixy}{(x^2 + y^2)^2}\right).$$

Now, $\lim_{x,y \to 0} f(x+iy)$ depends on $|y/x|$; the function tends to infinity, if $|y/x| > 1$. Thus, $f(z)$, as the function of a complex variable, is even not continuous in the point $z = 0$.

Below, we assume a solution of an ODE in the form of a power series and substitute that series into a differential equation in order to find the coefficients of the series. We make use of the following theorem on differentiation of a power series.

Theorem *Let*

$$\sum_{n=0}^{\infty} a_n(x - x_0)^n = f(x)$$

be a power series with the radius of convergence R. Then, the series

$$\sum_{n=1}^{\infty} n a_n(x - x_0)^{n-1}$$

converges to $f'(x)$ with the same radius of convergence R.

Problems

For the following functions, construct the Taylor series about $x = x_0$ and find its radius of convergence:

1. $f(x) = \sin x$;   (a) $x_0 = 0$;   (b) $x_0 = \pi/2$.
2. $f(x) = e^x$;   (a) $x_0 = 1$;   (b) $x_0 = -1$.
3. $f(x) = \frac{1}{1-x}$;   (a) $x_0 = 0$;   (b) $x_0 = -1$.
4. $f(x) = \frac{1}{(1-x)^2}$; $x_0 = 0$.
5. $f(x) = \ln(1 - x)$; $x_0 = 0$.
6. $f(x) = \ln(1+x)$; $x_0 = 0$.
7. $f(x) = \ln x$; $x_0 = a > 0$.
8. $f(x) = \arctan x$; $x_0 = 0$.
9. $f(x) = (1+x)^{1/2}$; $x_0 = 0$.
10. $f(x) = (1+x)^{-1/2}$; $x_0 = 0$.

6.2   Series Solutions Near an Ordinary Point

In the present section, we construct a power series expansion of a solution about a point x_0, where the coefficients of y and all derivatives of y in a linear ODE are analytic. Such point is called an *ordinary point*. For IVPs, it is natural to take x_0 to be a point, where the initial conditions are assigned.

6.2.1   First-Order Equations

Example 6.5 Let us begin with a simple initial value problem:

$$y' = xy, \quad y(0) = y_0.$$

For comparison with a power series solution, the compact analytical solution of IVP can be easily obtained by separating variables. This solution is $y = y_0 e^{x^2/2}$.

First, we notice that the coefficient of y is x, which is analytic everywhere. Thus, the initial condition is at the ordinary point $x_0 = 0$, and we start a power series solution by assuming a power series expansion about this point for the unknown function:

$$y = a_0 + a_1 x + a_2 x^2 + a_3 x^3 + \ldots = \sum_{k=0}^{\infty} a_k x^k. \tag{6.8}$$

Within its radius of convergence (that will be found a posteriori), we can differentiate (6.8) to obtain

$$y' = a_1 + 2a_2 x + 3a_3 x^2 + 4a_4 x^3 + \ldots = \sum_{k=1}^{\infty} k a_k x^{k-1}. \tag{6.9}$$

Substituting expansions (6.8) and (6.9) into the equation, we find:

$$a_1 + 2a_2 x + 3a_3 x^2 + 4a_4 x^3 + 5a_5 x^4 + 6a_6 x^5 + \ldots$$
$$= a_0 x + a_1 x^2 + a_2 x^3 + a_3 x^4 + a_4 x^5 + a_5 x^6 + \ldots$$

Equating the coefficients of the same powers of x gives:

$$a_1 = 0, \quad a_2 = a_0/2, \quad a_3 = a_1/3 = 0, \quad a_4 = a_2/4 = a_0/(2 \cdot 4),$$
$$a_5 = a_3/5 = 0, \quad a_6 = a_4/6 = a_0/(2 \cdot 4 \cdot 6), \ldots.$$

Thus, the expansion (6.8) is

$$y = a_0 \left(1 + \frac{x^2}{2} + \frac{1}{2}\frac{x^4}{4} + \frac{1}{2 \cdot 3}\frac{x^6}{8} + \cdots \right) = a_0 \sum_{n=0}^{\infty} \frac{1}{n!} \left(\frac{x^2}{2} \right)^n.$$

It is shown in Example 6.3 that this series converges for any x ($R = \infty$), and the sum of the series is $a_0 e^{x^2/2}$. Therefore, the previous formula is $y = y_0 e^{x^2/2}$. Applying the initial condition $y(0) = y_0$, we obtain $a_0 = y_0$ and thus we recover the IVP solution:

$$y = y_0 e^{x^2/2}.$$

Example 6.6 Consider another initial value problem:

$$y' = 1 + xy, \quad y(0) = y_0.$$

$x_0 = 0$, again, is the ordinary point. Substituting expansions (6.8) and (6.9) into the equation, we obtain:

$$a_1 + 2a_2x + 3a_3x^2 + 4a_4x^3 + 5a_5x^4 + 6a_6x^6 + \cdots$$
$$= 1 + a_0x + a_1x^2 + a_2x^3 + a_3x^4 + a_4x^5 + \cdots .$$

Equating the coefficients of the same powers of x, we have:

$$a_1 = 1, \quad a_2 = a_0/2, \quad a_3 = a_1/3 = 1/(1 \cdot 3), \quad a_4 = a_2/4 = a_0/(2 \cdot 4),$$
$$a_5 = a_3/5 = 1/(1 \cdot 3 \cdot 5), \quad a_6 = a_4/6 = a_0/(2 \cdot 4 \cdot 6), \quad \ldots .$$

Substituting these coefficients in (6.8) and collecting the terms that are proportional to a_0 and a_1, we find the general solution of the equation:

$$y = a_0 \left(1 + \frac{x^2}{2} + \frac{x^4}{2 \cdot 4} + \frac{x^6}{2 \cdot 4 \cdot 6} + \cdots \right) + \left(\frac{x}{1} + \frac{x^3}{1 \cdot 3} + \frac{x^5}{1 \cdot 3 \cdot 5} + \frac{x^7}{1 \cdot 3 \cdot 5 \cdot 7} + \cdots \right).$$

This power series solution converges for any x, as can be easily checked using the d'Alembert convergence test.

The value $a_0 = y_0$ again is obtained using the initial condition $y(0) = y_0$. The series in the first parentheses has the sum $e^{x^2/2}$.

There is no elementary function that corresponds to the second series. Nevertheless, we determined the expression for the solution that allows a calculation of $y(x)$ for any x, with arbitrary accuracy. Thus, the power series approach works even when the solution cannot be expressed via elementary functions.

The class of elementary functions is not sufficiently large to present solutions of ODEs, including some equations that are important for applications. Therefore, it makes sense to extend the class of useful functions by defining new functions, termed special functions. For instance, the second series can be expressed via the *error function*, defined as

$$\mathrm{erf} z = \frac{2}{\sqrt{\pi}} \int_0^z e^{-t^2/2} dt.$$

One can show that the second series is the Maclaurin series for the function:

$$\sqrt{\pi/2}\, \mathrm{erf}\left(x/\sqrt{2}\right).$$

Examples of other special functions will be given in Chap. 10.

Note that if the initial condition is $y(x_0) = y_0$, where $x_0 \neq 0$, then the power series centered at that x_0 should be used instead of a Maclaurin series, e.g.,

$$y = a_0 + a_1(x - x_0) + a_2(x - x_0)^2 + a_3(x - x_0)^3 + \ldots = \sum_{k=0}^{\infty} a_k(x - x_0)^k. \quad (6.10)$$

The method described above is based on the assumption that the solution can be presented in the form of a convergent power series (Taylor or Maclaurin), that is, the solution is an analytic function. Analyticity may not hold at *singular points*, which are points where the coefficients of an ODE diverge. In that case, the approach has to be modified (see Sect. 6.3).

6.2.2   Second-Order Equations

Similar approach based on expansions (6.8) or (6.10) can be applied to second (and higher)-order differential equations.

Example 6.7 Consider IVP for harmonic oscillator:

$$y'' + y = 0, \quad (6.11)$$

$$y(0) = A, \quad y'(0) = B. \quad (6.12)$$

Obviously, the solution is

$$y = A\cos x + B\sin x.$$

Since $x = 0$ is the ordinary point, we construct the solution in the form of a power series:

$$y = \sum_{n=0}^{\infty} a_n x^n. \tag{6.13}$$

Because the coefficients in (6.11) are not singular, we can expect that the solution of IVP in the form (6.13) converges for any x. For the second derivative, we obtain the expression:

$$y'' = \sum_{n=2}^{\infty} a_n n(n-1) x^{n-2}. \tag{6.14}$$

Next, we replace n with $(m+2)$ in (6.14):

$$y'' = \sum_{m=0}^{\infty} a_{m+2}(m+2)(m+1) x^m. \tag{6.15}$$

Replacing n with m in (6.13) and substituting (6.13) and (6.15) into Eq. (6.11), we get:

$$\sum_{m=0}^{\infty} [a_{m+2}(m+2)(m+1) + a_m] x^m = 0.$$

The series expansion for the function at the right-hand side of that equation is just

$$0 = \sum_{m=0}^{\infty} 0 \cdot x^m.$$

Therefore, equating the coefficients of the powers of x at the left and right side, we find:

$$a_{m+2}(m+2)(m+1) + a_m = 0, \quad m = 0, 1, 2, \ldots;$$

hence, we obtain the recurrence formula:

$$a_{m+2} = -\frac{a_m}{(m+1)(m+2)}, \quad m = 0, 1, 2, \ldots. \tag{6.16}$$

From (6.16) we obtain:

$$a_2 = -a_0/12 = -a_0/2!, \quad a_4 = -a_2/34 = a_0/4!, \quad \ldots,$$
$$a_{2k} = (-1)^k a_0/(2k)!, \quad \ldots$$

$$a_3 = -a_1/(2 \cdot 3) = -a_1/3!, \quad a_5 = -a_3/45 = a_1/5!, \quad \ldots,$$
$$a_{2k+1} = (-1)^k a_1/(2k+1)!, \quad \ldots$$

It is seen that the coefficients with even subscripts are determined by a_0, whereas the coefficients with odd subscripts are determined by a_1. a_0 and a_1 are arbitrary constants.

Thus, we obtain the solution of Eq. (6.12) in the form of a sum of two power series:

$$y(x) = a_0 \sum_{k=0}^{\infty} (-1)^k \frac{x^{2k}}{(2k)!} + a_1 \sum_{k=0}^{\infty} (-1)^k \frac{x^{2k+1}}{(2k+1)!}. \tag{6.17}$$

According to the d'Alembert criterion, both series converge for all x.

To satisfy the initial conditions (6.12), first we calculate $y'(x)$ using the term-by-term differentiation:

$$y'(x) = a_0 \sum_{k=1}^{\infty} (-1)^k \frac{x^{2k-1}}{(2k-1)!} + a_1 \sum_{k=0}^{\infty} (-1)^k \frac{x^{2k}}{(2k)!} \tag{6.18}$$

then substitute (6.17) and (6.18) into (6.12). This gives:

$$y(0) = a_0 = A, \quad y'(0) = a_1 = B.$$

Thus,

$$y(x) = A \sum_{k=0}^{\infty} (-1)^k \frac{x^{2k}}{(2k)!} + B \sum_{k=0}^{\infty} (-1)^k \frac{x^{2k+1}}{(2k+1)!}. \tag{6.19}$$

The series in (6.19) are the Maclaurin series for $\cos x$ and $\sin x$:

$$\cos x = 1 - x^2/2 + \ldots, \quad \sin x = x - x^3/6 + \ldots.$$

These functions are linearly independent. Therefore, we arrive to the same IVP solution, as was stated in the beginning of the example without using the series expansion (6.13):

$$y(x) = A \cos x + B \sin x.$$

Example 6.8 (*Airy Equation*)
Consider equation

$$y'' - xy = 0, \quad -\infty < x < \infty, \tag{6.20}$$

which is called Airy equation (named after George Airy (1801–1892), a British astronomer).
    Using the series (6.13)

$$y = \sum_{n=0}^{\infty} a_n x^n,$$

we find:

$$xy = \sum_{n=0}^{\infty} a_n x^{n+1} = \sum_{m=1}^{\infty} a_{m-1} x^m. \tag{6.21}$$

Now, we take the expression (6.15) for y'' and write separately the first term, i.e., the one that corresponds to $m = 0$:

$$y'' = 2a_2 + \sum_{m=1}^{\infty} (m+2)(m+1) a_{m+2}. \tag{6.22}$$

Substituting (6.21) and (6.22) into (6.20), we find:

$$2a_2 + \sum_{m=1}^{\infty} [(m+2)(m+1)a_{m+2} - a_{m-1}] x^m = 0. \tag{6.23}$$

Equating to zero the coefficient of x^0 in (6.23), we get $2a_2 = 0$; hence,

$$a_2 = 0. \tag{6.24}$$

Setting to zero the coefficients of x^m, $m = 1, 2, \ldots$, we find:

$$a_{m+2} = \frac{a_{m-1}}{(m+2)(m+1)}, \quad m = 1, 2, \ldots. \tag{6.25}$$

It follows from (6.24) and (6.25) that $a_5 = 0$, $a_8 = 0$, $\ldots$, $a_{3k+2} = 0$ for any natural k.
    The coefficients a_0 and a_1 can be chosen arbitrarily, and all other coefficients are calculated using (6.25). For the general solution, we have to find two linearly independent particular solutions represented by Maclaurin series.
    For the first solution, $y_1(x)$, we take $a_0 = 1$, $a_1 = 0$. Then, $a_4 = a_7 = \ldots = a_{3k+1} = \ldots = 0$. The nonzero coefficients are

$$a_3 = a_0/(3 \cdot 2) = 1/(3 \cdot 2); \quad a_6 = a_3/(6 \cdot 5) = 1/(6 \cdot 5 \cdot 3 \cdot 2); \ \ldots$$

$$a_{3k} = \frac{1}{[3k(3k-1)(3k-3)(3k-4) \cdot \ldots \cdot 3 \cdot 2]};$$

hence,

$$y_1(x) = 1 + \sum_{k=1}^{\infty} \frac{x^{3k}}{3k(3k-1)(3k-3)(3k-4) \cdot \ldots \cdot 3 \cdot 2}.$$

In order to find the second solution, $y_2(x)$, we take $a_0 = 0$, $a_1 = 1$; then $a_3 = a_6 = \ldots = a_{3k} = \ldots = 0$, whereas

$$a_4 = a_1/(4 \cdot 3) = 1/(4 \cdot 3); \quad a_7 = a_4/(7 \cdot 6) = 1/(7 \cdot 6 \cdot 4 \cdot 3); \ \ldots$$

$$a_{3k+1} = \frac{1}{[(3k+1)3k(3k-2)(3k-3) \cdot \ldots \cdot 4 \cdot 3]};$$

thus,

$$y_2(x) = x + \sum_{k=1}^{\infty} \frac{x^{3k+1}}{(3k+1)3k(3k-2)(3k-3) \cdot \ldots \cdot 4 \cdot 3}.$$

Because the expansions for $y_1(x)$ and $y_2(x)$ contain different powers of x, the linear combination $C_1 y_1(x) + C_2 y_2(x)$ is identically equal to 0 only at $C_1 = C_2 = 0$, that is, solutions $y_1(x)$ and $y_2(x)$ are linearly independent. The general solution of Eq. (6.20) is

$$y(x) = C_1 y_1(x) + C_2 y_2(x).$$

The properties of $y_1(x)$ and $y_2(x)$ are considered in more detail in Sect. 10.2.9.

Example 6.9 (*Hermite Equation*)
(named after Charles Hermite (1822–1901), a French mathematician)
The Hermite equation is the second-order linear ODE:

$$y'' - 2xy' + 2ny = 0, \quad n = 0, 1, 2, \ldots . \tag{6.26}$$

We again assume the solution $y(x)$ in the form of a power series:

$$y(x) = \sum_{k=0}^{\infty} a_k x^k. \tag{6.27}$$

Then,

$$y'(x) = \sum_{k=1} k a_k x^{k-1} = \sum_{k=0} k a_k x^{k-1},$$

$$y''(x) = \sum_{k=2} k(k-1) a_k x^{k-2} = \sum_{k=0} (k+2)(k+1) a_{k+2} x^k.$$

Substituting these expressions into (6.26), we obtain:

$$\sum_{k=0}^{\infty} [(k+2)(k+1) a_{k+2} - 2k a_k + 2n a_k] x^k = 0. \tag{6.28}$$

This identity holds for all x only when the coefficients of all powers x^k are zero; thus, the *recurrence relation* follows:

$$a_{k+2} = a_k \frac{2(k-n)}{(k+2)(k+1)}. \tag{6.29}$$

From there it follows that for a given value of n, $a_{k+2} = 0$ as well as all the following coefficients: $a_{k+2} = a_{k+4} = \ldots = 0$; thus, the series (6.27) reduces to a polynomial of degree n. For n even, this polynomial contains only even powers of x; for n odd it contains only odd powers of x. In the first case, the coefficient a_0 is arbitrary; in the second case a_1 is arbitrary; in both cases all the following coefficients are determined by the recurrence relation (6.29). These even and odd polynomials provide two linearly independent solutions of Eq. (6.26).

Polynomial solutions of the Hermite equation are called Hermite polynomials and labeled as $H_n(x)$. For instance, $H_0(x) = a_0$, $H_1(x) = a_1 x$, $H_2(x) = a_0(1 - 2x^2)$, $H_3(x) = a_1(1 - 2x^3/3)$, and $H_2(x) = a_0(1 - 4x^2 + 4x^4/3)$.

The Hermite polynomials and their applications are considered in more detail in Chap. 10.

6.3   Series Solutions Near a Regular Singular Point

In the previous section, we considered equations of the type

$$y'' + p(x)y' + q(x)y = 0, \tag{6.30}$$

where the functions $p(x)$ and $q(x)$ are analytic in an interval containing the point x_0, and constructed solutions in the form of a power series

$$y(x) = \sum_{n=0}^{\infty} a_n (x - x_0)^n.$$

Let us consider now the equation in a more general form:

$$P(x)y'' + Q(x)y' + R(x)y = 0, \tag{6.31}$$

where the functions $P(x)$, $Q(x)$, and $R(x)$ are analytic in an interval containing x_0. If $P(x_0) \neq 0$, we can rewrite (6.31) in the form (6.30) with

$$p(x) = \frac{Q(x)}{P(x)}, \quad q(x) = \frac{R(x)}{P(x)}, \tag{6.32}$$

where the functions $p(x)$ and $q(x)$ are also analytic near $x = x_0$. However, if $P(x_0) = 0$, the functions (6.32) are not analytic at x_0, and therefore the approach described in Sect. 6.2 may lead to a nonanalytic solution of Eq. (6.31) at x_0.

In fact, we already considered equations of this kind in Sect. 3.8. Recall that the *Euler equation*

$$x^2 y'' + pxy' + qy = 0, \tag{6.33}$$

has solutions of the form $y_1(x) = x^{r_1}$, $y_2(x) = x^{r_2}$, where $r_{1, 2}$ are two simple roots of the quadratic equation

$$r(r - 1) + pr + q = 0. \tag{6.34}$$

If r is complex, $r = a + ib$, then

$$x^r = x^a e^{ib \ln x} = x^a [\cos(b \ln x) + i \sin(b \ln x)].$$

If r is the double root of (6.34), then $y_1(x) = x^r$, and $y_2(x) = x^r \ln x$. Thus, except the case where both roots $r_{1, 2}$ are simple and integer, solutions of the Euler equation are not analytic functions in the point $x = 0$.

In the present section, we consider a generalization of Eq. (6.33):

$$L[y] = (x - x_0)^2 y'' + (x - x_0)p(x)y' + q(x)y = 0, \tag{6.35}$$

or

$$y'' + \frac{p(x)}{x - x_0} y' + \frac{q(x)}{(x - x_0)^2} y = 0, \tag{6.36}$$

where the functions

$$p(x) = \sum_{j=0}^{\infty} p_j (x - x_0)^j, \quad q(x) = \sum_{j=0}^{\infty} q_j (x - x_0)^j$$

are analytic at $x = x_0$ (rather than just constant as in the case of the Euler equation). The point $x = x_0$ is called a *regular singular point* of Eq. (6.35). Note that a point which is neither ordinary nor regular singular is called *irregular singular point*. We do not consider such points in this book.

Because the Euler equation is a particular case of (6.35), it is reasonable to expect that the nonanalyticity of the solutions of (6.35) will be similar to that of the solutions of the Euler equation, and to search the solutions of (6.35) in the form:

$$y = (x - x_0)^r \sum_{k=0}^{\infty} a_k (x - x_0)^k, \quad a_0 \neq 0 \tag{6.37}$$

(below we obtain the equation for r).

We start the solution by calculating the auxiliary quantity:

$$L[(x - x_0)^c] = (x - x_0)^c [c(c-1) + cp(x) + q(x)] = (x - x_0)^c \sum_{j=0}^{\infty} f_j(c)(x - x_0)^j,$$

where

$$f_0(c) = c(c-1) + p_0 c + q_0; \quad f_j(c) = p_j c + q_j, \quad j \neq 0.$$

Now, substituting (6.37) into (6.35), we find:

$$L\left[\sum_{k=0}^{\infty} a_k (x - x_0)^{r+k}\right] = \sum_{k=0}^{\infty} a_k L\left[(x - x_0)^{r+k}\right]$$

$$= \sum_{k=0}^{\infty} a_k (x - x_0)^{r+k} \sum_{j=0}^{\infty} f_j(k+j)(x - x_0)^j \tag{6.38}$$

$$= (x - x_0)^r \sum_{k=0}^{\infty} \sum_{j=0}^{\infty} a_k f_j(k+j)(x - x_0)^{k+j} = 0.$$

The coefficients of all powers of $x - x_0$ in (6.38) must be equal to zero. There is only one term with $k+j = 0$: both $k = j = 0$. Thus, we find:

$$a_0 f_0(r) = 0. \tag{6.39}$$

There are two terms corresponding to $k+j = 1$: $k = 0, j = 1$ and $k = 1, j = 0$. We find:

$$a_0 f_1(r) + a_1 f_0(r+1) = 0. \tag{6.40}$$

Similarly, at higher orders we find:

$$k+j=2, \quad a_0 f_2(r) + a_1 f_1(r+1) + a_2 f_0(r+2) = 0, \tag{6.41}$$

$$\cdots$$

$$k+j=m, \quad a_0 f_m(r) + a_1 f_{m-1}(r+1) + \dots a_m f_0(r+m) = 0 \tag{6.42}$$

$$\cdots$$

Recall that $a_0 \neq 0$; therefore, Eq. (6.39) is *the indicial equation* that determines r:

$$f_0(r) = r(r-1) + p_0 r + q_0 = 0. \tag{6.43}$$

Because we solve a homogeneous linear equation, a_0 is arbitrary. For each root r of (6.43), we have the following recursion formulas:

$$a_1(r) = -\frac{a_0 f_1(r)}{f_0(r+1)}, \quad \dots,$$

$$a_m(r) = -\frac{a_0 f_m(r) + a_1(r) f_{m-1}(r+1) + \dots a_{m-1}(r) f_1(r+m-1)}{f_0(r+m)}, \quad \dots.$$

The obtained formulas can be applied if all the values $f_0(r+1), \dots, f_0(r+m), \dots$ are nonzero.

Quadratic Eq. (6.43) has either two different roots, $r_1 \neq r_2$ or one double real root, $r_1 = r_2$.

If $r_1 \neq r_2$ and $r_1 - r_2$ is not a positive integer number, we can apply recursion formulas obtained above for finding both independent particular solutions of Eq. (6.35):

$$y_1(x) = (x - x_0)^{r_1} \sum_{m=0}^{\infty} a_m(r_1)(x - x_0)^m, \tag{6.44}$$

$$y_2(x) = (x - x_0)^{r_2} \sum_{m=0}^{\infty} a_m(r_2)(x - x_0)^m. \tag{6.45}$$

Specifically, those conditions are always satisfied when $r_{1,2} = \lambda \pm i\mu$ are complex. Instead of complex solutions

$$y_{1,2}(x) = (x - x_0)^{\lambda} \exp[i\mu \ln(x - x_0)] \sum_{m=0}^{\infty} a_m(\lambda \pm i\mu)(x - x_0)^m,$$

one can use their real and imaginary parts.

If $r_1 - r_2 = N > 0$ is an integer number, only one solution, $y_1(x)$, can be obtained using formula (6.44). The algorithm fails for the second solution, because $f_0(r_2+N) = f_0(r_1) = 0$.

Also, in the case $r_1 = r_2$, the algorithm described above allows to find only one solution.

In both cases, the extension of the algorithm is as follows: after finding solution $y_1(x)$ in the form (6.44), one has to search the second solution in the form:

$$y_2(x) = (x - x_0)^{r_2} \sum_{m=0}^{\infty} b_m (x - a)^m + A \ln |x - x_0| \, y_1(x).$$

The difference between two cases is as follows. In the case $r_1 = r_2$, the coefficient $A \neq 0$ and it can be chosen arbitrarily; also, one can take $b_0 = 0$ because the term proportional to $(x - x_0)^{r_2} = (x - x_0)^{r_1}$ is present in the expansion for $y_1(x)$. In the case $r_1 - r_2 = N > 0$, the coefficient $b_0 \neq 0$ and it can be chosen arbitrarily, while it may happen that $A = 0$. We shall not derive the rules formulated above.

As an example, consider the *Bessel equation* (named after Friedrich Bessel (1784–1846), a German astronomer, mathematician, physicist, and geodesist):

$$y''(x) + \frac{y'(x)}{x} + \left(1 - \frac{p^2}{x^2}\right) y(x) = 0. \tag{6.46}$$

In order to understand the behavior of the solution in the vicinity of the singular point $x = 0$, we construct the indicial equation for Eq. (6.46):

$$F(r) = r(r-1) + r - p^2 = r^2 - p^2 = 0;$$

its roots are $r_1 = p \geq 0$ and $r_2 = -p \geq 0$.

As we discussed, there always exists the solution:

$$y_1(x) = \sum_{n=0}^{\infty} a_n x^{p+n}, \quad a_0 \neq 0, \tag{6.47}$$

which is not singular in the point $x = 0$. As to the second linearly independent solution, there are the following options:

(i) If $r_1 - r_2 = 2p$ is not an integer, then there exists another solution:

$$y_2(x) = \sum_{n=0}^{\infty} a'_n x^{-p+n}, \quad a'_0 \neq 0, \tag{6.48}$$

which is singular at the point $x = 0$;

(ii) If $p = 0$, the second solution has the structure:

$$y_2(x) = ay_1(x) \ln x + \sum_{n=1}^{\infty} b_n x^n, \quad a \neq 0; \tag{6.49}$$

(iii) If $2p = N$ is a positive integer, then the general theory predicts the following structure of the second solution:

$$y_2(x) = ay_1(x) \ln x + \sum_{n=0}^{\infty} c_n x^{-p+n}, \quad c_0 \neq 0; \tag{6.50}$$

note that in the latter case, a can be equal to zero.

Let us find non-singular solutions (6.47) known as *Bessel functions of the first kind* and consider the case of integer values of $p = 0, 1, 2, \ldots$, which is the most important case in applications. Substituting (6.47) into (6.46), we obtain:

$$\sum_{n=0}^{\infty} n(2p + n)a_n x^{p+n} + \sum_{n=2}^{\infty} a_{n-2} x^{p+n} = 0.$$

Equating the terms with the same powers of x, we find that a_0 is arbitrary, $a_1 = 0$, and all other coefficients are determined by the *recurrence relation*:

$$a_n = -\frac{a_{n-2}}{n(2p+n)}, \quad n = 2, 3, \ldots. \tag{6.51}$$

We find that all a_n with odd subscripts vanish, while for the coefficients with even subscripts, we get:

$$a_2 = -\frac{a_0}{2^2(p+1)}, \quad a_4 = -\frac{a_2}{2^3(p+2)} = -\frac{a_0}{2^4 \cdot 2(p+2)}, \ldots$$

By induction, one can prove that for any even $n = 2k$

$$a_{2k} = \frac{(-1)^k a_0}{2^{2k} k!(p+k)(p+k-1)\ldots(p+1)}, \quad k = 1, 2, \ldots$$

Thus, the solution can be written as

$$y_1(x) = a_0 2^p p! J_p(x),$$

where the function

$$J_p(x) = \sum_{k=0}^{\infty} \frac{(-1)^k}{k!(p+k)!} \left(\frac{x}{2}\right)^{p+2k} \tag{6.52}$$

is called *Bessel function of the first kind* of order p (here we define $0! = 1$).

Thus, we have obtained the solution (6.47) of Eq. (6.46) that is finite at $x = 0$ (still for integer values of p only). Other cases are considered in detail in Chap. 10.

The first few terms of the expansion in Eq. (6.52) near zero for the first three functions are

$$J_0(x) = 1 - \frac{x^2}{2^2} + \frac{x^4}{2^4 \cdot 2! \cdot 2!} - \cdots, \qquad J_1(x) = \frac{x}{2} - \frac{x^3}{2^3 \cdot 2!} + \frac{x^5}{2^5 \cdot 2! \cdot 3!} - \cdots,$$

$$J_2(x) = \frac{x^2}{2^2 \cdot 2!} - \frac{x^4}{2^4 \cdot 3!} + \frac{x^6}{2^6 \cdot 2! \cdot 4!} - \cdots . \tag{6.53}$$

The coefficients in the Bessel function expansion decay very rapidly because of the factorial factor in the denominator in (6.52).

In problems below search particular solutions in the form of series expansion and find the terms up to x^4. You can use programs *ODE 1st order* and *ODE 2nd order* to plot the graphs of series and numerical solutions of IVP on some interval which includes the initial point x_0. (Input a formula for series solution into the Edit window "Analytical solution.")

Problems

1. $y' = y^2 - x$, $y(0) = 1$.
2. $y' = y^3 + x^2$, $y(1) = 1$.
3. $y' = y + xe^y$, $y(0) = 0$.
4. $y' = 2x + \cos y$, $y(0) = 0$.
5. $y'' + xy = 0$, $y(0) = 1$, $y'(0) = -1$.
6. $y'' = xy' - y^2$, $y(0) = 1$, $y'(0) = 2$.
7. $y'' = y'^2 + xy$, $y(0) = 4$, $y'(0) = -2$.
8. $x^2 y'' - xy' + (1+x)y = 0$, $y(1) = 1$, $y'(1) = 2$.
9. $x^2 y'' - x(1+x)y' - 3y = 0$, $y(1) = 1$, $y'(1) = 2$.

Chapter 7
Laplace Transform

7.1 Introduction

In this chapter we consider the *Laplace transform* (named after Pierre-Simon Laplace (1749–1827), a French scientist) which allows to transform some types of differential equations into algebraic equations. It is especially useful for solving linear ODEs with constant or piecewise constant coefficients, as well as integral equations introduced in Chap. 11.

A Laplace Transform $L[x(t)]$ of a real function $x(t)$ defined for all real numbers $t \geq 0$ is given by the following formula:

$$\widehat{x}(p) = L[x(t)] = \int_0^\infty e^{-pt} x(t) dt, \tag{7.1}$$

where p is, generally, a complex parameter. Function $\widehat{x}(p)$ is often called the *image* of the *original* function $x(t)$.

The right-hand side of Eq. (7.1) is called the *Laplace integral*. For the convergence of integral (7.1), the growth of function $x(t)$ should be limited: for sufficiently large t, $x(t)$ should satisfy the inequality

$$|x(t)| \leq M e^{at}, \quad M > 0.$$

In that case, the Laplace transform (LT) is defined for $\mathrm{Re}\, p > a$. Also, function $x(t)$ should be integrable on $(0, \infty)$. For instance, the LT of $x(t) = 1/t^\beta$ exists only if $\beta < 1$.

As an example, we find Laplace transform of two functions:

Let $x(t) = 1$; then $L[x(t)] = \int_0^\infty e^{-pt} dt = -\frac{1}{p} e^{-pt}\big|_0^\infty = \frac{1}{p}$.

Let $x(t) = e^{at}$; then

© The Author(s), under exclusive license to Springer Nature Switzerland AG 2023
V. Henner et al., *Ordinary Differential Equations*,
https://doi.org/10.1007/978-3-031-25130-6_7

Table 7.1 Basic Laplace transforms

$x(t)$	$L[x(t)]$	Convergence condition
1	$\dfrac{1}{p}$	$\mathrm{Re}\,p > 0$
t^n	$\dfrac{n!}{p^{n+1}}$	$n \geq 0$ is integer, $\mathrm{Re}\,p > 0$
t^a	$\dfrac{\Gamma(a+1)}{p^{a+1}}$	$a > -1$, $\mathrm{Re}\,p > 0$
e^{at}	$\dfrac{1}{p-a}$	$\mathrm{Re}\,p > \mathrm{Re}\,a$
$\sin ct$	$\dfrac{c}{p^2+c^2}$	$\mathrm{Re}\,p > |\mathrm{Im}\,c|$
$\cos ct$	$\dfrac{p}{p^2+c^2}$	$\mathrm{Re}\,p > |\mathrm{Im}\,c|$
$\sinh bt$	$\dfrac{b}{p^2-b^2}$	$\mathrm{Re}\,p > |\mathrm{Re}\,b|$
$\cosh bt$	$\dfrac{p}{p^2-b^2}$	$\mathrm{Re}\,p > |\mathrm{Re}\,b|$
$t^n e^{at}$	$\dfrac{n!}{(p-a)^{n+1}}$	$\mathrm{Re}\,p > \mathrm{Re}\,a$
$t \sin ct$	$\dfrac{2pc}{(p^2+c^2)^2}$	$\mathrm{Re}\,p > |\mathrm{Im}\,c|$
$t \cos ct$	$\dfrac{p^2-c^2}{(p^2+c^2)^2}$	$\mathrm{Re}\,p > |\mathrm{Im}\,c|$
$e^{at} \sin ct$	$\dfrac{c}{(p-a)^2+c^2}$	$\mathrm{Re}\,p > (\mathrm{Re}\,a + |\mathrm{Im}\,c|)$
$e^{at} \cos ct$	$\dfrac{p-a}{(p-a)^2+c^2}$	$\mathrm{Re}\,p > (\mathrm{Re}\,a + |\mathrm{Im}\,c|)$

$$L[x(t)] = \int_0^\infty e^{-pt} e^{at} \, dt = \int_0^\infty e^{(a-p)t} \, dt = \frac{1}{a-p} e^{(a-p)t}\big|_0^\infty = \frac{1}{p-a}.$$

It is seen that for convergence of the integral, $\mathrm{Re}\,p$ has to be larger than a.

Laplace transforms of some functions can be found in Table 7.1.

To determine the original function from the image, $\widehat{x}(p)$, one has to perform the *inverse Laplace transform*, which is denoted as

$$x(t) = L^{-1}[\widehat{x}(p)]. \tag{7.2}$$

Generally, the inverse Laplace transform can be evaluated using the approach of complex variable calculus that is described, for instance, in [6]. In this book, we shall use the table of LT presented above and the properties of the Laplace transform discussed below.

7.2   Properties of the Laplace Transform

Property 7.1   Laplace Transform is linear:

$$L[ax(t) + by(t)] = aL[x(t)] + bL[y(t)].\qquad(7.3)$$

This follows from the linearity of the integral (7.1).
Inverse transform is also linear:

$$L^{-1}[a\widehat{x}(p) + b\widehat{y}(p)] = aL^{-1}[\widehat{x}(p)] + bL^{-1}[\widehat{y}(p)].\qquad(7.4)$$

Property 7.2   Let $L[x(t)] = \widehat{x}(p)$. Then if c is a constant

$$L[x(ct)] = \frac{1}{c}\widehat{x}\left(\frac{p}{c}\right).\qquad(7.5)$$

This can be proven using the change of a variable, $t = u/k$:

$$\int_0^\infty e^{-pt}x(ct)dt = \frac{1}{c}\int_0^\infty e^{-\frac{p}{c}u}x(u)du = \frac{1}{c}\widehat{x}\left(\frac{p}{c}\right).$$

Property 7.3   Let $L[x(t)] = \widehat{x}(p)$, and

$$x_a(t) = \begin{cases} 0, & t < a, \\ x(t-a), & t \geq a, \end{cases}$$

where $a > 0$. Then,

$$\widehat{x}_a(p) = e^{-pa}\widehat{x}(p).\qquad(7.6)$$

This property is also known as the *Delay Theorem*.

Property 7.4   Let $L[x(t)] = \widehat{x}(p)$. Then for any complex constant c:

$$L[e^{ct}x(t)] = \int_0^\infty e^{-pt}e^{ct}x(t)dt = \widehat{x}(p-c).$$

This property is also known as the *Shift Theorem*.

Property 7.5   Let $h(t)$ be a *convolution* of functions $f(t)$ and $g(t)$:

$$h(t) = \int_0^t f(t-\tau)g(\tau)d\tau \tag{7.7}$$

(or $h(t) = \int_0^t f(\tau)g(t-\tau)d\tau$).

Convolution is symbolically denoted as

$$h(t) = f(t) * g(t).$$

The property states that the Laplace transform of $h(t)$ is (the proof can be found in [6])

$$\widehat{h}(p) = \widehat{f}(p)\widehat{g}(p); \tag{7.8}$$

in other words

$$L[h(t)] = L\left[\int_0^t f(t-\tau)g(\tau)d\tau\right] = \widehat{f}(p)\widehat{g}(p). \tag{7.9}$$

Property 7.6 Laplace transform of a first derivative is

$$L[x'(t)] = p\widehat{x}(p) - x(0). \tag{7.10}$$

Indeed, integration by parts gives:

$$\int_0^\infty e^{-pt}\widehat{x}(t)dt = x(t)e^{-pt}\big|_0^\infty + p\int_0^\infty e^{-pt}x(t)dt = -x(0) + p\widehat{x}(p).$$

Analogously, for the second derivative, we obtain:

$$L[x''(t)] = p^2\widehat{x}(p) - px(0) - x'(0). \tag{7.11}$$

And for the n-th derivative

$$L\left[x^{(n)}(t)\right] = p^n\widehat{x}(p) - p^{n-1}\widehat{x}(0) - p^{n-2}x'(0) - \ldots - x^{(n-1)}(0). \tag{7.12}$$

Property 7.7 Let $L[x(t)] = \widehat{x}(p)$; then the Laplace transform of the integral can be represented as

$$L\left[\int_0^t x(t)dt\right] = \frac{1}{p}\hat{x}(p). \tag{7.13}$$

This property can be proven by writing the LT of the integral as a double integral and then interchanging the order of the integrations.

Up to this point, we considered the Laplace transform of "usual" functions $x(t)$ defined by their values in the region $0 \leq t < \infty$. Below we consider the Laplace transform of a "generalized" function, the so-called *Dirac delta function* $\delta(t)$ (Paul Dirac (1902–1984), an English physicist), which is defined only by values of the integrals of its products with usual functions.

Let us define function

$$\Delta(t;a) = \frac{1}{a\sqrt{2\pi}}e^{-t^2/2a^2}. \tag{7.14}$$

For any $t \neq 0$, $\lim_{a\to 0}\Delta(t;a) = 0$, while $\lim_{a\to 0}\Delta(0;a) = \infty$. Thus, $\Delta(t;0)$ is not a "usual" function defined for any t. Consider the integral

$$I(x(t);a) = \int_{-\infty}^{\infty} x(t)\Delta(t;a)dt, \tag{7.15}$$

where $x(t)$ is an arbitrary continuous function. Formula (7.15) determines a number $I(x(t);a)$ that depends on the function $x(t)$ as a whole; such a dependence is called functional. Define $\tau = t/a$ and rewrite (7.15) as

$$I(x(t);a) = \int_{-\infty}^{\infty} \frac{x(at)}{\sqrt{2\pi}}e^{-\tau^2/2}d\tau.$$

In the limit $a \to 0$, we obtain a functional

$$I(x(t);0) = \lim_{a\to 0} I(x(t);a) = x(0)\frac{1}{\sqrt{2\pi}}e^{-\tau^2/2}d\tau = x(0). \tag{7.16}$$

In a contradistinction to functional (7.15), functional (7.16) cannot be presented as the integral of a product of function $x(t)$ with a certain function $\Delta(t;0)$. Nevertheless, we shall write that functional in a symbolic form as

$$I(x(t);0) = \int\limits_{-\infty}^{\infty} x(t)\delta(t)dt = x(0), \tag{7.17}$$

where $\delta(t)$, the delta function, is a "generalized function" defined by relation (7.17). The basic properties of $\delta(t)$ are

$$\int\limits_{-\infty}^{\infty} x(t)\delta(t-a)dt = x(a), \qquad \int\limits_{-\infty}^{\infty} \delta(t-a)dt = 1. \tag{7.18}$$

Note that the integrals are not necessarily taken from minus infinity to plus infinity, but they can be over any interval containing the point $t = a$.

Other properties of the δ function are

$$\delta(-t) = \delta(t), \quad \delta(at) = \frac{1}{|a|}\delta(t).$$

Note that function $\delta(t)$, which is formally equal to 0 for $t \neq 0$, is infinite at $t = 0$, and satisfies the relation $\int\limits_{-\infty}^{\infty} \delta(t)dt = 1$, can be considered as a "derivative" of the *Heaviside function* (Oliver Heaviside (1850–1925), an English mathematician and physicist):

$$\frac{d}{dt}H(t) = \delta(t),$$

where $H(t) = \begin{cases} 1, & \text{for } t > 0, \\ 0, & \text{for } t < 0. \end{cases}$

We shall consider the applications of the delta function in other chapters.

Property 7.8 Laplace transforms related to the delta function, $\delta(t)$, are

$$L[\delta(t)] = 1, \quad L[\delta(t-a)] = e^{-pa}, \quad L[\delta(t-a)x(t)] = e^{-pa}x(a). \tag{7.19}$$

Reading Exercise Prove the properties (7.19) using the relations (7.18) with $a > 0$.

7.3 Applications of the Laplace Transform for ODEs

One of the important applications of the Laplace transform is the solution of ODE IVPs, since often an equation for the transform of an unknown function can be solved much easier than the original equation. A transform must be inverted after it has been found, to return to the original function (i.e., the solution of a differential equation). Often this step requires a calculation of integrals of a complex variable, but in many situations, for example, the case of linear equations with constant coefficients, the inverted transforms can be found in the table of Laplace transforms. Some partial differential equations also can be solved using the method of Laplace transform. In this section we consider solutions of ordinary differential equations and systems of ODEs.

Let the task be to solve the IVP for the second-order linear equation with constant coefficients (below we denote an unknown function as $x(t)$):

$$ax'' + bx' + cx = f(t), \tag{7.20}$$

$$x(0) = \beta, \quad x'(0) = \gamma. \tag{7.21}$$

Let us apply the LT to the both sides of Eq. (7.20). Using linearity of the transform and Property 7.6 gives the *algebraic equation* for the transform function $\widehat{x}(p)$:

$$a\left(p^2\widehat{x} - \beta p - \gamma\right) + b(p\widehat{x} - \beta) + c\widehat{x} = \widehat{f}(p). \tag{7.22}$$

We have used the initial conditions (7.21). Solving this algebraic equation, one finds:

$$\widehat{x}(p) = \frac{\widehat{f}(p) + a\beta p + a\gamma + b\beta}{ap^2 + bp + c}. \tag{7.23}$$

Note that when the initial conditions are zero, the transform takes a simple form:

$$\widehat{x}(p) = \frac{\widehat{f}(p)}{ap^2 + bp + c}. \tag{7.24}$$

The inverse LT gives the function $x(t)$, which is the solution of the IVP. As we already mentioned, a general way involves the integration of a function of a complex variable. Fortunately, many LT can be inverted using tables and partial fractions, like in examples below.

Example 7.1 Solve differential equation:

$$x'' + 9x = 6\cos 3t$$

with zero initial conditions.

Solution Applying the LT to both sides of the equation and taking into the account that $x(0) = 0$, $x'(0) = 0$ gives:

$$p^2\widehat{x}(p) + 9\widehat{x}(p) = \frac{6p}{p^2 + 9}.$$

Then,

$$\widehat{x}(p) = \frac{6p}{(p^2 + 9)^2} = \frac{2 \cdot 3p}{\left(p^2 + 3^2\right)^2}.$$

The original function $x(t)$ is read out directly from the Laplace Transform table:

$$x(t) = t \sin 3t.$$

This function is the solution of the IVP.

Example 7.2 Solve the IVP:

$$x'' + 4x = e^t, \quad x(0) = 4, \quad x'(0) = -3.$$

Solution Applying the LT to both sides of the equation and taking into the account the initial conditions, we obtain:

$$p^2\widehat{x}(p) - 4p + 3 + 4\widehat{x}(p) = \frac{1}{p - 1}.$$

Solving for $\widehat{x}(p)$ gives

$$\widehat{x}(p) = \frac{4p^2 - 7p + 4}{(p^2 + 4)(p - 1)}.$$

Next, using the partial fractions, we can write:

$$\frac{4p^2 - 7p + 4}{(p^2 + 4)(p - 1)} = \frac{A}{p - 1} + \frac{Bp + C}{p^2 + 4}.$$

From there,

$$4p^2 - 7p + 4 = Ap^2 + 4A + Bp^2 + Cp - Bp - C,$$

and equating the coefficients of the second, first, and zeroth degrees of p, we have:

$$4 = A + B, \quad -7 = C - B, \quad 4 = 4A - C.$$

The solution to this system of equations is

$$A = 1/5, \quad B = 19/5, \quad C = -16/5.$$

Then,

$$x(t) = \frac{1}{5}L^{-1}\left[\frac{1}{p-1}\right] + \frac{19}{5}L^{-1}\left[\frac{p}{p^2+4}\right] - \frac{16}{5}L^{-1}\left[\frac{1}{p^2+4}\right].$$

Using the inverse transform from the table gives the solution of the IVP:

$$x(t) = (e^t + 19\cos 2t - 8\sin 2t)/5.$$

It is easy to check that this solution satisfies the equation and the initial conditions.

Example 7.3 Solve system of differential equations:

$$\begin{cases} x' = 2y + t, \\ y' = x + y, \end{cases}$$

with the initial conditions $x(0) = 0$, $y(0) = 0$.

Solution Applying the LT to each equation, we obtain an algebraic system for the functions $\widehat{x}(p)$ and $\widehat{y}(p)$:

$$\begin{cases} p\widehat{x} = 2\widehat{y} + \dfrac{1}{p^2}, \\ p\widehat{y} = \widehat{x} + \widehat{y}. \end{cases}$$

From the second equation, $\widehat{x} = (p-1)\widehat{y}$. Substitute it into the first equation:

$$p(p-1)\widehat{y} - 2\widehat{y} = \frac{1}{p^2},$$

from which

$$\widehat{y} = \frac{1}{p^2(p^2-p-2)} = \frac{1}{p^2(p-2)(p+1)} = \frac{1}{3p^2}\left[\frac{1}{p-2} - \frac{1}{p+1}\right].$$

Now use partial fractions:

1. $\dfrac{1}{p^2(p-2)} = \dfrac{A}{p} + \dfrac{B}{p^2} + \dfrac{C}{p-2}$,

which gives

$$1 = Ap(p-2) + B(p-2) + Cp^2 \text{ and } A+C = 0, \quad -2A+B = 0, \quad -2B = 1.$$
Thus, $A = -1/4$, $B = -1/2$, $C = 1/4$.

2. $\dfrac{1}{p^2(p+1)} = \dfrac{A}{p} + \dfrac{B}{p^2} + \dfrac{C}{p+1}$,

which gives

$1 = Ap(p+1) + B(p+1) + Cp^2$ and $A + C = 0,\ A + B = 0,\ B = 1$.

Thus, $A = -1,\ B = 1,\ C = 1$.

Then,

$$\widehat{y} = \frac{1}{3}\left[-\frac{1}{4p} - \frac{1}{2p^2} + \frac{1}{4(p-2)} + \frac{1}{p} - \frac{1}{p^2} - \frac{1}{p+1}\right] =$$

$$= \frac{1}{4p} - \frac{1}{2p^2} + \frac{1}{12(p-2)} - \frac{1}{3(p+1)}.$$

The inverse LT gives:

$$y(t) = \frac{1}{4} - \frac{t}{2} + \frac{1}{12}e^{2t} - \frac{1}{3}e^{-t}.$$

Function $x(t)$ can be found from the second equation of the initial system:

$$x(t) = y' - y = -\frac{3}{4} + \frac{t}{2} + \frac{1}{12}e^{2t} + \frac{2}{3}e^{-t}.$$

Thus, the solution of IVP is

$$\begin{cases} x(t) = -\dfrac{3}{4} + \dfrac{t}{2} + \dfrac{1}{12}e^{2t} + \dfrac{2}{3}e^{-t}, \\[2mm] y(t) = \dfrac{1}{4} - \dfrac{t}{2} + \dfrac{1}{12}e^{2t} - \dfrac{1}{3}e^{-t}. \end{cases}$$

It is easy to check that these $x(t)$ and $y(t)$ satisfy the equations and the initial conditions.

This next Example shows how a nonhomogeneous equation can be solved in terms of a convolution integral.

Example 7.4 Consider an IVP for a harmonic oscillator:

$$x'' + \omega_0^2 x = g(t) \tag{7.25}$$

with $x(0) = x_0,\ x'(0) = v_0$.

Solution Applying the LT to the equation, we obtain:

$$\widehat{x}(p) = \frac{x_0 p + v_0}{p^2 + \omega_0^2} + \frac{\widehat{g}(p)}{p^2 + \omega_0^2}.$$

The inverse of $\widehat{x}(p)$ is

$$x(t) = x_0 \cos \omega_0 t + \frac{v_0}{\omega_0} \sin \omega_0 t + \frac{1}{\omega_0} \int\limits_0^t g(t-u) \sin \omega_0 u \, du. \qquad (7.26)$$

Example 7.5 Find the current in the electrical circuit consisting of sequentially connected resistor R, inductor L, capacitor C, and a voltage source $E(t)$, provided that at the initial time $t = 0$, the switch is closed and the capacitor is not charged.
   Consider the cases when:

1. The voltage source is constant: $E(t) = E_0$ for $t \geq 0$.
2. At $t = 0$ the voltage source produces a short impulse: $E(t) = E_0 \delta(t)$.

Solution   The equation for the current in the circuit is

$$L\frac{di}{dt} + Ri + \frac{q}{C} = E(t), \qquad (7.27)$$

or taking into account that electric charge on a capacitor is $q = \int\limits_0^t i(\tau)d\tau$, we can write this equation as

$$L\frac{di}{dt} + Ri + \frac{1}{C} \int\limits_0^t i(\tau)d\tau = E(t).$$

Applying the LT to both sides of the equation gives:

$$L p \widehat{i} + R \widehat{i} + \frac{1}{C} \frac{\widehat{i}}{p} = \widehat{E}(p).$$

To transform the integral, we used Property 7.7 of the Laplace transform. From the last equation, the function $\widehat{i}(p)$ is

$$\widehat{i}(p) = \frac{\widehat{E}(p)}{R + Lp + 1/pC}.$$

1. If voltage is constant, then $\widehat{E}(p) = E_0/p$ and thus

$$\widehat{i}(p) = \frac{1}{p} \frac{E_0}{R + Lp + 1/pC} = \frac{E_0}{L} \frac{1}{p^2 + (R/L)p + 1/LC}.$$

Next, complete the square in the denominator and introduce notations:

$$\omega^2 = \frac{1}{LC} - \frac{R^2}{4L^2} \quad \left(\text{assume that } \omega^2 > 0\right), \quad \gamma = \frac{R}{2L}.$$

Then,

$$\widehat{i}(p) = \frac{E_0}{L} \frac{1}{(p+\gamma)^2 + \omega^2}.$$

From the table of LT, we find:

$$\frac{1}{\omega} L[e^{-\gamma t} \sin \omega t] = \frac{1}{(p+\gamma)^2 + \omega^2};$$

thus, the current in case of constant voltage is

$$i(t) = \frac{E_0}{\omega L} e^{-\gamma t} \sin \omega t.$$

2. If $E(t) = E_0 \delta(t)$, then $\widehat{E}(p) = E_0$ and

$$\widehat{i}(p) = \frac{E_0}{R + Lp + 1/pC} = \frac{E_0}{L} \frac{1}{p + (R/L) + 1/pLC}.$$

As in the previous solution, we complete the square in the denominator and write the expression for $\widehat{i}(p)$ in the form convenient for the application of the inverse transform:

$$\widehat{i}(p) = \frac{E_0}{L} \frac{p}{(p+\gamma)^2 + \omega^2} = \frac{E_0}{L} \left[\frac{p+\gamma}{(p+\gamma)^2 + \omega^2} - \frac{\gamma}{\omega} \frac{\omega}{(p+\gamma)^2 + \omega^2} \right].$$

Now from the table of LT, the original function $i(t)$ is

$$i(t) = \frac{E_0}{L} e^{-\gamma t} \left[\cos \omega t - \frac{\gamma}{\omega} \sin \omega t \right] \tag{7.28}$$

– the current is damping due to resistance R.

Example 7.6 The support points of two pendulums are connected by a spring that has the stiffness k. Each pendulum has length l and mass m. Find the motions of the pendulums, if both pendulums at the initial time are at equilibrium, and the one of them is given the velocity v, and the second one has zero velocity.

Solution The equations of motion and the initial conditions for the pendulums are

$$\begin{cases} m\ddot{x}_1 = -\dfrac{mg}{l}x_1 + k(x_2 - x_1), \\ m\ddot{x}_2 = -\dfrac{mg}{l}x_2 + k(x_1 - x_2), \end{cases}$$

$$x_1(0) = x_2(0) = 0, \dot{x}_1(0) = v, \dot{x}_2(0) = 0.$$

Applying the LT to each of these equations and using formula (7.11) and the initial conditions, we obtain for the left sides of equations:

$$L[m\ddot{x}_1] = m\left(p^2\widehat{x}_1 - v\right),$$

$$L[m\ddot{x}_2] = mp^2\widehat{x}_2.$$

Thus, after the transformation the system takes the form:

$$\begin{cases} m(p^2\widehat{x}_1 - v) = -\dfrac{mg}{l}\widehat{x}_1 + k(\widehat{x}_2 - \widehat{x}_1), \\ mp^2\widehat{x}_2 = -\dfrac{mg}{l}\widehat{x}_2 + k(\widehat{x}_1 - \widehat{x}_2). \end{cases}$$

From this algebraic system, we find $\widehat{x}_1$:

$$\widehat{x}_1 = \frac{v(p^2 + g/l + k/m)}{(p^2 + g/l + 2k/m)(p^2 + g/l)} = \frac{v}{2}\left(\frac{1}{p^2 + g/l + 2k/m} + \frac{1}{p^2 + g/l}\right).$$

Inverting the transform gives:

$$x_1(t) = \frac{v}{2}\left(\frac{1}{\omega_1}\sin\omega_1 t + \frac{1}{\omega_2}\sin\omega_2 t\right),$$

where $\omega_1 = \sqrt{g/l + 2k/m}$, $\omega_2 = \sqrt{g/l}$ are the frequencies. Equation for $x_2(t)$ and its solution can be found analogously (do it as an *Exercise*).

Problems

Solve the IVPs for differential equations and systems of differential equations using the Laplace Transform:

1. $x'' + 4x = e^t$, $x(0) = 0$, $x'(0) = 0$.
2. $x'' + 25x = e^{-2t}$, $x(0) = 1$, $x'(0) = 0$.
3. $x'' + x = 4\sin t$, $x(0) = 0$, $x'(0) = 0$.
4. $x'' + 16x = 3\cos 2t$, $x(0) = 0$, $x'(0) = 0$.
5. $x'' + x' = te^t$, $x(0) = 0$, $x'(0) = 0$.
6. $x'' + 2x' + 2x = te^{-t}$, $x(0) = 0$, $x'(0) = 0$.
7. $x'' - 9x = e^{3t}\cos t$, $x(0) = 0$, $x'(0) = 0$.
8. $x'' - x = 2e^t - t^2$, $x(0) = 0$, $x'(0) = 0$.

9. $\begin{cases} x' = 5x - 3y, \\ y' = x + y + 5e^{-t}, \end{cases}$ $\begin{cases} x(0) = 0, \\ y(0) = 0. \end{cases}$

10. $\begin{cases} x' + 4y = \sin 2t, \\ y' + x = 0, \end{cases}$ $\begin{cases} x(0) = 0, \\ y(0) = 0. \end{cases}$

11. $\begin{cases} x' + 4y = \cos 2t, \\ y' + 4x = \sin 2t, \end{cases}$ $\begin{cases} x(0) = 0, \\ y(0) = 0. \end{cases}$

12. $\begin{cases} x' = 2x - y, \\ y' = 2y - x - 5e^t \sin t, \end{cases}$ $\begin{cases} x(0) = 0, \\ y(0) = 0. \end{cases}$

13. $\begin{cases} x' = 2y + \cos t, \\ y' = x + y, \end{cases}$ $\begin{cases} x(0) = 0, \\ y(0) = 0. \end{cases}$

14. Solve Eq. (7.27) for a voltage source $E(t) = \begin{cases} E_0 = \text{const}, 0 < t < T \\ 0, \qquad\qquad t > T. \end{cases}$

Chapter 8
Fourier Series

In this chapter we consider the representations of functions in the form of series in trigonometric functions. This approach will be used for solving ODEs in the following chapters.

8.1 Periodic Processes and Periodic Functions

Periodic phenomena are very common. Periodicity means that some process repeats after some time interval T, called the period. Alternating electric current, an object in circular motion and wave phenomena are the examples of periodic physical phenomena. Such processes can be associated with periodic mathematical functions in time, t, which have the property:

$$\varphi(t + T) = \varphi(t).$$

The simplest periodic function is the sine (or cosine) function, $A \sin(\omega t + \alpha)$ (or $A \cos(\omega t + \alpha)$), where ω is the *angular frequency* related to the period by the relationship:

$$\omega = \frac{2\pi}{T} \tag{8.1}$$

(the quantity $f = 1/T$ is the frequency and $\omega t + \alpha$ is the phase).

Using these simple periodic functions, more complex periodic functions can be constructed, as was noted by the French mathematician *Joseph Fourier* (1768–1830).

V. Henner et al., *Ordinary Differential Equations*,
https://doi.org/10.1007/978-3-031-25130-6_8

For example, if we add the functions

$$y_0 = A_0, \quad y_1 = A_1 \sin(\omega t + \alpha_1), \quad y_2 = A_2 \sin(2\omega t + \alpha_2),$$
$$y_3 = A_3 \sin(3\omega t + \alpha_3), \quad \ldots \tag{8.2}$$

with multiple frequencies ω, 2ω, 3ω, ..., i.e. with the periods T, $T/2$, $T/3$,... we obtain a periodic function (with the period T), which, when graphed, can have an appearance very distinct from the graphs of any of the functions in Eq. (8.2).

It is natural to also investigate the inverse problem. Is it possible to represent a given arbitrary periodic function, $\varphi(t)$, with period T by a sum of simple functions such as those in Eq. (8.2)? As we shall see, for a very wide class of functions, the answer to this question is positive, but to do so may require infinitely many functions in Eq. (8.2). In these cases, the periodic function $\varphi(t)$ can be represented by the infinite *trigonometric series*:

$$\varphi(t) = A_0 + A_1 \sin(\omega t + \alpha_1) + A_2 \sin(2\omega t + \alpha_2) + \ldots =$$
$$= A_0 + \sum_{n=1}^{\infty} A_n \sin(n\omega t + \alpha_n), \tag{8.3}$$

where A_n and α_n are constants and $\omega = 2\pi/T$. Each term in Eq. (8.3) is called a *harmonic*, and the decomposition of periodic functions into harmonics is called *harmonic analysis*.

In many cases it is useful to introduce the variable:

$$x = \omega t = \frac{2\pi t}{T}$$

and to work with the functions

$$f(x) = \varphi\left(\frac{x}{\omega}\right),$$

which are also periodic, but with the *standard period* 2π: $f(x + 2\pi) = f(x)$. Using this shorthand, Eq. (8.3) becomes:

$$f(x) = A_0 + A_1 \sin(x + \alpha_1) + A_2 \sin(2x + \alpha_2) + \ldots =$$
$$= A_0 + \sum_{n=1}^{\infty} A_n \sin(nx + \alpha_n). \tag{8.4}$$

With the trigonometric identity $\sin(\alpha + \beta) = \sin\alpha\cos\beta + \cos\alpha\sin\beta$ and the notation

$$A_0 = 2a_0, \quad A_n \sin\alpha_n = a_n, \quad A_n \cos\alpha_n = b_n \quad (n = 1, 2, 3, \ldots),$$

we obtain a standard form for the harmonic analysis of a periodic function $f(x)$ as

$$f(x) = \frac{a_0}{2} + (a_1 \cos x + b_1 \sin x) + (a_2 \cos 2x + b_2 \sin 2x) + \ldots =$$
$$= \frac{a_0}{2} + \sum_{n=1}^{\infty} (a_n \cos nx + b_n \sin nx), \tag{8.5}$$

which is referred to as the *trigonometric Fourier expansion*.

8.2   Fourier Coefficients

To determine the limits of validity for the representation in Eq. (8.5) of a given function $f(x)$ with period 2π, and to find the coefficients a_n and b_n, we follow the approach that was originally elaborated by Fourier. We first assume that the function $f(x)$ can be integrated over the interval $[-\pi, \pi]$. If $f(x)$ is discontinuous at any point, we assume that the integral of $f(x)$ converges, and in this case we also assume that the integral of the absolute value of the function, $|f(x)|$, converges. A function with these properties is said to be *absolutely integrable*. Integrating the expression (8.5) term by term, we obtain:

$$\int_{-\pi}^{\pi} f(x)dx = \pi a_0 + \sum_{n=1}^{\infty} \left[a_n \int_{-\pi}^{\pi} \cos nx dx + b_n \int_{-\pi}^{\pi} \sin nx dx \right].$$

Since

$$\int_{-\pi}^{\pi} \cos nx dx = \frac{\sin nx}{n} \bigg|_{-\pi}^{\pi} = 0 \quad \text{and} \quad \int_{-\pi}^{\pi} \sin nx dx = -\frac{\cos nx}{n} \bigg|_{-\pi}^{\pi} = 0, \tag{8.6}$$

all the terms in the sum are zero, and we obtain:

$$a_0 = \frac{1}{\pi} \int_{-\pi}^{\pi} f(x)dx. \tag{8.7}$$

To find coefficients a_n, we multiply (8.5) by $\cos mx$ and then integrate term by term over the interval $[-\pi, \pi]$:

$$\int_{-\pi}^{\pi} f(x) \cos mx dx = a_0 \int_{-\pi}^{\pi} \cos mx dx + \sum_{n=1}^{\infty}$$

$$\times \left[a_n \int_{-\pi}^{\pi} \cos nx \cos mx dx + b_n \int_{-\pi}^{\pi} \sin nx \cos mx dx \right].$$

The first term is zero as was noted in Eq. (8.6). For any n and m, we also have:

$$\int_{-\pi}^{\pi} \sin nx \cos mx dx = \frac{1}{2} \int_{-\pi}^{\pi} [\sin(n+m)x + \sin(n-m)x] dx = 0, \qquad (8.8)$$

and if $n \neq m$, we obtain:

$$\int_{-\pi}^{\pi} \cos nx \cos mx dx = \frac{1}{2} \int_{-\pi}^{\pi} [\cos(n+m)x + \cos(n-m)x] dx = 0. \qquad (8.9)$$

Using these formulas along with the identity

$$\int_{-\pi}^{\pi} \cos^2 mx dx = \int_{-\pi}^{\pi} \frac{1 + \cos 2mx}{2} dx = \pi \qquad (8.10)$$

we see that all integrals in the sum are zero, except the one with the coefficient a_m. We thus have:

$$a_m = \frac{1}{\pi} \int_{-\pi}^{\pi} f(x) \cos mx dx \quad (m = 1, 2, 3, \ldots). \qquad (8.11)$$

The usefulness of introducing the factor 1/2 in the first term in (8.5) is now apparent since it allows the same formulas to be used for all a_n, including $n = 0$.

Similarly, multiplying Eq. (8.5) by $\sin mx$ and using, along with Eqs. (8.6) and (8.8), two other simple integrals

$$\int_{-\pi}^{\pi} \sin nx \sin mx dx = 0, \quad \text{if } n \neq m, \qquad (8.12)$$

and

$$\int_{-\pi}^{\pi} \sin^2 mx dx = \pi, \quad \text{if } n = m, \qquad (8.13)$$

we obtain the second coefficient:

$$b_m = \frac{1}{\pi} \int_{-\pi}^{\pi} f(x) \sin mx \, dx \quad (m = 1, 2, 3, \ldots). \tag{8.14}$$

Reading Exercise Obtain the same result as in Eqs. (8.8), (8.9) and (8.12) using *Euler's formula*:

$$e^{imx} = \cos mx + i \sin mx.$$

Equations (8.8), (8.9), and (8.12) also indicate that the system of functions

$$1, \cos x, \sin x, \cos 2x, \sin 2x, \ldots, \cos nx, \sin nx, \ldots \tag{8.15}$$

is *orthogonal* on $[-\pi, \pi]$.

It is important to notice that the above system is not orthogonal on the reduced interval $[0, \pi]$ because for n and m with different parity (one odd and the other even), we have:

$$\int_0^{\pi} \sin nx \cos mx \, dx \neq 0.$$

However, the system consisting of cosine functions only

$$1, \cos x, \cos 2x, \ldots, \cos nx, \ldots \tag{8.16}$$

is orthogonal on $[0, \pi]$, and the same is true for

$$\sin x, \sin 2x, \ldots, \sin nx, \ldots \tag{8.17}$$

A second observation, which we will need later, is that on an interval $[0, l]$ of arbitrary length l, both systems of functions

$$1, \cos \frac{\pi x}{l}, \cos \frac{2\pi x}{l}, \ldots, \cos \frac{n\pi x}{l}, \ldots \tag{8.18}$$

and

$$\sin \frac{\pi x}{l}, \sin \frac{2\pi x}{l}, \ldots, \sin \frac{n\pi x}{l}, \ldots \tag{8.19}$$

are orthogonal.

Reading Exercise Prove the above three statements.

Equations (8.7), (8.11), and (8.14) are known as the *Fourier coefficients*, and the series (8.5) with these definitions is called the *Fourier series*. Equation (8.5) is also referred to as the *Fourier expansion* of the function $f(x)$.

Notice that for the function $f(x)$ having period 2π, the integral

$$\int_{\alpha}^{\alpha+2\pi} f(x)dx$$

does not depend on the value of α. As a result, we may also use the following expressions for the Fourier coefficients:

$$a_m = \frac{1}{\pi}\int_0^{2\pi} f(x)\cos mx\,dx \quad \text{and} \quad b_m = \frac{1}{\pi}\int_0^{2\pi} f(x)\sin mx\,dx. \tag{8.20}$$

It is important to realize that to obtain the results above, we used a term-by-term integration of the series, which is justified only if the series converges uniformly. Until we know for sure that the series converges, we can only say that the series (8.5) *corresponds* to the function $f(x)$, which usually is denoted as

$$f(x) \sim \frac{a_0}{2} + \sum_{n=1}^{\infty}(a_n\cos nx + b_n\sin nx).$$

At this point we should remind the reader what is meant by *uniform convergence*. The series $\sum_{n=1}^{\infty} f_n(x)$ converges to the sum $S(x)$ uniformly on the interval $[a,b]$ if, for any arbitrarily small $\varepsilon > 0$, we can find a number N such that for all $n \geq N$, the remainder of the series $\left|\sum_{n=N}^{\infty} f_n(x)\right| \leq \varepsilon$ for all $x \in [a,b]$. This indicates that the series approaches its sum uniformly with respect to x.

The most important features of a uniformly converging series are:

(i) If $f_n(x)$ for any n is a continuous function, then $S(x)$ is also a continuous function.

(ii) The equality $\sum_{n=1}^{\infty} f_n(x) = S(x)$ can be integrated term by term along any subinterval within the interval $[a,b]$.

(iii) If the series $\sum_{n=1}^{\infty} f'_n(x)$ converges uniformly, then its sum is equal to $S'(x)$; i.e., the formula $\sum_{n=1}^{\infty} f_n(x) = S(x)$ can be differentiated term by term.

There is a simple and very practical criterion for convergence (which has been already discussed in Chap. 6) established by *Karl Weierstrass* (a German mathematician (1815–1897)), that says that if $|f_n(x)| < c_n$ for each term $f_n(x)$ in the series defined on the interval $x \in [a, b]$ (i.e. $f_n(x)$ is limited by c_n), where $\sum\limits_{n=1}^{\infty} c_n$ is a converging number series, then the series $\sum\limits_{n=1}^{\infty} f_n(x)$ converges uniformly on $[a, b]$. For example, the number series $\sum\limits_{n=1}^{\infty} 1/n^2$ is known to converge, so any trigonometric series with terms such as $\sin nx/n^2$ or similar will converge uniformly for all x because $|\sin nx/n^2| \leq 1/n^2$.

8.3 Convergence of Fourier Series

In this section we study the range of validity of Eq. (8.5) with Fourier coefficients given by Eqs. (8.11) and (8.14). To start, it is clear that if the function $f(x)$ is finite on $[-\pi, \pi]$, then the Fourier coefficients are bounded. This is easily verified, for instance, for a_n, since

$$| a_n | = \frac{1}{\pi} | \int_{-\pi}^{\pi} f(x) \cos nx dx | \leq \frac{1}{\pi} \int_{-\pi}^{\pi} |f(x)| \cdot | \cos nx | dx \leq \frac{1}{\pi} \int_{-\pi}^{\pi} |f(x)| dx.$$

$$(8.21)$$

The same result is valid in cases where $f(x)$ is not finite but is absolutely integrable, i.e., the integral of its absolute value converges:

$$\int_{-\pi}^{\pi} |f(x)| dx < \infty. \tag{8.22}$$

The *necessary condition* for the series convergence is that its terms tend to zero as $n \to \infty$. Because the absolute values of sine and cosine functions are bounded by the negative one and the positive one, the necessary condition for convergence of the trigonometric series in Eq. (8.5) is that coefficients of expansion a_n and b_n tend to zero as $n \to \infty$. This condition is valid for functions that are absolutely integrable, which is clear from the following lemma, proved by the famed German mathematician Georg Riemann (1826–1866):

Riemann's Lemma

If the function f(t) is absolutely integrable on [a, b], *then*

$$\lim_{\alpha \to \infty} \int_a^b f(t) \sin \alpha t\, dt = 0 \quad \text{and} \quad \lim_{\alpha \to \infty} \int_a^b f(t) \cos \alpha t\, dt = 0. \qquad (8.23)$$

We will not prove this lemma rigorously, but its meaning should be obvious. In the case of very fast oscillations, the sine and cosine functions change their sign very quickly as $\alpha \to \infty$. Thus, these integrals vanish for "reasonable" (i.e., absolutely integrable) functions because they do not change substantially as the sine and cosine alternate with opposite signs in their semi-periods.

Thus, for absolutely integrable functions, the necessary condition of convergence of Fourier series is satisfied. Before we discuss the problem of convergence of Fourier series in more detail, let us notice that almost any function significant for applications can be expanded in a converging Fourier series.

It is important to know how quickly the terms in (8.5) decrease as $n \to \infty$. If they decrease rapidly, the series converges rapidly. In this case, using very few terms, we have a good trigonometric approximation for $f(x)$, and the partial sum of the series, $S_n(x)$, is a good approximation to the sum $S(x) = f(x)$. If the series converges more slowly, a larger number of terms are needed to have a sufficiently accurate approximation.

Assuming that the series (8.5) converges, the speed of its convergence to $f(x)$ depends on the behavior of $f(x)$ over its period, or, in the case of nonperiodic functions, on the way it is extended from the interval [a, b] to the entire axis x, as we will discuss below. Convergence is most rapid for very smooth functions (functions which have continuous derivatives of higher order). Discontinuities in the derivative of the function, $f'(x)$, substantially reduce the rate of convergence, whereas discontinuities in $f(x)$ reduce the convergence rate even more with the result that many terms in the Fourier series must be used to approximate the function $f(x)$ with the needed precision. This should be fairly obvious, since the "smoothness" of $f(x)$ determines the rate of decrease of the coefficients a_n and b_n.

It can be shown [1] that the coefficients decrease:

1. Faster than $1/n^2$ (e.g., $1/n^3$) when $f(x)$ and $f'(x)$ are continuous but $f''(x)$ has a discontinuity;
2. At about the same rate as $1/n^2$ when $f(x)$ is continuous but $f'(x)$ has discontinuities;
3. At a rate similar to the rate of convergence of $1/n$ if $f(x)$ is discontinuous.

It is important to note that in the first two cases, the series converges uniformly, which follows from the Weierstrass criterion (named after German mathematician Karl Weierstrass, 1815–1897), because each term of Eq. (8.5) is bounded by the corresponding term in the converging numeric series $\sum_{n=1}^{\infty} \frac{1}{n^2} < \infty$.

The following very important Dirichlet theorem (German mathematician Peter Gustav Dirichlet (1805–1859) provided a rigorous foundation for the theory of Fourier series) describes the convergence of the Fourier series given in Eq. (8.5) for a function $f(x)$ at a point x_0, where $f(x)$ is continuous or where it may have a discontinuity (the proof can be found in the book [7]).

The Dirichlet Theorem

If the function $f(x)$ with period 2π is piecewise continuous in $[-\pi, \pi]$, then its Fourier series converges to $f(x_0)$ when x_0 is a continuity point, and to

$$\frac{f(x_0 + 0) + f(x_0 - 0)}{2}$$

if x_0 is a point of discontinuity.
At the ends of the interval $[-\pi, \pi]$, the Fourier series converges to

$$\frac{f(-\pi + 0) + f(\pi - 0)}{2}.$$

The definition of a function $f(x)$ *piecewise continuous* in $[a, b]$ mentioned in the theorem above is as follows:

(i) It is continuous on $[a, b]$ except perhaps at a finite number of points.
(ii) If x_0 is one such point, then the left and right limits of $f(x)$ at x_0 exist and are finite.
(iii) Both the limit from the right of $f(x)$ at a and the limit from the left at b exist and are finite.

Stated more briefly, the Fourier series of a piecewise continuous function $f(x)$ converge to $f(x)$ in the points of continuity, but in the points of discontinuity, it converges to another value.

8.4 Fourier Series for Nonperiodic Functions

We assumed above that the function $f(x)$ is defined on the entire x-axis and has the period 2π. But very often we need to deal with nonperiodic functions defined only on the interval $[-\pi, \pi]$. The theory elaborated above can still be used if we extend $f(x)$ periodically from $(-\pi, \pi)$ to all x. In other words we assign the same values of $f(x)$ to all the intervals $(\pi, 3\pi)$, $(3\pi, 5\pi)$, ..., $(-3\pi, \pi)$, $(-5\pi, -3\pi)$, ... and then use Eqs. (8.11) and (8.14) for the Fourier coefficients of this new function, which is periodic. Many examples of such extensions will be given below. If $f(-\pi) = f(\pi)$, we can include the end points, $x = \pm \pi$ and the Fourier series converges to $f(x)$

everywhere on $[-\pi, \pi]$. Over the entire axis, the expansion gives a periodic extension of the given function $f(x)$ defined originally on $[-\pi, \pi]$. In many cases $f(-\pi) \neq f(\pi)$, and the Fourier series at the ends of the interval $[-\pi, \pi]$ converges to

$$\frac{f(-\pi) + f(\pi)}{2},$$

which differs from both $f(-\pi)$ and $f(\pi)$.

The rate of convergence of the Fourier series depends on the discontinuities of the function and derivatives of the function after its extension to the entire axis. Some extensions do not increase the number of discontinuities of the original function, whereas others do increase this number. In the latter case, the rate of convergence is reduced. Among the examples given later in this chapter, Examples 8.2 and 8.3 are of the Fourier series of $f(x) = x$ on the interval $[0, \pi]$. In the first expansion, the function is extended to the entire axis as an even function and remains continuous, so that the coefficients of the Fourier series decrease as $1/n^2$. In the second example, this function is extended as an odd function and has discontinuities at $x = k\pi$ (integer k), in which case the coefficients decrease slower, as $1/n$.

8.5   Fourier Expansions on Intervals of Arbitrary Length

Suppose that a function $f(x)$ is defined on some interval $[-l, l]$ of arbitrary length $2l$ (where $l > 0$). Using the substitution

$$x = \frac{ly}{\pi} \qquad (-\pi \leq y \leq \pi),$$

we obtain the function $f\left(\frac{yl}{\pi}\right)$ of the variable y on the interval $[-\pi, \pi]$, which can be expanded using the standard Eqs. (8.5), (8.11), and (8.14) as

$$f\left(\frac{yl}{\pi}\right) = \frac{a_0}{2} + \sum_{n=1}^{\infty} (a_n \cos ny + b_n \sin ny),$$

with

$$a_n = \frac{1}{\pi} \int_{-\pi}^{\pi} f\left(\frac{yl}{\pi}\right) \cos ny\, dy \quad \text{and} \quad b_n = \frac{1}{\pi} \int_{-\pi}^{\pi} f\left(\frac{yl}{\pi}\right) \sin ny\, dy.$$

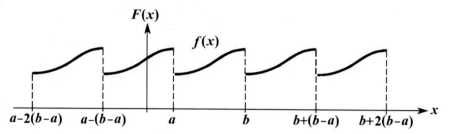

Fig. 8.1 Arbitrary function $f(x)$ defined on the interval $[a, b]$ extended to the x-axis as the function $F(x)$

Returning to the variable x, we obtain:

$$f(x) = \frac{a_0}{2} + \sum_{n=1}^{\infty} \left(a_n \cos \frac{n\pi x}{l} + b_n \sin \frac{n\pi x}{l} \right) \tag{8.24}$$

with

$$a_n = \frac{1}{l} \int_{-l}^{l} f(x) \cos \frac{n\pi x}{l} dx, \qquad n = 0, 1, 2, \ldots,$$

$$b_n = \frac{1}{l} \int_{-l}^{l} f(x) \sin \frac{n\pi x}{l} dx, \qquad n = 1, 2, \ldots. \tag{8.25}$$

If the function is given on an arbitrary interval of length $2l$, for instance $[0, 2l]$, the formulas for the coefficients of the Fourier series (8.24) become:

$$a_n = \frac{1}{l} \int_{0}^{2l} f(x) \cos \frac{n\pi x}{l} dx \quad \text{and} \quad b_n = \frac{1}{l} \int_{0}^{2l} f(x) \sin \frac{n\pi x}{l} dx. \tag{8.26}$$

In both cases, the series in Eqs. (8.24) (5.1) gives a periodic function with the period $T = 2l$.

If the function $f(x)$ is given on an interval $[a, b]$ (where a and b may have the same or opposite sign, i.e., the interval $[a, b]$ can include or exclude the point $x = 0$), different periodic continuations onto the entire x-axis may be constructed (see Fig. 8.1). As an example, consider the periodic continuation $F(x)$ of the function $f(x)$, defined by the condition:

$$F(x + n(b - a)) = f(x), \qquad n = 0, \pm 1, \pm 2, \ldots \qquad \text{for all } x.$$

In this case the Fourier series is given by Eq. (8.24) where $2l = b - a$. Clearly, instead of Eq. (8.25), the following formulas for the Fourier coefficients should be used:

$$a_n = \frac{2}{b-a} \int_a^b f(x) \cos \frac{2n\pi x}{b-a} dx, \quad b_n = \frac{2}{b-a} \int_a^b f(x) \sin \frac{2n\pi x}{b-a} dx. \quad (8.27)$$

The series in Eq. (8.24) gives a periodic function with the period $T = 2l = b - a$; however, the original function was defined only on the interval $[a, b]$ and is not periodic in general.

8.6   Fourier Series in Cosine or in Sine Functions

Suppose that $f(x)$ is an *even* function on $[-\pi, \pi]$, so that $f(x) \sin nx$ is odd. For this case

$$b_n = \frac{1}{\pi} \int_{-\pi}^{\pi} f(x) \sin nx dx = 0,$$

since the integral of an odd function over a symmetric interval equals zero. Coefficients a_n can be written as

$$a_n = \frac{1}{\pi} \int_{-\pi}^{\pi} f(x) \cos nx dx = \frac{2}{\pi} \int_0^{\pi} f(x) \cos nx dx, \quad (8.28)$$

since the integrand is even. Thus, for an even function $f(x)$, we may write:

$$f(x) = \frac{a_0}{2} + \sum_{n=1}^{\infty} a_n \cos nx. \quad (8.29)$$

Similarly, if $f(x)$ is an *odd* function, we have:

$$a_n = \frac{1}{\pi} \int_{-\pi}^{\pi} f(x) \cos nx dx = 0 \quad \text{and} \quad b_n = \frac{2}{\pi} \int_0^{\pi} f(x) \sin nx dx, \quad (8.30)$$

in which case we have

$$f(x) = \sum_{n=1}^{\infty} b_n \sin nx. \tag{8.31}$$

Thus, an even function on $[-\pi, \pi]$ is expanded in the set (8.17):

$$1, \cos x, \cos 2x, \ldots, \cos nx, \ldots$$

An odd function on $[-\pi, \pi]$ is expanded in the set (8.16):

$$\sin x, \sin 2x, \ldots, \sin nx, \ldots$$

Any function can be presented as a sum of even and odd functions:

$$f(x) = f_1(x) + f_2(x),$$

where

$$f_1(x) = \frac{f(x) + f(-x)}{2} \quad \text{and} \quad f_2(x) = \frac{f(x) - f(-x)}{2},$$

in which case $f_1(x)$ can be expanded into a cosine Fourier series and $f_2(x)$ into a sine series.

If the function $f(x)$ is defined only on the interval $[0, \pi]$, we can extend it to the interval $[-\pi, 0)$. This extension may be done in different ways, corresponding to different Fourier series. In particular, such extension can make $f(x)$ even or odd on $[-\pi, \pi]$, which leads to a cosine or sine series with a period 2π. In the first case on the interval $[-\pi, 0)$, we have:

$$f(-x) = f(x), \tag{8.32}$$

and in the second case

$$f(-x) = -f(x). \tag{8.33}$$

The points $x = 0$ and $x = \pi$ need special consideration because the sine and cosine series behave differently at these points. If $f(x)$ is continuous at these points, because of Eqs. (8.32) and (8.33), the cosine series converges to $f(0)$ at $x = 0$ and to $f(\pi)$ at $x = \pi$. The situation is different for the sine series, however. At $x = 0$ and $x = \pi$, the sum of the sine series in Eq. (8.31) is zero; thus, the series is equal to the functions $f(0)$ and $f(\pi)$, respectively, only when these values are zero.

If $f(x)$ is given on the interval $[0, l]$ (where $l > 0$), the cosine and sine series are

$$\frac{a_0}{2} + \sum_{n=1}^{\infty} a_n \cos \frac{n\pi x}{l} \tag{8.34}$$

and

$$\sum_{n=1}^{\infty} b_n \sin \frac{n\pi x}{l} \tag{8.35}$$

with the coefficients

$$a_n = \frac{2}{l} \int_0^l f(x) \cos \frac{n\pi x}{l} dx, \tag{8.36}$$

or

$$b_n = \frac{2}{l} \int_0^l f(x) \sin \frac{n\pi x}{l} dx. \tag{8.37}$$

To summarize the above discussion, we see that the Fourier series provides a way to obtain an *analytic formula* for functions defined by different formulas on different intervals by combining these intervals into a larger one. Such analytic formulas replace a discontinuous function by a continuous Fourier series expansion, which is often more convenient in a given application. As we have seen above, there are often many different choices of how to extend the original function, defined initially on an interval, to the entire axis. A specific choice of an extension depends on the application wherein the expansion is to be used. Many examples and problems demonstrating these points will be presented in the examples at the end of each of the following sections and the problems at the end of this chapter.

8.7   Examples

All the functions given below are differentiable or piecewise differentiable and can be represented by Fourier series.

Example 8.1   Find the cosine series for $f(x) = x^2$ on the interval $[-\pi, \pi]$.

Solution   The coefficients are

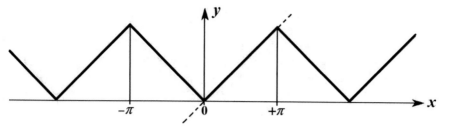

Fig. 8.2 The function $f(x) = x$ extended to the x-axis

$$\frac{1}{2}a_0 = \frac{1}{\pi} \int\limits_0^\pi x^2 dx = \frac{\pi^2}{3},$$

$$a_n = \frac{2}{\pi} \int\limits_0^\pi x^2 \cos nx dx = \frac{2}{\pi} x^2 \frac{\sin nx}{n} \Big|_0^\pi - \frac{4}{n\pi} \int\limits_0^\pi x \sin nx dx =$$

$$= \frac{4}{n\pi} x \frac{\cos nx}{n} \Big|_0^\pi - \frac{4}{n^2\pi} \int\limits_0^\pi \cos nx dx = (-1)^n \frac{4}{n^2}.$$

Thus,

$$x^2 = \frac{\pi^2}{3} + 4 \sum_{n=1}^\infty (-1)^n \frac{\cos nx}{n^2} \quad (-\pi \le x \le \pi). \tag{8.38}$$

In the case where $x = \pi$, we obtain a famous expansion:

$$\frac{\pi^2}{6} = \sum_{n=1}^\infty \frac{1}{n^2}. \tag{8.39}$$

Example 8.2 Let the function $f(x) = x$ be given on the interval $[0, \pi]$. Find the Fourier cosine series.

Solution Figure 8.2 gives an even periodic extension of $f(x) = x$ from $[0, \pi]$ onto the entire axis, and it also represents the sum of the series, Eq. (8.42), obtained below. For coefficients we have:

$$\frac{1}{2}a_0 = \frac{1}{\pi} \int\limits_0^\pi x dx = \frac{\pi}{2},$$

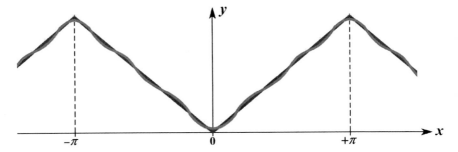

Fig. 8.3 Original function extended to the x-axis plotted together with the partial sum of the first five terms

$$a_n = \frac{2}{\pi} \int_0^\pi x \cos nx\, dx = \frac{2}{\pi} x \frac{\sin nx}{n}\Big|_0^\pi - \frac{2}{n\pi} \int_0^\pi \sin nx\, dx =$$

$$= 2\frac{\cos n\pi - 1}{n^2\pi} = 2\frac{(-1)^n - 1}{n^2\pi} \qquad (n > 0);$$

that is,

$$a_{2k} = 0, \qquad a_{2k-1} = -\frac{4}{(2k-1)^2\pi}, \qquad (k = 1, 2, 3, \ldots),$$

and thus,

$$x = \frac{\pi}{2} - \frac{4}{\pi} \sum_{k=1}^\infty \frac{\cos(2k-1)x}{(2k-1)^2} \qquad (0 \le x \le \pi). \tag{8.40}$$

Figure 8.3 shows the graph of the partial sum:

$$y = S_5(x) = \frac{\pi}{2} - \frac{4}{\pi}\left(\cos x + \frac{1}{3^2}\cos 3x + \frac{1}{5^2}\cos 5x\right)$$

together with the graph of the extended function.

Example 8.3 Find the Fourier series for $f(x) = \frac{\pi - x}{2}$ on the interval $(0, 2\pi)$.

Solution The coefficients are

$$a_0 = \frac{1}{\pi} \int_0^{2\pi} \frac{\pi - x}{2}\, dx = \frac{1}{2\pi}\left(\pi x - \frac{1}{2}x^2\right)\Big|_0^{2\pi} = 0,$$

$$a_n = \frac{1}{\pi} \int_0^{2\pi} \frac{\pi - x}{2} \cos nx\, dx = \frac{1}{2\pi}(\pi - x)\frac{\sin nx}{n}\Big|_0^{2\pi} - \frac{1}{2n\pi}\int_0^{2\pi}\sin nx\, dx = 0,$$

$$b_n = \frac{1}{\pi} \int_0^{2\pi} \frac{\pi - x}{2} \sin nx\, dx = -\frac{1}{2\pi}(\pi - x)\frac{\cos nx}{n}\Big|_0^{2\pi} - \frac{1}{2n\pi}\int_0^{2\pi}\cos nx\, dx = \frac{1}{n}.$$

Thus, we get the interesting result:

$$\frac{\pi - x}{2} = \sum_{n=1}^{\infty} \frac{\sin nx}{n} \quad (0 < x < 2\pi). \tag{8.41}$$

This equation is not valid at $x = 0$ and $x = 2\pi$ because the sum of the series equals zero. The equality also does not hold beyond $(0, 2\pi)$.

For $x = \frac{\pi}{2}$ we have an interesting result that was obtained by *Gottfried Leibniz* (1646–1716) using another technique:

$$\frac{\pi}{4} = 1 - \frac{1}{3} + \frac{1}{5} - \frac{1}{7} + \cdots \tag{8.42}$$

And for $x = \frac{\pi}{6}$, we obtain another representation of π:

$$\frac{\pi}{4} = 1 + \frac{1}{5} - \frac{1}{7} - \frac{1}{11} + \frac{1}{13} + \frac{1}{17} - \cdots \tag{8.43}$$

8.8   The Complex Form of the Trigonometric Series

For a real function $f(x)$ with a period 2π, the Fourier expansion

$$f(x) = \frac{a_0}{2} + \sum_{n=1}^{\infty}(a_n \cos nx + b_n \sin nx) \tag{8.44}$$

with

$$a_n = \frac{1}{\pi} \int\limits_{-\pi}^{\pi} f(x) \cos nx dx, \qquad (n = 0, 1, 2, \ldots),$$

$$b_n = \frac{1}{\pi} \int\limits_{-\pi}^{\pi} f(x) \sin nx dx, \qquad (n = 1, 2, 3, \ldots)$$

(8.45)

can be rewritten in a complex form. From Euler's formula

$$e^{iax} = \cos ax + i \sin ax$$

(8.46)

we have:

$$\cos nx = \frac{1}{2} \left(e^{inx} + e^{-inx} \right),$$

$$\sin nx = \frac{1}{2i} \left(e^{inx} - e^{-inx} \right) = \frac{i}{2} \left(e^{-inx} - e^{inx} \right),$$

from which we obtain

$$f(x) = \frac{a_0}{2} + \sum_{n=1}^{\infty} \left[\frac{1}{2} (a_n + b_n i) e^{-inx} + \frac{1}{2} (a_n - b_n i) e^{inx} \right].$$

Using the notations

$$c_0 = \frac{1}{2} a_0, \qquad c_n = \frac{1}{2} (a_n - b_n i), \qquad c_{-n} = \frac{1}{2} (a_n + b_n i),$$

we have:

$$f(x) = \sum_{n=-\infty}^{\infty} c_n e^{inx}.$$

(8.47)

With the Fourier equations for a_n and b_n (8.45), it is easy to see that the coefficients c_n can be written as

$$c_n = \frac{1}{2\pi} \int\limits_{-\pi}^{\pi} f(x) e^{-inx} dx \quad (n = 0, \pm 1, \pm 2, \ldots).$$

(8.48)

For functions with a period of $2l$, equations similar to (8.47) and (8.48) have the form:

$$f(x) = \sum_{n=-\infty}^{\infty} c_n e^{\frac{in\pi x}{l}} \tag{8.49}$$

and

$$c_n = \frac{1}{2l} \int_{-l}^{l} f(x) e^{-\frac{in\pi x}{l}} dx \qquad (n = 0, \pm 1, \pm 2, \ldots). \tag{8.50}$$

For periodic functions in time t with a period T, the same formulas can be written as

$$f(t) = \sum_{n=-\infty}^{\infty} c_n e^{\frac{2in\pi t}{T}} = \sum_{n=-\infty}^{\infty} c_n e^{in\omega t}, \qquad \omega = 2\pi/T \tag{8.51}$$

and

$$c_n = \frac{1}{T} \int_{-T/2}^{T/2} f(t) e^{-\frac{2in\pi t}{T}} dt = \frac{1}{T} \int_{-T/2}^{T/2} f(t) e^{-in\omega t} dt \qquad (n = 0, \pm 1, \pm 2, \ldots). \tag{8.52}$$

Several useful properties of these results can be easily verified:

(i) Because $f(x)$ is real, c_n and c_{-n} are complex conjugate and we have $c_{-n} = c_n^*$;
(ii) If $f(x)$ is even, all c_n are real;
(iii) If $f(x)$ is odd, $c_0 = 0$ and all c_n are pure imaginary.

If a function $f(x)$ of the real variable x is complex, we have:

$$f(x) = f_1(x) + if_2(x),$$

where $f_1(x), f_2(x)$ are real functions, in which case the Fourier series for $f(x)$ is the sum of Fourier series for $f_1(x)$ and $f_2(x)$, where the second series is multiplied by the imaginary number i. Equations (8.47) and (8.48) remain unchanged, but the above three properties of the coefficients are not valid (in particular, the coefficients c_n and c_{-n} are not complex conjugate). Instead of the above properties in this case, we have:

(i) If $f(x)$ is even, then $c_{-n} = c_n$;
(ii) If $f(x)$ is odd, then $c_{-n} = -c_n$.

Example 8.4 Represent the function:

$$f(x) = \begin{cases} 0, & -\pi \le x \le 0, \\ 1, & 0 < x \le \pi, \end{cases}$$

by a complex Fourier series.

Solution The coefficients are

$$c_0 = \frac{1}{2\pi} \int_0^\pi dx = \frac{1}{2},$$

$$c_n = \frac{1}{2\pi} \int_0^\pi e^{-inx} dx = \frac{1 - e^{-in\pi}}{2\pi ni} = \begin{cases} 0, & n = \text{even}, \\ \dfrac{1}{\pi ni}, & n = \text{odd}. \end{cases}$$

Thus,

$$f(x) = \frac{1}{2} + \frac{1}{\pi i} \sum_{\substack{n = -\infty \\ n = \text{odd}}}^{+\infty} \frac{1}{n} e^{inx}.$$

Reading Exercise Using Euler's formula, check that from this expression it follows that

$$\text{Im} f(x) = 0 \quad \text{(as expected) and} \quad \text{Re} f(x) = \frac{1}{2} + \frac{2}{\pi} \sum_{n = 1, 3, \ldots}^\infty \frac{\sin nx}{n}.$$

The same result can be obtained if we apply the real form of the Fourier series from the beginning.

Example 8.5 Find the Fourier series of the function $f(x) = e^{-x}$ on the interval $(-\pi, \pi)$.

Solution First use the complex Fourier series with the coefficients:

$$c_n = \frac{1}{2\pi} \int_{-\pi}^\pi e^{-x} e^{-inx} dx = \frac{1}{2\pi} \int_{-\pi}^\pi e^{-(1+in)x} dx = \frac{e^\pi e^{in\pi} - e^{-\pi} e^{-in\pi}}{2\pi(1 + in)}.$$

Then, with $e^{\pm in\pi} = \cos n\pi \pm i \sin n\pi = (-1)^n$, we have $c_n = \frac{(-1)^n(e^\pi - e^{-\pi})}{2\pi(1+in)}$; thus,

$$e^{-x} = \sum_{n=-\infty}^{\infty} c_n e^{\frac{in\pi x}{l}} = \frac{e^{\pi} - e^{-\pi}}{2\pi} \sum_{n=-\infty}^{\infty} \frac{(-1)^n e^{inx}}{1+in}.$$

On the interval $(-\pi, \pi)$, this series converges to e^{-x}, and at the points $x = \pm \pi$, its sum is $(e^{\pi} + e^{-\pi})/2$.

Reading Exercise Apply Euler's formula and check that this series in real form becomes

$$e^{-x} = \frac{e^{\pi} - e^{-\pi}}{\pi} \left[\frac{1}{2} + \sum_{n=1}^{\infty} \frac{(-1)^n}{1+n^2} (\cos nx + n \sin nx) \right].$$

The same result is obtained if we apply the real form of the Fourier series from the beginning.

8.9 Fourier Series for Functions of Several Variables

In this section we extend the previous ideas to generate the Fourier series for functions of two variables, $f(x, y)$, which has a period 2π in both variables x and y. Analogous to the development of Eq. (8.47), we write a double Fourier series for the function $f(x, y)$ as

$$f(x, y) = \sum_{n, m = -\infty}^{+\infty} \alpha_{nm} e^{i(nx+my)} \tag{8.53}$$

in the domain $(D) = (-\pi \le x \le \pi, -\pi \le y \le \pi)$.

The coefficients α_{nm} can be obtained by multiplying Eq. (8.53) by $e^{-i(nx + my)}$ and integrating over the domain (D), performing this integration for the series term by term. Because the functions e^{inx} are orthogonal on the interval $[-\pi, \pi]$ (same is true for e^{imy}), we obtain:

$$\alpha_{nm} = \frac{1}{4\pi^2} \iint_{(D)} f(x, y) e^{-i(nx+my)} dx dy \quad (n, m = 0, \pm 1, \pm 2, \ldots). \tag{8.54}$$

The previous two formulas give the Fourier series for $f(x, y)$ in the complex form. For the real Fourier series instead of Eq. (8.53), we have:

$$f(x, y) = \sum_{n,m=0}^{+\infty} [a_{nm} \cos nx \cos my + b_{nm} \cos nx \sin my$$
$$+ c_{nm} \sin nx \cos my + d_{nm} \sin nx \sin my], \tag{8.55}$$

where

$$a_{00} = \frac{1}{4\pi^2} \iint\limits_{(D)} f(x, y) dxdy, \qquad a_{n0} = \frac{1}{2\pi^2} \iint\limits_{(D)} f(x, y) \cos nx dxdy,$$

$$a_{0m} = \frac{1}{2\pi^2} \iint\limits_{(D)} f(x, y) \cos my dxdy, \quad a_{nm} = \frac{1}{\pi^2} \iint\limits_{(D)} f(x, y) \cos nx \cos my dxdy,$$

$$b_{0m} = \frac{1}{2\pi^2} \iint\limits_{(D)} f(x, y) \sin my dxdy, \quad b_{nm} = \frac{1}{\pi^2} \iint\limits_{(D)} f(x, y) \cos nx \sin my dxdy,$$

$$c_{n0} = \frac{1}{2\pi^2} \iint\limits_{(D)} f(x, y) \sin nx dxdy, \quad c_{nm} = \frac{1}{\pi^2} \iint\limits_{(D)} f(x, y) \sin nx \cos my dxdy,$$

$$d_{nm} = \frac{1}{\pi^2} \iint\limits_{(D)} f(x, y) \sin nx \sin my dxdy \quad \text{for all} \quad n, m = 1, 2, 3, \ldots$$

$$\tag{8.56}$$

8.10 Generalized Fourier Series

Consider expansions similar to trigonometric Fourier series using a set of orthogonal functions as a basis for the expansion. Recall that two complex functions, $\varphi(x)$ and $\psi(x)$, of a real variable x are orthogonal on the interval $[a, b]$ (which can be an infinite interval) if

$$\int_a^b \varphi(x)\psi^*(x)dx = 0, \tag{8.57}$$

where $\psi^*(x)$ is the complex conjugate of $\psi(x)$ (when $\psi(x)$ is real, $\psi^* = \psi$).

Let us expand some function $f(x)$ into a set of orthogonal functions $\{\varphi_n(x)\}$:

$$f(x) = c_1\varphi_1(x) + c_2\varphi_2(x) + \ldots + c_n\varphi_n(x) + \ldots = \sum_{n=1}^{\infty} c_n\varphi_n(x). \tag{8.58}$$

Multiplying by $\varphi_n(x)$, integrating, and using the orthogonality condition, we obtain the coefficients:

$$c_n = \frac{\int_a^b f(x)\varphi_n^*(x)dx}{\int_a^b \varphi_n(x)\varphi_n^*(x)dx} = \frac{1}{\lambda_n}\int_a^b f(x)\varphi_n^*(x)dx, \tag{8.59}$$

where

$$\|\varphi_n\|^2 = \int_a^b |\varphi_n(x)|^2 dx$$

are real numbers, e.g., the squared *norms* of the functions $\varphi_n(x)$.

Series (8.58) with coefficients (8.59) is called a *generalized Fourier series*.

If the set $\{\varphi_n(x)\}$ is normalized, then $\|\varphi_n\|^2 = 1$, and the previous formula becomes:

$$c_n = \int_a^b f(x)\varphi_n^*(x)dx. \tag{8.60}$$

Let us multiply Eq. (8.58) by its complex conjugate, $f^*(x) = \sum_{n=1}^{\infty} c_n^*\varphi_n^*(x)$, and integrate over the interval $[a, b]$ (or the entire axis). This gives, due to the orthogonality of the functions $\{\varphi_n(x)\}$:

$$\int_a^b |f(x)|^2 dx = \sum_{n=1}^{\infty} |c_n|^2 \int_a^b |\varphi_n(x)|^2 dx = \sum_{n=1}^{\infty} |c_n|^2. \tag{8.61}$$

Equation (8.61) is known as the *completeness equation* or *Parseval's identity*, which the French mathematician Marc-Antoine Parseval stated in 1799. If this equation is satisfied, the set of functions $\{\varphi_n(x)\}$ is *complete*. Omitting any function from that set will make it incomplete.

Equation (8.61) is an extension of the Pythagorean theorem to a space with an infinite number of dimensions; the square of the diagonal of an (infinite dimensional) parallelepiped is equal to the sum of the squares of its sides.

The completeness of the set $\{\varphi_n(x)\}$ means that any function $f(x)$ (for which $\int_a^b |f(x)|^2 dx < \infty$) can be expanded in this set (formula (8.58)) and no other functions except $\{\varphi_n(x)\}$ need to be included.

Note that from formula (8.61), it follows that $c_n \to 0$ as $n \to \infty$.

In the case of a trigonometric Fourier series (8.5) on $[-\pi, \pi]$, the complete orthogonal set of functions $\varphi_n(x)$ is

$$1, \ \cos x, \ \sin x, \ \cos 2x, \ \sin 2x, \ \ldots, \ \cos nx, \ \sin nx, \ \ldots,$$

and Eq. (8.61) becomes:

$$\int_{-\pi}^{\pi} f^2(x)dx = \frac{1}{2}\pi a_0^2 + \pi \sum_{n=1}^{\infty} \left(a_n^2 + b_n^2\right). \tag{8.62}$$

For a series on the interval $[-l, l]$, Eq. (8.61) reads:

$$\int_{-l}^{l} f^2(x)dx = \frac{1}{2}la_0^2 + l \sum_{n=1}^{\infty} \left(a_n^2 + b_n^2\right). \tag{8.63}$$

Other sets of complete orthogonal functions are a system of sines (8.16), or a system of cosines (8.17) on the interval $[0, \pi]$.

Parseval's identity has a clear meaning in the context of a *spectral resolution (or a harmonic analysis)*.

Every oscillatory process can be presented as a superposition of or expansion in monochromatic oscillations with various frequencies. A periodic oscillation with a period T is a superposition of a discrete set of monochromatic oscillations with frequencies that are integer multiples of a "fundamental" frequency $\omega = 2\pi/T$, where T is the period. We can write it in the form (8.51):

$$f(t) = \sum_{n=-\infty}^{\infty} c_n e^{in\omega t}.$$

The instantaneous intensity, or energy of the oscillator, is $|f(t)|^2$. When this series is squared and averaged over the time, $\left|\bar{f}(t)\right|^2 = \int_0^{\infty} |f(t)|^2 dt$, the products of the terms with different frequencies vanish, since they contain oscillating factors. Therefore, the average intensity is the sum of the intensities, or energies, of its monochromatic components:

$$|\bar{f}(t)|^2 = \sum_{n=1}^{\infty} |c_n|^2.$$

From a mathematical point of view, this is equivalent to Parseval's identity (8.61).

8.11 The Gibbs Phenomenon

In this section we take a closer look at the behavior of the Fourier series of a function $f(x)$ near a point of discontinuity (a finite jump) of the function. At these points the series cannot converge uniformly, and, in addition, partial sums exhibit specific defects.

Let us begin with an example. The Fourier series for the function

$$f(x) = \begin{cases} -\pi/2, & \text{if} \quad -\pi < x < 0, \\ 0, & \text{if} \quad x = 0, \pm\pi, \\ \pi/2, & \text{if} \quad 0 < x < \pi, \end{cases}$$

is

$$2\sum_{n=1}^{\infty} \frac{\sin(2n-1)x}{2n-1} = 2\left[\sin x + \frac{\sin 3x}{3} + \frac{\sin 5x}{5} + \dots\right]. \tag{8.64}$$

This expansion gives (an odd) continuation of the function $f(x)$ from the interval $(-\pi/2, \pi/2)$ to the entire x-axis. Because of the periodicity, we can restrict the analysis to the interval $(0, \pi/2)$. The partial sums, shown in Fig. 8.4, have bumps near the points $x = 0$ and $x = \pi$.

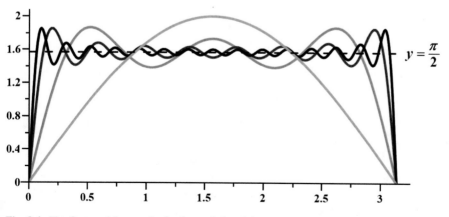

Fig. 8.4 The first partial sums S_1, S_3, S_{15}, and S_{30} of the expansion of (8.64), demonstrating the Gibbs phenomenon

We may isolate these points within infinitely small regions, $[0, \varepsilon)$ and $(\pi - \varepsilon, \pi]$, so that on the rest of the interval, $[\varepsilon, \pi - \varepsilon]$, this series converges uniformly. In Fig. 8.4 this corresponds to the fact that the graphs of the partial sums, for large enough n, are very close to the line $y = \pi/2$ along the interval $[\varepsilon, \pi - \varepsilon]$. Close to the points $x = 0$ and $x = \pi$, it is clear that the uniformity of the approximation of $f(x)$ with partial sums is violated because of the jump discontinuity in $f(x)$.

Next we point out another phenomenon that can be observed near the points $x = 0$ and $x = \pi$. Near $x = 0$, approaching the origin from the right, the graphs of the partial sums (shown in Fig. 8.4) oscillate about the line $y = \pi/2$. The significant thing to note is that the amplitudes of these oscillations do not diminish to zero as $n \to \infty$. On the contrary, the height of the first bump (closest to $x = 0$) approaches the value of $\delta = 0.281$ above the $y = \pi/2$ line. This corresponds to an additional $\delta/(\pi/2) \approx 0.18$ of the height of the partial sum above the "expected" value. The situation is similar when x approaches the value π from the left. Such a defect of the convergence is known as the *Gibbs phenomenon* (discovered by Henry Wilbraham (1848) and rediscovered by Josiah Willard Gibbs (1899), an American scientist known for significant theoretical contributions to physics, chemistry, and mathematics). In general, if the function $f(x)$ has a finite jump $|D|$ at some point x, the maximum elevation of the partial sum value near x when $n \to \infty$ is bigger than $|D|$ by $2\,|\,D\,|$ $\delta/\pi \approx 0.115$.

8.12   Fourier Transforms

A Fourier series is a representation of a function which uses a *discrete* system of orthogonal functions. This idea may be expanded to a *continuous* set of orthogonal functions. The corresponding expansion in this case is referred to as a *Fourier transform* (FT).

Let us start with the complex form of the Fourier series for a function $f(x)$ on the interval $[-l, l]$:

$$f(x) = \sum_{n=-\infty}^{\infty} c_n e^{\frac{in\pi x}{l}}, \tag{8.65}$$

with the coefficients

$$c_n = \frac{1}{2l} \int_{-l}^{l} f(x) e^{-\frac{in\pi x}{l}} dx \quad (n = 0, \pm 1, \pm 2, \ldots). \tag{8.66}$$

In physics terminology, Eq. (8.66) gives a discrete spectrum of a function $f(x)$ with *wave numbers* $k_n = \frac{n\pi}{l}$. Here $c_n e^{ik_n x}$ is a harmonic with a complex amplitude c_n defined by Eq. (8.66), or

$$c_n = \frac{1}{2l} \int_{-l}^{l} f(x)e^{-ik_n x}dx. \tag{8.67}$$

Suppose now that l is very large; thus, the distance between two neighboring wave numbers, $\Delta k = \frac{\pi}{l}$, is very small. Using the notation

$$\widehat{f}(k) = \int_{-\infty}^{\infty} f(x)e^{-ikx}dx, \tag{8.68}$$

we may write Eq. (8.67) in the form:

$$c_n = \frac{1}{2\pi} \int_{-\infty}^{\infty} f(x)e^{-ik_n x}dx \cdot \frac{\pi}{l} = \frac{1}{2\pi}\widehat{f}(k_n)\Delta k. \tag{8.69}$$

Using this definition Eq. (8.65) can be written as

$$f(x) = \sum_{n} c_n e^{ik_n x} = \frac{1}{2\pi}\sum_{n}\widehat{f}(k_n)e^{ik_n x}\Delta k \quad (-l < x < l). \tag{8.70}$$

In the limit $l \to \infty$, this becomes the integral:

$$f(x) = \frac{1}{2\pi} \int_{-\infty}^{\infty} \widehat{f}(k)e^{ikx}dk \quad (-\infty < x < \infty). \tag{8.71}$$

In this limit the wave number takes all values from $-\infty$ to ∞, i.e., when $l \to \infty$ the spectrum is continuous. The amplitudes are distributed continuously, and for each infinitesimal interval from k to $k + dk$, there is an infinitesimal amplitude:

$$dc = \frac{1}{2\pi}\widehat{f}(k)dk. \tag{8.72}$$

With this equation as a definition, $\widehat{f}(k)$ is called the *Fourier transform* of $f(x)$.

Equation (8.68) is called the direct Fourier transform, and Eq. (8.71) is referred to as the inverse Fourier transform. These formulas are valid if the function $f(x)$ is absolutely integrable on $(-\infty, \infty)$:

$$\int\limits_{-\infty}^{\infty} |f(x)| \, dx < \infty. \tag{8.73}$$

Function $f(x)$ is, generally speaking, a complex function of a real variable x. Its inverse Fourier transform can be complex even if the function $f(x)$ is real.

By the *Riemann-Lebesgue lemma* [6]

$$\widehat{f}(k) \to 0 \quad \text{as} \quad |k| \to \infty.$$

(Henri Lebesgue (1875–1941) was a French mathematician known for his theory of integration.)

As in the case of a Fourier series, at a point x_0 where function $f(x)$ is continuous, the inverse transform (8.71) gives $f(x_0)$. At a point where $f(x)$ has a finite discontinuity, the inverse transform gives a value:

$$[f(x_0 + 0) + f(x_0 - 0)]/2.$$

Example 8.6

Find the Fourier transform of the *Gaussian function*:

$$f(x) = \sqrt{\frac{b}{\pi}} e^{-bx^2}, \quad b = const. \tag{8.74}$$

(German mathematician Carl Friedrich Gauss (1777–1855) is known for many significant contributions to mathematics and science. He is ranked among history's most influential mathematicians.)

Let $f(x)$ be normalized by the condition $\int\limits_{-\infty}^{\infty} f(x)dx = 1$.

FT of the function (8.74) is

$$\widehat{f}(k) = \sqrt{\frac{b}{\pi}} \int\limits_{-\infty}^{\infty} e^{-bx^2} e^{-ikx} dx.$$

To evaluate the integral, complete the square:

$$-bx^2 - ikx = -\left(\sqrt{b}x + ik/2\sqrt{b}\right)^2 - k^2/4b,$$

and introduce a new variable $z = \sqrt{b}x + ik/2\sqrt{b}$.

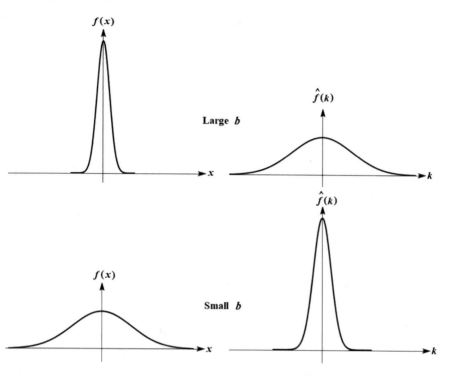

Fig. 8.5   Function $f(x)$ and its Fourier transform $\widehat{f}(k)$ for large and small values of b

Then, we obtain:

$$\widehat{f}(k) = e^{-k^2/4b} \frac{1}{\sqrt{\pi}} \int\limits_{-\infty}^{\infty} e^{-z^2}\, dz = e^{-k^2/4b}. \tag{8.75}$$

Function $\widehat{f}(k)$ is also a Gaussian function with a peak at the origin, decreasing as $k \to \pm\, \infty$. Note that if $f(x)$ is sharply peaked (large b), then $\widehat{f}(k)$ is flat, and vice versa; see Fig. 8.5.

This is typical of the Fourier transforms. In quantum mechanics when $f(x)$ is related with the probability of the location of a particle and $\widehat{f}(k)$ is related with the probability of the momentum, this feature is a manifestation of the *Heisenberg uncertainty principle*. In spectral analysis this reciprocal relation connects a time signal $f(t)$ and the spectral bandwidth $\widehat{f}(\omega)$.

In Chap. 7, the delta function was defined as the limit of function:

$$\Delta(t;a) = \frac{1}{a\sqrt{2\pi}} e^{-t^2/2a^2} \quad \text{as} \quad a \to 0.$$

Applying the result of Example 8.6, we find that

$$\widehat{\Delta}(k;a) = e^{-k^2 a^2/2}.$$

In the limit $a \to 0$, we find that the Fourier transform of delta function $\widehat{\delta}(k) = 1$, i.e.,

$$\delta(x) = \frac{1}{2\pi} \int_{-\infty}^{\infty} e^{ikx} dk. \tag{8.76}$$

This integral exists only in the sense of the principal value (v.p.), i.e.,

$$\delta(x) = \frac{1}{2\pi} \lim_{L\to\infty} \int_{-L}^{L} e^{ikx} dk. \tag{8.77}$$

Note that there are different ways to deal with the factor $1/2\pi$ in the formulas for direct and inverse transforms, but the product of both factors must be $1/2\pi$. Often, this factor is placed in the direct transform formula, while sometimes it is split into two identical factors, $1/\sqrt{2\pi}$, one in each equation. Using the definition given by Eqs. (8.68) and (8.71) has the advantage that the FT of $\delta(x)$ equals one, $\widehat{\delta}(k) = 1$.

Next consider the direct and the inverse Fourier transforms, (8.68) and (8.71), in the cases of even and odd functions.

If the function $f(x)$ is even, we have:

$$\widehat{f}(k) = \int_{-\infty}^{\infty} f(x)\cos kx dx - i \int_{-\infty}^{\infty} f(x)\sin kx dx = 2 \int_{0}^{\infty} f(x)\cos kx dx. \tag{8.78}$$

From this we see that $\widehat{f}(k)$ is also even, and with Eq. (8.71) we obtain:

$$f(x) = \frac{1}{\pi} \int_{0}^{\infty} \widehat{f}(k)\cos kx dk. \tag{8.79}$$

These formulas give what is known as the *Fourier cosine transform*. Similarly, if $f(x)$ is odd, we obtain the *Fourier sine transform*:

$$\widehat{if}(k) = 2 \int_0^\infty f(x) \sin kx dx, \quad f(x) = \frac{1}{\pi} \int_0^\infty \widehat{if}(k) \sin kx dk. \tag{8.80}$$

In this case usually $\widehat{if}(k)$ (rather than $\widehat{f}(k)$) is called the Fourier transform. We leave it to the reader to obtain Eq. (8.86) as a *Reading Exercise*.

If $f(x)$ is given on the interval $0 < x < \infty$, it can be extended to $-\infty < x < 0$ in either an even or odd way, and we may use either sine or cosine transforms.

In general case, we may write the FT as

$$f(x) = \frac{1}{2\pi} \int_{-\infty}^\infty \left[\widehat{a}(k) \cos kx + \widehat{b}(k) \cos kx \right] dk, \tag{8.81}$$

where

$$\widehat{a}(k) = \int_{-\infty}^\infty f(x) \cos kx dx \quad \text{and} \quad \widehat{b}(k) = \int_{-\infty}^\infty f(x) \sin kx dx. \tag{8.82}$$

Equations (8.81) and (8.82) give the representation of the function $f(x)$ in the form of the *Fourier integral*. In contrast to the Fourier series, which gives a function written as a discrete spectrum with frequencies $n\pi/l$ depending on the integer number n (harmonic number), the Fourier integral represents a function as a continuum of harmonics with frequencies distributed continuously from zero to infinity.

Reading Exercises

Below we state a few important properties of Fourier transform. We leave proofs to the reader:

(i) Prove that the Fourier transform of $f(-x)$ is equal to $\widehat{f}(-k)$.

(ii) Prove that if function $f(x)$ is real, then $\widehat{f}(-k)$ is equal to $\widehat{f}^*(k)$.

(iii) Prove that the Fourier transform of $f'(x)$ is equal to $ik\widehat{f}(k)$.

   Hint: The Fourier transform of $f'(x)$ is $\int_{-\infty}^\infty f'(x)e^{-ikx}dx$; differentiate this by parts and take into account that $\int_{-\infty}^\infty |f(x)| dx < \infty$; thus, $f(\pm\infty) = f'(\pm\infty) = 0$.

   Similarly, the Fourier transform of $f''(x)$ is equal to $-k^2\widehat{f}(k)$.

(iv) Prove that the Fourier transform of $f(x - x_0)$ is equal to $e^{-ikx_0}\widehat{f}(k)$.

(v) Prove that the Fourier transform of $f(ax)$ (where a is constant) is equal to $\frac{1}{a}\widehat{f}\left(\frac{k}{a}\right)$. This property shows that if we stretch the dimension of an "object" along the x-axis, then the dimension of a Fourier "image" compresses by the same factor.

For many practical applications, it is useful to present FT formulas for another pair of physical variables, *time and frequency*. Using Eqs. (8.68) and (8.75), we may write the direct and inverse transforms as

$$f(t) = \frac{1}{2\pi} \int_{-\infty}^{\infty} \widehat{f}(\omega) e^{i\omega t} d\omega \quad \text{and} \quad \widehat{f}(\omega) = \int_{-\infty}^{\infty} f(t) e^{-i\omega t} dt. \tag{8.83}$$

Fourier transform equations can be generalized to cases of higher dimensions. For instance, for an application with the spatial variables represented as vectors, Eqs. (8.72) and (8.75) become:

$$f\left(\vec{x}\right) = \frac{1}{2\pi} \int_{-\infty}^{\infty} \widehat{f}\left(\vec{k}\right) e^{i\vec{k}\cdot\vec{x}} d\vec{k} \quad \text{and}$$

$$\widehat{f}\left(\vec{k}\right) = \int_{-\infty}^{\infty} f\left(\vec{x}\right) e^{-i\vec{k}\cdot\vec{x}} d\vec{x}. \tag{8.84}$$

Guided by the proof of *Parseval's identity* (8.64) for Fourier series, we can prove the formula:

$$\int_{-\infty}^{\infty} |f(x)|^2 dx = \frac{1}{2\pi} \int_{-\infty}^{\infty} \left|\widehat{f}(k)\right|^2 dk, \tag{8.85}$$

known as the *Plancherel's identity* for Fourier transforms. A Swiss mathematician Michel Plancherel (1885–1967) proved the identity in 1910.

Proof
Using (8.71) we have:

$$\int_{-\infty}^{\infty} |f(x)|^2 dx = \frac{1}{2\pi} \int_{-\infty}^{\infty} \left\{ f(x) \int_{-\infty}^{\infty} \widehat{f}^*(k) e^{-ikx} dk \right\} dx =$$

$$= \frac{1}{2\pi} \int_{-\infty}^{\infty} \left\{ \widehat{f}^*(k) \int_{-\infty}^{\infty} f(x) e^{-ikx} dx \right\} dk = \frac{1}{2\pi} \int_{-\infty}^{\infty} \left|\widehat{f}(k)\right|^2 dk.$$

Quantity $\left|\widehat{f}(k)\right|^2$ is called *the spectral density* of a function $f(x)$.

According to (8.85), if $|f(x)|^2$ is absolutely integrable, then $\left|\widehat{f}(k)\right|^2$ is also absolutely integrable.

Plancherel's identity has a clear sense in terms of a *spectral resolution*. Consider a wave or a field which can be expanded in a Fourier integral containing a continuous sequence of different frequencies. Such expansion has the form (8.83); therefore, Plancherel's identity has the form:

$$\int\limits_{-\infty}^{\infty} |f(t)|^2 dt = \frac{1}{2\pi} \int\limits_{-\infty}^{\infty} \left|\widehat{f}(\omega)\right|^2 d\omega, \qquad (8.86)$$

or

$$\int\limits_{0}^{\infty} |f(t)|^2 dt = \frac{1}{2\pi} \int\limits_{0}^{\infty} \left|\widehat{f}(\omega)\right|^2 d\omega,$$

where $\left|\widehat{f}(\omega)\right|^2$ is the spectral density of a function $f(t)$.

This result means that the energy of a field or a signal $f(t)$ is equal to the integration over all frequency components of a field's energy spectral density.

Example 8.7
Let $f(x) = 1$ on $-1 < x < 1$ and zero outside this interval. This is an even function; thus, with the cosine Fourier transform, we have:

$$\widehat{f}(k) = 2\left(\int\limits_{0}^{1} 1 \cdot \cos kx dx + \int\limits_{1}^{\infty} 0 \cdot \cos kx dx\right) = \frac{2\sin k}{k}.$$

The inverse transform gives:

$$f(x) = 2 \int\limits_{0}^{\infty} \frac{\sin k}{\pi k} \cos kx dk. \qquad (8.87)$$

For instance, substituting $x = 0$ in (8.87) gives:

$$1 = 2 \int\limits_{0}^{\infty} \frac{\sin k}{\pi k} dk,$$

and we obtain the interesting result:

$$\int\limits_0^\infty \frac{\sin k}{k}\, dk = \frac{\pi}{2}. \tag{8.88}$$

Example 8.8. Harmonic Oscillations

The motion of a harmonic oscillator is governed by the linear second-order differential equation:

$$x''(t) + 2\lambda x'(t) + \omega_0^2 x(t) = f(t), \tag{8.89}$$

where constants ω_0 and λ are the natural frequency of a system and the damping coefficient. This motion has been analyzed in Sect. 3.9 in the case of a periodic external force $f(t) = f_{\max} \sin \omega t$. Using Fourier transforms we can solve this equation when a function $f(t)$ is arbitrary.

In all practical cases, an FT of $f(t)$ and of a solution $x(t)$ exist, and we can write Eq. (8.83) along with

$$x(t) = \frac{1}{2\pi} \int\limits_{-\infty}^\infty \widehat{x}(\omega) e^{i\omega t}\, d\omega.$$

Using property (*iii*), the transformed Eq. (8.89) reads:

$$-\omega^2 \widehat{x}(\omega) + 2\lambda i \omega \widehat{x}(\omega) + \omega_0^2 \widehat{x}(\omega) = \widehat{f}(\omega). \tag{8.90}$$

Solving (8.90) for $\widehat{x}(\omega)$ gives:

$$\widehat{x}(\omega) = \frac{\widehat{f}(\omega)}{(\omega_0^2 - \omega^2) + 2i\lambda\omega}; \tag{8.91}$$

thus, the solution of the Eq. (8.89) is

$$x(t) = \frac{1}{2\pi} \int\limits_{-\infty}^\infty \frac{\widehat{f}(\omega) e^{i\omega t}}{(\omega_0^2 - \omega^2) + 2i\lambda\omega}\, d\omega. \tag{8.93}$$

This integral can be evaluated using the complex variable calculus (see, for instance [8], which is beyond the scope of this book. As in Sect. 3.9, a general solution vanishes for large t because of a dissipation, and only the particular solution (8.93) describing the forced oscillations remains).

Example 8.9. RLC Circuit

In Sect. 3.10.2 we discussed an *RLC* circuit driven by a sinusoidal electromotive force. As a more general situation, consider a periodic *emf* $E(t)$, e.g., $E(t + T) = E(t)$, but suppose that $E(t)$ is not necessarily sinusoidal. Find the response of the system, e.g., the current $i(t)$.

The function $i(t)$ satisfies the differential equation:

$$L\frac{d^2i}{dt^2} + R\frac{di}{dt} + \frac{1}{C}i = \frac{dE}{dt}. \tag{8.94}$$

Under steady-state conditions $i(t)$ is also periodic with the period T, and we can write a Fourier expansions for $E(t)$ and $i(t)$:

$$E(t) = \sum_{n=-\infty}^{+\infty} E_n e^{in\omega t}, \quad i(t) = \sum_{n=-\infty}^{+\infty} c_n e^{in\omega t}, \quad \omega = 2\pi/T.$$

Next, differentiate $E(t)$ and $i(t)$:

$$\frac{dE}{dt} = \sum_{n=-\infty}^{+\infty} in\omega E_n e^{in\omega t}, \frac{di}{dt} = \sum_{n=-\infty}^{+\infty} in\omega c_n e^{in\omega t}, \frac{d^2i}{dt^2} = \sum_{n=-\infty}^{+\infty} (-n^2\omega^2) c_n e^{in\omega t},$$

substitute these series into (8.94) and equate the coefficients of the same exponent $e^{in\omega t}$ on both sides (on the basis of the completeness and orthogonality properties). It gives:

$$(-n^2\omega^2 L + in\omega R + 1/C)c_n = in\omega E_n;$$

therefore,

$$c_n = \frac{i(n\omega/L)}{(\omega_0^2 - n^2\omega^2) + 2\lambda n\omega i} E_n, \tag{8.95}$$

where $\omega_0^2 = 1/CL$ is the *natural frequency* of the circuit and $2\lambda = R/L$ is the *attenuation factor* of the circuit.

Fourier coefficients of the function $E(t)$ are

$$E_n = \frac{1}{T} \int_{-T/2}^{+T/2} E(t)e^{-in\omega t}dt,$$

and using them we can evaluate the Fourier coefficients c_n (8.95) of the function $i(t)$, which concludes the determination of the current.

Example 8.10. Power Radiation

Consider the quasiperiodic electric field decaying in time:

$$E(t) = \begin{cases} 0, & t < 0 \\ e^{-t/\tau} \sin \omega_0 t, & t > 0 \end{cases}$$

and find the power radiated by this field, $\int_0^\infty |E(t)|^2 dt$, in the frequency interval $(\omega, \omega + d\omega)$. Both $\omega > 0$ and $\omega_0 > 0$.

The Fourier transform is

$$\widehat{E}(\omega) = \int_0^\infty e^{-t/\tau} e^{-i\omega t} \sin \omega_0 t \, dt = \frac{1}{2i} \int_0^\infty e^{-t/\tau} e^{-i\omega t} \left(e^{i\omega_0 t} - e^{-i\omega_0 t} \right) dt$$

$$= \frac{1}{2} \left(\frac{1}{\omega + \omega_0 - i/\tau} - \frac{1}{\omega - \omega_0 - i/\tau} \right).$$

The total radiated power can be evaluated with the help of Plancherel's identity (8.90):

$$\int_0^\infty |E(t)|^2 dt = \frac{1}{2\pi} \int_0^\infty \left| \widehat{E}(\omega) \right|^2 d\omega. \tag{8.96}$$

For $\omega > 0$ the second term in $\widehat{E}(\omega)$ dominates, then

$$\left| \widehat{E}(\omega) \right|^2 d\omega \approx \frac{1}{8\pi} \frac{d\omega}{(\omega - \omega_0)^2 + 1/\tau^2}. \tag{8.97}$$

This *Lorentzian function*, named after a Dutch physicist and a Nobel Prize winner (1902) Hendrik Lorentz (1853–1928), falls off by ½ when $\omega = \omega_0 \pm 1/\tau$; thus, the width of the power curve can be defined as $2/\tau$. Therefore, the width of the energy spectrum is inversely proportional to the duration of the signal, in agreement with the uncertainty principle.

Example 8.11 Loaded Beam

A horizontal beam is supported at its ends at points $x = 0$ and $x = L$. The deflection of the beam, $y(x)$, is known to satisfy the equation:

$$\frac{d^4 y}{dx^4} = Rq(x), \tag{8.98}$$

where $q(x)$ is the load per unit length and R in the rigidity of the beam.

Since the function $y(x)$ must vanish at the points $x = 0$ and $x = L$, we can expand it into a Fourier sine series:

$$y(x) = \sum_{n=1}^{\infty} b_n \sin \frac{n\pi x}{L}.$$

Assuming that term-by-term differentiation is valid, next we obtain:

$$\frac{d^4 y(x)}{dx^4} = \sum_{n=1}^{\infty} \left(\frac{n\pi}{L}\right)^4 b_n \sin \frac{n\pi x}{L}.$$

Consider a uniformly loaded beam ($q = $ const.) and expand q into Fourier sine series:

$$q = \sum_{n=1}^{\infty} q_n \sin \frac{n\pi x}{L},$$

where

$$q_n = \frac{2}{L} \int_0^L q \sin \frac{n\pi x}{L} dx = \begin{cases} \dfrac{4q}{n\pi} & (n = \text{odd}), \\ 0 & (n = \text{even}). \end{cases}$$

Substitute both series into the DE and equate the coefficients of $\sin(n\pi x/L)$. This yields:

$$b_n = \frac{1}{R} \frac{L^4}{n^4 \pi^4} q_n = \begin{cases} \dfrac{4qL^4}{R\pi^5} \dfrac{1}{n^5} & (n = \text{odd}) \\ 0 & (n = \text{even}) \end{cases}$$

so that

$$y(x) = \frac{4qL^4}{R\pi^5} \sum_{n=1,3,5,\dots}^{\infty} \frac{1}{n^5} \sin \frac{n\pi x}{L}. \tag{8.99}$$

Due to the fifth power of n in the denominator, this series rapidly converges.

Reading Exercise Equation (8.98) can be solved in closed form by simple integrations. Compare the two solutions for some set of parameters. For this purpose, you can use the program *Fourier Series* of the book.

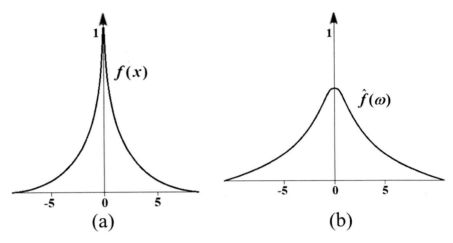

Fig. 8.6 The Fourier transform pair: (a) $f(x) = e^{-a|x|}$ and (b) $\widehat{f}(\omega) = \frac{2a}{\omega^2 + a^2}$

Problems

Determine a trigonometric Fourier of a function using:

(a) The expansion into a cosine and sine series;
(b) The expansion into a cosine series, e.g., keeping only *even terms*;
(c) The expansion into a sine series, e.g., keeping only *odd terms*.

1. $f(x) = x^2$ on the interval $[0, \pi]$
2. $f(x) = e^{ax}$ on the interval $[0, \pi]$
3. $f(x) = \cos ax$ on the interval $[-\pi, \pi]$
4. $f(x) = \sin ax$ on the interval $[-\pi, \pi]$
5. $f(x) = \pi - 2x$ on the interval $[0, \pi]$
   Find Fourier transform of a function $f(x)$:

6. $f(x) = \begin{cases} 0, & -\infty < x < 1, \\ 1, & 1 < x < 2, \\ 0, & 2 < x. \end{cases}$

7. $f(x) = e^{-a|x|}, \quad -\infty \leq x \leq \infty, \quad a > 0.$
   Check that (Fig. 8.6)

$$\widehat{f}(\omega) = \int\limits_{-\infty}^{+\infty} e^{-a|x|} e^{-i\omega x} dx = 2 \int\limits_{0}^{+\infty} e^{-ax} \cos \omega x \, dx = \frac{2a}{\omega^2 + a^2}.$$

8. $f(x) = \begin{cases} e^{-ax}, & x > 0, \\ 0, & x \leq 0. \end{cases}$
9. $f(x) = xe^{-a|x|} \quad (a > 0).$

10. $f(x) = e^{-x^2/2} \cos ax$.
11. $f(x) = 1/(x^2 + a^2)$.
12. Solve Eq. (8.89) using Fourier series for an arbitrary function $f(t)$.
     In particular, obtain an explicit series for $x(t)$ if $f(t)$ is given by:

(a) $f(t) = \begin{cases} 0, & -\pi \le t \le 0, \\ 1, & 0 < t \le \pi; \end{cases}$

(b) The triangle function in Fig. 8.2 on the interval $0 < t < 2\pi$

13. Find the deflection of the beam in Example 8.11 if its load per unit length is
    given by $q(x) = \frac{A}{L} x$; $A = const$.

    Answer: $y(x) = \frac{2AL^4}{R\pi^4} \sum_{n=1}^{\infty} \frac{(-1)^{n+1}}{n^5} \sin \frac{n\pi x}{L}$.

14. In Example 8.10 consider the solution for the current in the form (8.3):

$$I(t) = A_0 + \sum_{n=1}^{\infty} A_n \sin(n\omega t + \alpha_n).$$

Find coefficients A_n and $\tan \alpha_n$ for

$$E(t) = \begin{cases} E_0(1 + 4t/T), & -T/2 \le t < 0, \\ E_0(1 - 4t/T), & 0 \le t < T/2 \end{cases} \qquad E_0 = const.$$

If you decided to use the program *Fourier Series*, then to evaluate A_n and α_n, you can
set different values of the parameters R, L, C, E_0, T. For a particular set of these
parameters, one of the coefficients A_n can dominate over other coefficients – this is
the case of a resonance in one of the harmonics (modes). For instance, for $R = 1.6\Omega$,
$L = 6.35$ mH, $C = 4.0$ μF, $E_0 = 12$ V, and $T = 5 \times 10^{-3}$ s, the fifth, $n = 5$, mode
dominates [8].

Chapter 9
Boundary-Value Problems
for Second-Order ODEs

In the previous chapters, we considered initial value (Cauchy) problems for ordinary differential equations: all the conditions were imposed at the same point. That formulation of the problem is appropriate for phenomena evolving in time. For instance, we determine the initial location of the particle and its initial velocity, and then, by solving the Newton's equations, we determine the particle trajectory.

In the present chapter, we consider another kind of problems, where the conditions are imposed in different points, actually, on the boundaries of the region where the solution is defined. That is appropriate for the description of objects finite in space. For instance, we want to determine the distribution of temperature in a rod heated or cooled at its ends. It is clear that conditions characterizing heating have to be imposed simultaneously at both ends.

We will return to the problem of the heated rod in Chap. 13 where we will solve second-order linear partial differential equations (PDEs) using the method of separation of variables (or Fourier expansion method). We will see that this method allows to reduce that PDE problem to a certain kind of boundary-value problems for ordinary differential equations, the Sturm-Liouville problem that is considered below.

9.1   The Sturm-Liouville Problem

Let us consider a linear second-order ordinary differential equation:

$$a(x)y''(x) + b(x)y'(x) + c(x)y(x) + \lambda d(x)y(x) = 0 \tag{9.1}$$

with a *parameter* λ multiplied by the function $y(x)$. Generally, λ can be a complex number, and the solution $y(x)$ can be a complex function. Previously we have not considered complex-valued ODEs (note that the final solution of a physical problem

© The Author(s), under exclusive license to Springer Nature Switzerland AG 2023
V. Henner et al., *Ordinary Differential Equations*,
https://doi.org/10.1007/978-3-031-25130-6_9

must be real), but it is often useful to generalize the problem by allowing $y(x)$ to be complex.

Equation (9.1) can be written in the form:

$$\frac{d}{dx}[p(x)y'(x)] + [q(x) + \lambda r(x)]y(x) = 0, \tag{9.2}$$

where

$$p(x) = e^{\displaystyle\int \frac{b(x)}{a(x)}dx}, \quad q(x) = \frac{p(x)c(x)}{a(x)}, \quad r(x) = \frac{p(x)d(x)}{a(x)}. \tag{9.3}$$

Reading Exercise Verify that substitution of (9.3) into (9.2) gives Eq. (9.1).

The solution $y(x)$ is defined on a finite interval $[a, b]$ and obeys *homogeneous* (with zero right-hand sides) *boundary conditions* of the form:

$$\begin{aligned} \alpha_1 y' + \beta_1 y|_{x=a} = 0, \\ \alpha_2 y' + \beta_2 y|_{x=b} = 0. \end{aligned} \tag{9.4}$$

This kind of conditions, imposed in two different points, strongly differs from the set of initial conditions $y(a) = y_0$, $y'(a) = y_1$ imposed in the same point that we used formerly.

Equation (9.2) with boundary conditions (9.4) forms a special kind of boundary-value problems (BVPs) called the Sturm-Liouville problem (SLP). (Charles-Francois Sturm (1803–1855) and Joseph Liouville (1809–1882), Swiss and French mathematicians and friends, published a series of works on boundary-value problems for differential equations.)

Equation (9.2) has two linearly independent solutions, $y(x) = y^{(1)}(x, \lambda)$ and $y(x) = y^{(2)}(x, \lambda)$, and its general solution is

$$y(x) = C_1 y^{(1)}(x, \lambda) + C_2 y^{(2)}(x, \lambda), \tag{9.5}$$

where C_1 and C_2 are constants. Substitution of (9.4) into (9.2) gives a system of algebraic equations:

$$\begin{aligned} C_1\left[\alpha_1 y^{(1)'}(a, \lambda) + \beta_1 y^{(1)}(a, \lambda)\right] + C_2\left[\alpha_1 y^{(2)'}(a, \lambda) + \beta_1 y^{(2)}(a, \lambda)\right] = 0, \\ C_1\left[\alpha_2 y^{(1)'}(b, \lambda) + \beta_2 y^{(1)}(b, \lambda)\right] + C_2\left[\alpha_2 y^{(2)'}(b, \lambda) + \beta_2 y^{(2)}(b, \lambda)\right] = 0. \end{aligned} \tag{9.6}$$

If

$$\Delta(\lambda) = \begin{vmatrix} \alpha_1 y^{(1)\prime}(a,\lambda) + \beta_1 y^{(1)}(a,\lambda) & \alpha_1 y^{(2)\prime}(a,\lambda) + \beta_1 y^{(2)}(a,\lambda) \\ \alpha_2 y^{(1)\prime}(b,\lambda) + \beta_2 y^{(1)}(b,\lambda) & \alpha_2 y^{(2)\prime}(b,\lambda) + \beta_2 y^{(2)}(b,\lambda) \end{vmatrix} \neq 0, \qquad (9.7)$$

system (9.6) has only a trivial solution $C_1 = C_2 = 0$, i.e., the only solution of the BVP is $y(x) = 0$. One can prove that under definite conditions, $\Delta(\lambda) = 0$ has an infinite set of simple roots (*eigenvalues*) $\lambda = \lambda_n$, $n = 1, 2, \ldots, \infty$ (see Theorem 9.1 below). For those values of λ, $\Delta(\lambda_n) = 0$; therefore, the BVP has a set of nontrivial solutions $y(x) = C y_n(x)$, where $y_n(x) \neq 0$ is the *eigenfunction* and C is an arbitrary constant.

We emphasize that Eq. (9.2) and boundary conditions are homogeneous; otherwise, the BVP is not a Sturm-Liouville problem. It is clear that the constants α_1 and β_1 should not both be equal to zero simultaneously, nor the constants α_2 and β_2. However, it is allowed that some of them are equal to zero. If $\alpha_k = 0$ for $k = 1$ or $k = 2$, the corresponding boundary condition becomes $y = 0$ (known as the Dirichlet boundary condition). If $\beta_k = 0$, i.e., $y' = 0$, that is the Neumann (Carl Neumann (1832–1925), a German mathematician) boundary condition. In any case, the constants α_k and β_k correspond to definite physical restrictions; therefore, they are real.

Note that the boundary-value problems can be formulated also for nonlinear equations which are not the topic of our book.

The Sturm-Liouville problems appear by solving PDEs that are considered in Chap. 13. In various problems governed by PDEs, there are some restrictions on the signs of coefficients. If α_k and β_k are nonzero, typically α_1 and β_1 have opposite signs, whereas α_2 and β_2 have the same sign. When the above conditions on the signs are violated, the solutions of PDEs can be unbounded. Below we consider "normal" physical situations where the above-mentioned conditions on the signs hold.

Later on we let $p(x)$, $q(x)$, $r(x)$, and $p'(x)$ be continuous, real functions on the interval $[a, b]$ and let $p(x) > 0$ and $r(x) > 0$ on the finite interval (a, b). The coefficients α_k and β_k in Eq. (9.4) are assumed to be real and independent of λ.

The properties of the boundary-value problem (9.2), (9.4) are strongly different from the properties of the initial value problems that we considered formerly. The solution of the initial value problem is unique, when the coefficients are continuous. In the case of the SLP, that is not the case. The problem always has the trivial solution $y(x) = 0$. But for some special values of λ (called *eigenvalues*), the solution is not unique, and there are also nontrivial solutions (called *eigenfunctions*) that correspond to the eigenvalues λ. If $y(x)$ is an eigenfunction corresponding to a certain eigenvalue lambda, then the function $C y(x)$ is also an eigenfunction for any constant C. Thus, the eigenfunction is defined up to an arbitrary constant coefficient.

If we introduce the differential operator (called the *Sturm-Liouville operator*)

$$\begin{aligned} Ly(x) &= -\frac{d}{dx}[p(x)y'(x)] - q(x)y(x) = \\ &= -p(x)y''(x) - p'(x)y'(x) - q(x)y(x), \end{aligned} \qquad (9.8)$$

or

$$L = -p(x)\frac{d^2}{dx^2} - p'(x)\frac{d}{dx} - q(x),$$

then Eq. (9.2) becomes:

$$Ly(x) = \lambda r(x)y(x). \tag{9.9}$$

As it is seen from Eq. (9.8), L is a real *linear operator*. When $r(x) = const$ (in this case one can take $r(x) = 1$), this equation appears as an ordinary *eigenvalue problem*, $Ly(x) = \lambda y(x)$, for which we have to determine λ and $y(x)$. For $r(x) \neq 1$, we have a modified problem where the function $r(x)$ is called a *weight function*. As we stated above, the only requirement on $r(x)$ is that it is real and positive. Equations (9.2) and (9.9) are equivalent.

Now we discuss the properties of eigenvalues and eigenfunction of the SLP. Let us write Eq. (9.9) for two eigenfunctions, $y_n(x)$ and $y_m(x)$, corresponding to eigenvalues λ_n and λ_m, and take the complex conjugate of the equation for $y_m(x)$. Notice that in spite of the fact that $p(x)$, $q(x)$, and $r(x)$ are real, we cannot assume from the very beginning that λ and $y(x)$ are real – that has to be checked. We have:

$$Ly_n(x) = \lambda_n r(x)y_n(x)$$

and

$$Ly_m^*(x) = \lambda_m^* r(x)y_m^*(x).$$

Multiplying the first of these equations by $y_m^*(x)$ and the second one by $y_n(x)$, we then integrate both equations from a to b and subtract the two results to obtain

$$\int_a^b y_m^*(x)Ly_n(x)dx - \int_a^b y_n(x)Ly_m^*(x)dx = (\lambda_n - \lambda_m^*)\int_a^b r(x)y_m^*(x)y_n(x)dx. \tag{9.10}$$

Using the definition of L given by Eq. (9.8), the left side of Eq. (9.10) is

$$\left\{ p(x)\left[\frac{dy_m^*}{dx}y_n(x) - y_m^*(x)\frac{dy_n}{dx}\right] \right\}_a^b. \tag{9.11}$$

Reading Exercise Verify the previous statement.

Then, using the boundary conditions (9.4), it can be easily proved that the expression (9.11) equals zero.

Reading Exercise Verify that the expression in Eq. (9.11) equals zero.

Thus, we are left with

$$\int\limits_a^b y_m^*(x)Ly_n(x)dx = \int\limits_a^b y_n(x)Ly_m^*(x)dx. \qquad (9.12)$$

An operator, L, which satisfies Eq. (9.12), is known as a *Hermitian* or *self-adjoint operator*. Thus, we may say that *the Sturm-Liouville linear operator* satisfying homogeneous boundary conditions is *Hermitian*. Many important operators in physics, especially in quantum mechanics, are Hermitian. (Charles Hermite (1822–1901) was a French mathematician, who made serious contributions mainly to number theory and algebra.)

Let us show that Hermitian operators have *real eigenvalues* and their eigenfunctions are *orthogonal*. The right side of Eq. (9.10) gives:

$$\left(\lambda_n - \lambda_m^*\right) \int\limits_a^b r(x)y_m^*(x)y_n(x)dx = 0. \qquad (9.13)$$

When $m = n$, the integral cannot be zero (recall that $r(x) > 0$); thus, $\lambda_n^* = \lambda_n$, and we have proved that *the eigenvalues of a Sturm-Liouville problem are real*. Then, for $\lambda_m \neq \lambda_n$, Eq. (9.13) is

$$\int\limits_a^b r(x)y_m^*(x)y_n(x)dx = 0, \qquad (9.14)$$

and we conclude that *the eigenfunctions corresponding to different eigenvalues of a Sturm-Liouville problem are orthogonal (with the weight function $r(x)$)*. The squared *norm* of the eigenfunction $y_n(x)$ is defined to be

$$\|y_n\|^2 = \int\limits_a^b r(x)|y_n(x)|^2 dx. \qquad (9.15)$$

Note that the eigenfunctions of Hermitian operators always *can be chosen to be real*. Indeed, let us take complex conjugates of Eqs. (9.2) and (9.4). Because λ_n, as well as functions $p(x)$, $q(x)$, and $r(x)$ and coefficients α_1, β_1, α_2, and β_2, are real, we conclude that if $y_n(x)$ is an eigenfunction corresponding to the eigenvalue λ_n, then the complex conjugate function $y_n^*(x)$ is also an eigenfunction corresponding to the same eigenvalue. Therefore, if $y_n(x)$ is not real, it can be replaced by the real function $\mathrm{Re}\, y_n(x) = \left(y_n + y_n^*\right)/2$ or $\mathrm{Im}\, y_n(x) = \left(y_n - y_n^*\right)/2i$.

Real eigenfunctions are more convenient to work with because it is easier to match them to boundary conditions which are intrinsically real since they represent physical restrictions.

The above proof of orthogonality fails if $\lambda_m = \lambda_n$ for some $m \neq n$ (in other words, there exist different eigenfunctions belonging to the same eigenvalue). In the latter case, we cannot conclude that the corresponding eigenfunctions, $y_m(x)$ and $y_n(x)$, are orthogonal (although in some cases they are). If there are f eigenfunctions that have the same eigenvalue, we have an f-*fold degeneracy* of the eigenvalue (we work only with the second-order equations; hence, only $f = 2$ is possible). In general, a degeneracy reflects a symmetry of the underlying physical system (examples can be found below in the text). Within the set of eigenfunction $C_1 y_n(x) + C_2 y_m(x)$ corresponding to the degenerate eigenvalue, it is always possible to construct linear combinations of the eigenfunctions belonging to the same eigenvalue so that these new functions are orthogonal. For instance, eigenfunctions $y_n(x)$ and

$$\widetilde{y}_m(x) = y_m(x) - y_n(x)(y_n \cdot y_m)/\|y_n\|^2,$$

where $(y_n \cdot y_m) = \int\limits_a^b r(x) y_n(x) y_m^*(x) dx$, are orthogonal.

If $p(a) \neq 0$ and $p(b) \neq 0$, then $p(x) > 0$ on the closed interval $[a, b]$, and we have the so-called regular Sturm-Liouville problem. If $p(a) = 0$, then we do not impose the first of the boundary conditions in Eq. (9.4); instead, we require $y(x)$ and $y'(x)$ to be finite at $x = a$. Similar situations occur if $p(b) = 0$, or if both $p(a) = 0$ and $p(b) = 0$. All these cases correspond to the so-called singular Sturm-Liouville problem.

In some problems, x is a circular variable (for instance, the polar angle φ), and we can consider *the periodic Sturm-Liouville problem* with periodic boundary conditions: $y(\varphi) = y(\varphi + 2\pi)$, $y'(\varphi) = y'(\varphi + 2\pi)$. (In that case, $p(\varphi)$, $q(\varphi)$, and $r(\varphi)$ are periodic functions, because the points with coordinates φ and $\varphi + 2\pi$ coincide.)

In addition to problems defined on a finite interval, we shall consider problems defined on a semi-infinite or infinite region. In that case, the corresponding condition (9.4) is usually replaced by the condition of a physically relevant behavior on the infinity, e.g., it is assumed that the solution is bounded. In the latter case, the SLP is also considered as singular.

As we have seen above, because the eigenvalues are real, the eigenfunctions can also be chosen real. In the non-degenerate case, an arbitrary eigenfunction is just a real function multiplied by an arbitrary, generally complex, coefficient. In the degenerate case, when there are two linearly independent eigenfunctions for the same eigenvalue, it may be convenient to use truly complex eigenfunctions with linearly independent real and imaginary parts. Some examples will be given in the following sections.

The following items summarize the types of Sturm-Liouville problems:

(i) For $p(x) > 0$ and $r(x) > 0$ at a finite interval $x \in [a, b]$, and the boundary conditions are imposed separately at $x = a$ and $x = b$, we have the *regular* problem.

(ii) If $p(x)$ or $r(x)$ is equal to zero or discontinuous at least at one of the ends, or of the interval is infinite or semi-infinite, we have the *singular* problem.

(iii) For periodic functions $p(x)$, $q(x)$, and $r(x) > 0$, and periodic boundary conditions at the ends of interval $[a, b]$ we have the *periodic* problem.

The following theorem gives a list of several important *properties of the regular
and periodic Sturm-Liouville problems*:

Theorem 9.1

(i)	Each regular and each periodic Sturm-Liouville problem has an infinite number of discrete eigenvalues $\lambda_1 < \lambda_2 < \ldots < \lambda_n < \ldots$ such that $\lambda_n \to \infty$ as $n \to \infty$. All eigenvalues are real numbers.
(ii)	All eigenvalues of a regular Sturm-Liouville problem are simple, i.e., two eigenfunctions that belong to the same eigenvalue are linearly dependent; for a periodic Sturm-Liouville problem, this property does not hold.
(iii)	For each of the types of Sturm-Liouville problems, the eigenfunctions corresponding to different eigenvalues are linearly independent.
(iv)	For each of the types of Sturm-Liouville problems, the set of eigenfunctions is orthogonal with the weight function $r(x)$ on the interval $[a, b]$.
(v)	All $\lambda_n \geq 0$, if the following conditions are satisfied simultaneously: (1) $q(x) \leq 0$ on $[a, b]$; (2) $\alpha_1 = 0$ or $\beta_1/\alpha_1 \leq 0$; (3) $\alpha_2 = 0$ or $\beta_2/\alpha_2 \leq 0$. These are sufficient conditions: all the eigenvalues can be nonnegative even when some of those conditions are not satisfied.

Some of these properties have been proven previously, such as property (*iv*) and
part of property (*i*). The remaining part of property (*i*) will be shown in several
examples below, as well as property (*v*).

Let us prove property (*ii*) in the case of a regular SLP.

Proof

Let two eigenfunctions $y_1(x)$ and $y_2(x)$ correspond to the same eigenvalue λ. Then,
the Wronskian at $x = 0$

$$W[y_1, y_2](0) = \begin{vmatrix} y_1(0) & y_2(0) \\ y_1'(0) & y_2'(0) \end{vmatrix} = y_1(0)y_2'(0) - y_2(0)y_1'(0) = 0.$$

Indeed, from the b.c. (9.4), we obtain:

If $\alpha_1 = 0$, then $y_j(0) = 0$, $j = 1, 2$; hence, $W[y_1, y_2](0) = 0$.
If $\alpha_1 \neq 0$, then $y_j'(0) = -(\beta_1/\alpha_1)y_j(0)$, $j = 1, 2$; hence,

$$W[y_1, y_2](0) = -\frac{\beta_1}{\alpha_1}y_1(0)y_2(0) + \frac{\beta_1}{\alpha_1}y_2(0)y_1(0) = 0.$$

As we know from Chap. 3, if the Wronskian equals zero in some point, it equals
zero on the whole interval $[a, b]$; thus, $y_1(x)$ and $y_2(x)$ are linearly dependent, i.e.,
$y_2(x) = Cy_1(x)$, where C is a certain constant.

In Example 9.4 we will obtain a degenerate solution of periodic Sturm-Liouville problems.

Property (*iii*) can be proved similarly, and we leave this proof to the reader as a *Reading Exercise.*

Let us emphasize that the properties listed above are obtained for the regular and periodic Sturm-Liouville problems. In Chap. 10, we shall consider some special functions, e.g., Bessel functions and the orthogonal polynomials, such as Legendre polynomials that arise from singular Sturm-Liouville problems; thus, the first statement in the above theorem is not directly applicable to these important cases. In spite of that, singular Sturm-Liouville problems may also have an infinite sequence of discrete eigenvalues which we discuss in Chap. 13 for Bessel functions and for the orthogonal polynomials.

Eigenfunctions $y_n(x)$ form a *complete orthogonal set* on $[a, b]$. This means that any reasonable well-behaved function, $f(x)$, defined on $[a, b]$ can be expressed as a series (called a *generalized Fourier series*) of eigenfunctions of a SLP in which case we may write:

$$f(x) = \sum_{n}^{\infty} c_n y_n(x), \tag{9.16}$$

where it is convenient, in some cases, to start the summation with $n = 1$, in other cases with $n = 0$. An expression for the coefficients c_n can be found by multiplying both sides of Eq. (9.16) by $r(x)y_n^*(x)$ and integrating over $[a, b]$ to give

$$c_n = \frac{\int_a^b r(x)f(x)y_n^*(x)dx}{\|y_n\|^2}. \tag{9.17}$$

Substituting (9.17) into (9.16), we obtain:

$$f(x) = \sum_{n}^{\infty} \frac{1}{\|y_n\|^2} \int_a^b r(\xi)f(\xi)y_n(x)y_n^*(\xi)d\xi.$$

That means that

$$\sum_{n}^{\infty} \frac{1}{\|y_n\|^2} \int_a^b r(\xi)y_n(x)y_n^*(\xi)d\xi = \delta(x - \xi).$$

This is *the completeness property* for eigenfunctions of a SLP.

The Sturm-Liouville theory provides a theorem for convergence of the series in Eq. (9.16) at every point x of $[a, b]$:

Theorem 9.2

Let $\{y_n(x)\}$ be the set of eigenfunctions of a regular Sturm-Liouville problem, and let $f(x)$ and $f'(x)$ be piecewise continuous on a closed interval. Then, the series expansion (9.16) converges to $f(x)$ at every point where $f(x)$ is continuous and to the value $[f(x_0 + 0) + f(x_0 - 0)]/2$ if x_0 is a point of discontinuity.

The theorem is also valid for the orthogonal polynomials and Bessel functions related to singular SLP. This theorem, which is extremely important for applications, is similar to the theorem for trigonometric Fourier series.

Formulas in Eqs. (9.14), (9.15), (9.16), and (9.17) can be written in a more convenient way if we define a *scalar product of* functions φ and ψ as the number given by

$$\varphi \cdot \psi = \int_a^b r(x)\varphi(x)\psi^*(x)dx \qquad (9.18)$$

(another notation of the scalar product is (φ, ψ)).

This definition of the scalar product has properties identical to those for vectors in linear Euclidian space, a result which can be easily proved:

$$\begin{aligned}
\varphi \cdot \psi &= (\psi \cdot \varphi)^*, \\
(a\varphi) \cdot \psi &= a\varphi \cdot \psi, \qquad (a \text{ is a number}) \\
\varphi \cdot (a\psi) &= a^*\varphi \cdot \psi, \qquad\qquad\qquad (9.19) \\
\varphi \cdot (a\psi + b\phi) &= a\varphi \cdot \psi + b\varphi \cdot \phi, \\
\varphi \cdot \varphi &= |\varphi|^2 \geq 0.
\end{aligned}$$

The last property relies on the assumption made for the Sturm-Liouville equation that $r(x) \geq 0$. If φ is continuous on $[a, b]$, then $\varphi \cdot \varphi = 0$ only if φ is zero.

Reading Exercise Prove the relations given in Eq. (9.19).

In terms of the scalar product, the orthogonality of eigenfunctions (defined by Eq. (9.14)) means that

$$y_n \cdot y_m = 0, \quad \text{if} \quad n \neq m \qquad (9.20)$$

and the formula for the Fourier coefficients in Eq. (9.14) becomes:

$$c_n = \frac{f \cdot y_n^*}{y_n \cdot y_n}. \qquad (9.21)$$

Functions satisfying the condition

$$\varphi \cdot \varphi = \int_a^b r(x)|\varphi(x)|^2 dx < \infty. \tag{9.22}$$

Note that the definition of the scalar product resembles the corresponding definition defined in the linear algebra for vectors in an n-dimensional vector space.
However, now the "vector" has infinitely many components (the values of the
function in each point), and the space of functions has infinitely many basic vectors
(all the eigenfunctions). The space of functions with the scalar product (9.18) is the
infinite-dimensional *Hilbert space*, named after David Hilbert (1862–1943), who
was one of the most influential mathematicians of the late nineteenth and early
twentieth century. The complete orthogonal set of functions $\{y_n(x)\}$ serves as the
orthogonal basis in that space.

9.2 Examples of Sturm-Liouville Problems

Example 9.1 Solve the equation:

$$y''(x) + \lambda y(x) = 0 \tag{9.23}$$

on the interval $[0, l]$ with boundary conditions

$$y(0) = 0 \quad \text{and} \quad y(l) = 0. \tag{9.24}$$

Solution First, comparing Eq. (9.23) with Eqs. (9.8) and (9.9), we conclude that
(9.23) and (9.24) form a regular SLP with linear operator $L = -d^2/dx^2$ and functions
$q(x) = 0$, $p(x) = r(x) = 1$. (Note that L can be taken with either negative or positive
sign as is clear from Eqs. (9.2), (9.8), and (9.9).) As a *Reading Exercise*, verify that
L is Hermitian.

Let us discuss the cases $\lambda = 0$, $\lambda < 0$, and $\lambda > 0$ separately. If $\lambda = 0$, then the
general solution to Eq. (9.23) is

$$y(x) = C_1 x + C_2$$

and from boundary conditions (9.24), we have $C_1 = C_2 = 0$, i.e., there exists only the
trivial solution $y(x) = 0$. If $\lambda < 0$, then

$$y(x) = C_1 e^{\sqrt{-\lambda}x} + C_2 e^{-\sqrt{-\lambda}x}$$

and the boundary conditions (9.24) again give $C_1 = C_2 = 0$ and therefore the trivial
solution $y(x) = 0$. Thus, we have only the possibility $\lambda > 0$, in which case we write
$\lambda = \mu^2$ with μ real, and we have a general solution of Eq. (9.23) given by

$$y(x) = C_1 \sin \mu x + C_2 \cos \mu x.$$

The boundary condition $y(0) = 0$ requires that $C_2 = 0$, and the boundary condition $y(l) = 0$ gives $C_1 \sin \mu l = 0$. From this we must have $\sin \mu l = 0$ and $\mu_n = \frac{n\pi}{l}$ since the choice $C_1 = 0$ again gives the trivial solution. Thus, the eigenvalues are

$$\lambda_n = \mu_n^2 = \left(\frac{n\pi}{l}\right)^2, \quad n = 1, 2, \ldots \tag{9.25}$$

and the eigenfunctions are $y_n(x) = C_n \sin \frac{n\pi x}{l}$, where for $n = 0$ we have the trivial solution $y_0(x) = 0$. It is obvious that we can restrict ourselves to positive values of n since negative values do not give new solutions. These eigenfunctions are orthogonal over the interval $[0, l]$ since we can easily show that

$$\int_0^l \sin \frac{n\pi x}{l} \cdot \sin \frac{m\pi x}{l} \, dx = 0 \quad \text{for } m \neq n. \tag{9.26}$$

The orthogonality of eigenfunctions follows from the fact that the Sturm-Liouville operator, L, is Hermitian for the boundary conditions given in Eq. (9.24). The eigenfunctions may be normalized by writing

$$C_n^2 \int_0^l \sin^2 \frac{n\pi x}{l} \, dx = C_n^2 \cdot \frac{l}{2} = 1,$$

which results in the orthonormal eigenfunctions

$$y_n(x) = \sqrt{\frac{2}{l}} \sin n\pi x, \quad n = 1, 2, \ldots \tag{9.27}$$

Thus, we have shown that the boundary-value problem consisting of Eqs. (9.23) and (9.24) has eigenfunctions that are *sine* functions. It means that *the expansion in eigenfunctions of the Sturm-Liouville problem for solutions to Eqs. (9.23) and (9.24) is equivalent to the trigonometric Fourier sine series.*

Reading Exercise Suggest alternatives to boundary conditions (9.24) which will result in cosine functions as the eigenfunctions for Eq. (9.23).

Example 9.2 Determine the eigenvalues and corresponding eigenfunctions for the SLP:

$$y''(x) + \lambda y(x) = 0, \tag{9.28}$$

$$y'(0) = 0, \quad y(l) = 0. \tag{9.29}$$

Solution As in the previous example, the reader may check as a *Reading Exercise* that the parameter λ must be positive in order to have nontrivial solutions. Thus, we may write $\lambda = \mu^2$, so that we have oscillating solutions given by

$$y(x) = C_1 \sin \mu x + C_2 \cos \mu x.$$

The boundary condition $y'(0) = 0$ gives $C_1 = 0$, and the boundary condition $y(l) = 0$ gives $C_2 \cos \mu l = 0$. If $C_2 = 0$ we have a trivial solution; otherwise, we have $\mu_n = (2n - 1)\pi/2l$, for $n = 1, 2, \ldots$. Therefore, the eigenvalues are

$$\lambda_n = \mu_n^2 = \left[\frac{(2n-1)\pi}{2l} \right]^2 \tag{9.30}$$

and the eigenfunctions are

$$y_n(x) = C_n \cos \frac{(2n-1)\pi x}{2l}, \qquad n = 1, 2, \ldots \tag{9.31}$$

We leave it to the reader to prove that the eigenfunctions in Eq. (9.31) are orthogonal on the interval $[0, l]$. The reader may also normalize these eigenfunctions to find the normalization constant C_n which is equal to $\sqrt{2/l}$.

Example 9.3 Determine the eigenvalues and eigenfunctions for the SLP:

$$y''(x) + \lambda y(x) = 0, \tag{9.32}$$
$$y(0) = 0, \quad y'(l) + hy(l) = 0. \tag{9.33}$$

Solution As in the previous examples, nontrivial solutions exist only when $\lambda > 0$ (the reader should verify this as a *Reading Exercise*). Letting $\lambda = \mu^2$ we obtain a general solution as

$$y(x) = C_1 \sin \mu x + C_2 \cos \mu x.$$

From the boundary condition $y(0) = 0$, we have $C_2 = 0$. The other boundary condition gives $\mu \cos \mu l + h \sin \mu l = 0$. Thus, the eigenvalues are given by the equation:

$$\tan \mu_n l = -\mu_n/h. \tag{9.34}$$

We can obtain these eigenvalues by plotting $\tan \mu_n l$ and $-\mu_n/h$ on the same graph as in Fig. 9.1. The graph is plotted for positive μ, because negative μ do not bring new solutions.

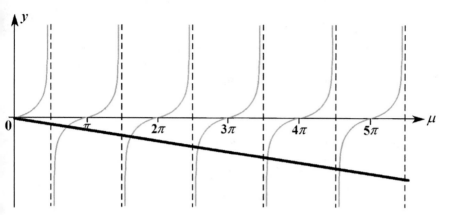

Fig. 9.1 The functions $\tan\mu_n l$ and $-\mu_n/h$ (for $h = 5$) plotted against μ. The eigenvalues of the Sturm-Liouville problem in Example 9.3 are given by the intersections of these lines

From the figure it is directly seen that there is an infinite number of discrete eigenvalues. The eigenfunctions

$$y_n(x) = C_n \sin\mu_n x, \quad n = 1, 2, \ldots \tag{9.35}$$

are orthogonal so that

$$\int_0^l \sin\mu_n x \cdot \sin\mu_m x\, dx = 0 \quad \text{for } m \neq n. \tag{9.36}$$

The orthogonality condition shown in Eq. (9.36) follows from the general theory as a direct consequence of the fact that the operator $L = -\frac{d^2}{dx^2}$ is Hermitian for the boundary conditions (9.33). We leave it to the reader to verify the previous statement as a *Reading Exercise*.

The normalized eigenfunctions are

$$y_n(x) = \sqrt{\frac{2(\mu_n^2 + h^2)}{l(\mu_n^2 + h^2) + h}} \sin\mu_n x. \tag{9.37}$$

Example 9.4 Solve the SLP:

$$y'' + \lambda y = 0, \quad 0 < x < l, \tag{9.38}$$

on the interval $[0, l]$ with periodic boundary conditions

$$y(0) = y(l), \quad y'(0) = y'(l). \tag{9.39}$$

Solution Again, verify as a *Reading Exercise* that nontrivial solutions exist only when $\lambda > 0$ (for which we will have oscillating solutions as before). Letting $\lambda = \mu^2$ we can write a general solution in the form:

$$y(x) = C_1 \cos \mu x + C_2 \sin \mu x.$$

The boundary conditions in Eq. (9.36) give:

$$\begin{cases} C_1(\cos \mu l - 1) + C_2 \sin \mu l = 0, \\ - C_1 \sin \mu l + C_2(\cos \mu l - 1) = 0. \end{cases} \tag{9.40}$$

This system of homogeneous algebraic equations for C_1 and C_2 has nontrivial solution only when its determinant is equal to zero:

$$\begin{vmatrix} \cos \mu l - 1 & \sin \mu l \\ - \sin \mu l & \cos \mu l - 1 \end{vmatrix} = 0, \tag{9.41}$$

which yields

$$\cos \mu l = 1. \tag{9.42}$$

The roots of Eq. (9.42) are

$$\lambda_n = \left(\frac{2\pi n}{l} \right)^2, \quad n = 0, 1, 2, \ldots \tag{9.43}$$

With these values of λ_n, Eq. (9.40) for C_1 and C_2 has two linearly independent nontrivial solutions given by

$$C_1 = 1, C_2 = 0, \quad \text{and} \quad C_1 = 0, C_2 = 1. \tag{9.44}$$

Substituting each set into the general solution, we obtain the eigenfunctions:

$$y_n^{(1)}(x) = \cos \sqrt{\lambda_n} x \quad \text{and} \quad y_n^{(2)}(x) = \sin \sqrt{\lambda_n} x. \tag{9.45}$$

In particular, when $l = 2\pi$ we have

$$y_n^{(1)}(x) = \cos nx \quad \text{and} \quad y_n^{(2)}(x) = \sin nx. \tag{9.46}$$

Therefore, for the eigenvalue $\lambda_0 = 0$, we have the eigenfunction $y_0(x) = 1$ (and a trivial solution $y(x) \equiv 0$). Each nonzero eigenvalue, λ_n, has two linearly independent eigenfunction so that for this example we have twofold degeneracy:

$$\|y_0\|^2 = l, \quad \|y_n\|^2 = \frac{l}{2} \quad \text{(for } n = 1, 2, \ldots\text{)}. \tag{9.47}$$

Example 9.5 Obtain the expansion of the function $f(x) = x^2(1 - x)$ using the eigenfunctions of the SLP:

$$y'' + \lambda y = 0, \quad 0 \le x \le \pi/2, \tag{9.48}$$

$$y'(0) = y'(\pi/2) = 0. \tag{9.49}$$

Solution   First, prove as a *Reading Exercise* that the eigenvalues and eigenfunctions of this boundary-value problem are

$$\lambda = 4n^2, \quad y_n(x) = \cos 2nx, \quad n = 0, 1, 2, \ldots \tag{9.50}$$

A Fourier series expansion, given in Eq. (9.16), of the function $f(x)$ using the eigenfunctions above is

$$x^2(1-x) = \sum_{n=0}^{\infty} c_n y_n(x) = \sum_{n=0}^{\infty} c_n \cos 2nx. \tag{9.51}$$

Since f and f' are continuous functions, this expansion will converge to $x^2(1 - x)$ for $0 < x < \pi/2$ as was stated previously. In Eq. (9.48) we see that the function $r(x) = 1$; thus, the coefficients of this expansion obtained from Eq. (9.17) are

$$c_0 = \frac{\displaystyle\int_0^{\pi/2} x^2(1-x)\,dx}{\displaystyle\int_0^{\pi/2} dx} = \frac{\pi^2}{4}\left(\frac{1}{3} - \frac{\pi}{8}\right),$$

$$c_n = \frac{\displaystyle\int_0^{\pi/2} x^2(1-x)\cos 2nx\,dx}{\displaystyle\int_0^{\pi/2} \cos^2 2nx\,dx} = \frac{(-1)^n}{n^4}\left[1 - \frac{3\pi}{4} + \frac{3}{2\pi n^2}\right] - \frac{3}{2\pi n^4},$$

$$n = 1, 2, 3, \ldots$$

Figure 9.2 shows the partial sum ($n = 10$) of this series, compared with the original function $f(x) = x^2(1 - x)$.

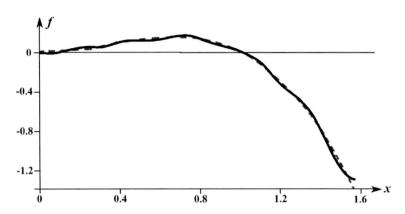

Fig. 9.2 Graphs of the function $f(x) = x^2(1 - x)$ (dashed line) and partial sum with $n = 10$ of the Fourier expansion of $f(x)$ (solid line)

Two important classes of *special functions*, Legendre and Bessel functions, are discussed in detail in Chap. 10. In the two following examples, they serve simply as illustrations of Sturm-Liouville problems.

Example 9.6 (*Fourier-Legendre Series*)
    The Legendre equation is

$$\frac{d}{dx}\left[\left(1- x^2\right)y'\right] + \lambda y = 0 \tag{9.52}$$

for x on the closed interval $[-1, 1]$. There are no boundary conditions in a straight form because $p(x) = 1 - x^2$ vanishes at the endpoints. However, we seek a finite solution, a condition which in this case acts as a boundary condition.

Solution The Legendre polynomials, $P_n(x)$, are the only solutions of Legendre's equation that are bounded on the closed interval $[-1, 1]$. The set of functions $\{P_n(x)\}$ where $n = 0, 1, 2, \ldots$, is orthogonal with respect to the weight function $r(x) = 1$ on the interval $[-1, 1]$ in which case the orthogonality relation is

$$\int_{-1}^{1} P_n(x)P_m(x)dx = 0 \quad \text{for} \ \ m \neq n. \tag{9.53}$$

The eigenfunctions for this problem are thus $P_n(x)$ with eigenvalues $\lambda = n(n + 1)$ for $n = 0, 1, 2, \ldots$ (see Sect. 10.3).

If $f(x)$ is a piecewise smooth on $[-1, 1]$, the series

$$\sum_{n=0}^{\infty} c_n P_n(x) \tag{9.54}$$

converges to

$$\frac{1}{2}[f(x_0 + 0) + f(x_0 - 0)] \tag{9.55}$$

at any point x_0 on $(-1, 1)$. Because $r(x) = 1$ in Eq. (9.17), the coefficients c_n are

$$c_n = \frac{\displaystyle\int_{-1}^{1} f(x) P_n(x) dx}{\displaystyle\int_{-1}^{1} P_n^2(x) dx}, \tag{9.56}$$

or written in terms of the scalar product

$$c_n = \frac{f(x) \cdot P_n(x)}{P_n(x) \cdot P_n(x)}. \tag{9.57}$$

Example 9.7 (*Fourier-Bessel Series*)
Consider the Sturm-Liouville problem:

$$(xy')' + \left(\lambda x - \frac{\nu^2}{x}\right) y = 0, \quad 0 \leq x \leq 1 \tag{9.58}$$

with boundary conditions such that $y(0)$ is finite and $y(1) = 0$. Here ν is a constant.

Solution   The eigenvalues for this problem are $\lambda = j_n^2$ for $n = 1, 2, \ldots$, where $j_1, j_2, j_3,$ $\ldots$ are the positive zeros of the functions $J_\nu(x)$ which are Bessel functions of order ν (see Sect. 10.2). If $f(x)$ is a piecewise smooth function on the interval $[0, 1]$, then for $0 < x < 1$ it can be resolved in the series:

$$\sum_{n=1}^{\infty} c_n J_\nu(j_n x), \tag{9.59}$$

which converges to

$$\frac{1}{2}[f(x_0 + 0) + f(x_0 - 0)]. \tag{9.60}$$

Since, in this SLP, $r(x) = x$, the coefficients c_n are

$$c_n = \frac{\displaystyle\int_0^1 xf(x)J_\nu(j_n x)dx}{\displaystyle\int_0^1 xJ_\nu^2(j_n x)dx}, \tag{9.61}$$

or in terms of the scalar product

$$c_n = \frac{f(x) \cdot J_\nu(j_n x)}{J_\nu(j_n x) \cdot J_\nu(j_n x)}. \tag{9.62}$$

Problems

In problems 1 through 7, find eigenvalues and eigenfunctions of the SLP for the equation:

$$y''(x) + \lambda y(x) = 0$$

with the following boundary conditions:

1. $y(0) = 0$, $y'(l) = 0$.
2. $y'(0) = 0$, $y(\pi) = 0$.
3. $y'(0) = 0$, $y'(1) = 0$.
4. $y'(0) = 0$, $y'(1) + y(1) = 0$.
5. $y'(0) + y(0) = 0$, $y(\pi) = 0$.
6. $y(-l) = y(l)$, $y'(-l) = y'(l)$ (periodic boundary conditions). As noted above, if the boundary conditions are periodic, then the eigenvalues can be degenerate. Show that in this problem two linearly independent eigenfunctions exist for each eigenvalue.
7. For the operator $L = -d^2/dx^2$ acting on functions $y(x)$ defined on the interval $[0, 1]$, find its eigenvalues and eigenfunctions (assume $r(x) = 1$):

    (a) $y(0) = 0$, $y'(1) + y(1) = 0$.
    (b) $y'(0) - y(0) = 0$, $y(1) = 0$.
    (c) $y(0) = y(1)$, $y'(0) = y'(1)$.

9.3 Nonhomogeneous BVPs

9.3.1 Solvability Condition

In the present section, we analyze the *nonhomogeneous* boundary-value problem:

$$[p(x)y'(x)]' + [q(x) + \lambda r(x)]y(x) = f(x), \quad a < x < b, \tag{9.63}$$

$$(\alpha_1 y' + \beta_1 y)|_{x=a} = 0, \quad (\alpha_2 y' + \beta_2 y)|_{x=b} = 0. \tag{9.64}$$

Let us begin with some simple analytically solvable problems.

Example 9.8 As the first example, we solve the following problem:

$$y''(x) + \lambda y(x) = 1, \tag{9.65}$$

$$y(0) = y(l) = 0, \tag{9.66}$$

where $\lambda > 0$ *is not* an eigenvalue of the corresponding homogeneous Sturm-Liouville problem, i.e.,

$$\lambda \neq \left(\frac{n\pi}{l}\right)^2, \quad n = 1, 2, \ldots \tag{9.67}$$

(see Example (9.1)). The solution of the nonhomogeneous problem (9.65) is

$$y(x) = y_h(x) + y_p(x), \tag{9.68}$$

where

$$y_h(x) = C_1 \sin \sqrt{\lambda}x + C_2 \cos \sqrt{\lambda}x \tag{9.69}$$

is the general solution of the corresponding homogeneous problem and

$$y_p(x) = \frac{1}{\lambda} \tag{9.70}$$

is the obvious particular solution of the nonhomogeneous problem. Substituting (9.68) into (9.66), we find:

$$C_2 + \frac{1}{\lambda} = 0, \quad C_1 \sin \sqrt{\lambda}l + C_2 \cos \sqrt{\lambda}l + \frac{1}{\lambda} = 0. \tag{9.71}$$

Because the determinant of the algebraic system (9.71)

$$\Delta(\lambda) = \begin{vmatrix} 0 & 1 \\ \sin \sqrt{\lambda}l & \cos \sqrt{\lambda} \end{vmatrix} = -\sin \sqrt{\lambda}l \neq 0,$$

due to (9.67), system (9.68) has a unique solution:

$$C_2 = -\frac{1}{\lambda}, \quad C_1 = \frac{\cos \sqrt{\lambda}l - 1}{\lambda \sin \sqrt{\lambda}l}.$$

Finally, we obtain:

$$y(x) = \frac{\left(\cos \sqrt{\lambda}l - 1\right) \sin \sqrt{\lambda}x - \left(\cos \sqrt{\lambda}x - 1\right) \sin \sqrt{\lambda}l}{\lambda \sin \sqrt{\lambda}l}.$$

Thus, in the case where λ is not an eigenvalue of the homogeneous problem, the nonhomogeneous problem has a unique solution.

Example 9.9 Let us consider again the BVP (9.65), (9.66), but this time with

$$\lambda = \lambda_1 = \left(\frac{\pi}{l}\right)^2.$$

Because $\sin \pi = 0$, $\cos \pi = -1$, system (9.71) becomes:

$$C_2 + \left(\frac{l}{\pi}\right)^2 = 0, \quad -C_2 + \left(\frac{l}{\pi}\right)^2 = 0. \tag{9.72}$$

Obviously, system (9.72) has no solutions. Thus, in the case where the homogeneous system has a nontrivial solution, the nonhomogeneous problem can have no solutions.

Example 9.10 Let us take now

$$\lambda = \lambda_2 = \left(\frac{2\pi}{l}\right)^2$$

with

$$\sin 2\pi = 0, \quad \cos 2\pi = 1.$$

Now system (9.71) becomes:

$$C_2 + \left(\frac{l}{2\pi}\right)^2 = 0, \quad C_2 + \left(\frac{l}{2\pi}\right)^2 = 0. \tag{9.73}$$

System (9.73) has an infinite number of solutions: $C_2 = -(l/2\pi)^2$ and C_1 is arbitrary. We can conclude that in the case where λ is an eigenvalue, BVP (9.65), (9.66) can have infinitely many solutions.

A question appears: when does nonhomogeneous problem (9.63), (9.64) with λ equal to an eigenvalue λ_n have no solutions, and when does it have infinitely many solutions? To answer that question, let us multiply both sides of (9.63) with the corresponding eigenfunction $y_n(x)$ and integrate them from a to b. In the right-hand side, we obtain:

$$\int_a^b f(x)y_n(x)dx. \tag{9.74}$$

In the left-hand side, using the integration by parts, we find:

$$\int_a^b y_n(x)\left\{[p(x)y'(x)]' + [q(x) + \lambda_n r(x)]y(x)\right\}dx =$$

$$= \int_a^b \left\{[p(x)y'(x)]' + [q(x) + \lambda_n r(x)]y(x)\right\}dx$$

$$+ y_n(x)p(x)y'(x)\big|_a^b - y_n'(x)p(x)y(x)\big|_a^b.$$

Because $y_n(x)$ is the solution of the homogeneous BVP (9.2), (9.4), the integral term is equal to zero. The rest of the terms can be written as

$$p(b)\left[y_n(b)y'(b) - y_n'(b)y(b)\right] - p(a)\left[y_n(a)y'(a) - y_n'(a)y(a)\right]. \tag{9.75}$$

Because both $y_n(x)$ and $y(x)$ satisfy the same boundary conditions, expression (9.75) is equal to zero. Thus, we obtain the following *solvability criterion* for the nonhomogeneous boundary-value problem (9.63), (9.64) at $\lambda = \lambda_n$:

$$\int_a^b f(x)y_n(x)dx = 0. \tag{9.76}$$

Note that the integral (9.76) does not contain the weight function $r(x)$.

Now we can understand why we came to different conclusions when considering Examples 9.9 and 9.10. In Example 9.9, for $\lambda = \lambda_1 = (\pi/l)^2$, we have:

$$y_1(x) = \sin \frac{\pi x}{l}. \tag{9.77}$$

Calculating the integral (9.74) with (9.77), $f = 1$, $a = 0$, and $b = l$, we obtain:

$$\int_0^l \sin \frac{\pi x}{l} dx = -\frac{l}{\pi} \cos \frac{\pi x}{l} \Big|_0^l = \frac{2l}{\pi} \neq 0.$$

Thus, condition (9.76) is not satisfied, and the problem has no solutions. In Example 9.10, for $\lambda = \lambda_2 = (2\pi/l)^2$ and $y_2(x) = \sin(2\pi x/l)$, expression (9.74) is

$$\int_0^l \sin \frac{2\pi x}{l} dx = -\frac{l}{2\pi} \cos \frac{2\pi x}{l} \Big|_0^l = 0.$$

Criterion (9.76) is satisfied, and the problem is solvable. Actually, in that case it has infinitely many solutions, because an arbitrary solution of the homogeneous problem $Cy_2(x)$ can be added without violation of the equation and boundary conditions.

We can summarize our observations as the following *Fredholm alternative*:

1. If λ is not an eigenvalue of the Sturm-Liouville problem (9.2), (9.4), i.e., the homogeneous problem has no nontrivial solutions, then the nonhomogeneous boundary-value problem (9.63), (9.64) has a unique solution for arbitrary continuous function $f(x)$.

2. If $\lambda = \lambda_n$ is an eigenvalue of the Sturm-Liouville problem (9.2), (9.4), i.e., the homogeneous problem has a nontrivial solution, then the nonhomogeneous boundary-value problem (9.63), (9.64):

    (i) Either has no solutions, when the right-hand side of the equation, $f(x)$ is not orthogonal to the eigenfunction $y_n(x)$:

    $$\int_a^b f(x)y_n(x)dx \neq 0,$$

    (ii) Or has infinitely many solutions, if $f(x)$ is orthogonal to $y_n(x)$:

    $$\int_a^b f(x)y_n(x)dx = 0.$$

Example 9.11 Solve the BVP:

$$y'' + y' = 1, \quad x \in (0, 1);$$
$$y'(0) = 0, y(1) = 1 - \text{the boundary conditions.}$$

In this problem the value of the function is given at one end of the interval and the value of the function's derivative at another end. Note that one of the boundary conditions is not homogeneous.

First, let us consider the homogeneous problem with homogeneous boundary conditions. The solution of the homogeneous equation $y'' + y' = 0$ is $Y(x) = C_1 + C_2 e^{-x}$. Substituting that solution into boundary conditions, we find $C_2 = 0$ and $C_1 + C_2 e^{-1} = 0$; hence, $C_1 = C_2 = 0$. Thus, the homogeneous problem has no nontrivial solution. In other words, the SLP

$$y'' + y' + \lambda y = 0, \quad 0 < x < 1; \quad y'(0) = 0, y(1) = 0$$

has no eigenvalue $\lambda = 0$.

Therefore, the solution of the BVP exists and is unique, and we can find it explicitly.

Searching a particular solution of the nonhomogeneous equation in the form $\bar{y}(x) = ax$, we obtain $a = 1$; thus, $\bar{y}(x) = x$. Therefore, a general solution of the given equation is

$$y(x) = C_1 + C_2 e^{-x} + x.$$

From there, $y'(x) = -C_2 e^{-x} + 1$. Substitution it in the boundary conditions gives:

$$\begin{cases} -C_2 + 1 = 0, \\ C_1 + C_2 e^{-1} + 1 = 1, \end{cases}$$

and we obtain $C_2 = 1$, $C_1 = -1/e$; thus, the solution of the BVP is

$$y(x) = -e^{-1} + e^{-x} + x.$$

Example 9.12 Solve the BVP:

$$y'' + y = 2x - \pi, \quad x \in [0, \pi];$$
$$y(0) = 0, y(\pi) = 0.$$

The homogeneous problem has the nontrivial solution $y_h(x) = \sin x$. Thus, we have to check the solvability condition and calculate:

$$I = \int\limits_0^\pi (2x - \pi) \sin x \, dx.$$

We find that $I = 0$; hence, we can expect that there will be a family of solutions. Indeed, the general solution of the given equation is

$$y(x) = C_1 \cos x + C_2 \sin x + 2x - \pi.$$

Either boundary condition gives $C_1 = \pi$, but the value of C_2 cannot be determined. Thus, the solution of the BVP is

$$y(x) = \pi \cos x + C \sin x + 2x - \pi$$

with an arbitrary C.

Example 9.13 Solve the BVP:

$$y'' + y = x, \quad x \in [0, \pi];$$
$$y(0) = y(\pi) = 0.$$

The nontrivial solution of the homogeneous problem is $y_h(x) = \sin x$. To check the solvability condition, we calculate:

$$I = \int\limits_0^\pi x \sin x \, dx$$

and find that $I \neq 0$; hence, the BVP has no solutions.

Indeed, the general solution of the equation is $y(x) = C_1 \cos x + C_2 \sin x + x$. Substituting that solution into boundary conditions, we find $C_1 = 0$ and $-C_1 + \pi = 0$, which is impossible.

Example 9.14 Solve the BVP:

$$y'' - y = 1, \quad x \in [0, \infty),$$
$$y(0) = 0, y \text{ is bounded at } x \to \infty.$$

The second boundary condition describes the behavior of the function we search at $x \to \infty$.

The solution of the homogeneous equation is $Y(x) = C_1 e^x + C_2 e^{-x}$. There are no solutions of the homogeneous equation satisfying the imposed boundary conditions; therefore, the solution of the inhomogeneous problem exists and is unique.

Searching a particular solution of the nonhomogeneous equation in the form $\bar{y}(x) = a$, we obtain $a = -1$; thus, $\bar{y}(x) = -1$. Therefore, a general solution of the given equation is $y(x) = C_1 e^x + C_2 e^{-x} - 1$.

The second boundary condition can be satisfied only if $C_1 = 0$. Then, the condition $y(0) = 0$ gives $C_2 - 1 = 0$. The solution of the BVP is

$$y(x) = e^{-x} - 1.$$

Example 9.15 Solve the BVP:

$$y'' - y = 1, \quad x \in (-\infty, \infty)$$

with the boundary conditions formulated in the following way: function $y(x)$ is bounded when $x \to \infty$ and $x \to -\infty$.

As in the previous example, the homogeneous problem has no solutions, and the solution of the inhomogeneous problem is unique. From a general solution $y(x) = C_1 e^x + C_2 e^{-x} - 1$, we see that both $C_1 = C_2 = 0$; thus, the solution of the BVP is $y(x) = -1$.

Problems

Solve the BVP:

1. $y'' + y = 1$, $y(0) = 0$, $y(\pi/2) = 0$.
2. $y'' + 4y' = 1$, $y'(0) = 0$, $y(1) = 1$.
3. $y'' - y' = 0$, $y(0) = -1$, $y'(1) - y(1) = 2$.
4. $y'' + y = 1$, $y(0) = 0$, $y(\pi) = 0$.
5. $y'' + 4y = x$, $y(0) = 0$, $y(\pi/4) = 1$.
6. $y'' - y' - 2y = 0$, $y'(0) = 2$, $y(+\infty) = 0$.
7. $y'' - y = 1$, $y(0) = 0$, $y(x)$ limited at $x \to +\infty$.
8. $y'' - y' = 0$, $y(0) = 3$, $y(1) - y'(1) = 1$.
9. $y'' + y = 1$, $y(0) = 0$, $y'(\pi/2) = 1$.
10. $y'' - 2y' - 3y = 0$.(a) $y(0) = 1$, $\lim\limits_{x \to \infty} y(x) = 0$, (b) $y'(0) = 2$, $\lim\limits_{x \to \infty} y'(x) = 0$.

9.3.2 The General Solution of Nonhomogeneous Linear Equation

Let us consider the case where the homogeneous problem (9.2), (9.4) has no nontrivial solutions; hence, the solution of the problem (9.63), (9.64) is unique, in more detail. Denote $Q(x) = q(x) + \lambda r(x)$ and rewrite problem (9.63), (9.64) as

$$L[y(x)] \equiv [p(x)y'(x)]' + Q(x)y(x) = f(x), \quad a < x < b, \tag{9.78}$$

$$U_1(y) \equiv (\alpha_1 y' + \beta_1 y)|_{x=a} = 0, \quad U_2(y) \equiv (\alpha_2 y' + \beta_2 y)|_{x=b} = 0. \tag{9.79}$$

To distinguish between the solution of the nonhomogeneous BVP (9.78), (9.79) and solutions of the homogeneous Eq. (9.2), denote the latter solutions as $u(x)$:

$$L[u(x)] \equiv [p(x)u'(x)]' + Q(x)u(x) = 0. \tag{9.80}$$

First, let us derive a convenient integral form for the general solution of the nonhomogeneous Eq. (9.78). According to the general scheme of the method of parameter variation described in Chap. 3, the general solution of (9.78) can be presented as

$$y(x) = c_1(x)u_1(x) + c_2(x)u_2(x), \tag{9.81}$$

where $u_1(x)$ and $u_2(x)$ are linearly independent solutions of (9.80). Functions $c_1(x)$ and $c_2(x)$ satisfy the system of equations:

$$c_1'u_1 + c_2'u_2 = 0, \quad c_1'u_1' + c_2'u_2' = f/p \tag{9.82}$$

(recall that it is assumed that $p(x) \neq 0$ in $[a, b]$). Solving system (9.82), we find:

$$c_1' = -\frac{u_2 f}{pW}, \quad c_2' = \frac{u_1 f}{pW}, \tag{9.83}$$

where

$$W = u_1 u_2' - u_2 u_1'$$

is the Wronskian.

Let us rewrite Eq. (9.80) as

$$y'' + \frac{p'(x)}{p(x)} y' + \frac{Q(x)}{p(x)} y = 0$$

and apply formula (3.24) in Chap. 3. We find:

$$W'(x) + \frac{p'(x)}{p(x)} W(x) = 0$$

i.e.,

$$(pW')' = 0.$$

Therefore,

$$p(x)W(x) \equiv c = const. \tag{9.84}$$

Thus, formulas (9.83) can be rewritten as

$$c_1'(x) = -\frac{u_2(x)f(x)}{c}, \quad c_2'(x) = \frac{u_1(x)f(x)}{c};$$

hence,

$$c_1(x) = c_1(x_0) - \frac{1}{c} \int_{x_0}^{x} f(\xi)u_2(\xi)d\xi, \quad c_2(x) = c_2(x_0) + \frac{1}{c} \int_{x_0}^{x} f(\xi)u_1(\xi)d\xi,$$

where x_0 is an arbitrary point in the closed interval $a \leq x \leq b$. Thus, the general solution of Eq. (9.78) is

$$y(x) = c_1(x_0)u_1(x) + c_2(x_0)u_2(x) + \int_{x_0}^{x} \frac{u_1(\xi)u_2(x) - u_1(x)u_2(\xi)}{c} f(\xi)d\xi. \tag{9.85}$$

Relation (9.85) is valid for any $x_0 \in [a, b]$. Applying that relation for $x_0 = a$ and $x_0 = b$, we obtain:

$$y(x) = c_1(a)u_1(x) + c_2(a)u_2(x) + \int_{a}^{x} \frac{u_1(\xi)u_2(x) - u_1(x)u_2(\xi)}{c} f(\xi)d\xi, \tag{9.86}$$

$$y(x) = c_1(b)u_1(x) + c_2(b)u_2(x) + \int_{b}^{x} \frac{u_1(\xi)u_2(x) - u_1(x)u_2(\xi)}{c} f(\xi)d\xi$$

$$= c_1(b)u_1(x) + c_2(b)u_2(x) + \int_{x}^{b} \frac{u_1(x)u_2(\xi) - u_1(\xi)u_2(x)}{c} f(\xi)d\xi. \tag{9.87}$$

Let us add (9.86) and (9.87) and then divide the sum by 2. We obtain:

$$y(x) = \bar{c}_1 u_1(x) + \bar{c}_2 u_2(x) + \int_{a}^{b} K(x, \xi)f(\xi)d\xi, \tag{9.88}$$

where

$$K(x, \xi) = \frac{u_1(\xi)u_2(x) - u_1(x)u_2(\xi)}{2c} \operatorname{sign}(x - \xi),$$

$$\overline{c}_1 = \frac{c_1(a) + c_1(b)}{2}, \quad \overline{c}_2 = \frac{c_2(a) + c_2(b)}{2}.$$

(9.89)

For $x \neq \xi$, the kernel $K(x, \xi)$ is the linear combination of functions $u_1(x)$ and $u_2(x)$; hence, it is a solution of the homogeneous Eq. (9.80). At $x = \xi$, function $K(x, \xi)$ is continuous:

$$K(x, \xi)|_{x=\xi^-} = K(x, \xi)|_{x=\xi^+} = 0.$$

However, that function is not smooth at $x = \xi$. Indeed,

$$K_x(x, \xi) = \frac{u_1(\xi)u_2'(x) - u_1'(x)u_2(\xi)}{2c} \operatorname{sign}(x - \xi), \quad x \neq \xi;$$

therefore,

$$K_x(x, \xi)|_{x=\xi^+} = \lim_{\varepsilon \to +0} K_x(\xi + \varepsilon, \xi) = \frac{1}{2c} W(\xi),$$

$$K_x(x, \xi)|_{x=\xi^-} = \lim_{\varepsilon \to +0} K_x(\xi - \varepsilon, \xi) = -\frac{1}{2c} W(\xi).$$

Thus, the derivative $K_x(x, \xi)$ has a jump at $x = \xi$:

$$K_x(x, \xi)|_{x=\xi^+} - K_x(x, \xi)|_{x=\xi^-} = \frac{W(\xi)}{c} = \frac{1}{p(\xi)}.$$

9.3.3   The Green's Function

Now, let us take into account the boundary conditions (9.79). Substituting (9.86) and (9.87) into (9.79), we find:

$$U_1(y) = \overline{c}_1 U_1(u_1) + \overline{c}_2 U_1(u_2) + \int_a^b U_1(K(x, \xi))f(\xi)d\xi = 0, \quad x = a; \quad (9.90)$$

$$U_2(y) = \overline{c}_1 U_2(u_1) + \overline{c}_2 U_2(u_2) + \int_a^b U_2(K(x, \xi))f(\xi)d\xi = 0, \quad x = b. \quad (9.91)$$

Recall that we consider the case where the homogeneous BVP has no nontrivial solutions. Therefore,

$$\begin{vmatrix} U_1(u_1) & U_1(u_2) \\ U_2(u_1) & U_2(u_2) \end{vmatrix} \neq 0; \tag{9.92}$$

otherwise, the homogeneous problem would have nontrivial solutions.

Relations (9.90), (9.91), and (9.92) are valid for any pair of linearly independent solutions of (9.80). Let us choose now a specific pair of functions $u_1(x)$ and $u_2(x)$ satisfying the *Cauchy problems* with the following initial conditions:

$$u_1(a) = \alpha_1, \quad u_1' = -\beta_1 \quad \text{and} \quad u_2(b) = \alpha_2, \quad u_2' = -\beta_2.$$

Obviously,

$$U_1(u_1) = 0, \quad U_2(u_2) = 0. \tag{9.93}$$

Let us emphasize that

$$U_1(u_2) \neq 0, \quad U_2(u_1) \neq 0; \tag{9.94}$$

otherwise, the homogeneous BVP would have solutions. Taking into account (9.93) and (9.94), we find that

$$U_1(K(x,\xi)) = -\frac{u_1(\xi)U_1(u_2)}{2c}, \quad U_2(K(x,\xi)) = -\frac{u_2(\xi)U_2(u_1)}{2c}. \tag{9.95}$$

Substituting (9.93), (9.95) into (9.90), (9.91), we obtain:

$$\bar{c}_2 U_1(u_2) - \int_a^b \frac{u_1(\xi)U_1(u_2)}{2c} f(\xi) d\xi = 0,$$

$$\bar{c}_1 U_2(u_1) - \int_a^b \frac{u_2(\xi)U_2(u_1)}{2c} f(\xi) d\xi = 0.$$

Taking into account (9.94), we find that

$$\bar{c}_1 = \int_a^b \frac{u_2(\xi)f(\xi)}{2c} d\xi, \quad \bar{c}_2 = \int_a^b \frac{u_1(\xi)f(\xi)}{2c} d\xi. \tag{9.96}$$

Substituting (9.96) into (9.88), we obtain the following integral expression for the solution of nonhomogeneous BVP (9.78), (9.79):

$$
y(x) = \int_a^x \frac{u_1(\xi)u_2(x)}{c} f(\xi)d\xi + \int_x^b \frac{u_2(\xi)u_1(x)}{c} f(\xi)d\xi =
$$

$$
= \int_a^b G(x,\xi)f(\xi)d\xi,
$$

(9.97)

where

$$
G(x,\xi) = \frac{u_1(\xi)u_2(x)}{c}, \quad a \le \xi \le x; \quad G(x,\xi) = \frac{u_1(x)u_2(\xi)}{c}, \quad x \le \xi \le b.
$$

Here $u_1(x)$ and $u_2(x)$ are solutions of Eq. (9.80) satisfying conditions (9.93), and c is determined by the relation (9.84). The function $G(x,\xi)$ is the *Green's function* of problem (9.78), (9.79). The integral operator (9.97) that transforms $f(x)$ into $y(x)$ is the inverse operator to the differential operator (9.78) that transforms $y(x)$ into $f(x)$.

Let us list the basic properties of the Green's function $G(x,\xi)$, named after a remarkable self-taught British mathematical physicist George Green (1793–1841).

1.	In each of the regions $a < x < \xi$ and $\xi < x < b$, function $G(x,\xi)$, as a function of x, is a solution of the homogeneous problem: $$LG(x,\xi) \equiv [p(x)G_x(x,\xi)]_x + Q(x)G(x,\xi) = 0 \qquad (9.98)$$ (here the subscript x means a partial derivative with respect to x). It satisfies the homogeneous boundary conditions: $$x = a, \quad U_1[G(x,\xi)] = \frac{U_1[u_1(x)]u_2(\xi)}{c} = 0, \qquad (9.99)$$ $$x = b, \quad U_2[G(x,\xi)] = \frac{U_2[u_2(x)]u_1(\xi)}{c} = 0. \qquad (9.100)$$
2.	Function $G(x,\xi)$ is continuous in the point $x = \xi$: $$G(\xi^+,\xi) = G(\xi^-,\xi) = \frac{u_1(\xi)u_2(\xi)}{c}. \qquad (9.101)$$
3.	Function $G(x,\xi)$ is not smooth in the point $x = \xi$: $$G_x(x,\xi)\vert_{x=\xi^+} - G_x(x,\xi)\vert_{x=\xi^-} = \frac{W(\xi)}{c} = \frac{1}{p(\xi)}. \qquad (9.102)$$

One can prove that the solution of the problem (9.98), (9.99), (9.100), (9.101), and (9.102) and that solution can be used for solving the nonhomogeneous BVP by means of formula (9.97).

Example 9.16 As an example, let us consider the problem:

$$
y''(x) + \lambda y(x) = f(x), \quad 0 < x < l; \quad y(0) = y(l) = 0, \qquad (9.103)
$$

where $\lambda > 0$, $\lambda \ne (n\pi/l)^2$, $n = 1, 2, \dots$

The Green's function $G(x, \xi)$ is the solution of the following problem:

$$G_{xx}(x, \xi) + \lambda G(x, \xi) = 0 \quad \text{for} \quad 0 < x < \xi \text{ and } \xi < x < l; \qquad (9.104)$$

$$G(0, \xi) = 0; \quad G(l, \xi) = 0; \qquad (9.105)$$

$$G(\xi^+, \xi) = G(\xi^-, \xi); \qquad (9.106)$$

$$G_x(\xi^+, \xi) - G_x(\xi^-, \xi) = 1. \qquad (9.107)$$

First, let us consider the problem in the region $0 < x < \xi$:

$$G_{xx}(x, \xi) + \lambda G(x, \xi) = 0; \quad G(0, \xi) = 0.$$

We obtain:

$$G(x, \xi) = C_1(\xi) \sin \sqrt{\lambda} x, \qquad (9.108)$$

where $C_1(\xi)$ is a still unknown function of ξ. Similarly, in the region $\xi < x < l$, we have the problem:

$$G_{xx}(x, \xi) + \lambda G(x, \xi) = 0; \quad G(l, \xi) = 0.$$

The solution can be written as

$$G(x, \xi) = C_2(\xi) \sin \sqrt{\lambda}(x - l). \qquad (9.109)$$

Calculating the derivatives

$$G_x(x, \xi) = C_1(\xi) \sqrt{\lambda} \cos \sqrt{\lambda} x, \quad 0 < x < \xi;$$
$$G_x(x, \xi) = C_2(\xi) \sqrt{\lambda} \cos \sqrt{\lambda}(x - l) \qquad (9.110)$$

and substituting (9.108), (9.109), (9.110), and (9.110) into (9.106) and (9.107), we find:

$$C_1(\xi) \sin \sqrt{\lambda} \xi - C_2(\xi) \sin \sqrt{\lambda}(\xi - l) = 0;$$
$$- C_1(\xi) \sqrt{\lambda} \cos \sqrt{\lambda} \xi + C_2(\xi) \sqrt{\lambda} \cos \sqrt{\lambda}(\xi - l) = 1. \qquad (9.111)$$

Solving system (9.111), we find:

$$C_1(\xi) = \frac{\sin \sqrt{\lambda}(\xi - l)}{\sqrt{\lambda} \sin \sqrt{\lambda} l}, \quad C_2(\xi) = \frac{\sin \sqrt{\lambda} \xi}{\sqrt{\lambda} \sin \sqrt{\lambda} l};$$

hence,

$$G(x,\xi) = \frac{\sin \sqrt{\lambda}(\xi - l)\sin \sqrt{\lambda}x}{\sqrt{\lambda}\sin \sqrt{\lambda}l}, \quad 0 \le x \le \xi \le l$$

and

$$G(x,\xi) = \frac{\sin \sqrt{\lambda}\xi \sin \sqrt{\lambda}(x - l)}{\sqrt{\lambda}\sin \sqrt{\lambda}l}, \quad 0 \le \xi \le x \le l.$$

Thus, the solution of problem is

$$y(x) = \frac{1}{\sqrt{\lambda}\sin \sqrt{\lambda}l} \times$$

$$\times \left[\sin \sqrt{\lambda}(\xi - l) \int_0^x f(\xi) \sin \sqrt{\lambda}\xi d\xi + \sin \sqrt{\lambda}x \int_x^l f(\xi) \sin \sqrt{\lambda}(\xi - l)d\xi \right].$$

Specifically, for $f(\xi) = 1$, we calculate integrals:

$$\int_0^x \sin \sqrt{\lambda}\xi d\xi = \frac{1 - \cos \sqrt{\lambda}x}{\sqrt{\lambda}},$$

$$\int_x^l \sin \sqrt{\lambda}(\xi - l)d\xi = \frac{-1 + \cos \sqrt{\lambda}(x - l)}{\sqrt{\lambda}}$$

and find

$$y(x) = \frac{1}{\sqrt{\lambda}\sin \sqrt{\lambda}l}\left[\sin \sqrt{\lambda}(x - l)\left(1 - \cos \sqrt{\lambda}x\right) + \sin \sqrt{\lambda}x\left(-1 + \cos \sqrt{\lambda}(x - l)\right)\right] =$$
$$= \frac{1}{\sqrt{\lambda}\sin \sqrt{\lambda}l}\left[\sin \sqrt{\lambda}(x - l) - \sin \sqrt{\lambda}\xi + \sin \sqrt{\lambda}l\right] =$$
$$= \frac{1}{\sqrt{\lambda}\sin \sqrt{\lambda}l}\left[\sin \sqrt{\lambda}x\left(\cos \sqrt{\lambda}l - 1\right) - \sin \sqrt{\lambda}l\left(\cos \sqrt{\lambda}x - 1\right)\right],$$

which coincides with the result of Example 9.8.

Note that the problem (9.98), (9.99), (9.100), (9.101), and (9.102) can be reformulated in a more compact form using the notion of the δ-function (see Sect. 8.12). Let us define the *Heaviside function* $H(x)$ by the following relations:

$$H(x) = 0, \quad x < 0; \quad H(0) = 1/2; \quad H(x) = 1, \quad x > 0.$$

(Oliver Heaviside (1850–1925) was another British self-taught mathematical phys-icist, whose primary contribution was a re-formulation of Maxwell's equations. His form of these equations is commonly used today.)

Function $G_x(x, \xi)$, which has a jump equal to $1/p(\xi)$ at $x = \xi$, can be considered as a sum of a continuous function and the function $H(x - \xi)/p(\xi)$. Using the definition of the δ-function, we can write that $H(x) = \int\limits_{-\infty}^{x} \delta(y)dy$; therefore, $\delta(x) = dH(x)/dx$.

Thus, $G_{xx}(x, \xi)$ is a sum of a continuous function and the function $\delta(x - \xi)/p(\xi)$. That allows us to rewrite formulas (9.98) and (9.102) as

$$LG(x, \xi) = [p(x)G_x(x, \xi)]_x + Q(x)G(x, \xi) = \delta(x - \xi). \qquad (9.112)$$

Equation (9.112) is solved on the interval $x \in (a, b)$ with boundary conditions (9.99), (9.100).

Problems

1. The following problem is given:

$$\left[(1 - x^2)y'\right]' + 6y = f(x), \quad -\frac{1}{\sqrt{3}} < x < \frac{1}{\sqrt{3}},$$

$$y\left(-\frac{1}{\sqrt{3}}\right) = y\left(\frac{1}{\sqrt{3}}\right) = 0.$$

Find the solvability condition for that problem. Give an example of function $f(x)$ that satisfies the solvability condition.

2. Using the appropriate Green's function, find the solution of the following problem:

$$(xy')' + xy = f(x), \quad 0 < x < 1,$$

$$y'(0) = 0, \quad y'(1) = 1.$$

3. Using the Green's function found in problem 2, find the solution of the problem:

$$(xy')' + xy = 2x, \quad 0 < x < 1,$$

$$y'(0) = 0, \quad y'(1) = 1.$$

Chapter 10
Special Functions

As we have seen in the previous chapters, typically, solutions of differential equations cannot be presented using elementary functions. Generally, the solution is a function determined by a certain integral or power series. However, there are some classes of functions, named special functions, which are ubiquitous in different branches of science. Those classes of functions have been studied thoroughly, and often they bear the names of scientists who discovered them.

In the present chapter, we describe some especially important special functions.

10.1 Gamma Function

We start our description of special functions with the *Gamma function* $\Gamma(x)$ (that function was invented and investigated by Daniel Bernoulli, Euler, and Legendre).

The Gamma function is a generalization of the *factorial*:

$$n! = n(n-1) \cdot \ldots \cdot 2 \cdot 1$$

defined only for natural numbers $n = 1, 2, 3, \ldots$, to non-integer values of the argument.

First, we define the Gamma function for arbitrary *positive* numbers by the following formula:

$$\Gamma(x) = \int_0^\infty t^{x-1} e^{-t} dt, \quad x > 0. \tag{10.1}$$

The basic property of the Gamma function is the relation:

$$\Gamma(x+1)=x\Gamma(x). \tag{10.2}$$

In order to derive that relation, we apply the integration by parts:

$$\Gamma(x+1)=\int_0^\infty t^x e^{-t}dt=-\int_{t=0}^{t=\infty} t^x d(e^{-t})=-t^x e^{-t}\Big|_{t=0}^{t=\infty}+x\int_0^\infty t^{x-1}e^{-t}dt=x\Gamma(x).$$

Starting with

$$\Gamma(1)=\int_0^\infty e^{-t}dt=1, \tag{10.3}$$

we obtain:

$$\Gamma(2)=1, \quad \Gamma(3)=2\cdot 1, \quad \Gamma(4)=3\cdot 2\cdot 1, \quad \dots.$$

Thus, for natural $x \geq 2$, we have:

$$\Gamma(x)=(x-1)! \tag{10.4}$$

As an example of a non-integer value of x, let us calculate integral (10.1) for $x=1/2$ using the substitution $t=y^2$:

$$\Gamma(1/2)=\int_0^\infty t^{-1/2}e^{-t}dt=2\int_0^\infty e^{-y^2}dy=\sqrt{\pi}.$$

Using Eq. (10.2), we find:

$$\Gamma(3/2)=(1/2)\Gamma(1/2)=\sqrt{\pi}/2.$$

Reading Exercise Show that for any integer $n \geq 1$:

$$\Gamma\left(n+\frac{1}{2}\right)=\frac{1\cdot 3\cdot 5\cdot\dots\cdot(2n-1)}{2^n}\Gamma\left(\frac{1}{2}\right). \tag{10.5}$$

Now, let us define the function $\Gamma(x)$ for negative values of x *postulating* relation (10.2), which can be written as

$$\Gamma(x) = \frac{\Gamma(x+1)}{x}, \tag{10.6}$$

for *arbitrary* x. We may use this equation to find, for example, $\Gamma(-1/2)$ in the following way:

$$\Gamma\left(-\frac{1}{2}\right) = \frac{\Gamma(1/2)}{-1/2} = -2\sqrt{\pi}.$$

It is clear that using Eqs. (10.1) and (10.2), we can find $\Gamma(x)$ for all values of x *except* 0 *and negative integers*. Indeed, integral (10.1) diverges at $x = 0$; hence, $\Gamma(0)$ is not defined. Then, from Eq. (10.2), we see that $\Gamma(-1)$ is also not defined because it involves $\Gamma(0)$. Thus, $\Gamma(x)$ diverges for any negative integer value of x. From Eq. (10.6), we find that for small x

$$\Gamma(x) = \frac{\Gamma(1+x)}{x} \sim \frac{1}{x}, \quad \Gamma(-1+x) = \frac{\Gamma(x)}{-1+x}$$

$$\sim -\frac{1}{x}, \quad \Gamma(-2+x) = \frac{\Gamma(-1+x)}{-2+x} \sim \frac{1}{2x}, \quad \dots$$

Thus,

$$\Gamma(-n+x) \sim \frac{(-1)^n}{n!x}, \quad |x| \ll 1.$$

A graph of $\Gamma(x)$ is plotted in Fig. 10.1.

Equations (10.2) and (10.6) allow us to find the value of $\Gamma(x)$ for any real x using the value of $\Gamma(x)$ on the interval $1 \le x < 2$.

For example, $\Gamma(3.4) = 2.4 \cdot \Gamma(2.4) = 2.4 \cdot 1.4 \cdot \Gamma(1.4)$.

Based on this fact, a table of values for $\Gamma(x)$ only has to include the interval $[1,2)$ for values of x (Table 10.1). In that interval, the minimum value of $\Gamma(x)$ is reached at $x = 1.46116321\dots$.

As an application of the relation between the factorial and the Gamma function, let us consider the asymptotic behavior of the factorial at large n. Using relation (10.4), we can write the following integral expression for the factorial:

$$n! = \Gamma(n+1) = \int_0^{\infty} t^n e^{-t} dt = \int_0^{\infty} e^{n \ln t - t} dt.$$

Changing the integration variable, $t = ny$, we find:

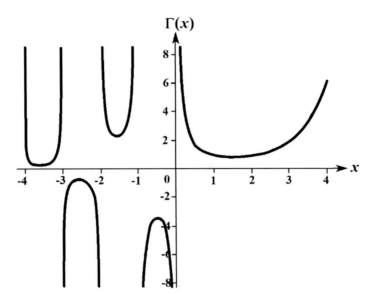

Fig. 10.1 Graph of the Gamma function, $\Gamma(x)$

Table 10.1 Values of $\Gamma(x)$ for $x \in [1,\ 2]$

x	$\Gamma(x)$	x	$\Gamma(x)$	x	$\Gamma(x)$
1	1	1.35	0.891151	1.7	0.908639
1.05	0.973504	1.4	0.887264	1.75	0.919063
1.1	0.951351	1.45	0.885661	1.8	0.931384
1.15	0.933041	1.5	0.886227	1.85	0.945611
1.2	0.918169	1.55	0.888868	1.9	0.961766
1.25	0.906402	1.6	0.893515	1.95	0.979881
1.3	0.897471	1.65	0.900116	2	1

$$n! = e^{n \ln n} n \int_0^\infty e^{n(\ln y - y)} dy.$$

One can see that the integrand has a maximum at $y = 1$; at large n, it is very sharp, i.e., the main contribution into the integral is given by the vicinity of the point $y = 1$. That circumstance allows to approximate the function in the exponent by the first terms of its Taylor expansion around the point $y = 1$:

$$n(\ln y - y) = -n - \frac{n}{2}(y - 1)^2 + \ldots .$$

Introducing the new variable

$$z = \sqrt{\frac{n}{2}}(y-1),$$

we find for large n:

$$n! \approx e^{n\ln n}ne^{-n}\int_0^\infty e^{-n(y-1)^2/2}dy = e^{n\ln n}n\sqrt{\frac{2}{n}}\int_{-\sqrt{n/2}}^\infty e^{-z^2}dz \approx \left(\frac{n}{e}\right)^n\sqrt{2n}\int_{-\infty}^\infty e^{-z^2}dz.$$

Thus,

$$n! \approx \left(\frac{n}{e}\right)^n\sqrt{2n\pi}. \tag{10.7}$$

This is the famous *Stirling's approximation* of the factorial, named after the Scottish mathematician James Stirling (1692–1770). The formula agrees very well with the precise value of $n!$ even for values of n which are not very large. For instance, for $n = 10$ the relative error of Eq. (10.7) is less than 0.8%. Most of the applications of Eq. (10.7) belong to statistical physics where it is often necessary to evaluate factorials of very large numbers.

Example 10.1 Show that

$$\int_0^\infty e^{-x^4}dx = \Gamma\left(\frac{5}{4}\right).$$

Solution With the substitution $x^4 = u$, we have:

$$\int_0^\infty e^{-x^4}dx = \int_0^\infty e^{-u}\frac{1}{4}u^{-3/4}du = \frac{1}{4}\int_0^\infty e^{-u}u^{\frac{1}{4}-1}du = \frac{1}{4}\Gamma\left(\frac{1}{4}\right) = \Gamma\left(\frac{1}{4}+1\right).$$

Example 10.2 Using the Gamma function show that

$$\int_0^1 x^k \ln x\, dx = -\frac{1}{(k+1)^2}, \quad k > -1.$$

Solution The substitution $\ln x = t$ gives:

$$\int_0^1 x^k \ln x\, dx = \int_{-\infty}^0 te^{(k+1)t}\, dt.$$

With another substitution $(k + 1)t = -u$, because $k + 1 > 0$, we obtain:

$$\int_0^1 x^k \ln x\, dx = -\frac{1}{(k+1)^2}\int_0^\infty ue^{-u}\, du = -\frac{1}{(k+1)^2}\Gamma(2) = -\frac{1}{(k+1)^2}.$$

Example 10.3 Show that the Gamma function can be defined by the integrals

$$\Gamma(z) = \int_0^1 \left[\ln\frac{1}{t}\right]^{z-1} dt, \quad \mathrm{Re}\, z > 0.$$

Solution Do the substitution $t = e^{-u}$; then, $dt = -e^{-u}du$, $\ln\frac{1}{t} = u$; u varies from ∞ to 0, t from 0 to 1; thus,

$$\int_0^1 \left[\ln\frac{1}{t}\right]^{z-1} dt = \int_\infty^0 u^{z-1}(-e^{-u})du = \int_0^\infty e^{-u}u^{z-1}du = \Gamma(z).$$

Problems

1. Evaluate $\Gamma(1)$, $\Gamma(2)$, $\Gamma(3)$, $\Gamma(4)$, $\Gamma(n)$.
2. Evaluate $\Gamma\left(\frac{1}{2}\right)$, $\Gamma\left(\frac{3}{2}\right)$, $\Gamma\left(\frac{5}{2}\right)$, $\Gamma\left(\frac{7}{2}\right)$, $\Gamma\left(n + \frac{1}{2}\right)$.
3. Evaluate the integral $\int_0^\infty e^{-x^2}dx$.
4. Evaluate the integral $\int_0^\infty e^{-x^n}dx \quad (n > 0)$.
5. Evaluate the integral $\int_0^\infty x^4 e^{-x^2}dx$.
6. Evaluate the integral $\int_0^\infty x^{2n-1}e^{-x^2}dx \quad (n > 0)$.
7. Evaluate the integral $\int_0^\infty x^m e^{-x^n}dx \quad ((m+1)/n > 0)$.
8. Evaluate the integral $\int_0^\infty \frac{dx}{\sqrt{xe^x}}$.

9. Evaluate the integral $\int_0^\infty x^p e^{-ax} \ln x \, dx$, $\quad a > 0$ using the relation:

$$\frac{d}{dp}\Gamma(p) = \int_0^\infty x^{p-1} e^{-x} \ln x \, dx.$$

10. Evaluate the integral $\int_{-\infty}^\infty x^{2n} e^{-ax^2} \, dx$, $\quad a > 0$.

10.2 Bessel Functions

10.2.1 Bessel Equation

Most of the applications of special functions are in the field of linear partial differential equations that are considered in Chap. 13. The powerful approach of *separation of variables* allows to obtain the solution of a partial differential equation by solving some ordinary differential equations. A basic element of that approach is finding *eigenvalues* and *eigenfunctions* of a certain Sturm-Liouville problem and the expansion of functions into a *generalized Fourier series* over the set of those eigenfunctions.

A number of important physical problems are governed by the *Bessel equation*. Solving that equation in cylindrical or spherical coordinates leads to the universal equation:

$$r^2 y''(r) + r y'(r) + \left(\lambda r^2 - p^2\right) y(r) = 0, \tag{10.8}$$

which is defined within a certain interval, with some boundary conditions on its ends. Here $r \geq 0$ is the radial coordinate, p^2 is a positive real number, and λ is the eigenvalue.

For example, the function $y(r)$, defined on a closed interval $[0, \ l]$, could be restricted to a specified behavior at the points $r = 0$ and $r = l$. For instance, at $r = l$ the value of the solution, $y(l)$, could be prescribed, or its derivative $y'(l)$, or their linear combination $\alpha y'(l) + \beta y(l)$. Generally, Eq. (10.8) has nontrivial solutions which correspond to a given set of boundary conditions only for some values of the parameter λ, which are called *eigenvalues* (see Sect. 9.3). The goal then becomes to find eigenvalues for these boundary conditions and the corresponding solutions, $y(r)$, which are called *eigenfunctions*.

The point $r = 0$ is a *regular singular point* (see Sect. 6.3). That becomes clear if we write Eq. (10.8) in the form:

$$y''(r) + \frac{y'(r)}{r} + \left(\lambda - \frac{p^2}{r^2}\right)y(r) = 0. \tag{10.9}$$

Assuming that $\lambda > 0$, we can apply the change of the variable $x = \sqrt{\lambda}r$ and transform Eqs. (10.8) and (10.9) to their standard forms:

$$x^2 y''(x) + xy'(x) + \left(x^2 - p^2\right)y(x) = 0 \tag{10.10}$$

and

$$y''(x) + \frac{y'(x)}{x} + \left(1 - \frac{p^2}{x^2}\right)y(x) = 0 \tag{10.11}$$

that we use below.

10.2.2   Bessel Functions of the First Kind

Equation (10.11) has been discussed already in Sect. 6.3 (see Eq. (6.47)). It was shown that for integer values of $p = 0, 1, 2, \ldots$, this equation has a nonsingular solution that can be presented in the form of a series:

$$y_1(x) = \sum_{n=0}^{\infty} a_n x^{p+n}, \quad a_0 \neq 0, \tag{10.12}$$

where coefficients a_n satisfy the recurrence relation

$$a_n = -\frac{a_{n-2}}{n(2p+n)}, \quad n = 2, 4, \ldots. \tag{10.13}$$

That solution can be written as

$$y_1(x) = a_0 2^p p! J_p(x),$$

where

$$J_p(x) = \sum_{k=0}^{\infty} \frac{(-1)^k}{k!(p+k)!}\left(\frac{x}{2}\right)^{p+2k} \tag{10.14}$$

is called *Bessel function of the first kind* of order p (we define $0! = 1$).

The first few terms of the expansion in Eq. (10.12) near zero for the first three functions are

$$J_0(x) = 1 - \frac{x^2}{2^2} + \frac{x^4}{2^4 \cdot 2! \cdot 2!} - \cdots, \qquad J_1(x) = \frac{x}{2} - \frac{x^3}{2^3 \cdot 2!} + \frac{x^5}{2^5 \cdot 2! \cdot 3!} - \cdots$$

$$J_2(x) = \frac{x^2}{2^2 \cdot 2!} - \frac{x^4}{2^4 \cdot 3!} + \frac{x^6}{2^6 \cdot 2! \cdot 4!} - \cdots$$

$$(10.15)$$

Reading Exercise Show that $\lim\limits_{x \to 0} \frac{J_1(x)}{x} = \frac{1}{2}$.

Note that the functions $J_n(x)$ are even if n is an even integer and odd if n is an odd integer function of x. Note also that in applications of the Bessel functions, x is usually a radial coordinate which is positive by definition; hence, the region $x < 0$ is unphysical.

For future reference we present the useful fact that at $x = 0$ we have $J_0(0) = 1$ and $J_n(0) = 0$ for $n \geq 1$. Figure 10.2 shows graphs of functions $J_0(x)$, $J_1(x)$, and $J_2(x)$. (Graphs and figures in the text are plotted with book's program.) Table 10.2 lists a few first roots of the equation $J_n(\mu) = 0$ for Bessel functions of orders 0, 1, and 2.

A convenient notation for roots of equation $J_n(\mu) = 0$ is $\mu_k^{(n)}$, where n stands for the order of the Bessel function and k stands for the root number.

Now let us generalize the above case to *arbitrary real p*. Using relation (10.13), we find:

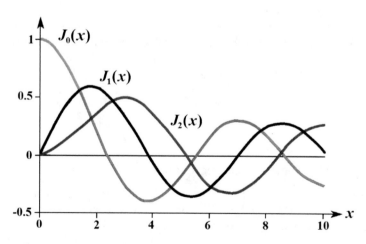

Fig. 10.2 Graphs of functions $J_0(x)$, $J_1(x)$, and $J_2(x)$

Table 10.2 Positive roots of $J_0(x)$, $J_1(x)$, and $J_2(x)$

Function	Roots				
	μ_1	μ_2	μ_3	μ_4	μ_5
$J_0(x)$	2.4048	5.5201	8.6537	11.7915	14.9309
$J_1(x)$	3.8317	7.0156	10.1735	13.3237	16.4706
$J_2(x)$	5.1356	8.4173	11.6198	14.7959	17.9598

$$a_2 = -\frac{a_0}{2(2+2p)} = -\frac{a_0}{2^2(1+p)} = -\frac{a_0\Gamma(p+1)}{2^2\Gamma(p+2)},$$

$$a_4 = -\frac{a_2}{2^3(p+2)} = \frac{a_0}{2!2^4(p+1)(p+2)} = \frac{a_0\Gamma(p+1)}{2!2^4\Gamma(p+3)},$$

$$\cdots\cdots\cdots\cdots\cdots\cdots\cdots\cdots\cdots\cdots\cdots$$

$$a_{2k} = \frac{(-1)^k a_0\Gamma(p+1)}{2^{2k}k!\Gamma(p+k+1)},$$

$$\cdots\cdots\cdots\cdots\cdots\cdots\cdots\cdots\cdots\cdots\cdots$$

Thus, we obtain the series:

$$
\begin{aligned}
J_p(x) &= \sum_{k=0}^{\infty} \frac{(-1)^k}{\Gamma(k+1)\Gamma(k+p+1)}\left(\frac{x}{2}\right)^{p+2k} \\
&= \left(\frac{x}{2}\right)^p \sum_{k=0}^{\infty} \frac{(-1)^k}{\Gamma(k+1)\Gamma(k+p+1)}\left(\frac{x}{2}\right)^{2k},
\end{aligned}
\tag{10.16}
$$

which *converges for any p.*

One can see that it is sufficient to replace the integer-valued function $p!$ in (10.14) by the Gamma function $\Gamma(p + 1)$, which is defined for arbitrary real values of p.

Recall that the Bessel equation contains p only as p^2. Therefore, we can obtain another solution of the Bessel equation replacing p by $-p$:

$$J_{-p}(x) = \left(\frac{x}{2}\right)^{-p} \sum_{k=0}^{\infty} \frac{(-1)^k}{\Gamma(k+1)\Gamma(k-p+1)}\left(\frac{x}{2}\right)^{2k}. \tag{10.17}$$

Reading Exercise Show that for integer, $p = n$, $J_{-n}(x) = (-1)^n J_n(x)$.

If p is not integer, solutions (10.16) and (10.17) are linearly independent, since the first terms in Eqs. (10.16) and (10.17) contain different powers of x, x^p and x^{-p}, respectively. Therefore, the *general solution* of Eq. (10.9) can be written in the form:

$$y = C_1 J_p(x) + C_2 J_{-p}(x). \tag{10.18}$$

Note that because the sum in (10.17) contains only even powers x^{2k}, that is correct also for half-integers:

$$p = (2m + 1)/2, \quad m = 0, 1, 2, \ldots. \tag{10.19}$$

Thus, for p from the set (10.19), the solution does not include a logarithmic term which is allowed by the general theory (see Sect. 6.3).

The functions $J_p(x)$, $p > 0$ are bounded as $x \to 0$. In fact, the function $J_p(x)$ is continuous for all x since it is a uniformly converging power series. For non-integer values of p, this follows from the properties of the Gamma function and the series (10.16). The point $x = 0$ does not belong to the domain of definition of the function (10.17), since x^{-p} for $p > 0$ diverges at this point.

For integer $p = n$, because of (10.17), $J_{-n}(x)$ and $J_n(x)$ are linearly dependent. Therefore, we still have to find the second linearly independent solution for integer p, including $p = 0$. That will be done in Sect. 10.2.4.

10.2.3   *Properties of Bessel Functions*

The basic properties of functions $J_p(x)$ are listed below.

1.	All Bessel functions are defined and continuous for $x > 0$ and have derivatives of all orders
2.	Bessel functions of even orders are even functions (since their expansion contains only even powers of the argument). Bessel functions of odd orders are odd functions
3.	Each Bessel function has an infinite number of real roots. Roots located on the positive semiaxis can be marked by integer numbers in the increasing order. Zeros of $J_p(x) = 0$ fall between the zeros of $J_{p+1}(x) = 0$
4.	The behavior of Bessel functions in the vicinity of zero is given by the first terms of the series in Eq. (10.14); for large x the asymptotic formula may be used: $$J_p(x) \approx \sqrt{\tfrac{2}{\pi x}} \cos\left(x - \tfrac{p\pi}{2} - \tfrac{\pi}{4}\right) \tag{10.20}$$ With increasing x the accuracy of this formula quickly increases. When $J_p(x)$ is replaced by the right-hand side of Eq. (10.17), the error is very small for large x and has the same order as $x^{-3/2}$ From Eq. (10.20) it follows, in particular, that the function $J_p(x)$ has roots that are close (for large x) to the roots of the equation: $$\cos\left(x - \tfrac{p\pi}{2} - \tfrac{\pi}{4}\right) = 0;$$ thus, the distance between two adjacent roots of the function $J_p(x)$ tends to π when roots tend to infinity. A graph of $J_p(x)$ has the shape of a curve which depicts decaying oscillation; the "wavelength" is almost constant (close to π), and the amplitude decays inversely proportional to square root of x. In fact we have $\lim\limits_{x \to \infty} J_p(x) = 0$
5.	*Recurrence formulas* $$J_{p+1}(x) = \tfrac{2p}{x} J_p(x) - J_{p-1}(x), \tag{10.21a}$$ $$J_{p+1}(x) = \tfrac{p}{x} J_p(x) - J'_p(x), \tag{10.21b}$$ $$J'_p(x) = -\tfrac{p}{x} J_p(x) + J_{p-1}(x) \tag{10.21c}$$

(continued)

6.	Integration formulas

$$\int x^{-p} J_{p+1}(x) dx = -x^{-p} J_p(x), \tag{10.22a}$$

$$\int x^p J_{p-1}(x) dx = x^p J_p(x), \tag{10.22b}$$

$$\int_0^x z J_0(z) dz = x J_1(x), \tag{10.22c}$$

$$\int_0^x z^3 J_0(z) dz = 2x^2 J_0(x) + x(x^2 - 4) J_1(x) \tag{10.22d}$$

7.	Differentiation formulas

$$\frac{d}{dx}\left[x^{-p} J_p(x)\right] \equiv -x^{-p} J_{p+1}(x), \quad p = 0, 1, 2, \ldots \tag{10.23a}$$

$$\frac{d}{dx}\left[x^p J_p(x)\right] \equiv x^p J_{p-1}(x), \quad p = 1, 2, 3, \ldots \tag{10.23b}$$

In particular, $J_0'(x) = -J_1(x), \quad J_1'(x) = J_0(x) - \dfrac{J_1(x)}{x}$

These identities are easily established by operating on the series which defines the function.

In many physical problems with the spherical symmetry, one encounters Bessel functions of half-integer orders where $p = (2n + 1)/2$ for $n = 0, 1, 2, \ldots$. For instance, using expansion (10.16) for $p = 1/2$ and $p = -1/2$

$$y(x) = x^{1/2} \sum_{k=0}^{\infty} a^k x^k \quad (a_0 \neq 0), \tag{10.24}$$

we obtain:

$$J_{1/2}(x) = \left(\frac{2}{\pi x}\right)^{1/2} \sum_{k=0}^{\infty} (-1)^k \frac{x^{2k+1}}{(2k+1)!} \tag{10.25}$$

and

$$J_{-1/2}(x) = \left(\frac{2}{\pi x}\right)^{1/2} \sum_{k=0}^{\infty} (-1)^k \frac{x^{2k}}{(2k)!}. \tag{10.26}$$

By comparing expansions in Eqs. (10.25) and (10.26) to the Maclaurin series expansions of $\sin x$ and $\cos x$, we obtain:

$$J_{1/2} = \left(\frac{2}{\pi x}\right)^{1/2} \sin x, \tag{10.27}$$

$$J_{-1/2} = \left(\frac{2}{\pi x}\right)^{1/2} \cos x. \tag{10.28}$$

Note that $J_{1/2}(x)$ is bounded for all x and function $J_{-1/2}(x)$ diverges at $x = 0$. Recall that Eq. (10.16) gives an expansion of $J_p(x)$ which is valid for any value of p.

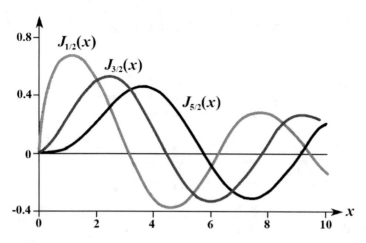

Fig. 10.3 Graphs of functions $J_{1/2}(x)$, $J_{3/2}(x)$, and $J_{5/2}(x)$

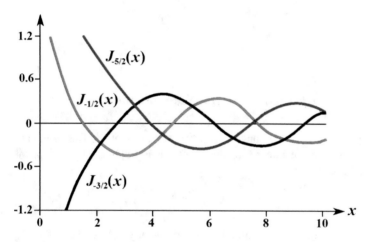

Fig. 10.4 Graphs of $J_{-1/2}(x)$, $J_{-3/2}(x)$ and $J_{-5/2}(x)$

Figure 10.3 shows graphs of functions $J_{1/2}(x)$, $J_{3/2}(x)$, and $J_{5/2}(x)$; Fig. 10.4 shows graphs for $J_{-1/2}(x)$, $J_{-3/2}(x)$, and $J_{-5/2}(x)$.

Reading Exercise Using the recurrence Eq. (10.21) and the expression for $J_{1/2}(x)$, obtain the functions $J_{3/2}(x)$, $J_{-3/2}(x)$, $J_{5/2}(x)$, and $J_{-5/2}(x)$. For instance, the answer for $J_{3/2}(x)$ is

$$J_{3/2} = \left(\frac{2}{\pi x}\right)^{1/2}\left(\frac{\sin x}{x} - \cos x\right). \tag{10.29}$$

Also let us mention an important result used in the theory of radiation:

$$e^{ix\cos\theta} = J_0(x) + 2\sum_{n=1}^{\infty} i^n J_n(x)\cos nx. \tag{10.30}$$

Reading Exercise

1. Assuming that the equality in Eq. (10.30) can be integrated term by term (explain why this is valid), obtain the integral form for the function $J_0(x)$:

$$J_0(x) = \frac{1}{2\pi} \int_0^{2\pi} e^{ix\sin\theta} d\theta = \frac{1}{2\pi} \int_0^{2\pi} e^{ix\cos\theta} d\theta. \tag{10.31}$$

2. Show that

$$e^{ix\sin\theta} = \sum_{n=-\infty}^{\infty} J_n(x)e^{in\theta}. \tag{10.32}$$

10.2.4   Bessel Functions of the Second Kind

In the previous section, we have found two solutions of the Bessel equation, $J_p(x)$ and $J_{-p}(x)$, which are linearly independent for any values of p except the natural ones. We have shown that

$$J_{-n}(x) = (-1)^n J_n(x), \quad n = 0, 1, 2, \ldots. \tag{10.33}$$

Therefore, still we have to find the second, linearly independent of $J_n(x)$, solution for natural n.

First, let us define the *Bessel functions of the second kind*, or *Neumann functions* $N_p(x)$ as

$$N_p(x) = \frac{J_p(x)\cos(p\pi) - J_{-p}(x)}{\sin(p\pi)}, \quad p \neq 0, 1, 2, \ldots. \tag{10.34}$$

Because of the linear independence of $J_p(x)$ and $J_{-p}(x)$ for non-integer p, $N_p(x)$ is a solution of the Bessel equation linearly independent of $J_p(x)$; therefore, the general solution of the Bessel equation can be written as

$$y(x) = C_1 J_p(x) + C_2 N_p(x). \tag{10.35}$$

For integer p, expression (10.34) is an indeterminate form 0/0, so we define

$$N_n(x) = \lim_{p \to n} N_p(x), \quad n = 0, 1, 2, \ldots$$

One can show that functions $N_n(x)$ are singular at $x = 0$: $N_0(x)$ diverges logarithmically, while for $n = 1, 2, \ldots$ the expansion for $N_n(x)$ at small x contains both a logarithmic term and a power-law divergence, $N_n(x) \sim x^{-n}$. Therefore, the function $N_n(x)$ is linearly independent of function $J_n(x)$ that is regular at $x = 0$. Thus, (10.35) is the general solution of the Bessel equation for any real p.

If the physical formulation of the problem requires regularity of the solution at zero, the coefficient C_2 in Eq. (10.35) must be zero.

For integer $p = n$

$$N_{-n}(x) = (-1)^n N_n(x). \tag{10.36}$$

At large x,

$$N_n(x) \approx \sqrt{\frac{2}{\pi x}} \sin\left(x - \frac{n\pi}{2} - \frac{\pi}{4}\right). \tag{10.37}$$

Thus, as $x \to \infty$, $N_n(x) \to 0$ is oscillating with decaying amplitude, see Fig. 10.5. Several roots of Neumann functions are listed in Table 10.3.

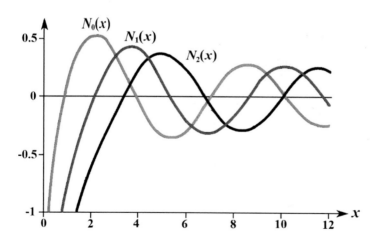

Fig. 10.5 Graphs of $N_0(x)$, $N_1(x)$, and $N_2(x)$

Table 10.3 Positive roots of the functions $N_0(x)$, $N_1(x)$, and $N_2(x)$

Function	Roots				
	μ_1	μ_2	μ_3	μ_4	μ_5
$N_0(x)$	0.8936	3.9577	7.0861	10.2223	13.3611
$N_1(x)$	2.1971	5.4297	8.5960	11.7492	14.8974
$N_2(x)$	3.3842	6.7938	10.0235	13.2199	16.3789

10.2.5   Bessel Functions of the Third Kind

The cylindrical functions of the third kind (or *Hankel functions*) can be expressed using functions of the first and second kind:

$$H_p^{(1)}(x) = J_p(x) + iN_p(x), \quad H_p^{(2)}(x) = J_p(x) - iN_p(x). \tag{10.38}$$

The Hankel functions $H_p^{(1)}(x)$ and $H_p^{(2)}(x)$ are linearly independent and are often used, for instance, in electrodynamics in the problems involving the theory of radiation. Their convenience follows, in particular, from the fact that at large x, their asymptotics is in the form of plane waves:

$$H_p^{(1)}(x) \approx \sqrt{\frac{2}{\pi x}} \exp\left\{ i\left(x - \frac{p\pi}{2} - \frac{\pi}{4} \right) \right\},$$
$$H_p^{(2)}(x) \approx \sqrt{\frac{2}{\pi x}} \exp\left\{ -i\left(x - \frac{p\pi}{2} - \frac{\pi}{4} \right) \right\}. \tag{10.39}$$

Note that differentiation formulas (10.23a and 10.23b) are also valid for Bessel functions of the second and third kind, $N_p(x)$ and $H_p^{(1,2)}(x)$.

Reading Exercise

1. Find expressions for the functions $N_{1/2}(x)$ and $H_{1/2}^{(1,2)}(x)$.

2. Using the recurrence relations and the expressions for $N_{1/2}(x)$ and $H_{1/2}^{(1,2)}(x)$, find expressions for the corresponding functions for $p = \pm 3/2, \pm 5/2$.

Reading Exercise There exist many relations connecting Bessel functions of different types. Prove the formula:

$$J_p(x)N_{p+1}(x) - J_{p+1}(x)N_p(x) = -\frac{2}{\pi x}. \tag{10.40}$$

10.2.6   Modified Bessel Functions

Next, we discuss the frequently encountered, modified Bessel equation given by

$$x^2 y''(x) + xy'(x) - (x^2 + p^2)y(x) = 0. \tag{10.41}$$

Using the variable transformation $x = -iz$, this equation can be reduced to the ordinary Bessel Eq. (10.10):

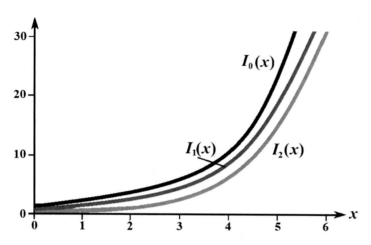

Fig. 10.6 Graphs of $I_0(x)$, $I_1(x)$, and $I_2(x)$

$$z^2 y''(z) + z y'(z) + (z^2 - p^2) y(z) = 0. \tag{10.42}$$

Making the change of a variable $x = iz$ in the expression for $J_p(z)$ (see (10.16)), we obtain the expression $i^p I_p(x)$, where

$$I_p(x) = \left(\frac{x}{2}\right)^p \sum_{k=0}^{\infty} \frac{1}{\Gamma(k+1)\Gamma(k+p+1)} \left(\frac{x}{2}\right)^{2k}. \tag{10.43}$$

This formula is correct for arbitrary real p, in particular, in order to obtain I_{-p} one needs only to replace p by $-p$ in Eq. (10.43).

Reading Exercise Show that the series (10.43) for the case $I_0(x)$ converges for all x.

Function $I_p(x)$ is called the *modified Bessel function of the first kind*. It is obvious that the relation between this function and the ordinary Bessel function of the first kind is given by

$$I_p(x) = (-i)^p J_p(ix). \tag{10.44}$$

Since the terms in the series (10.43) have constant signs, $I_p(x)$ increases monotonically with increasing x. Figure 10.6 shows graphs of the functions $I_0(x)$, $I_1(x)$, and $I_2(x)$. Note that $I_0(0) = 1$ and $I_n(0) = 0$ when $n = 1, 2, 3, \ldots$.

Reading Exercise Prove that, similarly to formulas (10.27) and (10.28) for functions $J_{\pm 1/2}(x)$, the functions $I_{\pm 1/2}(x)$ admit a simple representations given by

$$I_{1/2}(x) = \left(\frac{2}{\pi x}\right)^{1/2} \sinh(x), \quad I_{-1/2}(x) = \left(\frac{2}{\pi x}\right)^{-1/2} \cosh(x). \tag{10.45}$$

Another solution of Eq. (10.41) is defined as follows:

$$K_p(x) = \frac{\pi}{2} \frac{I_{-p}(x) - I_p(x)}{\sin p\pi}. \tag{10.46}$$

For integer $p = n$, $n = 0, \pm 1, \pm 2, \ldots$, we define $K_n(x) = \lim\limits_{p \to n} K_p(x)$. One can show that $K_p(x)$ and $I_p(x)$ are linearly independent for any real p.

The function $K_p(x)$ is called the *modified Bessel function of the second* kind, or Macdonald's function. In analogy to the relation in Eq. (10.44), there exists a connection between function $K_p(x)$ and the Hankel function:

$$K_p = \begin{cases} \dfrac{\pi}{2} i^{p+1} H_p^{(1)}(ix), & -\pi < \arg x \le \dfrac{\pi}{2} \\[2mm] \dfrac{\pi}{2} (-i)^{p+1} H_p^{(1)}(ix), & -\dfrac{\pi}{2} < \arg x \le \pi. \end{cases} \tag{10.47}$$

Reading Exercise Obtain the relations in Eq. (10.47).

The general solution of Eq. (10.41) is

$$y(x) = C_1 I_p(x) + C_2 K_p(x). \tag{10.48}$$

Using the approach described in Sect. 6.3, one can obtain solutions $K_p(x)$ in the form of series. In particular, function $K_0(x)$ may be written as

$$K_0(x) = -I_0(x)\left(\ln \frac{x}{2} + \gamma\right) + \sum_{k=1}^{\infty} \frac{x^{2k}}{2^{2k}(k!)^2} h_k. \tag{10.49}$$

Figure 10.7 shows graphs of functions $K_0(x)$, $K_1(x)$, and $K_2(x)$. For integer p the functions $K_n(x)$ diverge as $x \to 0$ and monotonically decrease with increasing x.

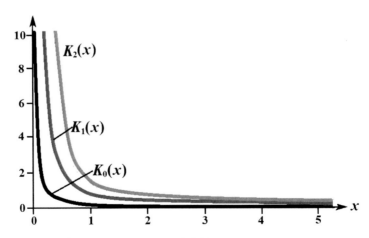

Fig. 10.7 Graphs of functions $K_0(x)$, $K_1(x)$, and $K_2(x)$

10.2.7 Boundary Value Problems and Fourier-Bessel Series

As mentioned in Sect. 10.2.1, the main application of Bessel functions is solving partial differential equations (usually in cylindrical or spherical coordinates) using the method of separation of variables (see Chap. 13). That method includes finding eigenvalues and eigenfunctions of a Sturm-Liouville problem, which consists of the Bessel equation with some boundary conditions. The next stage is the expansion of functions into a series over the set of eigenfunctions which are some Bessel functions. In the present section, we consider both stages.

For instance, assume that the solution $y(x)$ of the Bessel equations (10.8) defined on the interval $[0,\ l]$ is finite at $x = 0$, and at $x = l$, it obeys a homogeneous boundary condition $\alpha y'(l) + \beta y(l) = 0$ (notice that we use letter x instead of r considering Eq. (10.8)). When $\alpha = 0$, this mixed boundary condition becomes of the Dirichlet type, when $\beta = 0$ it is of the Neumann type. The Bessel equation and boundary condition(s) form the Sturm-Liouville problem; thus, it has the nontrivial solutions only for nonnegative discrete eigenvalues λ_k:

$$\lambda_1 < \lambda_2 < \ldots < \lambda_k < \ldots . \tag{10.50}$$

The corresponding eigenfunctions are

$$J_p\left(\sqrt{\lambda_1}x\right),\ J_p\left(\sqrt{\lambda_2}x\right),\ \ldots,\ J_p\left(\sqrt{\lambda_k}x\right),\ \ldots \tag{10.51}$$

(if function $y(x)$ has not to be finite at $x = 0$, we can also consider the set of eigenfunctions $N_p\left(\sqrt{\lambda_k}x\right)$). The boundary condition at $x = l$ gives:

$$\alpha\sqrt{\lambda}\,J'_p\left(\sqrt{\lambda}l\right) + \beta J_p\left(\sqrt{\lambda}l\right) = 0.$$

Setting $\sqrt{\lambda}l \equiv \mu$ we obtain a transcendental equation for μ:

$$\alpha\mu J'_p(\mu) + \beta l J_p(\mu) = 0, \tag{10.52}$$

which has an infinite number of roots which we label as $\mu_k^{(p)}$. From there the eigenvalues are $\lambda_k = (\mu_k/l)^2$; we see that we need only *positive roots*, $\mu_m^{(n)}$, because negative roots do not give new values of λ_k.

For instance, the condition $y(l) = 0$ leads to the equality $C_1 J_p\left(\sqrt{\lambda}l\right) = 0$. Since $C_1 \neq 0$ for nontrivial solutions (one can take $C_1 = 1$), we have $J_p\left(\sqrt{\lambda}l\right) = 0$. This means that $\sqrt{\lambda}l$ is the root of the equation $J_p(x) = 0$. Therefore, $\sqrt{\lambda}l = \mu_k^{(p)}$, where $\mu_1^{(p)}, \mu_2^{(p)}, \ldots$ are positive roots of the Bessel function $J_p(x)$. Thus, *for fixed p we have* the set of eigenvalues $\lambda_k = \left(\mu_k^{(p)}/l\right)^2$ and corresponding eigenfunctions:

$$J_p\left(\frac{\mu_1^{(p)}}{l}x\right), \ J_p\left(\frac{\mu_2^{(p)}}{l}x\right), \ \ldots, \ J_p\left(\frac{\mu_k^{(p)}}{l}x\right), \ \ldots \tag{10.53}$$

As follows from the Sturm-Liouville theory, these functions form a *complete set* and are pair-wise orthogonal (with weight x) on the interval $[0, \ l]$:

$$\int_0^l J_p\left(\frac{\mu_k^{(p)}}{l}x\right)J_p\left(\frac{\mu_k^{(p)}}{l}x\right)x\,dx = 0, \quad i \neq j. \tag{10.54}$$

A Fourier series expansion (or generalized Fourier series) of an arbitrary function $f(x)$ using the set of functions (10.53) is called *Fourier-Bessel series* and is given by the expression:

$$f(x) = \sum_{k=0}^{\infty} c_k J_p\left(\frac{\mu_k^{(p)}}{l}x\right). \tag{10.55}$$

The orthogonality property allows to find the coefficients of this series. We multiply Eq. (10.55) by $J_p\left(\frac{\mu_k^{(p)}}{l}x\right)$ and integrate term by term with weight x. This gives an expression for the coefficient as

$$c_k = \frac{\displaystyle\int_0^l f(x)J_p\left(\frac{\mu_k^{(p)}}{l}x\right)x\,dx}{\displaystyle\int_0^l x\left[J_p\left(\frac{\mu_k^{(p)}}{l}x\right)\right]^2 dx}. \tag{10.56}$$

The *squared norm* $\|R_{pk}\|^2 = \int_0^l xJ_p^2\left(\frac{\mu_k^{(p)}}{l}x\right)dx$ is:

1. For the *Dirichlet* boundary condition $\alpha = 0$ and $\beta = 1$, in which case eigenvalues are obtained from the equation

$$J_p(\mu) = 0$$

and we have:

$$\|R_{pk}\|^2 = \frac{l^2}{2}\left[J_p'\left(\mu_k^{(p)}\right)\right]^2. \tag{10.57}$$

2. For the *Neumann* boundary condition $\alpha = 1$ and $\beta = 0$, in which case eigenvalues are obtained from the equation

$$J'_p(\mu) = 0$$

and we have:

$$\left\| R_{pk} \right\|^2 = \frac{l^2}{2\left(\mu_k^{(p)}\right)^2} \left[\left(\mu_k^{(p)}\right)^2 - p^2\right] J_p^2\left(\mu_k^{(p)}\right). \tag{10.58}$$

3. For the *mixed* boundary condition $\alpha \equiv 1$ and $\beta \equiv h$, in which case eigenvalues
are obtained from the equation

$$\mu J'_p(\mu) + hlJ_p(\mu) = 0$$

and we have:

$$\left\| R_{pk} \right\|^2 = \frac{l^2}{2} \left[1 + \frac{l^2 h^2 - p^2}{\left(\mu_k^{(p)}\right)^2}\right] J_p^2\left(\mu_k^{(p)}\right). \tag{10.59}$$

The completeness of the set of functions $J_p\left(\frac{\mu_k^{(p)}}{l}x\right)$ on the interval $(0, l)$ means
that for any square integrable on $[0,\ l]$ function $f(x)$, the following is true:

$$\int_0^l xf^2(x)dx = \sum_k \left\| J_p\left(\frac{\mu_k^{(p)}}{l}x\right) \right\|^2 c_k^2. \tag{10.60}$$

This is Parseval's equality for the Fourier-Bessel series. It has the same significance of completeness as in the case of the trigonometric Fourier series with sines
and cosines as the basis functions where the weight function equals one instead of
x as in the Bessel series.

Regarding the convergence of the series (10.55), we note that the sequence of the
partial sums of the series, $S_n(x)$, converges on the interval $(0, l)$ on average (i.e., in
the mean) to $f(x)$ (with weight x), which may be written as

$$\int_0^l [f(x) - S_n(x)]^2 x dx \to 0, \quad \text{if} \quad n \to \infty.$$

This property is true for any function $f(x)$ from the class of piecewise-continuous functions because the orthogonal set of functions (10.56) is complete on the interval $[0, l]$. For such functions, $f(x)$, the series (10.55) converges absolutely and uniformly. We present the following theorem, which states a somewhat stronger result about the convergence of the series in Eq. (10.58) than convergence in the mean, without the proof:

Theorem
If the function $f(x)$ is piecewise-continuous on the interval $(0, l)$, then the Fourier-Bessel series converges to $f(x)$ at the points where the function $f(x)$ is continuous, and to

$$\frac{1}{2}[f(x_0 + 0) + f(x_0 - 0)],$$

if x_0 is a point of finite discontinuity of the function $f(x)$.

Below we consider several examples of the expansion of functions into the Fourier-Bessel series using the functions $J_p(x)$.

Example 10.4 Let us expand the function $f(x) = A$, $A = $ const., in a series using the Bessel functions $X_k(x) = J_0\left(\mu_k^{(0)} x/l\right)$ on the interval $[0, l]$, where $\mu_k^{(0)}$ are the positive roots of the equation $J_0(\mu) = 0$.

Solution First, we calculate the norm, $\|X_k\|^2 = \left\| J_0\left(\mu_k^{(0)} x/l\right) \right\|^2$ using the relation $J_0'(x) = -J_1(x)$ to obtain

$$\left\| J_0\left(\mu_k^{(0)} x/l\right) \right\|^2 = \frac{l^2}{2}\left[J_0'\left(\mu_k^{(0)}\right)\right]^2 = \frac{l^2}{2}J_1^2\left(\mu_k^{(0)}\right).$$

Using the substitution $z = \mu_k^{(0)} x/l$ and the second relation of (10.22c), we may calculate the integral:

$$\int_0^l J_0\left(\frac{\mu_k^{(0)}}{l}x\right)x dx = \frac{l^2}{\left(\mu_k^{(0)}\right)^2}\int_0^{\mu_k^{(0)}} J_0(z)z dz = \frac{l^2}{\left(\mu_k^{(0)}\right)^2}[zJ_1(z)]_0^{\mu_k^{(0)}} = \frac{l^2}{\mu_k^{(0)}}J_1\left(\mu_k^{(0)}\right).$$

For the coefficients c_k of the expansion (10.55), we have:

$$c_k = \frac{\int\limits_0^l AJ_0\left(\mu_k^{(0)}x/l\right)\,xdx}{\left\|J_0\left(\mu_k^{(0)}x/l\right)\right\|^2} = \frac{2A}{l^2\left[J_1\left(\mu_k^{(0)}\right)\right]^2}\frac{l^2}{\mu_k^{(0)}}J_1\left(\mu_k^{(0)}\right) = \frac{2A}{\mu_k^{(0)}J_1\left(\mu_k^{(0)}\right)}.$$

Thus, the expansion is

$$f(x) = 2A\sum_{k=0}^{\infty}\frac{1}{\mu_k^{(0)}J_1\left(\mu_k^{(0)}\right)}J_0\left(\frac{\mu_k^{(0)}}{l}x\right).$$

Figure 10.8 shows the function $f(x) = 1$ and the partial sum of its Fourier-Bessel series when $l = 1$. From this figure it is seen that the series converges very slowly (see Fig. 10.8d), and even when 50 terms are kept in the expansion (Fig. 10.8c), the difference from $f(x) = 1$ can easily be seen. This is because at the endpoints of the interval, the value of the function $f(x) = 1$ and the functions $J_0\left(\mu_k^{(0)}x\right)$ are different. The obtained expansion does not converge well near the endpoints.

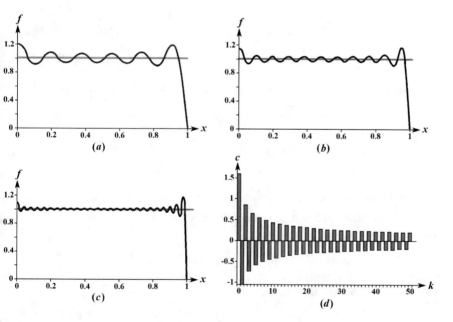

Fig. 10.8 The function $f(x) = 1$ and the partial sum of its Fourier-Bessel series. The graph of $f(x)$ is shown in gray, the graph of the series by the black line. (a) 11 terms are kept in the series ($N = 10$); (b) $N = 20$; (c) $N = 50$; (d) values of the coefficients c_k of the series

Example 10.5 Let us modify the previous problem. We expand the function $f(x) = A$, $A = $ const, given on the interval $[0, l]$, in a Fourier series in Bessel functions $X_k(x) = J_0\left(\mu_k^{(0)} x/l\right)$, where $\mu_k^{(0)}$ are now the positive roots of the equation $J_0'(\mu) = 0$.

Solution For Neumann boundary condition, $J_0'(\mu) = 0$, for $k = 0$, we have:

$$\mu_0^{(0)} = 0, \quad X_0(x) = J_0(0) = 1, \quad \|X_0\|^2 = \|J_0(0)\|^2 = \frac{l^2}{2}.$$

The first coefficient of the expansion (10.55) is

$$c_0 = \frac{A}{\|J_0(0)\|^2} \int_0^l J_0(0)x\,dx = \frac{2A}{l^2} \int_0^l x\,dx = A.$$

The next coefficients c_k can be evaluated by using the substitution $z = \mu_k^{(0)} x/l$ and using the integration formula:

$$\int_0^l J_0\left(\frac{\mu_k^{(0)}}{l} x\right) x\,dx = \frac{l^2}{\left(\mu_k^{(0)}\right)^2} [zJ_1(z)]_0^{\mu_k^{(0)}} = \frac{l^2}{\mu_k^{(0)}} J_1\left(\mu_k^{(0)}\right).$$

Applying the relation $J_0'(x) = -J_1(x)$ and then recalling that $J_0'\left(\mu_k^{(0)}\right) = 0$, we find:

$$c_k = \frac{2A}{\mu_k^{(0)}} J_1\left(\mu_k^{(0)}\right) = 0, \quad \text{when } k > 0.$$

Thus, we obtain a simple expansion, $f(x) = c_0 J_0\left(\mu_0^{(0)} x/l\right) = A$. In fact, this means that the given function is actually one of the functions from the set of eigenfunctions used for eigenfunction expansion.

10.2.8 Spherical Bessel Functions

Equation (10.11) and Bessel functions considered above appear by solving partial differential equations in cylindrical coordinates, hence they can be called "cylindrical Bessel equation" and "cylindrical Bessel functions". In this section we briefly consider spherical Bessel functions which are related to the solutions of certain boundary value problems in spherical coordinates.

Consider the following equation:

$$\frac{d^2R(x)}{dx^2} + \frac{2}{x}\frac{dR(x)}{dx} + \left[1 - \frac{l(l+1)}{x^2}\right]R(x) = 0. \tag{10.61}$$

Parameter l takes discrete nonnegative integer values: $l = 0, 1, 2, \ldots$ Eq. (10.61) is called the *spherical Bessel equation*. It differs from the cylindrical Bessel equation, Eq. (10.11), by the coefficient 2 in the second term. Equation (10.61) can be transformed to a Bessel cylindrical equation by the substitution $R(x) = y(x)/\sqrt{x}$.

Reading Exercise Check that equation for $y(x)$ is

$$\frac{d^2y(x)}{dx^2} + \frac{1}{x}\frac{dy(x)}{dx} + \left[1 - \frac{(l+1/2)^2}{x^2}\right]y(x) = 0. \tag{10.62}$$

If we introduce $s = l + 1/2$ in Eq. (10.62), we recognize this equation as the Bessel equation which has the general solution:

$$y(x) = C_1J_s(x) + C_2N_s(x), \tag{10.63}$$

where $J_s(x)$ and $N_s(x)$ are (cylindrical) Bessel and Neumann functions. Because $s = l + 1/2$, these functions are of half-integer order. Inverting the transformation, we find that the solution, $R(x)$, to Eq. (10.61) is

$$R(x) = C_1\frac{J_{l+1/2}(x)}{\sqrt{x}} + C_2\frac{N_{l+1/2}(x)}{\sqrt{x}}. \tag{10.64}$$

If we consider a regular at $x = 0$ solution, the coefficient $C_2 \equiv 0$.

The *spherical Bessel function* $j_l(x)$ is defined to be a solution finite at $x = 0$; thus, it is a multiple of $J_{l+1/2}(x)/\sqrt{x}$. The coefficient of proportionality is usually chosen to be $\sqrt{\pi/2}$ so that

$$j_l(x) = \sqrt{\frac{\pi}{2x}}J_{l+1/2}(x). \tag{10.65}$$

For $l = 0$, $J_{1/2}(x) = \sqrt{\frac{2}{\pi x}}\sin x$; thus,

$$j_0(x) = \frac{\sin x}{x}. \tag{10.66}$$

Analogously we may define the *spherical Neumann functions* as

$$n_l(x) = \sqrt{\frac{\pi}{2x}}N_{l+1/2}(x), \tag{10.67}$$

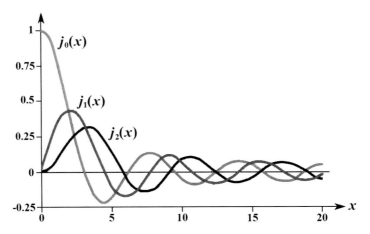

Fig. 10.9 Graphs of functions $j_0(x)$, $j_1(x)$, and $j_2(x)$

from where (using Eq. (10.29))

$$n_0(x) = \sqrt{\frac{\pi}{2x}} J_{-1/2}(x) = -\frac{\cos x}{x}. \qquad (10.68)$$

Expressions for the first few terms of the functions $j_l(x)$ and $n_l(x)$ are

$$j_1(x) = \frac{\sin x}{x^2} - \frac{\cos x}{x} + \ldots, \qquad j_2(x) = \left(\frac{3}{x^3} - \frac{1}{x}\right) \sin x - \frac{3}{x^2} \cos x + \ldots, \qquad (10.69)$$

$$n_1(x) = -\frac{\cos x}{x^2} - \frac{\sin x}{x} + \ldots, \qquad n_2(x) = -\left(\frac{3}{x^3} - \frac{1}{x}\right) \cos x - \frac{3}{x^2} \sin x + \ldots$$
$$(10.70)$$

The spherical Bessel functions with $l = 0, 1, 2$ are sketched in Figs. 10.9 and 10.10.

The following *recurrence relations* are valid:

$$f_{l-1}(x) + f_{l+1}(x) = (2l+1)x^{-1}f_l(x), \qquad (10.71)$$

$$lf_{l-1}(x) - (l+1)f_{l+1}(x) = (2l+1)\frac{d}{dx}f_l(x). \qquad (10.72)$$

Differentiation formulas

$$\frac{d}{dx}\left[x^{l+1}j_l(x)\right] = x^{l+1}j_{l-1}(x), \qquad \frac{d}{dx}\left[x^{-l}j_l(x)\right] = -x^{-l}j_{l+1}(x). \qquad (10.73)$$

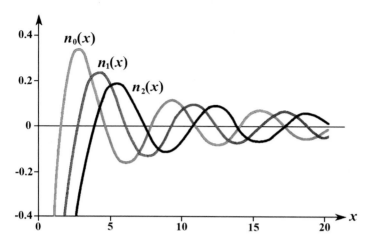

Fig. 10.10 Graphs of functions $n_0(x)$, $n_1(x)$, and $n_2(x)$

Asymptotic values

$$j_l(x) \sim \frac{1}{x} \cos\left[x - \frac{\pi}{2}(l+1)\right], \quad n_l(x) \sim \frac{1}{x} \sin\left[x - \frac{\pi}{2}(l+1)\right], \quad \text{as } x \to \infty \quad (10.74)$$

(the last expression has good precision for $x \gg l(l+1)$).

10.2.9 Airy Functions

The Airy functions are the particular solution of the equation (Airy equation):

$$y'' - xy = 0 \tag{10.75}$$

(this equation is named after the British astronomer George Biddell Airy (1801–1892)). This simple equation has a turning point where the character of the solution *changes from oscillatory to exponential*.

There are Airy function of the first kind, $Ai(x)$, and the second kind, $Bi(x)$; when $x \to -\infty$ both oscillate (with equal decaying amplitude, the phase shift between them is $\pi/2$; see Fig. 10.11). For $x \to +\infty$, function $Ai(x)$ monotonically decays and function $Bi(x)$ monotonically increases.

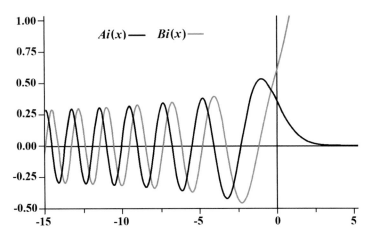

Fig. 10.11 Functions $Ai(x)$ (black) and $Bi(x)$ (gray)

For real x functions $Ai(x)$ and $Bi(x)$ can be presented by the integrals:

$$Ai(x) = \frac{1}{\pi} \int\limits_{0}^{\infty} \cos\left(\frac{t^3}{3} + xt\right) dt, \tag{10.76}$$

$$Bi(x) = \frac{1}{\pi} \int\limits_{0}^{\infty} \left[\exp\left(-\frac{t^3}{3} + xt\right) + \sin\left(\frac{t^3}{3} + xt\right)\right] dt. \tag{10.77}$$

This can be easily proven – differentiating these integrals we obtain the Airy equation in both cases.

Relation of Airy functions with Bessel functions of different kinds:

For positive values of the argument, the Airy functions are related to modified Bessel functions:

$$Ai(x) = \sqrt{\frac{x}{9}}\left[I_{-1/3}\left(\frac{2}{3}x^{3/2}\right) - I_{1/3}\left(\frac{2}{3}x^{3/2}\right)\right] = \frac{1}{\pi}\sqrt{\frac{x}{3}}K_{1/3}\left(\frac{2}{3}x^{3/2}\right), \tag{10.78}$$

$$Bi(x) = \sqrt{\frac{x}{3}}\left[I_{1/3}\left(\frac{2}{3}x^{3/2}\right) + I_{-1/3}\left(\frac{2}{3}x^{3/2}\right)\right]. \tag{10.79}$$

For negative values of the argument, the Airy functions are related to Bessel functions:

$$Ai(x) = \sqrt{\frac{|x|}{9}}\left[J_{1/3}\left(\frac{2}{3}|x|^{3/2}\right) + J_{-1/3}\left(\frac{2}{3}|x|^{3/2}\right)\right],$$ (10.80)

$$Bi(x) = \sqrt{\frac{|x|}{3}}\left[J_{-1/3}\left(\frac{2}{3}|x|^{3/2}\right) - J_{1/3}\left(\frac{2}{3}|x|^{3/2}\right)\right].$$ (10.81)

Problems
Solve the following equations:

1. $x^2 y'' + xy' + (3x^2 - 2)y = 0$.
2. $x^2 y'' + xy' - (4x^2 + \frac{1}{2})y = 0$.
3. $x^2 y'' + 2xy' + (x^2 - 1)y = 0$.
4. $y'' + \frac{y'}{x} + (1 - \frac{1}{9x^2})y = 0$.
5. $x^2 y'' + xy' + 4(x^4 - 2)y = 0$.
6. Prove the following integral formulas:

$$\int x^p J_{p-1}(x)dx = x^p J_p(x) + C;$$

$$\int x^{-p} J_{p+1}(x)dx = -x^{-p} J_p(x) + C;$$

$$\int J_{p+1}(x)dx = \int J_{p-1}(x)dx - 2J_p(x).$$

7. Show that $\int J_0(x)J_1(x)dx = -\frac{1}{2}J_0^2(x)$, $\int_0^1 xJ_0(ax)dx = \frac{1}{a}J_1(a)$.

8. Show that $\int_0^1 xJ_0^2(ax)dx = \frac{J_0^2(a)+J_1^2(a)}{2}$.

9. Show that $J_{p-1}(x) = J_{p+1}(x)$ at every value of the extrema of the function $J_p(x)$, and $J_{p-1}(x) = -J_{p+1}(x) = J_p'(x)$ for every positive root of the equation $J_p(x) = 0$. Using book's software plot graphs of functions $J_2(x)$, $J_3(x)$, and $J_4(x)$ on the same screen and verify that the above equalities at extremals are correct.

10. Show that $\int_0^\infty J_1(x)dx = \int_0^\infty J_3(x)dx = \ldots = \int_0^\infty J_{2n+1}(x)dx$ and $\int_0^\infty J_0(x)dx = \int_0^\infty J_2(x)dx = \ldots = \int_0^\infty J_{2n}(x)dx$.

11. Show that $N_0'(x) = -N_1(x)$.
12. Show that $\int x^{n+1}J_n(x)dx = x^{n+1}J_{n+1}(x) + C$ for integer $n \geq 1$.
13. Using a series expansion of the function $J_0(x)$, show that

$$I_0(x) = 1 + \frac{x^2}{2^2} + \frac{x^4}{2^2 \cdot 4^2} + \frac{x^6}{2^2 \cdot 4^2 \cdot 6^2} \cdots$$

14. Prove that $\int x I_0(ax)dx = \frac{x}{a}I_0'(ax) + C$.
15. Prove that $\int j_l(x)dx = -j_0(x)$, $\int j_0(x)x^2 dx = x^2 j_l(x)$.

In the following three problems, find the eigenfunctions of the Sturm-Liouville problems for the Bessel equation assuming that function $y(x)$ is finite at $x = 0$:

16. $x^2 y'' + x y' + (\lambda x^2 - 1)y = 0$, $y(1) = 0$.
17. $x^2 y'' + x y' + \lambda x^2 y = 0$, $y'(2) = 0$.
18. $x^2 y'' + x y' + (\lambda x^2 - 9)y = 0$, $y(3) + 2y'(3) = 0$.

Problems 19 through 25 can be solved with the program FourierSeries (some of these problems are in the Library section of the program), Maple, or Mathematica.

19. Expand the function $f(x)$, given on the interval $[0, 1]$, in a Fourier series in Bessel functions of the first kind, $X_k(x) = J_0(\mu_k x)$, where $\mu_k^{(0)}$ are positive roots of the equation $J_0(\mu) = 0$, if:

$$\begin{array}{ll} \text{(a)} \, f(x) = \sin \pi x; & \text{(b)} \, f(x) = x^2; \\ \text{(c)} \, f(x) = x(1-x); & \text{(d)} \, f(x) = 1-x^2; \\ \text{(e)} \, f(x) = \cos \frac{\pi x}{2}. \end{array}$$

20. Expand the function $f(x)$, given on the interval $[0, 1]$, in a Fourier series in Bessel functions $X_k(x) = J_1(\mu_k x)$, where $\mu_k^{(1)}$ are positive roots of the equation $J_1(\mu) = 0$, if:

$$\begin{array}{ll} \text{(a)} \, f(x) = x; & \text{(b)} \, f(x) = \sin \pi x; \\ \text{(c)} \, f(x) = \sin^2 \pi x; & \text{(d)} \, f(x) = x(1-x); \\ \text{(e)} \, f(x) = x(1-x^2). \end{array}$$

21. Expand the function:

$$f(x) = A\left(1 - \frac{x^2}{l^2}\right), \quad A = \text{const},$$

given on the interval $[0, l]$, in Fourier series in Bessel functions $X_k(x) = J_0\left(\mu_k^{(0)} x/l\right)$, where $\mu_k^{(0)}$ are positive roots of the equation $J_0(\mu) = 0$.

22. Expand the function:

$$f(x) = Ax, \quad A = \text{const},$$

given on the interval $[0, l]$, in a Fourier series in Bessel functions $X_k(x) = J_1\left(\mu_k^{(1)} x/l\right)$, where $\mu_k^{(1)}$ are positive roots of the equation $J_1'(\mu) = 0$.

23. Expand the function:

$$f(x) = Ax^2, \quad A = \text{const},$$

given on the interval $[0, l]$, in Fourier series in Bessel functions of the first kind $X_k(x) = J_0\left(\mu_k^{(0)} x/l\right)$, where $\mu_k^{(0)}$ are positive roots of the equation $J_0'(\mu) = 0$.

24. Expand the function:

$$f(x) = \begin{cases} x^2, & 0 \le x < 1, \\ x, & 1 \le x < 2, \end{cases}$$

given on the interval $[0, 2]$, in Fourier series in Bessel functions of the first kind $X_k(x) = J_2\left(\mu_k^{(2)} x/l\right)$ ($l = 2$), where $\mu_k^{(2)}$ are positive roots of the equation $\mu J_2'(\mu) + hlJ_2(\mu) = 0$.

10.3 Legendre Functions

10.3.1 Legendre Equation and Legendre Polynomials

In applications one often encounters an eigenvalue problem containing a second-order linear homogeneous differential equation:

$$\left(1 - x^2\right)y'' - 2xy' + \lambda y = 0, \quad -1 \le x \le 1, \tag{10.82}$$

where λ is a real parameter. Equation (10.85) can be rewritten in the obviously *self-adjoint* form given by

$$\frac{d}{dx}\left[\left(1 - x^2\right)\frac{dy}{dx}\right] + \lambda y = 0, \quad -1 \le x \le 1 \tag{10.83}$$

and is called *the Legendre equation*. Such an equation frequently arises after a separation of variables procedure in spherical coordinates (in those problems the variable x is $x = \cos\theta$, where θ is a polar angle) in many problems of mathematical physics.

Prominent examples include heat conduction in a spherical domain, vibrations of spherical solids and shells, as well as boundary value problems for the electric potential in spherical coordinates. Note that the condition $p(x) > 0$ characteristic for the regular Sturm-Liouville problem is violated for $p(x) = 1 - x^2$ at the ends of the interval $x = \pm 1$ where Eq. (10.83) has regular singular points. Therefore, solutions of Eq. (10.83) are generally singular in those points.

To define the eigenvalue problem, we have to supplement the equation with boundary conditions. In the case of the Legendre equation, it is sufficient to demand that the solution is bounded at the endpoints:

$$y(\pm 1) < \infty. \tag{10.84}$$

Because the boundary value problem (10.83) and (10.84) is a kind of Sturm-Liouville problem, we can expect that different eigenfunctions $y_m(x)$ and $y_n(x)$ are orthogonal on the interval $[-1, 1]$ with the weight $r(x) = 1$:

$$\int_{-1}^{1} y_m(x)y_n(x)dx = 0, \quad n \neq m. \tag{10.85}$$

The proof is similar to that in the case of the regular Sturm-Liouville problem (see Sect. 9.3). Indeed, from (10.83) we find:

$$y_m \frac{d}{dx}\left[(1-x^2)\frac{dy_n}{dx}\right] + \lambda y_m y_n = 0,$$

$$y_n \frac{d}{dx}\left[(1-x^2)\frac{dy_m}{dx}\right] + \lambda y_n y_m = 0.$$

Subtracting the second equation from the first one and integrating over x from -1 to 1, we find:

$$(\lambda_n - \lambda_m) \int_{-1}^{1} y_m(x)y_n(x)dx$$

$$= \int_{-1}^{1} \left\{ y_m \frac{d}{dx}\left[(1-x^2)\frac{dy_n}{dx}\right] - y_n \frac{d}{dx}\left[(1-x^2)\frac{dy_m}{dx}\right] \right\}dx.$$

Integrating by parts, we find that the right-hand side of this equation vanishes; therefore, (10.85) holds if $\lambda_n \neq \lambda_m$.

Let us solve Eq. (10.82) on the interval $x \in [-1, 1]$ assuming that the function $y(x)$ is finite at the points $x = -1$ and $x = 1$. Let us search for a solution in the form of a power series in x:

$$y(x) = \sum_{k=0}^{\infty} a_k x^k. \tag{10.86}$$

Then,

$$xy' = \sum_{k=1}^{\infty} k a_k x^k = \sum_{k=0}^{\infty} k a_k x^k,$$

$$y'' = \sum_{k=2}^{\infty} k(k-1) a_k x^k = \sum_{k=0}^{\infty} (k+2)(k+1) a_{k+2} x^k,$$

$$x^2 y'' = \sum_{k=2}^{\infty} k(k-1) a_k x^k = \sum_{k=0}^{\infty} k(k-1) a_k x^k.$$

In y'' series the index of summations was changed from k to $k+2$ to start the summation in all terms from $k = 0$.
Substituting in Eq. (10.82), we obtain:

$$\sum_{k=0}^{\infty} \left[(k+2)(k+1) a_{k+2} - \left(k^2 + k - \lambda \right) a_k \right] x^k = 0.$$

This identity can be true for all $x \in [-1, 1]$ only when all coefficients of x^k are zero; therefore, we obtain *the recurrence formula*:

$$a_{k+2} = \frac{k(k+1) - \lambda}{(k+1)(k+2)} a_k. \tag{10.87}$$

Now we can state an important fact. The series in Eq. (10.86) converges on an open interval $-1 < x < 1$, as it can be seen from a ratio test, $\lim_{k \to \infty} \left| \frac{a_{k+2} x^{k+2}}{a_k x^k} \right| = x^2$, but diverges at the points $x = \pm 1$. Therefore, this series is not a bounded solution of the differential equation on the entire interval $-1 \leq x \leq 1$ unless it *terminates as a polynomial* with a finite number of terms. This can occur if the numerator in Eq. (10.87) is zero for some index value, $k_{max} = n$, such that

$$\lambda = n(n+1). \tag{10.88}$$

This gives $a_{n+2} = 0$ and consequently $a_{n+4} = 0, \ldots$; thus $y(x)$ will contain a finite number of terms and thus turn out to be a polynomial of degree n; thus,

$$y(x) = \sum_{k=0}^{n} a_k x^k. \tag{10.89}$$

It follows from (10.87) that this polynomial is even when n is an even number, and odd for odd n.
Because λ can take only *nonnegative integer* values, $\lambda = n(n+1)$, Eq. (10.82) actually becomes:

$$(1-x^2)y'' - 2xy' + n(n+1)y = 0. \tag{10.90}$$

Let us consider several particular cases. If $n = 0$ (which means that the highest degree of the polynomial is 0), $a_0 \neq 0$ and $a_2 = 0$, $a_4 = 0$, etc. The value of λ for this case is $\lambda = 0$, and we have $y(x) = a_0$. If $n = 1$, $a_1 \neq 0$, and $a_3 = a_5 = \ldots = 0$, then $\lambda = 2$ and $y(x) = a_1 x$. If $n = 2$, the highest degree of the polynomial is 2, we have $a_2 \neq 0$, $a_4 = a_6 = \ldots = 0$, $\lambda = 6$, and from the recurrence relation, we obtain $a_2 = -3a_0$. This results in $y(x) = a_0(1-3x^2)$. If $n = 3$, $a_3 \neq 0$, and $a_5 = a_7 = \ldots = 0$, $\lambda = 12$ and from the recurrence relation, we obtain $a_3 = -5/3a_1$ and as the result, $y(x) = a_1(x-5x^3/3)$. Constants a_0 and a_1 remain arbitrary unless we impose some additional requirements. A convenient requirement is that the solutions (the polynomials) obtained in this way should have the value 1 when $x = 1$.

The polynomials obtained above are denoted as $P_n(x)$ and called *the Legendre polynomials*. Those polynomials were mentioned in Sect. 9.2 (Example 9.6). The first few, which we derived above, are

$$P_0(x) = 1, \quad P_1(x) = x, \quad P_2(x) = \frac{1}{2}(3x^2 - 1), \quad P_3(x) = \frac{1}{2}(5x^3 - 3x). \tag{10.91}$$

Let us list two more (which the reader may derive as a *Reading Exercise* using the above relationships):

$$P_4(x) = \frac{1}{8}(35x^4 - 30x^2 + 3), \quad P_5(x) = \frac{1}{8}(63x^5 - 70x^3 + 15x). \tag{10.92}$$

Reading Exercise Obtain $P_n(x)$ for $n = 6, 7$.

Reading Exercise Show by direct substitution that $P_2(x)$ and $P_3(x)$ satisfy Eq. (10.82).

Rodrigues' formula gives another way to calculate the Legendre polynomials:

$$P_n(x) = \frac{1}{2^n n!} \frac{d^n}{dx^n}(x^2 - 1)^n \tag{10.93}$$

(here a zero-order derivative means the function itself).

Reading Exercise Using Rodrigues' formula show that

$$P_n(-x) = (-1)^n P_n(x), \tag{10.94}$$

$$P_n(-1) = (-1)^n. \tag{10.95}$$

Let us state several useful properties of Legendre polynomials:

$$P_n(1) = 1, \quad P_{2n+1}(0) = 0, \quad P_{2n}(0) = (-1)^n \frac{1 \cdot 3 \cdot \ldots \cdot (2n-1)}{2 \cdot 4 \cdot \ldots \cdot 2n}. \tag{10.96}$$

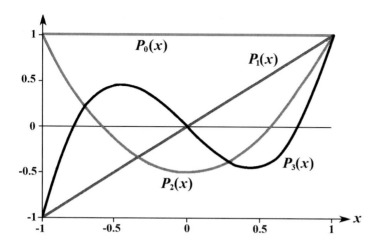

Fig. 10.12 First four polynomials, $P_n(x)$

The Legendre polynomial $P_n(x)$ has n real and simple (i.e., not repeated) roots, all lying in the interval $-1 < x < 1$. Zeros of the polynomials $P_n(x)$ and $P_{n+1}(x)$ alternate as x increases. In Fig. 10.12 the first four polynomials, $P_n(x)$, are shown, and their properties, as listed in Eqs. (10.96) through (10.98), are reflected in these graphs.

The following *recurrence formula* relates three polynomials:

$$(n+1)P_{n+1}(x) - (2n+1)xP_n(x) + nP_{n-1}(x) = 0. \qquad (10.97)$$

This formula gives a simple (and the most practical) way to obtain the Legendre polynomials of any order, one by one, starting with $P_0(x) = 1$ and $P_1(x) = x$.

There are recurrence relations for Legendre polynomials and their derivatives which are often useful:

$$P'_{n-1}(x) - xP'_n(x) + nP_n(x) = 0, \qquad (10.98)$$

$$P'_n(x) - xP'_{n-1}(x) - nP_{n-1}(x) = 0. \qquad (10.99)$$

To summarize, the solution of the Sturm-Liouville problem for Eq. (10.83) with boundary conditions stating that the solution is bounded on the closed interval $-1 \leq x \leq 1$ is a set of Legendre polynomials, $P_n(x)$, which are the eigenfunctions of the Sturm-Liouville operator. The eigenvalues are $\lambda = n(n+1)$, $n = 0, 1, 2, \ldots$. As a solution of the Sturm-Liouville problem, the Legendre polynomials, $P_n(x)$, form a *complete orthogonal* set of functions on the closed interval $[-1, 1]$, a property we will find very useful in the applications considered below.

If the points $x = \pm 1$ are excluded from a domain, the solution in the form of an infinite series is also acceptable. In this case logarithmically diverging at $x = \pm 1$ functions $Q_n(x)$ are also the solutions of Legendre equation (for details see Ref. [9]).

10.3.2   Fourier-Legendre Series in Legendre Polynomials

The Legendre polynomials are orthogonal on the interval $[-1, 1]$:

$$\int_{-1}^{1} P_n(x)P_m(x)dx = 0, \quad m \neq n. \tag{10.100}$$

The norm squared of Legendre polynomials is (see Ref. [9])

$$\|P_n\|^2 = \int_{-1}^{1} P_n^2(x)dx = \frac{2}{2n+1}. \tag{10.101}$$

Equations (10.100) and (10.101) can be combined and written as

$$\int_{-1}^{1} P_n(x)P_m(x)dx = \begin{cases} 0, & m \neq n, \\ \dfrac{2}{2n+1}, & m = n. \end{cases} \tag{10.102}$$

The Legendre polynomials form a complete set of functions on the interval $[-1, 1]$; thus, $\{P_n(x)\}$, $n = 0, 1, 2, \ldots$ provide a basis for an eigenfunction expansion for functions $f(x)$, bounded on the interval $[-1, 1]$:

$$f(x) = \sum_{n=0}^{\infty} c_n P_n(x). \tag{10.103}$$

Due to the orthogonality of the polynomials $P_n(x)$ of different degrees, the coefficients c_n are

$$c_n = \frac{1}{\|P_n\|^2} \int_{-1}^{1} f(x)P_n(x)dx = \frac{2n+1}{2} \int_{-1}^{1} f(x)P_n(x)dx. \tag{10.104}$$

Like in the case of the regular Sturm-Liouville problem, it can be shown that the sequence of the partial sums of this series (10.106), $S_N(x)$, converges on the interval $(-1, 1)$ on average (i.e., in the mean) to $f(x)$, which may be written as

$$\int\limits_{-1}^{1} [f(x) - S_N(x)]^2 dx \to 0 \quad \text{as} \quad N \to \infty. \tag{10.105}$$

The function $f(x)$ should be square integrable, that is, we require that the integral $\int_{-1}^{1} f^2(x)dx$ exists.

For an important class of piecewise-continuous functions, the series in Eq. (10.102) converges absolutely and uniformly. The following theorem states a stronger result about the convergence of the series (10.102) than convergence in the mean.

Theorem

If the function $f(x)$ is piecewise-continuous on the interval $(-1, 1)$, then the Fourier-Legendre series converges to $f(x)$ at the points where the function $f(x)$ is continuous and to

$$\frac{1}{2}[f(x_0 + 0) + f(x_0 - 0)], \tag{10.106}$$

if x_0 is a point of finite discontinuity of the function $f(x)$.

For any square integrable function, $f(x)$, we have:

$$\int\limits_{-1}^{1} f^2(x)dx = \sum_{n=0}^{\infty} \|P_n\|^2 c_n^2 = \sum_{n=0}^{\infty} \frac{2}{2n+1} c_n^2. \tag{10.107}$$

This is Parseval's equality (the completeness equation) for the Fourier-Legendre series. Clearly, for a partial sum on the right, we have Bessel's inequality:

$$\int\limits_{-1}^{1} f^2(x)dx \geq \sum_{n=0}^{N} \frac{2}{2n+1} c_n^2. \tag{10.108}$$

Below we consider several examples of the expansion of functions into the Fourier-Legendre series. In some cases the coefficients can be found analytically; otherwise, we may calculate them numerically using book's software, Maple, or Mathematica.

Example 10.6 Expand function $f(x) = A$, $A = $ const., in a Fourier-Legendre series in $P_n(x)$ on the interval $-1 \leq x \leq 1$.

Solution The series is

$$A = c_0 P_0(x) + c_1 P_1(x) + \ldots,$$

where coefficients c_n are

$$c_n = \frac{1}{\|P_n^2(x)\|} \int_{-1}^{1} A P_n(x) dx = \frac{(2n+1)A}{2} \int_{-1}^{1} P_n(x) dx.$$

From this formula it is clear that the only nonzero coefficient is $c_0 = A$.

Example 10.7 Expand the function $f(x) = x$ in a Fourier-Legendre series in $P_n(x)$ on the interval $-1 \leq x \leq 1$.

Solution The series is

$$x = c_0 P_0(x) + c_1 P_1(x) + \ldots,$$

where c_n are

$$c_n = \frac{1}{\|P_n^2(x)\|} \int_{-1}^{1} x P_n(x) dx = \frac{2n+1}{2} \int_{-1}^{1} x P_n(x) dx.$$

It is clear that the only nonzero coefficient is

$$c_1 = \frac{3}{2} \int_{-1}^{1} x P_1(x) dx = \frac{3}{2} \int_{-1}^{1} x^2 dx = 1.$$

As in the previous example, this result is apparent because one of the polynomials, $P_1(x)$ in this example, coincides with the given function, $f(x) = x$.

Example 10.8 Expand the function $f(x)$ given by

$$f(x) = \begin{cases} 0, & -1 < x < 0, \\ 1, & 0 < x < 1 \end{cases}$$

in a Fourier-Legendre series.

Solution The expansion $f(x) = \sum\limits_{n=0}^{\infty} c_n P_n(x)$ has coefficients:

$$c_n = \frac{1}{\|P_n\|^2} \int\limits_{-1}^{1} f(x)P_n(x)dx = \frac{2n+1}{2} \int\limits_{0}^{1} P_n(x)dx.$$

The first several coefficients are

$$c_0 = \frac{1}{2} \int\limits_{0}^{1} dx = \frac{1}{2}, \quad c_1 = \frac{3}{2} \int\limits_{0}^{1} x\,dx = \frac{3}{4}, \quad c_2 = \frac{5}{2} \int\limits_{0}^{1} \frac{1}{2}(3x^2-1)dx = 0.$$

Continuing, we find for the given function $f(x)$:

$$f(x) = \frac{1}{2}P_0(x) + \frac{3}{4}P_1(x) - \frac{7}{16}P_3(x) + \frac{11}{32}P_5(x) + \ldots.$$

The series converges slowly because of discontinuity of the given function $f(x)$ at point $x = 0$ (Fig. 10.13).

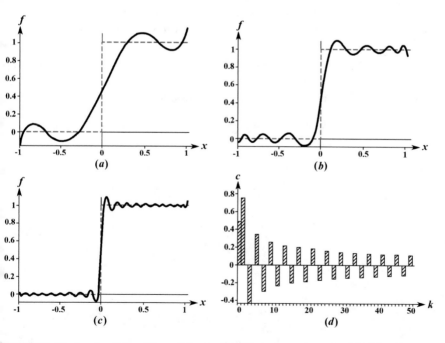

Fig. 10.13 The function $f(x)$ and the partial sum of its Fourier-Legendre series (10.106) with $N+1$ terms. The graph of $f(x)$ is shown by the dashed line, and the graph of the series is shown by the solid line. (a) $N = 5$, (b) $N = 15$, (c) $N = 50$; values of the coefficients c_n of the series are shown in panel (d)

10.3.3   Associate Legendre Functions

In this section we consider a generalization of Eq. (10.82):

$$\left(1-x^2\right)y'' - 2xy' + \left(\lambda - \frac{m^2}{1-x^2}\right)y = 0, \quad -1 \leq x \leq 1, \tag{10.109}$$

where m is a number. An example of the derivation of Eq. (10.109) from a real-world partial differential equation is given in Chap. 13 (Sect. 13.5). As it will be shown below, Eq. (10.109) has nontrivial solutions bounded at $x = \pm 1$ only for the values of $\lambda = n(n+1)$. In mathematical physics problems, the values of m are *integer*; also the values of m and n are related by inequality $|m| \leq n$. The solution of the original PDE problem depends on the sign of m, but Eq. (10.109) contains only m^2; hence, its solutions do not depend on its sign and later on we assume $m \geq 0$. Solutions for $m < 0$ can be obtained from those with $m > 0$ by the replacement of m with $|m|$.

Equation (10.109) is called the *associated Legendre equation of order m*. In the Sturm-Liouville form, this equation can be written as

$$\frac{d}{dx}\left[\left(1-x^2\right)\frac{dy}{dx}\right] + \left(\lambda - \frac{m^2}{1-x^2}\right)y = 0, \quad -1 \leq x \leq 1. \tag{10.110}$$

To solve Eq. (10.110), we can use a solution of Eq. (10.83). Let us introduce a new function, $z(x)$, to replace $y(x)$ in Eq. (10.110) using the substitution:

$$y(x) = \left(1 - x^2\right)^{\frac{m}{2}} z(x). \tag{10.111}$$

Substituting Eq. (10.114) into Eq. (10.113), we obtain:

$$\left(1-x^2\right)z'' - 2(m+1)xz' + [\lambda - m(m+1)]z = 0. \tag{10.112}$$

If $m = 0$, Eq. (10.112) reduces to Eq. (10.82); thus, its solutions are Legendre polynomials $P_n(x)$.

Let us solve Eq. (10.112) expanding $z(x)$ in a power series:

$$z = \sum_{k=0}^{\infty} a_k x^k. \tag{10.113}$$

With this we have:

$$z' = \sum_{k=1}^{\infty} k a_k x^{k-1} = \sum_{k=0}^{\infty} k a_k x^{k-1},$$

$$z'' = \sum_{k=2}^{\infty} k(k-1) a_k x^{k-2} = \sum_{k=0}^{\infty} (k+2)(k+1) a_{k+2} x^k,$$

$$x^2 z'' = \sum_{k=2}^{\infty} k(k-1) a_k x^k = \sum_{k=0}^{\infty} k(k-1) a_k x^k.$$

Substituting these expressions into Eq. (10.112), we obtain:

$$\sum_{k=0}^{\infty} \{(k+2)(k+1) a_{k+2} + [\lambda - (k+m)(k+m+1)] a_k\} x^k = 0.$$

Functions x^k are linearly independent; thus, the coefficients of each power of x^k must be zero which leads to a recurrence relation for coefficients a_k:

$$a_{k+2} = -\frac{\lambda - (k+m)(k+m+1)}{(k+2)(k+1)} a_k. \tag{10.114}$$

Reading Exercise Using this recurrence relation, check that the series in Eq. (10.113) converges for $-1 < x < 1$ and diverges at the ends of the intervals, $x = \pm 1$.

Below we will discuss only solutions which are regular on the closed interval, $-1 \le x \le 1$. This means that the series (10.113) should *terminate as a polynomial* of some maximum degree. Denoting this degree as q, we obtain $a_q \ne 0$ and $a_{q+2} = 0$ so that if $\lambda = (q + m)(q + m + 1)$, $q = 0, 1, \ldots$, then $a_{q+2} = a_{q+4} = \ldots = 0$. Introducing $n = q + m$, because q and m are nonnegative integers, we have $n = 0$, $1, 2, \ldots$ and $n \ge m$. Thus, we see that $\lambda = n(n + 1)$ as in the case of Legendre polynomials. Clearly if $n = 0$ the value of $m = 0$; thus, $\lambda = 0$ and the function $z(x) = a_0$ and $y(x) = P_0(x)$.

From the above discussion, we obtain that $z(x)$ is an even or odd polynomial of degree $(n-m)$:

$$z(x) = a_{n-m} x^{n-m} + a_{n-m-2} x^{n-m-2} + \ldots + \begin{cases} a_0, \\ a_1 x. \end{cases} \tag{10.115}$$

Let us present several examples for $m = 1$. If $n = 1$, then $q = 0$; thus, $z(x) = a_0$. If $n = 2$, then $q = 1$; thus, $z(x) = a_1 x$. If $n = 3$, $q = 2$, and $z(x) = a_0 + a_2 x^2$, and from the recurrence formula, we have $a_2 = -5 a_0$.

Reading Exercises
1. Find $z(x)$ for $m = 1$ and $n = 4$.
2. Find $z(x)$ for $m = 2$ and $n = 4$.
3. For all the above examples, check that (keeping the lowest coefficients arbitrary)
$z(x) = \frac{d^m}{dx^m} P_n(x)$.

Given that $\lambda = n(n + 1)$, we can obtain a solution of Eq. (10.112) using the solution of the Legendre Eq. (10.83). Let us differentiate Eq. (10.112) with respect to the variable x:

$$\left(1 - x^2\right)\left(z'\right)'' - 2[(m + 1) + 1]x(z')' + [n(n + 1) - (m + 1)(m + 2)]z' = 0.$$

$$(10.116)$$

It is seen that if in this equation we replace z' by z and $(m + 1)$ by m, the obtained equation becomes Eq. (10.112). In other words, if $P_n(x)$ is a solution of Eq. (10.112) for $m = 0$, then $P'_n(x)$ is a solution of Eq. (10.116) for $m = 1$. Repeating this we obtain that $P''_n(x)$ is a solution for $m = 2$, $P'''_n(x)$ is a solution for $m = 3$, etc. For arbitrary integer m, where $0 \leq m \leq n$, a solution of Eq. (10.112) is the function $\frac{d^m}{dx^m} P_n(x)$; thus,

$$z(x) = \frac{d^m}{dx^m} P_n(x), \quad 0 \leq m \leq n. \tag{10.117}$$

With Eqs. (10.117) and (10.111), we have a solution of Eq. (10.110) given by

$$y(x) = \left(1 - x^2\right)^{\frac{m}{2}} \frac{d^m}{dx^m} P_n(x), \quad 0 \leq m \leq n. \tag{10.118}$$

The functions defined in Eq. (10.118) are called the *associated Legendre functions* and denoted as $P_n^m(x)$:

$$P_n^m(x) = \left(1 - x^2\right)^{\frac{m}{2}} \frac{d^m}{dx^m} P_n(x). \tag{10.119}$$

Notice that $\frac{d^m}{dx^m} P_n(x)$ is a polynomial of degree $n-m$; thus,

$$P_n^m(-x) = (-1)^{n-m} P_n^m(x), \tag{10.120}$$

which is referred to as the parity property. From Eq. (10.119) it is directly seen that $P_n^m(x) = 0$ for $|m| > n$ because in this case m-th order derivatives of polynomial $P_n(x)$ of degree n are equal to zero. The graphs of several $P_n^m(x)$ are plotted in Fig. 10.14. Thus, from the above discussion, we see that Eq. (10.109) has eigenvalues λ:

$$m(m + 1), \quad (m + 1)(m + 2), \quad (m + 2)(m + 3) \tag{10.121}$$

with corresponding eigenfunctions, bounded on $[-1, 1]$, are

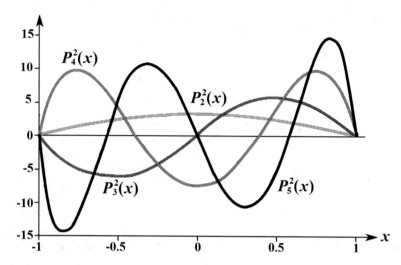

Fig. 10.14 Graphs of $P_2^2(x)$, $P_3^2(x)$, $P_4^2(x)$, and $P_5^2(x)$

$$P_m^m(x), \quad P_{m+1}^m(x), \quad P_{m+2}^m(x), \quad \ldots \tag{10.122}$$

Equation (10.109) (or (10.110)) does not change when the sign of m changes. Therefore, a solution of Eq. (10.109) for positive m is also a solution for negative values $-|m|$. Thus, we can define $P_n^m(x)$ as equal to $P_n^{|m|}(x)$ for $-n \leq m \leq n$:

$$P_n^{-|m|}(x) = P_n^{|m|}(x). \tag{10.123}$$

The first several associated Legendre functions $P_n^1(x)$ for $m = 1$ are

$$P_1^1(x) = \sqrt{1 - x^2} \cdot [P_1(x)]' = \sqrt{1 - x^2},$$
$$P_2^1(x) = \sqrt{1 - x^2} \cdot [P_2(x)]' = \sqrt{1 - x^2} \cdot 3x,$$
$$P_3^1(x) = \sqrt{1 - x^2} \cdot [P_3(x)]' = \sqrt{1 - x^2} \cdot \frac{3}{2}\left(5x^2 - 1\right)$$

and the first several associated Legendre functions, $P_n^2(x)$, for $m = 2$ are

$$P_2^2(x) = (1 - x^2) \cdot [P_2(x)]'' = (1 - x^2) \cdot 3,$$
$$P_3^2(x) = (1 - x^2) \cdot [P_3(x)]'' = (1 - x^2) \cdot 15x,$$
$$P_4^2(x) = (1 - x^2) \cdot [P_4(x)]'' = (1 - x^2) \cdot \frac{15}{2}\left(7x^2 - 1\right).$$

Associated Legendre function $P_n^m(x)$ has $(n-m)$ simple (not repeating) real roots on the interval $-1 < x < 1$.

The following *recurrence formula* is often useful:

$$(2n + 1)xP_n^m(x) - (n - m + 1)P_{n+1}^m(x) - (n + m)P_{n-1}^m(x) = 0. \qquad (10.124)$$

10.3.4   Fourier-Legendre Series in Associated Legendre Functions

Functions $P_n^m(x)$ *for fixed value of* $|m|$ and all possible values of the lower index

$$P_m^m(x), \quad P_{m+1}^m(x), \quad P_{m+2}^m(x), \quad \ldots \qquad (10.125)$$

form an *orthogonal* (with respect to the weight function $r(x) = 1$) and *complete* set of functions on the interval $[-1, 1]$ (because of relation (10.123), below we assume $m \geq 0$). In other words, for each value of m, there is an orthogonal and complete set of functions (10.125). This follows from the fact that these functions are also the solutions of a Sturm-Liouville problem. Thus, the set of functions (10.125) for any given m is a basis for an eigenfunction expansion for functions bounded on $[-1, 1]$, and we may write:

$$f(x) = \sum_{k=0}^{\infty} c_k P_{m+k}^m(x). \qquad (10.126)$$

The expansions in $P_n^m(x)$ are used for solving some PDE problems. The formula for the coefficients c_k ($k = 0, 1, 2, \ldots$) follows from the orthogonality of the functions in Eq. (10.128):

$$c_k = \frac{1}{\left\| P_{m+k}^m \right\|^2} \int_{-1}^{1} f(x)P_{m+k}^m(x)dx = \frac{[2(m+k) + 1]k!}{2(2m+k)!} \int_{-1}^{1} f(x)P_{m+k}^m(x)dx.$$

$$(10.127)$$

As previously, the sequence of the partial sums $S_N(x)$ of series (10.126) converges on the interval $(-1, 1)$ in the mean to the square integrable function $f(x)$, i.e.,

$$\int_{-1}^{1} [f(x) - S_N(x)]^2 dx \to 0 \quad \text{as} \quad N \to \infty.$$

For piecewise-continuous functions, the same theorem as in the previous section is valid.

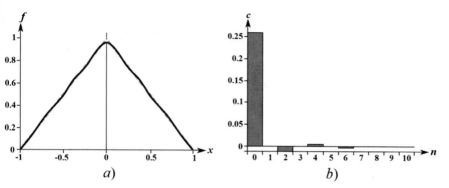

Fig. 10.15 The function $f(x)$ and the partial sum of its Fourier-Legendre series in terms of associated Legendre functions $P_n^2(x)$. (**a**) The graph of $f(x)$ is shown by the dashed line, the graph of the partial sum with $N = 10$ terms of series by the solid line; (**b**) values of the coefficients c_k of the series

Example 10.9 Expand the function:

$$f(x) = \begin{cases} 1 + x, & -1 \le x < 0, \\ 1 - x, & 0 \le x \le 1 \end{cases}$$

in terms of associated Legendre functions $P_n^m(x)$ of order $m = 2$.

Solution The series is

$$f(x) = c_0 P_2^2(x) + c_1 P_3^2(x) + c_2 P_4^2(x) + c_3 P_5^2(x) + \ldots,$$

where coefficients c_k are (Fig. 10.15)

$$c_k = \frac{2k + 5}{2} \frac{k!}{(k + 4)!} \left[\int_{-1}^{0} (1 + x) P_{k+2}^2(x)dx + \int_{0}^{1} (1 - x) P_{k+2}^2(x)dx \right].$$

Because $f(x)$ is an even function of x, $c_1 = c_3 = c_5 = \ldots = 0$.
The first two coefficients with even index are

$$c_0 = \frac{5}{48} \left[\int_{-1}^{0} (1 + x)3(1 - x^2)dx + \int_{0}^{1} (1 - x)3(1 - x^2)dx \right] = \frac{25}{96},$$

$$c_2 = \frac{1}{80} \left[\int_{-1}^{0} (1 + x)\frac{15}{2}(1 - x^2)(7x^2 - 1)dx + \int_{0}^{1} (1 - x)\frac{15}{2}(1 - x^2)(7x^2 - 1)dx \right]$$

$$= -\frac{1}{80}.$$

Problems

1. Using the recurrence formula in Eq. (10.87), show that the coefficient of x^n (the highest degree term) in $P_n(x)$ is $a_n = \frac{1 \cdot 3 \cdot 5 \cdots (2n-1)}{n!}$.

2. Prove Rodrigues' formula using the solution of the previous problem.
   Hint: Show that $z = (x^2 - 1)^n$ satisfies equation:

$$\left(1 - x^2\right)\frac{d^{n+2}z}{dx^{n+2}} - 2x\frac{d^{n+1}z}{dx^{n+1}} + n(n+1)\frac{d^n z}{dx^n} = 0,$$

   which means that $\frac{d^n z}{dx^n}$ is a solution of Legendre equation. Therefore, this solution must be a constant multiple of $P_n(x)$: $\frac{d^n z}{dx^n} = cP_n(x)$. Then, comparing the coefficients of highest degree on both sides of this relation, one can obtain $\frac{d^n}{dx^n}(x^2 - 1)^n = 2^n n! P_n(x)$ that is Rodrigues' formula.

3. Prove the following useful recurrence formulas:

$$P_{n-1}(x) = xP_n(x) + \frac{1 - x^2}{n}P_n'(x), \quad n \geq 1,$$

$$P_{n+1}(x) = xP_n(x) - \frac{1 - x^2}{n+1}P_n'(x), \quad n \geq 0.$$

4. The integral $I_{nm} = \int\limits_{-1}^{1} xP_n(x)P_m(x)dx$ is important in applications (e.g., in quantum mechanics problems). Show that

$$I_{nm} = \frac{2(n+1)}{(2n+1)(2n+3)}\delta_{m,n+1} + \frac{2n}{(2n+1)(2n-1)}\delta_{m,n-1}.$$

5. Prove that

$$\int\limits_{-1}^{1} xP_n(x)P_n'(x)dx = \frac{2n}{2n+1}.$$

6. Integrate the recurrence relation in Eq. (10.98) by parts to show that

$$I_n = \int\limits_{0}^{1} P_n(x)dx = \frac{P_{n-1}(0)}{n+1}, \quad n \geq 1.$$

   This integral is useful in applications. With Eq. (10.96) we can find that for odd values $n = 1, 3, 5, \ldots$:

$$I_n = (-1)^{(n-1)/2} \frac{(n-1)!}{2^n \left(\frac{n+1}{2}\right)! \left(\frac{n-1}{2}\right)!} = \frac{\Gamma(3/2)}{\Gamma(3/2 + n/2)\Gamma(1 - n/2)}$$

(clearly that $I_1 = 1$), $I_n = 0$ if n is even.

7. Let us present the Legendre equation as the eigenvalue problem $Ly(x) = \lambda y(x)$, where $L = - \frac{d}{dx}\left[(1-x^2)\frac{d}{dx}\right]$. Check that operator L is Hermitian on $[-1,1]$ i.e.,

$$\int_{-1}^{1} P_m(x)LP_n(x)dx = \int_{-1}^{1} P_n(x)LP_m(x)dx.$$

For each of problems 8 through 15, expand the function $f(x)$ in the Fourier-Legendre functions $P_n^m(x)$ on $[-1,1]$. Do expansion for (a) $m = 0$ – in this case the functions $P_n^m(x)$ are the Legendre polynomials $P_n(x)$; (b) for $m = 1$; (c) for $m = 2$.

Using the FourierSeries program, Maple, or Mathematica, obtain the pictures of several functions $P_n^m(x)$, plot the graphs of the given function $f(x)$, and of the partial sums $S_N(x)$ of the series, build the histograms of coefficients c_k of the series:

8. $f(x) = 5x - 2$.

9. $f(x) = 1 - 3x^2$.

10. $f(x) = \begin{cases} -1, & -5 < x < 0, \\ 1, & 0 < x < 5. \end{cases}$

11. $f(x) = \begin{cases} 0, & -5 < x < 0, \\ x, & 0 < x < 5. \end{cases}$

12. $f(x) = 5\cos\frac{\pi x}{2}$ for $-1 \leq x \leq 1$.

13. $f(x) = \begin{cases} 0, & -10 < x < 0, \\ \sqrt{10-x}, & 0 < x < 10. \end{cases}$

14. $f(x) = 7\sin \pi x$.

15. $f(x) = 3e^x$.

10.4   Elliptic Integrals and Elliptic Functions

Unlike the special functions considered in the previous sections that are useful for solving linear differential equations, the *elliptic functions* are used for solving some nonlinear ODEs and PDEs.

Let us define three *incomplete* elliptic integrals:

$$F(\varphi, k) = \int_0^{\varphi} \frac{d\theta}{\sqrt{1 - k^2 \sin^2 \theta}} \tag{10.128}$$

(incomplete elliptic integral of the first kind)

$$E(\varphi, k) = \int_0^{\varphi} \sqrt{1 - k^2 \sin^2 \theta} d\theta \tag{10.129}$$

(incomplete elliptic integral of the second kind), and

$$\Pi(\varphi, n, k) = \int_0^{\varphi} \frac{d\theta}{\left(1 - n \sin^2 \theta\right)\sqrt{1 - k^2 \sin^2 \theta}} \tag{10.130}$$

(incomplete elliptic integral of the third kind). Later on, we discuss only the case where the *modulus k* satisfies the condition $0 \le k^2 \le 1$; therefore, the elliptic integral is real. Substituting $x = \sin \theta$ into expressions (10.128)–(10.130), one obtains the following expressions for the elliptic integrals on the interval $-\pi/2 \le \varphi \le \pi/2$:

$$F(\varphi, k) = \int_0^{\sin \varphi} \frac{dx}{\sqrt{(1 - x^2)(1 - k^2 x^2)}}, \tag{10.131}$$

$$E(\varphi, k) = \int_0^{\sin \varphi} \sqrt{\frac{1 - k^2 x^2}{1 - x^2}} dx, \tag{10.132}$$

and

$$\Pi(\varphi, n, k) \int_0^{\sin \varphi} \frac{dx}{(1 - nx^2)\sqrt{(1 - x^2)(1 - k^2 x^2)}}. \tag{10.133}$$

Actually, any integrals of the kind

$$\int_a^x f\left(x, \sqrt{g(x)}\right) dx,$$

where f is a rational function and g is a polynomial of the fourth or third degree, can be transformed to the combinations of integrals (10.131)–(10.133).

Fig. 10.16 Plots of $F(\varphi, k)$
for $k^2 = 0$ (dashed line),
$0 < k^2 < 1$ (solid line), and
$k^2 = 1$ (dotted line)

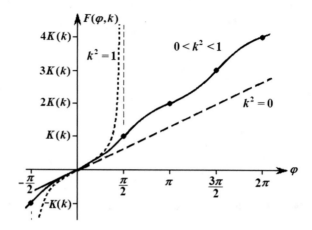

The plots of the function $F(\varphi, k)$ for different values of k are shown in Fig. 10.16.
Note that $F(\varphi, 0) = \varphi$ and

$$F(\varphi, 1) = \ln\left(\frac{1 + \sin\varphi}{\cos\varphi}\right), \quad |\varphi| < \frac{\pi}{2}.$$

The values at $\varphi = \pi/2$ of the integrals defined above are called *complete* elliptic
integrals:

$$K(k) = F\left(\frac{\pi}{2}, k\right), \quad E(k) = E\left(\frac{\pi}{2}, k\right), \quad \Pi(n, k) = \Pi\left(\frac{\pi}{2}, n, k\right).$$

Useful formulas for derivatives of complete elliptic integrals are

$$E'(k) = \frac{E(k) - K(k)}{k}, \quad K'(k) = \frac{E(k)}{k(1 - k^2)} - \frac{K(k)}{k}.$$

Let us present also *Legendre's relation*:

$$EK' + E'K - KK' = \frac{\pi}{2}.$$

The function *inverse* to $F(\varphi, k)$ is called *elliptic amplitude*: if $x = F(\varphi, k)$, then
$\varphi = \mathrm{am}(x, k)$. Define also the following *Jacobi elliptic functions*: *elliptic sine*,

$$\mathrm{sn}\, x = \sin\mathrm{am}(x, k),$$

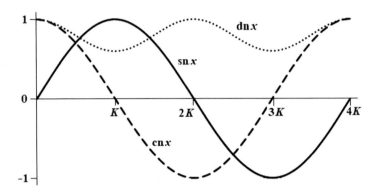

Fig. 10.17 Schematic plots of Jacobi elliptic functions: snx (solid line), cnx (dashed line), and dnx (dotted line)

elliptic cosine

$$\mathrm{cn} x = \cos \mathrm{am}(x, k),$$

and *delta amplitude*

$$\mathrm{dn} x = \sqrt{1 - k^2 \sin^2 x}.$$

Those functions are useful for description of nonlinear oscillations (see Chap. 5). Schematic plots of Jacobi elliptic functions are shown in Fig. 10.17.

The derivatives of those functions are (check these results as *Reading Exercises*):

$$\frac{d}{dx}\mathrm{sn} x = \mathrm{cn} x \mathrm{dn} x, \quad \frac{d}{dx}\mathrm{cn} x = -\mathrm{sn} x \mathrm{dn} x, \quad \frac{d}{dx}\mathrm{dn} x = -k^2 \mathrm{sn} x \mathrm{cn} x.$$

10.5   Hermite Polynomials

In Chap. 6 we already considered the Hermite equation:

$$y'' - 2xy' + 2ny = 0, \quad n = 0, 1, 2, \ldots \tag{10.134}$$

which solution are the Hermite polynomials

$$f(x) \equiv H_n(x) = \sum_{k}^{n} a_k x^k. \tag{10.135}$$

In more formal approach, the Hermite polynomials are the solutions of the boundary value problem:

$$\frac{d}{dx}\left(e^{-x^2}\frac{dy}{dx}\right) + \lambda e^{-x^2}y = 0, \quad -\infty < x < \infty, \tag{10.136}$$

with the boundary condition that function $y(x)$ can increase not faster than a polynomial when $x \to \pm\infty$. This problem has nontrivial eigenfunctions:

$$H_n(x) = (-1)^n e^{x^2}\frac{d^n}{dx^n}e^{-x^2} \tag{10.137}$$

and the eigenvalues $\lambda_n = 2n, \quad n = 0, 1, \ldots$.

Formula (10.137) is called Rodrigues' formula for the Hermite polynomials – it allows to obtain $H_n(x)$ by sequential differentiation of function $e^{-x^2/2}$. For instance

$$H_0(x) = (-1)^0 e^{x^2}e^{-x^2} = 1, \quad H_1(x) = (-1)^1 e^{x^2}\frac{d}{dx}e^{-x^2} = 2x.$$

From (10.137) follow

$$\frac{d^m}{dx^m}H_n(x) = 2^m n(n-1)\ldots(n-m+1)H_{n-m}(x) \tag{10.138}$$

and the recurrence formula

$$H_n(x) - 2xH_{n-1}(x) + 2(n-1)H_{n-2}(x) = 0, \quad n \geq 2. \tag{10.139}$$

The explicit expressions for $H_n(x)$ through the factorials are

$$H_n(x) = \sum_{k=0}^{\left[\frac{n}{2}\right]}\frac{(-1)^k n!}{k!(n-2k)!}(2x)^{n-2k}, \tag{10.140}$$

where $[n/2]$ is the largest integer of $n/2$, i.e. the largest integer that is not smaller than $n/2$.

Let us find the orthogonality condition for Hermite polynomials.

For $y_n = e^{-x^2/2}H_n(x)$, we have:

$$y_n'' + (2n+1-x^2)y_n = 0, \quad y_m'' + (2m+1-x^2)y_m = 0.$$

Multiplying the first of these equations by y_m, the second by y_n, then subtracting the obtained relations, we have:

$$y_n''y_m - y_m''y_n + 2(n-m)y_ny_m = 0, \quad \text{or} \quad \frac{d}{dx}(y_n'y_m - y_m'y_n) + 2(n-m)y_ny_m = 0.$$

Integrating the last relation on $-\infty < x < \infty$ and taking into account that $y_n(x) \to 0$ when $x \to \pm \infty$, we obtain:

$$2(n-m) \int_{-\infty}^{\infty} y_n y_m dx = 2(n-m) \int_{-\infty}^{\infty} e^{-x^2} H_n(x) H_m(x) dx.$$

For $n \neq m$ we see that H_n and H_m orthogonal with the weight e^{-x^2}.

To consider the case $n = m$, let us take the recurrence formula (10.139) and the one obtained from it by replacing n to $n + 1$:

$$H_n(x) - 2xH_{n-1}(x) + 2(n-1)H_{n-2}(x) = 0, \quad n = 2, 3, \ldots$$
$$H_{n+1}(x) - 2xH_n(x) + 2nH_{n-1}(x) = 0, \quad n = 1, 2, \ldots$$

Multiplying the first relation by H_{n-1}, the second by H_n and subtracting one from another, we obtain:

$$2nH_{n-1}^2(x) + H_{n+1}(x)H_{n-1}(x) - 2(n-1)H_{n-2}(x)H_n(x) - H_n^2(x) = 0, \quad n = 2, 3, \ldots$$

Next multiply this relation by e^{-x^2} and integrate from $-\infty$ to ∞; with the orthogonality of Hermite polynomials, it gives:

$$\int_{-\infty}^{\infty} e^{-x^2} H_n^2(x) dx = 2n \int_{-\infty}^{\infty} e^{-x^2} H_{n-1}^2 dx, \quad n = 2, 3, \ldots. \tag{10.141}$$

Repeating this formula, we arrive to

$$\int_{-\infty}^{\infty} e^{-x^2} H_n^2(x) dx = 2n \int_{-\infty}^{\infty} e^{-x^2} H_{n-1}^2 dx$$
$$= \ldots = 2^{n-1} n! \int_{-\infty}^{\infty} e^{-x^2} H_1^2 dx = 2^{n-1} n! 4 \int_{-\infty}^{\infty} e^{-x^2} x^2 dx = 2^n n! \sqrt{\pi}.$$

The last transition was obtained by integrating by parts with $\int_{-\infty}^{\infty} e^{-x^2} dx = \sqrt{\pi}$:

$$\int_{-\infty}^{\infty} e^{-x^2} x^2 dx = -\frac{1}{2} \int_{-\infty}^{\infty} x d\left(e^{-x^2}\right) = -\frac{1}{2} x e^{-x^2} \Big|_{-\infty}^{\infty} + \frac{1}{2} \int_{-\infty}^{\infty} e^{-x^2} dx = \frac{\sqrt{\pi}}{2}.$$

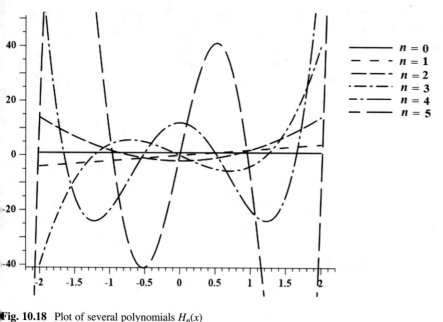

Fig. 10.18 Plot of several polynomials $H_n(x)$

Thus, the orthogonality and normalization conditions are

$$\int_{-\infty}^{\infty} e^{-x^2} H_n(x) H_m(x) dx = \begin{cases} 0, & \text{if } m \neq n; \\ 2^n n! \sqrt{\pi}, & \text{if } m = n. \end{cases} \qquad (10.142)$$

The first few polynomials are (Fig. 10.18)

$$H_0(x) = 1, \quad H_1(x) = 2x, \quad H_2(x) = 4x^2 - 2, \quad H_3(x) = 8x^3 - 12x,$$
$$H_4(x) = 16x^4 - 48x^2 + 12.$$

The Hermite polynomials form a complete orthogonal system of functions on $(-\infty, \infty)$; therefore, they can be used as a basis to expand functions $f(x)$ for which $\int_{-\infty}^{\infty} e^{-x^2} f^2(x) dx < \infty$: in points x where $f(x)$ is continuous

$$f(x) = \sum_{n=0}^{\infty} c_n H_n(x), \quad c_n = \frac{1}{2^n n! \sqrt{\pi}} \int_{-\infty}^{\infty} e^{-x^2} f(x) H_n(x) dx. \qquad (10.143)$$

As the application of the Hermite polynomials, let us consider the quantum oscillator problem – models based on this problem have found wide application in condensed matter physics and quantum field theory.

The *Schrödinger equation* for a particle with mass m in a potential $U = \frac{m\omega^2 x^2}{2}$ has the form (E is the particle energy, ω is oscillations frequency, $\hbar$ is the reduced Planck constant):

$$\psi''(x) + \frac{2m}{\hbar^2}\left(E - \frac{m\omega^2 x^2}{2}\right)\psi(x) = 0. \tag{10.144}$$

Let us introduce the dimensionless length and energy:

$$z = \sqrt{\frac{m\omega}{\hbar}}x \equiv \alpha x, \quad \lambda = \frac{2E}{\hbar\omega}. \tag{10.145}$$

With them Eq. (10.144) becomes:

$$\frac{d^2\psi}{dz^2} + \left(\lambda - z^2\right)\psi = 0. \tag{10.146}$$

When $z \to \infty$, one can neglect λ and it is easy to check that $\psi(z) \sim e^{-z^2/2}$. Using this asymptotic, let us search function $\psi(z)$ in the form:

$$\psi = f(z)e^{-z^2/2}. \tag{10.147}$$

Substituting (10.147) in Eq. (10.144) gives equation for $f(z)$:

$$f'' - 2zf' + (\lambda - 1)f = 0. \tag{10.148}$$

This is Eq. (10.134) with $\lambda-1$ in place of $2n$; its solution is $f(x) \equiv H_n(x)$. The relation $\lambda = 2n + 1$, with $\lambda = 2E/\hbar\omega$, leads to the energy quantization:

$$E_n = \hbar\omega\left(n + \frac{1}{2}\right). \tag{10.149}$$

The spectrum is equidistant (Fig. 10.19): $\Delta E = E_{n'} - E_n = \hbar\omega$. The wave function of the quantum harmonic oscillator is

$$\psi_n(z) = C_n H_n(z)e^{-z^2/2}. \tag{10.150}$$

Fig. 10.19 Functions $\psi_n(x)$ and energy levels of harmonic oscillator

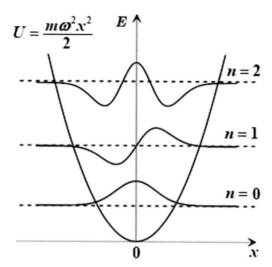

Coefficients C_n are determined from the normalization condition for the wave function (meaning that the total probability for a particle to be on $(-\infty, \infty)$ is 1):

$$\int_{-\infty}^{+\infty} |\psi_n|^2 dz = C_n^2 \int_{-\infty}^{+\infty} H_n^2(z) e^{-z^2} dz = C_n^2 2^n n! \sqrt{\pi} = 1,$$

$$C_n = \frac{1}{\sqrt{2^n n!} \, \pi^{1/4}}.$$

$$(10.151)$$

The final form of the wave functions of the harmonic oscillator (in dimensionless and dimensional form) is

$$\psi_n(z) = \frac{1}{\sqrt{2^n n!} \, \pi^{1/4}} H_n(z) e^{-z^2/2},$$

$$\psi_n(x) = \frac{1}{\sqrt{2^n n!}} \left(\frac{m\omega}{\hbar\pi}\right)^{1/4} H_n\left(\sqrt{\frac{m\omega}{\hbar}} x\right) e^{-\frac{m\omega}{2\hbar}x^2}.$$

$$(10.152)$$

An important property of the quantum oscillator is that its lowest energy is not zero:

$$E_0 = \frac{\hbar\omega}{2}.$$

The shape of the ground state wave function, $\psi_0(x)$, in Fig. 10.19 demonstrates that the quantum oscillator, unlike the classical one, is not rested at $x = 0$ when its energy has the minimum value. A particle in the ground state performs so-called zero oscillations. This result seems paradoxical, but it is in full agreement with the uncertainty principle – energy $E_0 = 0$ would mean that the momentum and coordinate of the particle simultaneously are zero, i.e., precisely certain values.

More applications of special functions are considering in other chapters, mostly in Chap. 13.

Problems

1. Show that

$$H_{2n}(0) = (-1)^n \frac{(2n)!}{n!}, \quad H_{2n+1}(0) = 0,$$

$$H'_{2n}(0) = 0, \quad H'_{2n+1}(0) = (-1)^n \frac{(2n+1)!}{n!} 2.$$

2. Find $H_2(x)$, $H_3(x)$, and $H_4(x)$ using Rodrigues' formula.
3. Prove that

$$e^{x^2} H_n(x) = -\frac{d}{dx} e^{x^2} H_{n-1}(x).$$

4. Prove that

$$e^{-x^2} H_n(x) = -\frac{d}{dx} e^{-x^2} H_{n-1}(x).$$

5. Prove that

$$\int_{-\infty}^{\infty} e^{-x^2} x^2 H_n^2(x) dx = 2^n n! \left(n + \frac{1}{2}\right) \sqrt{\pi}$$

(this integral occurs when calculating the mean square of the displacement of the quantum oscillator).
6. Expand function $f(x) = Ax^2 + Bx + C$ in Fourier-Hermite series.

Hint: Use the method of undefined coefficients to find c_i in expression:

$$Ax^2 + Bx + C = c_0 H_0(x) + c_1 H_1(x) + c_2 H_2(x).$$

Chapter 11
Integral Equations

11.1 Introduction

Integral equations appear in the descriptions of the processes in electrical circuits, in the models of materials deformation, and in the theory of radiation transport through matter. Many applications of integral equations are associated with partial differential equations of mathematical physics.

An integral equation is the equation that contains the unknown function under the integral sign (as an integrand).

Consider the following integral equation:

$$y(x) = y_0 + \int_{x_0}^{x} f(x, y)dx, \tag{11.1}$$

where $y(x)$ is the unknown function and $f(x, y)$ is the given function. From (11.1) it is clear that $y(x_0) = y_0$.

Differentiation of (11.1) with respect to x gives the first-order ODE:

$$\frac{dy}{dx} = f(x, y) \tag{11.2}$$

with the initial condition $y(x_0) = y_0$. Thus, the integral Eq. (11.1) reduces to the first-order Cauchy problem (first-order IVP). Conversely, the differential Eq. (11.2) together with the initial condition is equivalent to the integral Eq. (11.1).

Often it is easier to solve the integral Eq. (11.1) than the differential Eq. (11.2). Approximate solution of (11.1) can be obtained *iteratively*. Let set $y(x) = y_0$ to be initial approximation. Substituting it into (11.1), we obtain the first approximation (first iteration):

© The Author(s), under exclusive license to Springer Nature Switzerland AG 2023
V. Henner et al., *Ordinary Differential Equations*,
https://doi.org/10.1007/978-3-031-25130-6_11

$$y_1(x) = y_0 + \int_{x_0}^{x} f(x, y_0)dx. \tag{11.3}$$

The second approximation is obtained by substituting the first approximation in the original Eq. (11.1):

$$y_2(x) = y_0 + \int_{x_0}^{x} f(x, y_1(x))dx. \tag{11.4}$$

Continuing the iterative process, we obtain n-th approximation (iteration), which with some precision gives the solution of Eq. (11.1):

$$y_n(x) = y_0 + \int_{x_0}^{x} f(x, y_{n-1}(x))dx. \tag{11.5}$$

For smooth functions $f(x, y)$, the iterative process converges rapidly to the exact solution of (11.1) in the neighborhood of the point (x_0, y_0).

Example 11.1 Solve the following equation iteratively in the vicinity of the point $x_0 = 1$:

$$y(x) = 2 + \int_{1}^{x} \frac{y(x)}{x} dx.$$

Solution The initial approximation, as it seen from the equation, is $y_0(1) = 2$. The first approximation is

$$y_1(x) = 2 + \int_{1}^{x} \frac{2}{x} dx = 2 + 2\ln|x|.$$

And then the second approximation is

$$y_2(x) = 2 + \int_{1}^{x} \frac{2 + 2\ln|x|}{x} dx = 2 + 2\ln|x| + \ln^2|x|.$$

The original integral equation is equivalent to the differential equation $y' = y/x$ with the initial condition $y(1) = 2$. This IVP has the exact solution:

$$y = 2x.$$

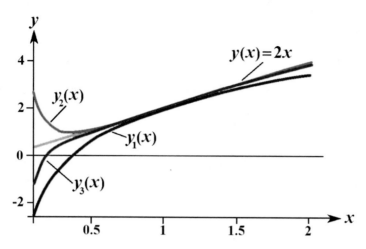

Fig. 11.1 Iterative solution of the equation in the Example 11.1 on the interval $x \in [0.1, 2]$

At the point $x_0 = 1$, the exact and the approximate solutions give the same result. Let us compare the exact solution and the first and second approximations at $x = 1.1$. The exact solution is $y(1.1) = 2 \cdot 1.1 = 2.2$, and the first and second approximations are

$$y_1(1.1) = 2.19062, \quad y_2(1.1) = 2.19970.$$

It is seen that the iterative solution is rapidly approaching the exact solution. However, as we move away from the point $x_0 = 1$, the difference of exact and approximate values increases, and more iterations are required to approximate the exact solution with acceptable accuracy.

Figure 11.1 shows the graphs of the exact solution $y = 2x$ and the first and second iterations $y_1(x)$, $y_2(x)$ on the interval $x \in [0.1, 2]$. These graphs allow to compare the results of the iterative solution with the exact solution, as one moves away from the point $y(1) = 2$.

Problems

Solve the following integral equations: (a) iteratively (find the first two approximations, not counting the initial one) and (b) by reducing them to differential equations (in these problems the differential equations can be solved exactly):

1. $y(x) = \int\limits_{\pi/2}^{x} e^{y(x)} \sin x \, dx.$ 2. $y(x) = \int\limits_{0}^{x} \sqrt{1 - y^2(x)} \, dx.$

3. $y(x) = \int\limits_{0}^{x} \cos^2 y(x) \, dx.$ 4. $y(x) = 1 - \int\limits_{1}^{x} \frac{xy(x) + 1}{x^2} \, dx.$

11.2   Introduction to Fredholm Equations

In physical problems integral equations often arise when one need to find function $y(x)$ acting on the system, when the net outcome is given by the integral over some interval $[a, b]$, and the integrand is the function $y(x)$ multiplied by some function $K(x, t)$, describing the influence of the system.

Consider the following two types of linear (with respect to unknown function) integral equations:

$$\lambda \int_a^b K(x,t)y(t)dt = g(x), \tag{11.6}$$

$$y(x) - \lambda \int_a^b K(x,t)y(t)dt = g(x). \tag{11.7}$$

Here $y(x)$ is the unknown function, $K(x, t)$ is the given function (called *the kernel* of the integral equation), $g(x)$ is the given function on $[a, b]$, and λ is the parameter. Variables x and t (here t is the integration variable) are real, but λ, $y(x)$, $K(x, t)$, $g(x)$ may be complex. The unknown function $y(x)$ also may be complex. $K(x, t)$ and $g(x)$ are continuous functions.

Equation (11.6) is *Fredholm integral equation of the first kind*; Eq. (11.7) is *Fredholm integral equation of the second kind*. When $g(x) = 0$, the integral equation is called *homogeneous*; otherwise, it is *nonhomogeneous*. For instance, the problem of finding the force density $y(x)$, acting on a string, such that the string takes on the given shape $g(x)$ on $a \le x \le b$, leads to Eq. (11.6).

Let us consider first Fredholm equations of the second kind with *degenerated kernels*. A kernel $K(x, t)$ is degenerated, if it can be written as a product, or a sum of products, of the functions of x and of t:

$$K(x,t) = \sum_{i=1}^n \varphi_i(x)\psi_i(t).$$

Then, Eq. (11.7) reads:

$$y(x) = \lambda \int_a^b K(x,t)y(t)dt + g(x) = \lambda \int_a^b \sum_{i=1}^n \varphi_i(x)\psi_i(t)y(t)dt + g(x), \tag{11.8}$$

and it can be reduced to a system of n linear algebraic equations for n unknowns. Next, we show how this reduction is carried out.

With the notation

$$A_i = \int_a^b \psi_i(t)y(t)dt \quad (i=1,2,\ldots,n) \tag{11.9}$$

Equation (11.8) reads:

$$y(x) = \lambda \sum_{i=1}^n A_i\varphi_i(x) + g(x). \tag{11.10}$$

Then, change the summation index to k and substitute this expression for $y(x)$ in (11.9), where the integration variable t is replaced by x:

$$A_i = \lambda \sum_{k=1}^n \int_a^b A_k\varphi_k(x)\psi_i(x)dx + \int_a^b \psi_i(x)g(x)dx \quad (i=1,2,\ldots,n). \tag{11.11}$$

Let introduce the notations:

$$\alpha_{ik} = \int_a^b \varphi_k(x)\psi_i(x)dx, \quad g_i = \int_a^b \psi_i(x)g(x)dx \quad (i,k=1,2,\ldots,n). \tag{11.12}$$

Now we can see that (11.11) results in a system of linear algebraic equations for A_i:

$$A_i - \lambda \sum_{k=1}^n \alpha_{ik}A_k = g_i \quad (i=1,2,\ldots,n). \tag{11.13}$$

Coefficients α_{ik} and free terms g_i are given by (11.12).

For the nonhomogeneous Eq. (11.8) ($g(x) \neq 0$), (11.13) constitutes a system of nonhomogeneous linear algebraic equations, which has a unique solution if its determinant is not zero. If for some values of the parameter λ the determinant is zero, then the nonhomogeneous Eq. (11.8) has no solutions, but at the same values of λ the corresponding homogeneous equation, obviously, has infinitely many solutions. From this discussion the usefulness of introducing the parameter λ becomes clear.

Values of λ at which the homogeneous equation has a solution are called the *eigenvalues* of the problem. These values constitute a *spectrum* of the problem. Functions $y_i(x)$ that are obtained from (11.10) for each value of λ from the spectrum are called the *eigenfunctions*.

Example 11.2 Solve integral equation:

$$y(x) - \lambda \int_0^1 \left(xt^2 - x^2 t\right) y(t) dt = x. \tag{11.14}$$

Solution Let $A = \int_0^1 t^2 y(t) dt$, $B = \int_0^1 t y(t) dt$. Then (11.14) has the form:

$$y(x) = \lambda Ax - \lambda Bx^2 + x.$$

Change x to t in $y(x)$, and plug $y(t)$ in the expressions for A and B:

$$A = \int_0^1 t^2 \left(\lambda At - \lambda Bt^2 + t\right) dt,$$

$$B = \int_0^1 t \left(\lambda At - \lambda Bt^2 + t\right) dt.$$

Integration gives the system of linear nonhomogeneous algebraic equations for the unknowns A and B:

$$\begin{cases} A = \lambda A/4 - \lambda B/5 + 1/4, \\ B = \lambda A/3 - \lambda B/4 + 1/3. \end{cases}$$

Solution of this system is

$$A = \frac{60 - \lambda}{240 + \lambda^2}, \quad B = \frac{80}{240 + \lambda^2}.$$

Substitution of these expressions in $y(x) = \lambda Ax - \lambda Bx^2 + x$ gives the solution of the integral Eq. (11.14) on $[0, 1]$. This can be verified as follows: when $y(x) = \lambda Ax - \lambda Bx^2 + x$ and $y(t) = \lambda At - \lambda Bt^2 + t$ with the above expressions for $A(\lambda)$ and $B(\lambda)$ are substituted in (11.14) and the integration is performed, the result is the equality $x = x$. In this solution λ is the arbitrary parameter. In Fig. 11.2 the graphs of the solutions of Eq. (11.14) are shown for several values of λ.

For the values of λ which are the roots of the quadratic equation

$$240 + \lambda^2 = 0,$$

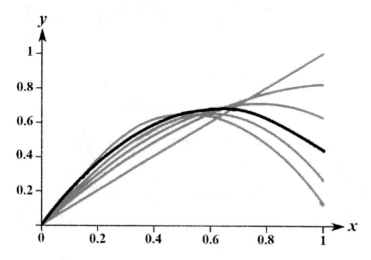

Fig. 11.2 Solutions of Eq. (11.14) for $\lambda = 0, 2, 4, 6, 8, 10$. The solution corresponding to value $\lambda = 6$ is shown by the black line

that is $\lambda_{1,2} = \pm i\sqrt{240}$ (the eigenvalues), the original nonhomogeneous equation has no solutions. But for $\lambda = \lambda_{1, 2}$, the homogeneous equation

$$y(x) - \lambda \int\limits_0^1 \left(xt^2 - x^2t\right)y(t)dt = 0, \qquad (11.15)$$

which corresponds to the original nonhomogeneous equation, has nontrivial solutions (homogeneous equations always have trivial zero solution) – the eigenfunctions of the homogeneous equation. Let us find them.

For Eq. (11.15) the system (11.13) is homogeneous, and for the coefficients A and B, we obtain the system:

$$\begin{cases} A = \lambda A/4 - \lambda B/5, \\ B = \lambda A/3 - \lambda B/4. \end{cases} \qquad (11.16)$$

It is not difficult to see that this system is linearly dependent (its determinant is zero), and it has nontrivial solution only for $\lambda = \lambda_1$ or $\lambda = \lambda_2$. Then, assigning to A an arbitrary real number α and finding B in terms of α, we obtain from the system (2.11): $A = \alpha, B = \frac{5\lambda - 20}{4\lambda}\alpha$. Thus, the eigenfunctions of Eq. (11.15) are

$$y_1(x) = \lambda_1\alpha x - \lambda_1 Bx^2, \quad y_2(x) = \lambda_2\alpha x - \lambda_2 Bx^2.$$

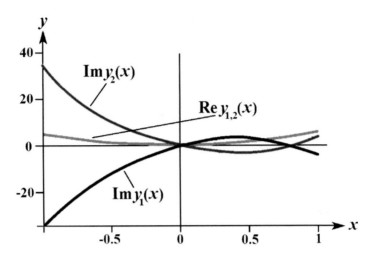

Fig. 11.3 Eigenfunctions of Eq. (11.15)

The spectrum of eigenvalues of the homogeneous equation, as we found above, is $\lambda_{1,2} = \pm i\sqrt{240}$. Value of α is obviously arbitrary, since both sides of Eq. (11.15) can be divided by this parameter.

The functions $y_1(x)$ and $y_2(x)$ are complex. Taking $\alpha = 1$, we obtain $\mathrm{Re}\,B = 5/4$, $\mathrm{Im}B = \pm 5/\sqrt{240}$. In Fig. 11.3 we show graphs of the real and imaginary parts of these solutions:

$$\mathrm{Re}\,y_{1,2}(x) = 5x^2, \quad \mathrm{Im}y_{1,2}(x) = \pm\sqrt{240}\left(x - \frac{5}{4}x^2\right).$$

Example 11.3 Solve the integral equation:

$$y(x) - \lambda \int\limits_0^1 \left(xe^t + x^2 t\right)y(t)dt = x.$$

Solution Let us denote $A = \int_0^1 e^t y(t)dt$, $B = \int_0^1 ty(t)dt$. Then, the original integral equation takes the form:

$$y(x) = \lambda Ax + \lambda Bx^2 + x.$$

Replacing the variable x to the variable t in this function $y(x)$, and substituting $y(t)$ in the expressions for A and B, and then integrating, results in the system of linear nonhomogeneous algebraic equations:

$$\begin{cases} (1-\lambda)A + (2-e)\lambda B = 1, \\ -\dfrac{1}{3}\lambda A + \left(1-\dfrac{\lambda}{4}\right)B = \dfrac{1}{3} \end{cases}.$$

Its solution is

$$A = \frac{12 - (11 - 4e)\lambda}{(11 - 4e)\lambda^2 - 15\lambda + 12}, \qquad B = \frac{4}{(11 - 4e)\lambda^2 - 15\lambda + 12}.$$

Substituting these A and B in $y(x) = \lambda Ax + \lambda Bx^2 + x$ gives the solution of the nonhomogeneous integral equation on the interval $[0, 1]$. The parameter λ is arbitrary, but for the values of λ that turn into zero, the denominator of the expressions for A and B:

$$\lambda_1 = \frac{15 + \sqrt{144e - 171}}{22 - 8e} \approx 117, \quad \lambda_2 = \frac{15 - \sqrt{144e - 171}}{22 - 8e} \approx 0.603,$$

the original nonhomogeneous integral equation does not have solutions. But in this case the corresponding homogeneous equation

$$y(x) - \lambda \int\limits_0^1 (xe^t + x^2 t)y(t)dt = 0$$

has solutions. When solving the homogeneous equation, the system of equations for the coefficients A and B is also homogeneous:

$$\begin{cases} (1-\lambda)A + (2-e)\lambda B = 0, \\ -\dfrac{1}{3}\lambda A + \left(1-\dfrac{\lambda}{4}\right)B = 0. \end{cases}$$

For $\lambda = \lambda_1$ or $\lambda = \lambda_2$, this system has a nontrivial solution. Letting $A = \alpha$, from the last equation, we obtain:

$$B = \frac{\lambda - 1}{(2 - e)\lambda}\alpha.$$

For $\lambda = \lambda_1$ we have $B_1 = -1.38\alpha$; for $\lambda = \lambda_2$, we have $B_2 = 0.917\alpha$. The eigenfunctions of the homogeneous equation are $y(x) = \lambda Ax + \lambda Bx^2$, that is,

$$y_1(x) = 117\alpha x - 161\alpha x^2, \ y_2(x) = 0.603\alpha x + 0.553\alpha x^2,$$

where α is the arbitrary parameter, which can be taken equal to one, for simplicity.

Notice that nonlinear equations with degenerated kernels can be solved in similar way.

Example 11.4 Solve homogeneous nonlinear integral equation:

$$y(x) = \lambda \int_0^1 xty^2(t)dt.$$

Solution Let

$$A = \int_0^1 ty^2(t)dt,$$

then

$$y(x) = A\lambda x.$$

Substitution $y(x) = A\lambda x$ in the expression for A gives:

$$A = \int_0^1 t\lambda^2 A^2 t^2 dt.$$

Integration results in

$$A = \frac{\lambda^2}{4}A^2.$$

This equation has two solutions:

$$A_1 = 0, \qquad A_2 = \frac{4}{\lambda^3}.$$

Consequently, the original integral equation on the interval $[0, 1]$ has two solutions (eigenfunctions) for any $\lambda \neq 0$:

$$y_1(x) = 0, \qquad y_2(x) = \frac{4}{\lambda}x.$$

The first one is a trivial solution of homogeneous equation. It is easy to verify that the substitution of these functions into the original equation transforms this equation into the identity. For the first function, the substitution obviously gives $0 = 0$. Let us substitute the second function in the equation:

$$\frac{4}{\lambda}x = \lambda \int\limits_0^1 xt \frac{16}{\lambda^2} t^2 dt.$$

Canceling common factors and calculating the integral again gives the identity.

The spectrum of the problem is continuous – it consists of all real, nonzero values of λ.

What if the kernel is not degenerated function? In this situation, we can use the fact that any continuous kernel (as well as kernels with finite discontinuities) can be represented as an infinite sum (Fourier series) of degenerate kernels. Using this, a Swedish mathematician Erik Ivar Fredholm (1866–1927), known for his work on the theory of integral equations, proved several theorems. We present one of these theorems (without a proof) for Eq. (11.7) (in fact, the scheme of the proof has practically discussed above). This theorem formulates the *Fredholm alternative*.

Theorem 11.1 *Either the nonhomogeneous Eq.* (11.7) *with a continuous kernel has a unique continuous solution for any function $g(x)$ (i.e., λ is not an eigenvalue), or the homogeneous equation corresponding to* (11.7) *has at least one nontrivial solution (an eigenfunction).*

Theory of Fredholm integral equations of the first kind is much more complicated than the one of equations of the second kind. Here we will describe only the general idea of a solution.

Solution of Eq. (11.6):

$$\lambda \int\limits_a^b K(x,t)y(t)dt = g(x)$$

can be found in the form of a Fourier series

$$y(x) = \sum_i c_i f_i(x), \tag{11.17}$$

where $f_1(x)$, $f_2(x)$, $f_3(x)$, ... is some complete set of functions on the interval (a, b), and c_i are the coefficients of the Fourier expansion. Completeness of the set means that the series (11.17) converges to $y(x)$ on (a, b), and the coefficients c_i are given by

$$c_i = \frac{\int\limits_a^b y(x)f_i(x)dx}{\int\limits_a^b f_i^2(x)dx} \tag{11.18}$$

(for simplicity this formula is written for the real functions $f(x)$).

Substitution of (11.18) in (11.6) gives:

$$g(x) = \lambda \sum_i c_i h_i(x), \tag{11.19}$$

where

$$h_i(x) = \int_a^b K(x,t) f_i(t) dt. \tag{11.20}$$

From (11.19) one can find the coefficients c_i, which finalizes the solution of Eq. (11.6).

Problems

Solve the following Fredholm integral equations of the second kind with degenerate kernels (in some problems $\lambda = 1$):

1. $y(x) = \sin x + \lambda \int_0^{\pi/2} \sin x \cos t\, y(t) dt.$

2. $y(x) = x + \lambda \int_0^{2\pi} (2\pi - t) y(t) \sin x\, dt.$

3. $y(x) = 2 \int_0^1 xt y^3(t) dt.$

4. $y(x) = \int_{-1}^1 (xt + x^2 t^2) y^2(t) dt.$

5. $y(x) = \int_0^1 x^2 t^2 y^2(t) dt.$

6. $y(x) = \lambda \int_0^{2\pi} y(t) \sin x \sin t\, dt.$

7. $y(x) = \lambda \int_0^1 (2xt - 4x^2) y(t) dt.$

8. $y(x) = \lambda \int_{-1}^1 (5xt^3 + 4x^2 t) y(t) dt.$

9. $y(x) = \lambda \int_{-1}^1 (5xt^3 + 4x^2 t + 3xt) y(t) dt.$

10. $y(x) = \lambda \int_0^\pi xy(t) \sin 2t\, dt + \cos 2x.$

11. $y(x) = \lambda \int_0^{2\pi} \sin(x + t) y(t) dt + 2.$

11.3 Iterative Method for the Solution of Fredholm Integral Equations of the Second Kind

Iterative method allows to obtain a sequence of functions that converges to the solution of the integral equation. It is natural to set function $y(x) = g(x)$ to be the zeroth approximation of the solution of Eq. (11.7):

$$y(x) = \lambda \int_a^b K(x,t)y(t)dt + g(x).$$

This function is substituted in the integrand:

$$y(x) = g(x) + \lambda \int_a^b K(x,t)g(t)dt. \tag{11.21}$$

Next, (11.21) is substituted in the integrand in Eq. (11.7) and so on. As a result, we obtain:

$$y(x) = g(x) + \lambda \int_a^b K(x,t)g(t)dt + \lambda^2 \int_a^b dt \int_a^b dt' K(x,t)K(t,t')g(t') + \dots$$

$$\tag{11.22}$$

Series (11.22) is called the *Neumann series* of Eq. (11.7). If the kernel $K(x,t)$ is bounded on (a,b), then the series converges for small values of λ (as elsewhere in this chapter, we are not concentrated on the questions of mathematical rigor).

Example 11.5 Find the solution of the integral equation:

$$y(x) = \lambda \int_0^1 \frac{x}{1+t^2} y(t)dt + 1 + x^2 \tag{11.23}$$

by the iterative method. Find the range of values of λ, for which the Neumann series converges.

Solution The zeroth approximation is $y_0(x) = 1 + x^2$. Substituting this in the integrand and integrating results in the first approximation:

$$y_1(x) = \lambda \int_0^1 \frac{x}{1+t^2} (1+t^2)dt + 1 + x^2 = \lambda \int_0^1 xdt + 1 + x^2 = \lambda x + 1 + x^2.$$

Substituting first approximation in the integrand of Eq. (11.23) gives the second approximation:

$$y_2(x) = \lambda \int\limits_0^1 \frac{x}{1+t^2}(\lambda t + 1 + t^2)dt + 1 + x^2 = \int\limits_0^1 \frac{\lambda^2 xt}{1+t^2}dt + \lambda \int\limits_0^1 xdt + 1 + x$$

$$= x\left(\lambda + \frac{\lambda^2 \ln 2}{2}\right) + 1 + x^2.$$

The third approximation is (check it as *Reading Exercise*)

$$y_3(x) = x\left(\lambda + \frac{\lambda^2 \ln 2}{2} + \frac{\lambda^3 \ln^2 2}{4}\right) + 1 + x^2.$$

It is clear that n-th approximation can be written as

$$y_n(x) = x\left(\lambda + \frac{\lambda^2 \ln 2}{2} + \frac{\lambda^3 \ln^2 2}{4} + \ldots + \frac{\lambda^n \ln^{n-1} 2}{2^{n-1}}\right) + 1 + x^2.$$

For $n \to \infty$ we obtain the series:

$$y(x) = x \sum_{n=1}^{\infty} \frac{\lambda^n \ln^{n-1} 2}{2^{n-1}} + 1 + x^2.$$

The series $\sum_{n=1}^{\infty} \frac{\lambda^n \ln^{n-1} 2}{2^{n-1}}$ is geometric with the common ratio $q = \frac{\lambda \ln 2}{2}$ and the first term $a_1 = \lambda$. This series converges to the sum $S = \frac{a_1}{1-q} = \frac{\lambda}{1 - \lambda\frac{\ln 2}{2}}$, if $\left|\frac{\lambda \ln 2}{2}\right| < 1$. Thus, the range of values of the parameter λ, such that the solution of the integral equation on the interval $[0, 1]$ exists, is $|\lambda| < \frac{2}{\ln 2}$. The solution is the function

$$y(x) = \frac{\lambda x}{1 - \frac{\ln 2}{2}\lambda} + 1 + x^2. \tag{11.24}$$

In Fig. 11.4 the graphs of the functions $y_1(x)$, $y_2(x)$, $y_3(x)$ and $y_0(x)$ are shown for two values of λ.

Problems
Solve the following Fredholm integral equations of the second kind by the iterative method, and find the range of values of λ for which the Neumann series converges. Plot the graphs of the zeroth, first, and second iterations on the corresponding intervals for several values of λ:

1. $y(x) = \lambda \int\limits_0^\pi \sin(x+t)y(t)dt + \cos x.$

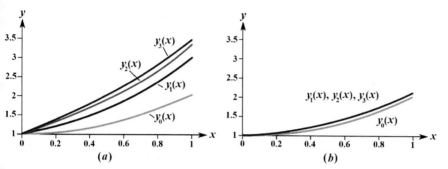

Fig. 11.4 Solutions of Eq. (11.23) obtained by the iterative method: (a) $\lambda = 1$; (b) $\lambda = 0.1$

2. $y(x) = \lambda \int_0^1 xty(t)dt + x.$

3. $y(x) = \lambda \int_0^1 y(t)dt + \sin \pi x.$

4. $y(x) = \lambda \int_0^{\pi/2} \sin x \cos x \, y(t)dt + 1 .$

5. $y(x) = \lambda \int_0^1 xe^t y(t)dt + e^{-x}.$

6. $y(x) = \lambda \int_0^{1/2} y(t)dt + x.$

7. $y(x) = \lambda \int_0^{\pi} x \sin 2t \, y(t)dt + \cos 2x.$

8. $y(x) = \lambda \int_0^{2\pi} \sin(x + t)y(t)dt + 2.$

11.4   Volterra Equation

Integral equations that contain the integral with a variable limit of integration are called *Volterra equations*, named after an Italian mathematician and physicist Vito Volterra (1860–1940):

$$\lambda \int_a^x K(x,t)y(t)dt + g(x) = 0, \qquad (11.25)$$

$$\lambda \int_a^x K(x,t)y(t)dt + g(x) = y(x). \qquad (11.26)$$

Equation (11.25) is called Volterra equation of the first kind and Eq. (11.26) of the second kind. Volterra equations are linear.

Solution of the initial value problem for linear differential equations leads to Volterra integral equations of the second kind, and vice versa – often the Volterra equation can be solved by reducing it to a differential equation. We consider this solution scheme with the example. Note that constant λ is often omitted.

Example 11.6 Solve Volterra equation of the second kind:

$$x + \int_0^x xty(t)dt = y(x),$$

by reducing it to a differential equation.

Solution Let

$$u(x) = \int_0^x ty(t)dt.$$

Next, we differentiate this expression and use the original equation in the form $y(x) = x + xu(x)$:

$$u' = xy(x) = x(x + xu(x)).$$

The solution of this differential equation is found using separation of variables:

$$u(x) = -1 + Ce^{x^3/3}.$$

Turning again to the equation $y(x) = x + xu(x)$, we obtain:

$$y(x) = xCe^{x^3/3}.$$

The initial condition follows from the integral equation, and it is $y(0) = 0$. Obviously, constant C cannot be obtained from this condition.

To find C, substitute the expression for $y(x)$ in the original equation:

$$x + \int_0^x xt^2 Ce^{t^3/3}dt = xCe^{x^3/3}.$$

Calculation of the integral gives:

$$xCe^{x^3/3} = xCe^{x^3/3} + x(1 - C),$$

from which $C = 1$. Thus, the final answer is

$$y(x) = xe^{x^3/3}.$$

Problems

Solve the following integral equations by reducing them to the differential equations:

1. $y(x) = x + \frac{x^2}{2} - \int\limits_0^x y(t)dt.$

2. $y(x) = e^x + \int\limits_0^x y(t)dt.$

3. $y(x) = \cos x + \int\limits_0^x e^{x-t}y(t)dt.$

4. $y(x) = 2 + \int\limits_0^x ty(t)dt.$

11.5   Solution of Volterra Equation with the Difference Kernel Using the Laplace Transform

The *Volterra equation of the second kind with the difference kernel* has the form:

$$\lambda \int\limits_0^x K(x-t)y(t)dt + g(x) = y(x). \qquad (11.27)$$

For instance, the initial value problem for linear differential equations with constant coefficients can be reduced to such equations. Equation (11.27) can be solved using Laplace transform (LT).

Recall that the Laplace transform $L[f(x)]$ of the real function $f(x)$ is

$$\widehat{f}(p) = L[f(x)] = \int\limits_0^\infty e^{-px}f(x)dx, \qquad (11.28)$$

where p is, generally, a complex parameter. $\widehat{f}(p)$ is the image of the original function $f(x)$.

The right side of Eq. (11.28) is called the *Laplace integral*. For its convergence in the case where p is complex, it is necessary that $\text{Re}p > 0$; if p is real, then for convergence $p > 0$. Also, a growth of function $f(x)$ should be restricted and $f(x)$ should be bounded on $(0, \infty)$.

Convolution of functions $f_1(x)$ and $f_2(x)$ of the real variable x is the function $f(x)$, which is defined as

$$f(x) = \int\limits_0^x f_1(x-t)f_2(t)dt. \qquad (11.29)$$

The integral in Eq. (11.27) is the convolution of functions $K(x)$ and $y(x)$; thus,

$$L\left[\int_0^x K(x-t)y(t)dt\right] = \widehat{K}(p)\widehat{y}(p).\tag{11.30}$$

It is clear that LT is linear:

$$L[C_1 f_1(x) + C_2 f_2(x)] = C_1 L[f_1(x)] + C_2 L[f_2(x)].\tag{11.31}$$

Applying the LT to Eq. (11.27) and using these properties, we obtain:

$$\widehat{y}(p) = \int_0^\infty \left[g(x) + \lambda \int_0^x K(x-t)y(t)dt\right] e^{-px}dx =$$

$$= \widehat{g}(p) + \lambda \int_0^\infty \left[\int_0^x K(x-t)y(t)dt\right] e^{-px}dx = \widehat{g}(p) + \lambda\widehat{K}(p)\widehat{y}(p).$$

Then,

$$\widehat{y}(p) = \frac{\widehat{g}(p)}{1 - \lambda\widehat{K}(p)} \quad \left(\widehat{K}(p) \neq \frac{1}{\lambda}\right),\tag{11.32}$$

where $\widehat{y}(p), \widehat{g}(p), \widehat{K}(p)$ are the Laplace transforms of the corresponding functions.

To determine the original function from the image, $\widehat{f}(p)$, one has to perform the *inverse Laplace transform*, which is denoted as

$$f(x) = L^{-1}\left[\widehat{f}(p)\right].\tag{11.33}$$

Inverse transform is also linear:

$$L^{-1}\left[C_1\widehat{f}_1(p) + c_2\widehat{f}_2(p)\right] = C_1 L^{-1}\left[\widehat{f}_1(p)\right] + C_2 L^{-1}\left[\widehat{f}_2(p)\right].\tag{11.34}$$

Laplace transforms for some functions can be found in the table in Chap. 7.

Example 11.7 Solve the integral equation:

$$y(x) = \sin x + 2 \int_0^x \cos(x-t)y(t)dt.$$

Solution Calculating directly from the definition (5.2), or using the table, we have:

$$L[\sin x] = \frac{1}{p^2 + 1}, \qquad L[\cos x] = \frac{p}{p^2 + 1}.$$

Applying the LT to both sides of the equation, taking into the account linearity of the transform and the convolution theorem, gives:

$$\widehat{y}(p) = \frac{1}{p^2 + 1} + \frac{2p}{p^2 + 1}\widehat{y}(p).$$

Then,

$$\widehat{y}(p)\left[1 - \frac{2p}{p^2 + 1}\right] = \frac{1}{p^2 + 1},$$

and simplifying:

$$\widehat{y}(p) = \frac{1}{(p-1)^2}.$$

Solution of the given integral equation is the inverse Laplace transform of this function. Using the table, we finally find:

$$y(x) = xe^x.$$

Example 11.8 Solve the integral equation:

$$y(x) = \cos x - \int\limits_0^x (x-t)\cos(x-t)y(t)dt.$$

Solution Applying the LT to this equation and using transforms

$$L[\cos x] = \frac{p}{p^2 + 1}, \qquad L[x\cos x] = \frac{p^2 - 1}{(p^2 + 1)^2},$$

gives

$$\widehat{y}(p) = \frac{p}{p^2 + 1} - \frac{p^2 - 1}{(p^2 + 1)^2}\widehat{y}(p).$$

Thus,

$$\widehat{y}(p) = \frac{p^2 + 1}{p^3 + 3p} = \frac{p}{p^2 + 3} + \frac{1}{p(p^2 + 3)}.$$

The second term in the above formula is not directly in the table. Using the method of partial fractions, we can decompose this term into the sum of two elementary fractions:

$$\widehat{y}(p) = \frac{p}{p^2 + 3} + \frac{1}{3p} - \frac{p}{3p^2 + 9}.$$

Finally, the inverse Laplace transform gives the solution of the integral equation:

$$y(x) = \frac{2}{3} \cos \sqrt{3}x + \frac{1}{3}.$$

Problems
Solve the following integral equations using the Laplace transform:

1. $y(x) = e^x - \int\limits_0^x e^{x-t} y(t) dt.$

2. $y(x) = x - \int\limits_0^x e^{x-t} y(t) dt.$

3. $y(x) = e^{2x} + \int\limits_0^x e^{t-x} y(t) dt.$

4. $y(x) = 2 + \frac{1}{2} \int\limits_0^x (x - t)^2 y(t) dt.$

11.6   Applications

Below we present some typical examples of the description of natural processes by integral equations.

11.6.1   Falling Object

Let us consider an object of the mass m falling without friction with zero initial velocity from the height $y = h$ to the height $y = 0$ along the surface determined by the equation $y = y(x)$ (see Fig. 11.5).

The conservation of the full energy, which is the sum of the kinetic energy $E_k = m(dl/dt)^2/2$, where l is the length of the trajectory along the curve $y = y(x)$, and the potential energy $E_p = mgy$, gives:

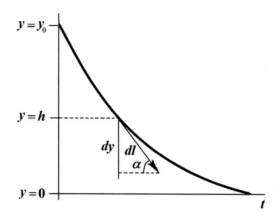

Fig. 11.5 Object moving freely from point $y_0 = y(0)$

$$\frac{1}{2}m\left(\frac{dl}{dt}\right)^2 + mgy = mgh.$$

The falling time $T(h)$ is given by the formula:

$$T(h) = \int\limits_{y=h}^{y=0} dt = \int\limits_{y=h}^{y=0} \frac{dl}{\sqrt{2g(h-y)}}. \tag{11.35}$$

Assume that we can measure the dependence $T(h)$ experimentally. Can we find the shape of the line $y = y(x)$ from those data, i.e., solve *the inverse problem*?

Let us define the function:

$$u(y) \equiv -\frac{dl}{dy} = \frac{1}{\sin \alpha(y)},$$

where $\alpha(y)$ is the inclination angle (see Fig. 11.5) and rewrite (11.35) as

$$\int\limits_0^h \frac{u(y)dy}{\sqrt{h-y}} = \sqrt{2g}T(h). \tag{11.36}$$

Equation (11.36), which is called *Abel's equation*, is Volterra equation of the first kind. (Niels Henrik Abel (1802–1829) was a Norwegian mathematician, who during his short life made fundamental contributions to several fields, primarily in algebra.) Solving that integral equation, we find $u(y)$. Taking into account that

$$\frac{dy}{dx} = -\tan \alpha(y) = -\frac{\sin \alpha(y)}{\sqrt{1 - \sin^2 \alpha(y)}} = -\frac{1}{\sqrt{u^2(y) - 1}},$$

we can determine the shape of the surface by solving equation:

$$\frac{dx}{dy} = -\sqrt{u^2(y) - 1}. \tag{11.37}$$

Let us solve Eq. (11.36) in the particular case when the motion time T from the height $y = h$ to the height $y = 0$ does not depend on h (in that case, the line $y = y(x)$ is called "tautochrone"). Using the LT of (11.36) and taking into account the rule for the LT of convolution, we find:

$$L\left[\frac{1}{\sqrt{y}}\right](p) \cdot L[u](p) = \sqrt{2g}\frac{T}{p}.$$

Taking into account that

$$L\left[\frac{1}{\sqrt{y}}\right](p) = \sqrt{\frac{\pi}{p}},$$

we find:

$$L[u](p) = \sqrt{\frac{2g}{\pi}}Tp^{-1/2};$$

hence,

$$u(y) = \sqrt{\frac{y_0}{y}}, \quad y_0 = \frac{2gT^2}{\pi^2}.$$

According to (11.37), the shape of the curve is determined by the following ODE:

$$\frac{dx}{dy} = -\sqrt{\frac{y_0 - y}{y}};$$

hence,

$$x(y) = -\int \sqrt{\frac{y_0 - y}{y}}dy + C. \tag{11.38}$$

In order to calculate the integral, let us define the new variable θ by the relation:

$$y = y_0 \cos^2 \frac{\theta}{2} = \frac{y_0}{2}(1 + \cos \theta), \quad \theta_0 \le \theta \le \pi, \tag{11.39}$$

where

$$\theta_0 = \arccos\left(\frac{2h}{y_0} - 1\right).$$

Substituting (11.39) into (11.38), we find:

$$x = y_0 \int \sin^2 \frac{\theta}{2} d\theta + C = y_0 \int \frac{1}{2}(1 - \cos \theta) d\theta + C = \frac{y_0}{2}(\theta - \sin \theta) + C.$$
$$\tag{11.40}$$

The constant C is arbitrary.

The curve determined by formulas (11.39) and (11.40) is cycloid. We shall return to the discussion of the tautochrone problem in Sect. 12.4.2.

Note that integral equations are widely used in the formulation of inverse problems for partial differential equations. That subject is beyond the scope of the present book.

11.6.2 Population Dynamics

Let us consider a population of animals. At a given time instant t, the population is characterized by function $n(x, t)$ that determines the number density of animals of the age x. Denote the probability of the survival from age 0 to age x as $s(x)$; then the probability of the survival from age x_1 to age x_2 is $s(x_2)/s(x_1)$. Let $f(x)$ be the fertility at age x; we assume that $f(x) \ne 0$ in the interval of ages $x_- < x < x_+$ and $f(x) = 0$ otherwise. Also, denote the total number of births during the time interval $(t, t + dt)$ as $b(t)dt$.

The "old" animals with the age $x > t$ already existed at the time instant $t = 0$, when their age was $x - t$. The probability of their survival from age $x - t$ to age x is $s(x)/s(x - t)$. Therefore, $n(x, t) = n(x - t, 0)s(x)/s(x - t)$ for $x > t$.

The "young" animals with $x < t$ were born at the time instant $t - x$ and survived to age x with probability $s(x)$. Thus, $n(x, t) = b(x - t)s(x)$ for $x < 0$.

The total number density of births at time instant t is

$$b(t) = \int_0^t n(x, t)f(x)dx + \int_t^\infty n(x, t)f(x)dx =$$

$$= \int_0^t b(t - x)s(x)f(x)dx + \int_t^\infty n(x - t, 0)\frac{s(x)}{s(x - t)}f(x)dx.$$

The obtained equation can be written as

$$b(t) = \int_0^t K(x)b(t-x)dx + g(t), \qquad (11.41)$$

where

$$K(x) = s(x)f(x)$$

and

$$g(t) = \int_t^\infty n(x-t,0)\frac{s(x)}{s(x-t)}f(x)dx = \int_0^\infty n(y,0)\frac{s(y+t)}{s(y)}f(y+t)dy,$$

where $y = x - t$. Equation (11.41), which is a Volterra integral equation of the second kind, is called *Lotka equation* (named after an American biophysicist and statistician Alfred James Lotka (1880–1949)).

Note that $K(x) > 0$ for $x_- < x < x_+$, and $K(x) = 0$ otherwise. Because $f(x) = 0$ for $x > x_+$, $g(t) = 0$ for $t > x_+$. Thus, for $t > x_+$ we obtain a Fredholm integral equation:

$$b(t) = \int_{x_-}^{x_+} K(x)b(t-x)dx. \qquad (11.42)$$

The linear homogeneous Eq. (11.42) has a set of exponential solutions of the kind:

$$b(t) = b_0 e^{\lambda t}.$$

The values of r are determined by the characteristic equation:

$$P(\lambda) = \int_{x_-}^{x_+} K(x)e^{-\lambda x}dx = 1. \qquad (11.43)$$

First, assume that λ is real. Recall that $K(x) > 0$ in the interval $x_- < x < x_+$. Therefore,

$$\frac{dP(\lambda)}{d\lambda} = - \int_{x_-}^{x_+} xK(x)e^{-\lambda x}dx < 0.$$

Thus, $P(\lambda)$ is a monotonically decreasing function that changes from $+\infty$ as $x \to -\infty$ and 0 as $x \to +\infty$. Obviously, Eq. (11.43) has exactly one real root that we shall denote as $\lambda = \lambda_0$.

Equation (11.43) has also complex roots, $\lambda_n = r_n + i\omega_n$, that satisfy the equations:

$$\int_{x_-}^{x_+} K(x)e^{-r_n x}\cos\omega_n x\, dx = 1 \tag{11.44}$$

and

$$\int_{x_-}^{x_+} K(x)e^{-r_n x}\sin\omega_n x\, dx = 1. \tag{11.45}$$

The structure of the general solution is

$$b(t) = b_0 e^{\lambda_0 t} + \sum_n b_n e^{\lambda_n t}. \tag{11.46}$$

Since $\cos\omega_n t < 1$ and $K(x) > 0$

$$1 = \int_{x_-}^{x_+} K(x)e^{-r_n x}\cos\omega_n x\, dx < \int_{x_-}^{x_+} K(x)e^{-r_n x}dx = P(r_n).$$

Because $P(\lambda_0) = 1$ and $P(x)$ is a decreasing function, $r_n < \lambda_0$. Therefore, the asymptotic of the solution (11.46) at large t is determined by the real growth rate λ_0, which is called "the Malthusian parameter" (see Chap. 1).

11.6.3 Viscoelasticity

Integral equations are used for the description of the dynamics of media *with memory* (e.g., polymer materials). The evolution of such system is determined by its whole history rather than its instantaneous state.

As the example, let us mention the *viscoelastic* materials. Recall that for the classical linear elastic string, the stress $\sigma(t)$ and the strain $\varepsilon(t)$ are proportional at any time instant t:

$$\sigma(t) = E\varepsilon(t),$$

where E is the stiffness of the string. In a shear viscous flow, the stress is proportional to the shear, i.e., the temporal derivative of the strain:

$$\sigma(t) = \eta \frac{d\varepsilon(t)}{dt}.$$

For a viscoelastic material with memory, the stress and strain are connected through a *hereditary integral*:

$$\sigma(t) = E(t)\varepsilon(0) + \int_0^t E(t-\tau)\frac{d\varepsilon(\tau)}{d\tau} d\tau.$$

The viscoelastic problems are governed mostly by partial differential equations, and we shall not discuss that subject in this book.

Chapter 12
Calculus of Variations

As we saw in previous chapters, the laws of nature are often expressed by ordinary differential equations with the appropriate initial or boundary conditions.

In this chapter, we move on to the next level. We shall consider problems where the governing ODE is a manifestation of the following underlying principle: the function describing a real process is optimal when compared to other functions in the same class.

12.1 Functionals: Introduction

In order to formulate a principle that we just mentioned, it is necessary to assign a number V to each function $y(x)$ from a chosen class of functions: $V = V[y(x)]$. Let us emphasize that V is not a composite function depending on the variable x: the argument of V is the function $y(x)$ as a whole. The variable V defined on the set of functions $y(x)$ is called a *functional*.

It turns out that the basic natural processes are described by functions that provide extrema (i.e., local or global minima or maxima) of functionals. A principle of this kind was first discovered by *Pierre de Fermat* (1662): the light spreads from one point to another along the path of a shortest time (see Sect. 12.4.3). Later, *Leonhard Euler* (1774) and *Pierre Louis Maupertuis* (1776) found that the trajectory of a particle moving in a force field corresponds to a minimum of a functional called the action. The principle of the least action (see Sect. 12.5.2) is one of the most basic principles of physics: Newton's second law and other elementary laws of physics are just its consequences. In the case of a free particle in the gravitational field, that principle is reduced to the minimization principle discovered by *Albert Einstein* (1916): the particle moves along the geodesic lines, i.e., the shortest curves connecting given points, in the curved space-time continuum.

The examples given above clearly demonstrate the importance of solving the problems of finding functionals' extrema, which are called variational problems. The

V. Henner et al., *Ordinary Differential Equations*,
https://doi.org/10.1007/978-3-031-25130-6_12

Fig. 12.1 The curve
$y = y(x)$ that motivates the
length functional $l[y(x)]$

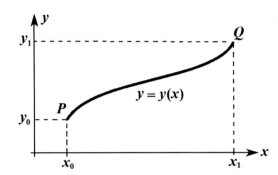

Fig. 12.2 The surface
$z = z(x, y)$ that motivates the
functional $S[z(x, y)]$

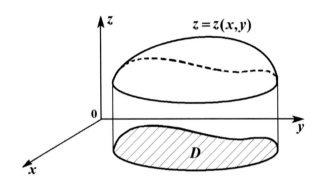

methods for solving those problems form the calculus of variations. That branch of
mathematics was initiated by *Johann Bernoulli* (1696) who formulated the problem
of finding the trajectory of fastest descent for an object in a uniform gravitational
field (the brachistochrone problem; see Sect. 12.4.1).

The crucial steps in the creation of the calculus of variation have been made by
Joseph-Louis Lagrange (1755) and *Leonhard Euler* (1756).

Let us start with several geometric examples of functionals.

1. *The length of a curve*

Consider a curve $y = y(x)$ connecting two given points $P(x_0, y_0)$ and $Q(x_1, y_1)$ in
the xy plane (Fig. 12.1).

The length of the curve is given by the functional:

$$l[y(x)] = \int_{x_0}^{x_1} \sqrt{1 + (y')^2}\, dx. \tag{12.1}$$

2. *Surface area*

Consider a surface $z = z(x, y)$ shown in Fig. 12.2. Let D be a projection of this
surface on the xy plane.

The area of the surface is given by the integral:

$$S = \iint\limits_{D} \sqrt{1 + \left(z'_x\right)^2 + \left(z'_y\right)^2}\, dx\, dy, \tag{12.2}$$

which is a functional $S[z(x, y)]$.

One can also consider functionals that depend on several functions of one or more variables, for instance:

$$V[y_1(x), y_2(x), \ldots, y_n(x)].$$

The next example illustrates that.

3. *The length of a curve in the three-dimensional space*

Consider a curve defined by the parametric equations $\begin{cases} x = x(t), \\ y = y(t), \\ z = z(t). \end{cases}$

The length of a curve,

$$l = \int\limits_{t_1}^{t_2} \sqrt{(x')^2 + (y')^2 + (z')^2}\, dt, \tag{12.3}$$

is a functional $l[x(t), y(t), z(t)]$.

The integral form of functionals appear in physical and geometrical applications, where the goal is to find functions, called *extremals*, which correspond to maximum or minimum values of the functional.

12.2   Main Ideas of the Calculus of Variations

12.2.1   Function Spaces

Each functional $V[y(x)]$ is defined on a certain class of functions. For instance, a functional in the form of an integral,

$$V[y(x)] = \int\limits_{x_0}^{x_1} F\left(x, y(x), y'(x), \ldots, y^{(n)}(x)\right) dx, \tag{12.4}$$

is defined if the integrand is an integrable (e.g., piecewise continuous) function of x on the interval $x_0 \le x \le x_1$. Therefore, $y(x)$ has to belong to a definite set of function

(function space) that we shall denote as M. Depending on the type of the functional, we shall demand typically that $y(x)$ is continuous or continuously differentiable n times ($n \geq 1$).

For functions that belong to a certain function space M, we can define the "distance" between two functions $y_1(x)$ and $y_2(x)$. When M is a set of continuous functions, we can define that distance ("the norm") as

$$\|y_1(x) - y_2(x)\| \equiv \max_{x_0 \leq x \leq x_1} |y_1(x) - y_2(x)|. \tag{12.5}$$

For continuously differentiable function, we can either adopt the norm (12.5) or define:

$$\|y_1(x) - y_2(x)\| \equiv \max_{x_0 \leq x \leq x_1} |y_1(x) - y_2(x)| \\ + \max_{x_0 \leq x \leq x_1} |y_1'(x) - y_2'(x)|. \tag{12.6}$$

In the latter case, two functions are considered as close if not only the values of functions but also the values of their derivatives are close to each other everywhere. Similarly, we can define:

$$\|y_1(x) - y_2(x)\| \equiv \sum_{m=0}^{n} \max_{x_0 \leq x \leq x_1} \left|y_1^{(m)}(x) - y_2^{(m)}(x)\right|, \tag{12.7}$$

if the corresponding derivatives exist and are continuous (here $y_j^{(0)}(x) \equiv y_j(x)$, $j = 1, 2$).

The function space M can be a subset of a wider set of functions. For instance, if $y(x)$ cannot be negative due to its physical meaning (i.e., $y(x)$ is the mass density), the space M can be a set of nonnegative functions continuous in the closed interval $x_0 \leq x \leq x_1$, which is a subset of the set of all functions continuous in $[x_0, x_1]$. If a certain function $\bar{y}(x)$ that belongs to M has some zeros in $[x_0, x_1]$, then there exist continuous functions arbitrary close to $\bar{y}(x)$ that are not nonnegative, i.e., they do not belong to M.

We shall call the set of nonnegative functions with zeros "the boundary" of the space M of nonnegative functions. For any strictly positive functions $\bar{y}(x)$, the minimum on the closed interval $[x_0, x_1]$ is positive. Therefore, all the functions sufficiently close to $\bar{y}(x)$ are also strictly positive, i.e., they belong to M. We shall call the functions in M, which are fully surrounded by functions belonging to M, "the inner points of the set M."

Now we are ready to introduce the notion of continuity of a functional. To do this, we introduce a *variation*, or *increment*, δy of the argument $y(x)$ of a functional as the difference between two functions, $y(x)$ and $\bar{y}(x)$:

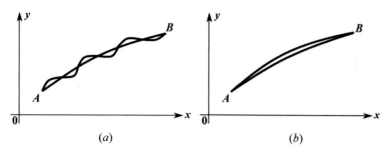

Fig. 12.3 Illustration of the concept of curves proximity

$$\delta y = y(x) - \bar{y}(x), \tag{12.8}$$

where $\bar{y}(x) \in M$ is a certain fixed function and $y(x) \in M$ is an arbitrary function.

A functional $V[y(x)]$ is called *continuous* on the interval $[x_0, x_1]$, if a small change of $y(x)$ corresponds to a small change in the functional $V[y(x)]$, i.e., if for any $\varepsilon > 0$, there exists $\delta > 0$, such that

$$|V[y(x)] - V[\bar{y}(x)]| < \varepsilon \quad \text{for} \quad \|y(x) - \bar{y}(x)\| < \delta, \tag{12.9}$$

where $\|\dots\|$ is one of the norms defined above. If the functional does not contain y', it is sufficient to use the norm (12.5), i.e., to demand $|y(x) - \bar{y}(x)| < \delta$ for $x \in [x_0, x_1]$, as shown in Fig. 12.3a.

For functionals that contain y', we shall use the norm (12.6), i.e., $|y'(x) - \bar{y}'(x)|$ for $x \in [x_0, x_1]$ also must be small, as shown in Fig. 12.3b. If a functional contains higher derivatives y'', y''', ..., it is necessary to consider as close the curves for which $|y''(x) - \bar{y}''(x)|$, $|y'''(x) - \bar{y}'''(x)|$, etc. also are small. All presented in Sect. 12.1 functionals continuously depend on their arguments.

Though in Fig. 12.3 the endpoints of all curves $y(x)$ are fixed at x_1 and x_2, in Sect. 12.7 we will consider "moving ends."

12.2.2 Variation of a Functional

Let us discuss now the behavior of $V[y(x)] - V[\bar{y}(x)]$ for close functions $y(x)$ and $\bar{y}(x)$. It is instructive to discuss first the behavior of usual functions for close values of the argument. In calculus, if the increment of a function $f(x)$,

$$\Delta f = f(x + \Delta x) - f(x),$$

can be presented as

$$\Delta f = A(x)\Delta x + \beta(x, \Delta x)\Delta x,$$

where $\beta(x, \Delta x) \to 0$ as $\Delta x \to 0$, then a function $y = f(x)$ is *differentiable at x*, and the *linear in Δx* part of the increment, $A(x)\Delta x$, is the *differential, df*; obviously, the function differentiable at x is continuous at that point.

Dividing by Δx and taking the limit as $\Delta x \to 0$, we obtain $A(x) = f'(x)$; therefore,

$$df = f'(x)\Delta x.$$

Notice that df can be also introduced as the derivative of $f(x+\alpha\Delta x)$ with respect of α evaluated at $\alpha = 0$:

$$\frac{d}{d\alpha}f(x + \alpha\Delta x)\Big|_{\alpha \to 0} = f'(x + \alpha\Delta x)\Delta x|_{\alpha=0} = f'(x)\Delta x = df.$$

Below we shall define the notion of the variation of a functional, which is somewhat similar to that of the differential of a function.

First, let us introduce the notion of a linear functional. A functional is called *linear*, and denoted as $L[y(x)]$, if

$$L[C_1 y_1(x) + C_2 y_2(x)] = C_1 L[y_1(x)] + C_2 L[y_2(x)], \qquad (12.10)$$

where C_1, C_2 are arbitrary constants.

If the increment of a functional,

$$\Delta V = V[\bar{y}(x) + \delta y] - V[\bar{y}(x)],$$

can be presented as

$$\Delta V = L[\bar{y}(x), \delta y] + \beta(\bar{y}(x), \delta y)|\delta y|_{\max}, \qquad (12.11)$$

where $L[\bar{y}(x), \delta y]$ is a *linear in $\delta y = y(x) - \bar{y}(x)$* functional, $|\delta y|_{\max}$ is the maximum value of $|\delta y|$ and $\beta(\bar{y}(x), \delta y) \to 0$ when $|\delta y|_{\max} \to 0$, then $L[\bar{y}(x), \delta y]$ is called *the variation of a functional* on the function $\bar{y}(x)$, and is denoted as

$$\delta V = L[\bar{y}(x), \delta y]. \qquad (12.12)$$

Thus, δV is the main part of ΔV which is linear in δy.

For a chosen function δy, let us consider a family of functions:

$$y(x, \alpha) = \bar{y}(x) + \alpha \delta y. \qquad (12.13)$$

where α is a parameter. It allows to treat the functional

$$V[y(x)] = V[\bar{y}(x) + \alpha \delta y] \equiv \varphi(\alpha) \tag{12.14}$$

as the function of α with fixed $\bar{y}(x)$ and δy. Then, variation of a functional also can be introduced as the derivative of $V[\bar{y}(x) + \alpha \delta y(x)]$ with respect of parameter α, evaluated at $\alpha = 0$:

$$\delta V = \frac{d}{d\alpha} V[\bar{y}(x) + \alpha \delta y]|_{\alpha = 0}. \tag{12.15}$$

Note that variation of the functional does not necessarily exist, for instance, $V[y] = \int_{x_0}^{x_1} |y - x| dx$ does not have a variation on the function $y = x$. Indeed, for $y(x) = x + \delta y$, the functional $\Delta V[y] = \int_{x_0}^{x_1} |\delta(y)| dx$ is not linear even for small δy.

The functionals given in Sect. 12.1 have variations if the curves and surfaces are smooth.

Later on we consider only functionals that have variations, and we shall not tell that each time.

Example 12.1 Find a variation of the functional $V[y] = \int_{x_0}^{x_1} y^2(x) dx$.

Solution

$$\Delta V = \int_{x_0}^{x_1} [\bar{y}(x) + \delta y(x)]^2 dx - \int_{x_0}^{x_1} \bar{y}^2(x) dx = \int_{x_0}^{x_1} 2\bar{y}(x)\delta y(x) dx + \int_{x_0}^{x_1} [\delta y(x)]^2 dx.$$

The first functional on the right-hand side is linear in $\delta y(x)$; thus, the variation is

$$\delta V = 2 \int_{x_0}^{x_1} \bar{y}(x)\delta y(x) dx.$$

We can also find a variation of the functional using the definition (12.15):

$$V[\bar{y}(x) + \alpha \delta y(x)] = \int_{x_0}^{x_1} [\bar{y}(x) + \alpha \delta y(x)]^2 dx.$$

Then $\frac{d}{d\alpha} V[\bar{y} + \alpha \delta y] = 2 \int_{x_0}^{x_1} [\bar{y} + \alpha \delta y] \delta y dx$; therefore,

$$\delta V = \frac{d}{d\alpha} V[\bar{y} + \alpha \delta y]\Big|_{\alpha=0} = 2 \int_{x_0}^{x_1} \bar{y}\delta y dx,$$

which is the same result as the one already obtained.

Example 12.2 Find a variation of the functional $V[y] = \int_a^b \left(y^2 - y'^2\right) dx$.

Solution

$$V[\bar{y}(x) + \alpha\delta y(x)] = \int_a^b \left\{ [\bar{y}(x) + \alpha\delta y(x)]^2 - [\bar{y}'(x) + \alpha\delta y'(x)]^2 \right\} dx.$$

Then,

$$\frac{d}{d\alpha} V[\bar{y} + \alpha\delta y] = \int_a^b [2(\bar{y} + \alpha\delta y)\delta y - 2(\bar{y}' + \alpha\delta y')\delta y'] dx;$$

therefore,

$$\delta V = \frac{d}{d\alpha} V[\bar{y} + \alpha\delta y]\Big|_{\alpha=0} = 2 \int_a^b (y\delta y - y'\delta y') dx.$$

12.2.3   *Extrema of a Functional*

As we had already said, variational problems are concerned with finding functions, on which a functional reaches an *extremum* within a certain function space, M.

A functional $V[y(x)]$ attains a *maximum* on $\bar{y}(x) \in M$, if for *any function* $y(x) \in M$ the difference between $V[y(x)]$ and $V[\bar{y}(x)]$ is nonpositive:

$$\Delta V = V[y(x)] - V[\bar{y}(x)] \leq 0.$$

If $\Delta V \leq 0$, and $\Delta V = 0$ only at $y = \bar{y}(x)$, then a *strict maximum* is attained on the function $y = \bar{y}(x)$.

Similarly, a functional attains a *minimum* on $\bar{y}(x) \in M$ if $\Delta V \geq 0$ on *any function* $y(x) \in M$. If $\Delta V \geq 0$, and $\Delta V = 0$ only at $y = \bar{y}(x)$, then a *strict minimum* is attained on the function $y = \bar{y}(x)$.

Example 12.3 Show that the functional

$$V[y] = \int\limits_0^1 (x^2 + y^2)\,dx$$

has a strict minimum on the function $y = 0$.

Solution

$$\Delta V = V[y(x)] - V[0] = \int\limits_0^1 (x^2 + y^2)\,dx - \int\limits_0^1 x^2\,dx = \int\limits_0^1 y^2\,dx \ge 0;$$

and the equality to zero is reached only for the function $y(x) = 0$. This means that the minimum is strict.

Generally, it is not easy to show that a definite function provides the minimum or maximum of the functional over the whole function space (a global minimum or maximum). Below we shall deal with a more restricted problem: we shall check whether it gives the minimum/maximum value of the functional among all the close functions. Obviously, that will give us a necessary condition for an extremum of a functional.

Theorem
If a functional $V[y(x)]$ attains a maximum (a minimum) on the function $y = \bar{y}(x)$, where $\bar{y}(x) \in M$ is an inner point of the definition space of the functional, then on that function the variation is zero:

$$\delta V[\bar{y}(x)] = 0. \tag{12.16}$$

(Similarly, the derivative or the differential of a function is zero at the interval's inner extremum point.)

Proof By the condition of the theorem, the function

$$\varphi(\alpha) = V[\bar{y}(x) + \alpha\delta y]$$

has an extremum at $\alpha = 0$; thus, the derivative $\varphi'(\alpha)|_{\alpha = 0} = \varphi'(0) = 0$. Therefore,

$$\frac{\partial}{\partial\alpha} V[\bar{y}(x) + \alpha\delta y]_{\alpha=0} = \delta V = 0 \text{ i.e., } \delta V = 0 \text{ on } \bar{y}(x).$$

Thus, on the curve on which a functional reaches the extremum, its variation equals zero.

Problems

Find a variation of the functionals:

1. $V[y] = \int\limits_a^b yy'dx.$

2. $V[y] = \int\limits_a^b (x + y)dx.$

3. $V[y] = y^2(0) + \int\limits_0^1 \left(xy + y'^2\right)dx.$

4. $V[y] = \int\limits_0^\pi y' \sin y dx.$

5. $V[y(x)] = \int\limits_0^1 xy^3(x)dx.$

6. $V[y(x)] = \int\limits_0^1 \left(xy + y'^2\right)ydx.$

7. $V[y(x)] = \int\limits_0^2 x^2\sqrt{1 + y^2}dx.$

8. $V[y(x)] = \int\limits_a^b (y'e^{-2y} - xy^3)dx.$

9. $V[y(x)] = \int\limits_0^1 \left(xy'^2 + (x + 1)y^2\right)dx.$

12.3   The Euler Equation and the Extremals

12.3.1   The Euler Equation

Consider a functional:

$$V[y(x)] = \int\limits_{x_0}^{x_1} F(x, y(x), y'(x))dx, \qquad (12.17)$$

and let all functions $y(x)$ in M be *fixed at the given endpoints* $A(x_0, y_0)$ and $B(x_1, y_1)$ (Fig. 12.4). We want to find the *extremal* of this variational problem. The following theorem gives the *necessary condition* that the extremal solution has to satisfy.

(From now on, we shall denote the extremal $\bar{y}(x)$ just as $y(x)$.)

Fig. 12.4 The extremal $y(x)$
(bold line) and other curves
$y(x, \alpha)$ connecting points
A and B

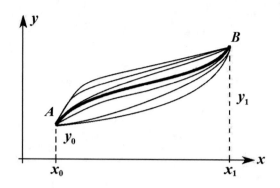

The Euler Theorem

If a functional $V[y(x)]$ reaches an extremum on a certain function $y(x)$, and

$$y(x_0) = y_0, \quad y(x_1) = y_1, \tag{12.18}$$

then $y(x)$ is a solution of the ordinary differential equation:

$$\frac{\partial F}{\partial y} - \frac{d}{dx}\frac{\partial F}{\partial y'} = 0 \tag{12.19}$$

which is called the Euler equation.

Note We assume that the function $F = F(x, y(x), y'(x))$ has a necessary number of continuous partial derivatives.

Proof Consider a family of functions:

$$y(x, \alpha) = y(x) + \alpha \delta y, \tag{12.20}$$

which for $\alpha = 0$ contains the extremal $y(x)$.

The functional $V[y(x, \alpha)]$ can be considered as a function of α:

$$V[y(x, \alpha)] \equiv \varphi(\alpha), \tag{12.21}$$

and $\varphi(\alpha)$ attains the extremum at $\alpha = 0$.

The necessary condition for the extremum of the function $\varphi(\alpha)$ at $\alpha = 0$ is

$$\varphi'(\alpha) = 0. \tag{12.22}$$

For

$$\varphi(\alpha) = \int_{x_0}^{x_1} F[x, y(x, \alpha), y'_x(x, \alpha)] \, dx, \tag{12.23}$$

we have:

$$\varphi'(\alpha) = \frac{\partial}{\partial \alpha} \int_{x_0}^{x_1} F dx = \int_{x_0}^{x_1} \left[\frac{\partial F}{\partial y} \frac{\partial}{\partial \alpha} y(x, \alpha) + \frac{\partial F}{\partial y'} \frac{\partial}{\partial \alpha} y'(x, \alpha) \right] dx.$$

With

$$\frac{\partial}{\partial \alpha} y(x, \alpha) = \frac{\partial}{\partial \alpha} [y(x) + \alpha \delta y(x)] = \delta y(x),$$

$$\frac{\partial}{\partial \alpha} y'(x, \alpha) = \frac{\partial}{\partial \alpha} [y'(x) + \alpha \delta y'(x)] = \delta y'(x),$$

we obtain:

$$\varphi'(\alpha) = \int_{x_0}^{x_1} \left[\frac{\partial F(x, y(x, \alpha), y'(x, \alpha))}{\partial y} \delta y(x) + \frac{\partial F(x, y(x, \alpha), y'(x, \alpha))}{\partial y'} \delta y'(x) \right]$$

and

$$\varphi'(0) = \int_{x_0}^{x_1} \left[\frac{\partial F(x, y(x), y'(x))}{\partial y} \delta y(x) + \frac{\partial F(x, y(x), y'(x))}{\partial y'} \delta y'(x) \right].$$

As we know, $\varphi'(0)$ is a variation of the functional:

$$\delta V = \frac{d}{d \alpha} V[y(x) + \alpha \Delta y]_{\alpha = 0} = \varphi'(\alpha)|_{\alpha = 0} = \varphi'(0) = 0.$$

Thus,

$$\int_{x_0}^{x_1} \left[\frac{\partial F}{\partial y} \delta y(x) + \frac{\partial F}{\partial y'} \delta y'(x) \right].$$

Evaluating second integral by means of the integration by parts and using $\delta y' = (\delta y)'$, we obtain:

$$\int_{x_0}^{x_1} \frac{\partial F}{\partial y'} (\delta y)' dx = \frac{\partial F}{\partial y'} \delta y \Big|_{x_0}^{x_1} - \int_{x_0}^{x_1} d \left(\frac{\partial F}{\partial y'} \right) \delta y = \int_{x_0}^{x_1} \frac{d}{dx} \left(\frac{\partial F}{\partial y'} \right) \delta y dx.$$

Here we used $\frac{\partial F}{\partial y'} \delta y \Big|_{x_0}^{x_1} = 0$, because $\delta y(x) = 0$ at the endpoints x_0 and x_1.

Thus,

$$\delta V = \int_{x_0}^{x_1} \left[\frac{\partial F}{\partial y} - \frac{d}{dx} \left(\frac{\partial F}{\partial y'} \right) \right] \delta y(x) dx = 0.$$

Since the expression in the brackets is a continuous function and a variation $\delta y(x)$ is an arbitrary function, we can conclude (the strict proof of this point is given by the Fundamental lemma of the calculus of variations) that the above integral can be zero only if

$$\frac{\partial F}{\partial y} - \frac{d}{dx} \frac{\partial F}{\partial y'} = 0. \tag{12.24}$$

Using brief notation, this equation is conveniently written as

$$F_y - \frac{d}{dx} F_{y'} = 0. \tag{12.25}$$

Differentiating $\frac{d}{dx} \frac{\partial F}{\partial y'}$, Eq. (12.24) can be written as

$$F_y - F_{y'x} - y' F_{y'y} - y'' F_{y'y'} = 0. \tag{12.26}$$

All forms, (12.24), (12.25), and (12.26) are equivalent.

The Euler equation gives the *necessary condition for extremum* of the functional (12.19) with the conditions (12.18). In other words, the extrema of this problem *can be attained* only on functions which are the solutions of the Euler equation (12.24).

Equation (12.24) (sometimes called *the Euler-Lagrange equation*) was obtained by Leonhard Euler in 1744. A function F in the integral functional (12.19) is often called the *Lagrange function*, or the *Lagrangian*.

Equation (12.24) is the second-order ordinary differential equation, and thus its general solution depends on two arbitrary constants: $y = y(x, C_1, C_2)$. These constants can be determined from the boundary conditions:

$$\begin{cases} y(x_0, C_1, C_2) = y_0, \\ y(x_1, C_1, C_2) = y_1, \end{cases}$$

i.e., in order to find the extremal $y(x)$, we have to solve the boundary value problem.

Recall that boundary value problems do not always have a solution, and if there is a solution, it may not be unique. But in many variational problems, the existence of a solution is obvious from the physical or geometric meaning of the problem. Note that the Euler theorem gives the necessary condition for the *extremum*; in Sect. 12.3.3 we formulate an additional necessary condition for the *minimum*.

For practical purposes, in order to determine whether a maximum or a minimum of a functional is attained on the extremal, often it is sufficient to compare the value of the functional on the extremal and on some close functions satisfying the same boundary conditions. If a value of the functional on the extremal is greater (smaller) than its values on the close curves, then the functional attains a maximum (minimum) on the extremal. Similar practical approach is used in calculus. However, note that a functional can increase for some function variations and decrease for other function variations. The example is given in Sect. 12.3.3.

Example 12.4 Find the extremal of the functional:

$$V[y(x)] = \int\limits_{1}^{2} \left(y'^2 - 2xy\right) dx, y(1) = 0, \ y(2) = -1.$$

Solution Here $F(x, y, y') = y'^2 - 2xy$, and the Euler equation is

$$\frac{\partial F}{\partial y} - \frac{d}{dx}\frac{\partial F}{\partial y'} = -2x - 2y'' = 0 \text{ or } y'' + x = 0.$$

Its general solution is

$$y(x) = -\frac{x^3}{6} + C_1 x + C_2,$$

and using the boundary conditions, we obtain $C_1 = 1/6$, $C_2 = 0$. Thus, the extremum *may be* attained on the extremal:

$$y(x) = \frac{x}{6}\left(1 - x^2\right).$$

Actually, that extremum is a minimum (see Sect. 12.3.3 for corresponding criteria).

12.3.2 Special Cases of Integrability of the Euler Equation

12.3.2.1 F Does Not Contain x, F = F(y, y')

In this case, the Euler equation (12.26) is reduced to

$$F_y + y' F_{y'y} - y'' F_{y'y'} = 0. \tag{12.27}$$

We multiply this equation by y' and add and subtract $F_y y''$ on the left-hand side. This gives the full derivative:

$$F_{yy}y' + F_{yy'}y'' - F_{y'y}y'' - F_{y'y}y'^2 - F_{y'y'}y''y' = 0, \quad \text{or} \quad \frac{d}{dx}\left(F - y'F_{y'}\right) = 0.$$

Therefore,

$$F - y'F_{y'} = \text{const.} \tag{12.28}$$

is the *first integral of the Euler equation*. In this case, instead of a second-order Euler equation, we can consider the first-order equation (12.28). This equation sometimes is easier to solve.

Example 12.5 Find the extremal of the functional:

$$V[y(x)] = \int\limits_0^1 e^y y'^2 dx, \quad y(0) = 0, y(1) = 4.$$

Solution The integrand does not contain variable x; thus, the first integral of the Euler equation is

$$F - y'F_{y'} = -e^y y'^2 = C_1.$$

Clearly, $C_1 < 0$ and replacing C_1 by $-C_1^2$, we arrive to $e^{y/2}y' = C_1$. Separating the variables and evaluating the integrals, we obtain $2e^{y/2} = C_1x + C_2$. The boundary conditions give $C_2 = 2$, $C_1 = 2(e^2 - 2)$; thus, the extremal is $y(x) = 2 \ln [(e^2 - 1)x + 1]$.

12.3.2.2   *F* Does Not Contain *y*, $F = F(x, y')$

In this case, the Euler equation is

$$\frac{d}{dx}F_{y'}(x, y') = 0,$$

and therefore, we arrive at the first-order differential equation:

$$F_{y'}(x, y') = \text{const.} \tag{12.29}$$

Example 12.6 Among all possible curves connecting the points $(1, 1)$ and $(2, 2)$, find the one on which an extremum of the functional:

$$V[y(x)] = \int\limits_1^2 y'(1 + x^2y')dx$$

can be attained.

Solution Function F does not depend on y, and we have:

$$\frac{d}{dx}F_{y'}(x,y') = 0,\text{ or }\frac{d}{dx}\left(1+2x^2y'\right) = 0.$$

Thus,

$$1+2x^2y' = C,\text{ or }y' = \frac{C-1}{2x^2}.$$

Then,

$$y(x) = -\frac{C-1}{2x}+C_2,\text{ or }y(x) = \frac{C_1}{x}+C_2.$$

With the boundary conditions $y(1) = 1$ and $y(2) = 2$, we have $C_1 = -2$, $C_2 = 3$. The extremal is $y(x) = -\frac{2}{x}+3$.

12.3.2.3   F Does Not Contain y', $F = F(x,y)$

In this case the Euler equation has the form:

$$F_y(x,y) = 0. \tag{12.30}$$

Equation (12.29) is algebraic. It defines either one curve, or a finite number of curves that may not satisfy the boundary conditions. Only in exceptional cases, when the curve (12.29) passes through the boundary points (x_0, y_0) and (x_1, y_1), the functional may attain an extreme value.

Example 12.7 Find the extremals of the functional:

$$V[y(x)] = \int\limits_{0}^{\pi/2} y(2x-y)dx,\quad y(0) = 0,\ y\left(\frac{\pi}{2}\right) = \frac{\pi}{2}.$$

Solution The Euler equation

$$F_y(x,y) = 2x-2y = 0,\text{ or }y = x.$$

Since the boundary conditions are satisfied, the functional $V[y(x)]$ may attain an extremum on the line $y = x$. For other boundary conditions, for example, $y(0) = 0$, $y\left(\frac{\pi}{2}\right) = 1$, the extremal does not pass through the boundary points, and the variational problem does not have a solution.

12.3.2.4   *F* Depends Only on y', $F = F(y')$

The Euler equation becomes:

$$y''F_{y'y'} = 0; \tag{12.31}$$

thus, either $y'' = 0$, which gives $y = C_1x+C_2$, or $F_{y'y'} = 0$. The roots of equation $F_{y'y'}(y') = 0$ (if there are any) are constants, $y' = k_i$. These roots also give straight-line solutions, $y = k_ix+C$. Thus, the extremals are $y = C_1x+C_2$, where C_1 and C_2 should be obtained from the boundary conditions $y_0 = y(x_0)$, $y_1 = y(x_1)$.

Example 12.8 Find the shortest path between two points in the plane.

Solution The answer, obviously, is the straight line connecting these points. Let us obtain this result using the Euler equation.

The length of the curve connecting two points in a plane is

$$l[y(x)] = \int_{x_0}^{x_1} \sqrt{1 + (y')^2}\,dx,$$

and

$$F_y = F_{y'x} = F_{y'y} = 0, F_{y'y'} = \frac{1}{\left[1 + (y')^2\right]^{\frac{3}{2}}} \neq 0.$$

The Euler equation gives $y''F_{y'y'} = 0$, and because $F_{y'y'} \neq 0$, we have:

$$y'' = 0.$$

Thus, $y = C_1x+C_2$.

Using the boundary conditions $y_0 = y(x_0)$, $y_1 = y(x_1)$, the line connecting the two points is

$$y(x) = \frac{y_1 - y_0}{x_1 - x_0}(x - x_0) + y_0.$$

Problems

Find the extremals of the functionals:

1. $V[y] = \int_{x_0}^{x_1} \left(y'^2 + 2yy' - 16y^2\right)dx.$

2. $V[y] = \int_{x_0}^{x_1} y'(x + y')dx.$

3. $V[y] = \int\limits_{x_0}^{x_1} \frac{y'^2}{x^3} dx.$

4. $V[y] = \int\limits_{-1}^{0} \left(12xy - y'^2\right) dx,\ y(-1) = 1,\ y(0) = 0.$

5. $V[y] = \int\limits_{1}^{2} \left(y'^2 + 2yy' + y^2\right) dx,\ y(1) = 1,\ y(2) = 0.$

6. $V[y] = \int\limits_{0}^{1} yy'^2 dx,\ y(0) = 1,\ y(1) = \sqrt[3]{4}.$

7. $V[y] = \int\limits_{0}^{\pi} \left(4y\cos x + y'^2 - y^2\right) dx,\ y(0) = 0,\ y(\pi) = 0.$

8. $V[y] = \int\limits_{0}^{1} \left(y'^2 - y^2 - y\right) e^{2x} dx,\ y(0) = 0,\ y(1) = e^{-1}.$

9. $V[y] = \int\limits_{-1}^{1} \left(y'^2 - 2xy\right) dx,\ y(-1) = -1,\ y(1) = 1.$

10. $V[y] = \int\limits_{-1}^{0} \left(y'^2 - 2xy\right) dx,\ y(-1) = 0,\ y(0) = 2.$

11. $V[y] = \int\limits_{1}^{e} \left(xy'^2 + yy'\right) dx,\ y(1) = 0,\ y(e) = 1.$

12. $V[y] = \int\limits_{x_0}^{x_1} [2xy + (x^2 + e^y)y'] dx,\ y(x_0) = y_0,\ y(x_1) = y_1.$

13. $V[y] = \int\limits_{0}^{1} (e^y + xy') dx,\ y(0) = 0,\ y(1) = \alpha.$

14. $V[y] = \int\limits_{0}^{\pi/4} \left(y'^2 - y^2\right) dx,\ y(0) = 1,\ y(\pi/4) = \frac{1}{\sqrt{2}}.$

15. $V[y] = \int\limits_{0}^{\pi} \left(y'^2 - y^2\right) dx,\ y(0) = 1,\ y(\pi) = -1.$

16. $V[y] = \int\limits_{0}^{1} \left(x + y'^2\right) dx,\ y(0) = 1,\ y(1) = 2.$

17. $V[y] = \int\limits_{0}^{1} \left(y^2 + y'^2\right) dx,\ y(0) = 0,\ y(1) = 1.$

18. $V[y] = \int\limits_{0}^{1} \left(y'^2 + 4y^2\right) dx,\ y(0) = e^2,\ y(1) = 1.$

19. $V[y(x)] = \int\limits_{x_0}^{x_1} (y^2 + 2xyy') dx.$

20. $V[y(x)] = \int\limits_{0}^{1} (y^2 + x^2 y') dx,\ y(0) = 0,\ y(1) = y_1.$

12.3.3   Conditions for the Minimum of a Functional

The solution of the Euler equation (12.19) with boundary conditions (12.18) can correspond to a maximum, a minimum, or a saddle point of the functional (12.17). In order to be sure that the obtained solution corresponds to the minimum of the functional, an additional analysis is needed, which is based on the calculation of the *second variation* of the functional.

Let us consider the family of functions (12.23) and expand them into the Maclaurin series up to the second order:

$$\varphi(\alpha) - \varphi(0) = \alpha \delta V[y; \delta y] + \alpha^2 \delta^2 V[y; \delta y] + \varepsilon, \tag{12.32}$$

where

$$\delta V[y; \delta y] = \int_{x_0}^{x_1} \left(F_y \delta y + F_{y'} \delta y' \right) dx \tag{12.33}$$

is the *first variation*

$$\delta^2 V[y; \delta y] = \frac{1}{2} \int_{x_0}^{x_1} \left[F_{yy} (\delta y)^2 + 2F_{yy'} \delta y \delta y' + F_{y'y'} (\delta y')^2 \right] dx \tag{12.34}$$

is the *second variation*, and ε denotes the terms of the higher order in the limit $|\alpha| \ll 1$. As we know, $\delta V[y; \delta y] = 0$ for the extremal solution. Thus, the positiveness of $\varphi(\alpha) - \varphi(0)$ demands $\delta^2 V[y; \delta y] > 0$ for any δy satisfying the boundary conditions:

$$\delta y(x_0) = \delta y(x_1) = 0. \tag{12.35}$$

Integrating by parts the second term in the integrand of (12.34)

$$\int_{x_0}^{x_1} 2F_{yy'} \delta y \delta y' dx = \int_{x_0}^{x_1} F_{yy'} \left[\delta y^2 \right]' = - \int_{x_0}^{x_1} \left(\frac{d}{dx} F_{yy'} \right) (\delta y)^2 dx,$$

we can rewrite (12.34) as

$$\delta^2 V[y; \delta y] = \frac{1}{2} \int_{x_0}^{x_1} \left[F_{y'y'} (\delta y')^2 + \left(F_{yy} - \frac{d}{dx} F_{yy'} \right) (\delta y)^2 \right] dx. \tag{12.36}$$

The functional $V[y(x)]$ has a minimum, if the expression (12.36) is positive definite for any $\delta y(x)$, $\delta y'(x)$ satisfying the conditions (12.35).

To find a *maximum*, we can apply the criteria for the minimum to the functional $-V[y(x)]$.

In some cases, a positiveness of the second variation is easy to obtain, like in the following example.

Example 12.9 Let us find the minimum of the problem:

$$V[y] = \frac{1}{2} \int_0^\pi \left(y'^2 + y^2 + 2y \sin x \right) dx, y(0) = y(\pi) = 0$$

and discuss if the necessary condition for the minimum is satisfied.

The Euler equation gives:

$$y'' - y = \sin x$$

and its solution with the boundary conditions is $y(x) = -0.5 \sin x$.

Let us take a function $y(x) = -0.5 \sin x + \delta y$ and substitute it in $V[y]$ to calculate the second variation of the functional:

$$\delta^2 V[y; \delta y] = V[y] - V[-0.5 \sin(x)] = \frac{1}{2} \int_0^\pi \left[(\delta y')^2 + (\delta y)^2 \right] dx.$$

We see that $\delta^2 V[y; \delta y] > 0$ for any nonzero differentiable function δy.

The rigorous analysis of the conditions for the positiveness of the second variation is rather complex, and here we present only the basic ideas of the derivation.

One of the necessary conditions is rather obvious: $F_{y'y'} > 0$. Indeed, we can take a small but rapidly oscillating variation $\delta y(x)$ such that $|\delta y'|$ is large while $|\delta y|$ is small, so that the first term in the right-hand side of (12.36) is much larger than the second one. Then, the second term can be neglected. We can see that (12.36) is positive for any $\delta y(x)$ only if $F_{y'y'} > 0$.

In order to obtain a second condition, let us consider expression (12.36) as a functional of $\delta y(x)$. The extremal solution $y(x)$ corresponds to the minimum of the functional $V[y(x)]$ if the functional (12.36) has the single minimum equal to 0 at $\delta y(x) = 0$. To be sure that (12.36) has no other minima, we use *the Euler equation for that functional*:

$$\left(F_{yy} - \frac{d}{dx} F_{yy'} \right) \delta y - \frac{d}{dx} \left(F_{y'y'} \delta y' \right) = 0. \tag{12.37}$$

Equation (12.37) has to have no nontrivial solutions.

The exact formulation of the criterion for positiveness of the second variation is as follows.

The functional (12.34) is positive definite for any $\delta y(x)$ and $\delta y'(x)$ satisfying the conditions (12.35), if

(i) $F_{y'y'} > 0$ for all $x_0 \le x \le x_1$.
(ii) For any $x_0 < x_* \le x_1$, the only solution of Eq. (12.37) with boundary conditions $\delta y(x_0) = \delta y(x_*) = 0$ is the trivial solution $\delta y(x) = 0$.

The conditions presented above are the necessary conditions for the extremal solution being a minimum of the functional $V[y(x)]$. At the same time, they are sufficient conditions for the "weak minimum" of the functional: the difference $V[y(x) + \delta y(x)] - V[y(x)]$ is positive if both maximum variations $|\delta y(x)|$ and $|\delta y'(x)|$ are sufficiently small on $x_0 \le x \le x_1$.

Example 12.10 Consider the extremal solution $y(x) = 0$ of the functional:

$$V[y(x)] = \int\limits_0^{x_1} \left(y'^2 - y^2 \right) dx. \qquad (12.38)$$

The second variation is

$$\delta^2 V[0; \delta y] = \int\limits_0^{x_1} \left[(\delta y')^2 - (\delta y)^2 \right] dx. \qquad (12.39)$$

Obviously, condition (i) is satisfied. The Euler equation for the functional of the second variation can be written as

$$\delta y'' + \delta y = 0. \qquad (12.40)$$

It has nontrivial solutions:

$$\delta y(x) = C \sin x, \qquad (12.41)$$

which satisfy the conditions $\delta y(0) = 0$ and $\delta y(\pi) = 0$. Thus, if $x_1 < \pi$, there are no nontrivial solutions of (12.40), such that

$$\delta y(0) = \delta y(x_*) = 0, \quad x_* \le x_1, \qquad (12.42)$$

and the function $y(x) = 0$ minimizes the functional $V[y(x)]$. However, if $x_1 \ge \pi$, there do exist nontrivial solutions (12.41) of Eq. (12.40) with the boundary conditions (12.41); therefore, $y(x) = 0$ is not the minimizer of the functional. Indeed, for the trial function $y(x) = C \sin (\pi x/x_1)$,

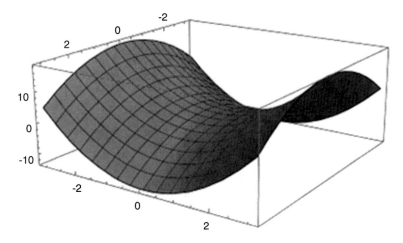

Fig. 12.5 The saddle in Example 12.10 ($-\pi \leq c_1, c_2 \leq \pi$)

$$V[y(x_1)] = \frac{C^2\left(\pi^2 - x_1^2\right)}{2x_1}.$$

If $x_1 = \pi$, the functional $V[y(x)]$ is equal to 0 for any function of the family (12.41). For $x_1 > \pi$, the functional is not bounded from below, and it can take arbitrary negative values.

Note that for the trial function

$$y(x) = \sum_{n=1}^{\infty} c_n \sin(n\pi x/x_1),\ V[y(x)] = \sum_{n=1}^{\infty} (c_n)^2 \left[(n\pi)^2 - (x_1)^2\right]/(2x_1).$$

This means that for $x_1 > \pi$, the extremum solution $y(x) = 0$ is always a "saddle point" of the functional: it decreases for c_n, such that $n < x_1/\pi$ and grows for other c_n.

Figure 12.5 shows a plot of function $V(c_1, c_2)$ for $\pi < x_1 < 2\pi$; all $c_n = 0$ for $n > 2$. If we take $x_1 = \frac{3\pi}{2}$, then $V(c_1, c_2) = -\frac{5\pi}{12}c_1^2 + \frac{7\pi}{12}c_2^2$ is a saddle around $c_1 = c_2 = 0$.

12.4   Geometrical and Physical Applications

12.4.1   The Brachistochrone Problem

Find a curve, along which an object slides in a vertical plane from point A to point B in the shortest time.

This problem historically is the first that led to the calculus of variations. The problem was stated in 1696 by Johann Bernoulli, and it is called *the brachistochrone problem*. (Brachistochrone translates as "the shortest time" from Greek.)

Fig. 12.6 The coordinate system for the brachistochrone problem

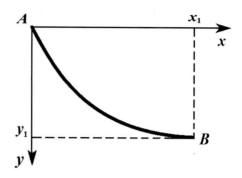

Let us align the origin of the coordinate system with the point A; see Fig. 12.6.

Neglecting the friction and the resistance of the medium, the kinetic energy of an object at point B equals the potential energy loss. Assuming zero initial velocity, we have:

$$\frac{mv^2}{2} = mgy,$$

which gives the object's speed

$$v = \frac{dl}{dt} = \sqrt{2gy}.$$

Using $l[y(x)] = \int\limits_0^{x_1} \sqrt{1 + (y')^2}\,dx$, we find the time it takes to get from $A(0,0)$ to $B(x_1, y_1)$:

$$T[y(x)] = \int\limits_0^{x_1} \frac{l}{v} = \frac{1}{\sqrt{2g}} \int\limits_0^{x_1} \frac{\sqrt{1 + (y')^2}}{\sqrt{y}}\,dx, \quad y(0) = 0, \quad y(x_1) = y_1. \quad (12.43)$$

Obviously, $T[y(x)]$ is a functional and we want to find the curve $y(x)$ on which it attains a minimum.

Since the integrand does not explicitly contain x, the first integral of the Euler equation $F - F_y y' = \text{const}$ gives:

$$\frac{\sqrt{1 + (y')^2}}{\sqrt{y}} - \frac{(y')^2}{\sqrt{y\left[1 + (y')^2\right]}} = C.$$

Simplifying, we have:

$$\frac{1}{\sqrt{y\left[1 + (y')^2\right]}} = C \quad \text{or} \quad y = \frac{C_1}{1 + (y')^2}$$

Let t be a parameter, such that $y' = \cot(t/2)$. Then,

$$y = \frac{C_1}{1 + \cot^2(t/2)} = C_1 \sin^2(t/2) = \frac{C_1}{2}(1 - \cos t).$$

From $\frac{dy}{dx} = \cot(t/2)$ it follows that

$$dx = \frac{dy}{\cot (t/2)} = \frac{C_1 \sin(t/2)\cos(t/2)}{\cot (t/2)} dt = C_1 \sin^2 t dt = \frac{C_1}{2}(1 - \cos t)dt;$$

thus, $x = \frac{C_1}{2}(t - \sin t) + C_2$. Next, $C_2 = 0$ since $y = 0$ at $x = 0$.
Now we see that the brachistochrone curve is the *cycloid*:

$$
\begin{cases}
x(t) = \dfrac{C_1}{2}(t - \sin t), \\[2mm]
y(t) = \dfrac{C_1}{2}(1 - \cos t).
\end{cases}
\tag{12.44}
$$

The radius $C_1/2$ of the rotating circle that generates the cycloid can be obtained by requiring that the point $B(x_1, y_1)$ belongs to the cycloid.

12.4.2   The Tautochrone Problem

The brachistochrone curve has one more interesting property: the time of the object's motion in the uniform gravity field toward the lowest point on the curve does not depend on the location of the initial point on the curve. Therefore, this curve is also called the "tautochrone," which means *equal time* in Greek.

We take equations of a cycloid in the form:

$$x(t) = \frac{y_1}{2}(t - \sin t), \quad y(t) = \frac{y_1}{2}(1 - \cos t), \quad 0 \le t \le \pi \tag{12.45}$$

and assume that the object starts its motion with zero velocity at a point (x_0, y_0):

$$x = x_0 = x(t_0), \quad y = y_0 = y(t_0).$$

Let us find the time of its arrival in the final point $x = x_1 = x(\pi)$, $y = y_1 = y(\pi)$. The conservation of energy (recall that the y-axis is directed downward)

$$\frac{mv^2}{2} - mgy = -mgy_0,$$

allows to find the velocity of the object:

$$v = \sqrt{2g(y - y_0)},$$

and the travel time between the initial and final points

$$T = \frac{1}{\sqrt{2g}} \int_{y_0}^{y_1} \frac{\sqrt{1 + (y')^2}\,dy}{y'\sqrt{y - y_0}} \tag{12.46}$$

(note that in order to obtain the expression (12.46), we took into account that $dx = dy/y'$.) Using the relations

$$y' = \cot\frac{t}{2}, \quad \sqrt{1 + y'^2} = \csc\frac{t}{2}$$

and taking into account that

$$y(t) = y_1 \sin^2\frac{t}{2},$$

we find that

$$\frac{\sqrt{1 + y'^2}}{y'} = \sec\frac{t}{2} = \sqrt{\frac{y_1}{y_1 - y}}.$$

Thus, the travel time

$$T = \sqrt{\frac{y_1}{2g}} \int_{y_0}^{y_1} \frac{dy}{\sqrt{(y_1 - y)(y - y_0)}} = \sqrt{\frac{2y_1}{g}}\pi \tag{12.47}$$

does not depend on y_0; in other words, all objects that start their motion at any point (x_0, y_0) on the curve will reach the final point (x_1, y_1) simultaneously. The curve (a cycloid) that possesses this amazing property was found by Christiaan Huygens in 1659. Interestingly, the motivation to find such curve was practical: there was the interest in building a clock, such that the period of the pendulum's oscillation does not depend on the oscillation amplitude.

12.4.3 The Fermat Principle

The Fermat principle proposed by the French mathematician Pierre de Fermat in 1662 states that optical rays in a medium propagate along the path with the least propagation time.

Time is the functional, $T[y] = \int_a^b \frac{dl}{v}$, where dl is the length differential, v is the speed of light in a medium $a(t_1)$ and $b(t_2)$ are the initial and final points. In two dimensions

$$T[y(x)] = \int_a^b \frac{\sqrt{1 + y'^2}\,dx}{v(x,y)}. \tag{12.48}$$

In a medium with the refraction index $n(x, y)$, the speed of light is $v(x, y) = c/n(x, y)$ (c is the speed of light in a vacuum); thus,

$$T[y(x)] = \frac{1}{c} \int_{t_1}^{t_2} n(x,y)\sqrt{1 + y'^2}\,dx. \tag{12.49}$$

For example, consider a situation when n depends only on y. The first integral of the Euler equation is

$$\frac{n(y)}{\sqrt{1 + y'^2}} = \text{const.} \tag{12.50}$$

The denominator is $\sqrt{1 + \tan^2\varphi} = 1/\cos\varphi$, where φ is the angle between the tangent line to the curve $y(x)$ and the x-axis. Thus, the last equation is

$$n(y)\cos\varphi = \text{const.} \tag{12.51}$$

For $n = \text{const}$ we see that $\varphi = \text{const}$; thus, the path is a straight line, as it should be.

Denote by θ the angle between an incident ray and the normal to the surface line. With $\theta + \varphi = \pi/2$ and $\cos(\pi/2 - \theta) = \sin\theta$, we obtain from Eq. (12.51) *Snell's law*, describing the refraction on the boundary of two media with the indexes of refraction n_1 and n_2:

$$n_1 \sin\theta_1 = n_2 \sin\theta_2. \tag{12.52}$$

Snell's law is named after a Dutch astronomer and mathematician Willebrord Snellius (1580–1626).

Reading Exercise Find a ray's trajectory in a medium with $n(y) = a/y$, $a = $ const.

Reading Exercise Find a ray's trajectory in a medium with $n(y) = ay$, $a = $ const.

12.4.4   The Least Surface Area of a Solid of Revolution

This problem reads:

Find the curve $y = y(x)$ with the endpoints at $A(x_0, y_0)$ and $B(x_1, y_1)$, such that, when the curve is revolved about the x-axis, the surface area of the resultant solid of revolution is minimized.

The curve that is sought (see Fig. 12.7) is the extremal of the functional:

$$S[y(x)] = \int 2\pi y ds = 2\pi \int_{x_0}^{x_1} y\sqrt{1 + (y')^2} dx, \quad y(x_0) = y_0, \ y(x_1) = y_1. \quad (12.53)$$

The integrand does not contain x; thus, the first integral of the Euler equation, $F - y'F_{y'} = C_1$, gives:

$$y\sqrt{1 + (y')^2} - \frac{yy'}{\sqrt{1 + (y')^2}} = C_1, \quad \text{or} \quad \frac{y}{\sqrt{1 + (y')^2}} = C_1.$$

Let us introduce the parameter t by requiring $y' = \sinh t$. Integration gives $y = C_1 \cosh t$, and then $dx = \frac{dy}{y'} = \frac{C_1 \sinh t dt}{\sinh t} = C_1 dt$. Thus, we arrive to $x = C_1 t + C_2$. The parametric equations

$$\begin{cases} x(t) = C_1 t + C_2 \\ y(t) = C_1 \cosh t \end{cases} \quad (12.54)$$

Fig. 12.7 A solid of revolution

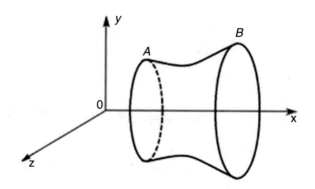

describe a *chain curve*. A chain curve that revolves about the *x*-axis generates a surface called a *catenoid* (it is not a circle, like could be naively expected).

Constants C_1 and C_2 are determined from the boundary conditions. For instance, if to take $A(0,0)$ and $B(x_1,0)$, then $x(0) = 0$ gives $C_2 = 0$, and $y(0) = 0$ gives $C_1 = 1$; thus,

$$\begin{cases} x(t) = t \\ y(t) = \cosh t. \end{cases}$$

12.4.5   The Shortest Distance Between Two Points on a Sphere

This problem reads:

Show that the shortest distance between two points on a sphere is attained when the points lie on a segment of a large circle of the sphere, i.e., on a circle whose center coincides with the center of the sphere.

The coordinates of a point on a sphere of radius a and the center at the origin are

$$x = a \sin \theta \cos \varphi, \quad y = a \sin \theta \sin \varphi, \quad z = a \cos \theta, \qquad (12.55)$$

where θ and φ are the spherical angles. The square of the differential of the arclength of a curve connecting the points is

$$dl^2 = dx^2 + dy^2 + dz^2 = a^2 d\theta^2 + a^2 \sin^2\theta \, d\varphi^2 \qquad (12.56)$$

(with $\theta = \theta(\varphi)$); thus, the length of the curve is

$$l = \int_{\varphi_1}^{\varphi_2} \sqrt{a^2 \sin^2\theta + a^2 \left(\frac{d\theta}{d\varphi}\right)^2} \, d\varphi. \qquad (12.57)$$

This is the functional $l[\theta(\varphi)] = \int_a^b F(\theta, \theta') d\varphi$, where $F(\theta, \theta')$ does not contain the variable φ. The first integral of the Euler equation is $F - \theta' F_{\theta'} = \text{const}$, which gives

$$\sqrt{a^2 \sin^2\theta + a^2 (d\theta/d\varphi)^2} - \frac{a^2 (d\theta/d\varphi)^2}{\sqrt{a^2 \sin^2\theta + a^2 (d\theta/d\varphi)^2}} = C_1.$$

After some algebra, we have:

$$C_1^2 \left(\frac{d\theta}{d\varphi}\right)^2 = a^2 \sin^4\theta - C_1^2 \sin^2\theta.$$

The solution of this first-order ODE is

$$\varphi = C_1 \int \frac{d\theta}{\sqrt{a^2 \sin^4\theta - C_1^2 \sin^2\theta}}.$$

We factorize inside the square root, as $a^2 \sin^4\theta - C_1^2 \sin^2\theta = C_1^2 \sin^4\theta(a^2/C_1^2 - \csc^2\theta)$, use the identity $\csc^2\theta = \cot^2\theta$, and introduce $\tilde{C}_1 = 1/\sqrt{(a/C_1)^2 - 1}$. Next, using the substitution

$$u = \tilde{C}_1 \cot\theta, \quad du = -\tilde{C}_1 \csc^2\theta d\theta,$$

we obtain:

$$\varphi = -\int \frac{du}{\sqrt{1-u^2}} = -\arcsin u + C_2 = -\arcsin\left(\tilde{C}_1 \cot\theta\right) + C_2,$$

or

$$\sin(\varphi - C_2) = -\tilde{C}_1 \cot\theta.$$

Finally, using the identity $\sin(\varphi - C_2) = \sin\varphi \cos C_2 - \cos\varphi \sin C_2$, we get:

$$a \sin C_2 \sin\theta \cos\varphi - a \cos C_2 \sin\theta \sin\varphi - a\tilde{C}_1 \cos\theta = 0.$$

In the Cartesian coordinates, this equation reads:

$$Ax + By + Cz = 0,$$

with $A = \sin C_2$, $B = -\cos C_2$, $C = -\tilde{C}_1$.

This is *the equation of a plane through the origin and, therefore, through the center of the sphere.* This means that the shortest distance between two points on the sphere is attained at the intersection of this plane with the surface of the sphere.

This example is the special case of determining the shortest curve on a given surface and connecting two given points. Such curves are called *geodesics*. In the general relativity, the *gravitational forces make masses move along the geodesic curves.*

Problems

1. Find a geodesic curve on the surface of a cone.
   Hint: the surface of a cone is given by equation $\sin\theta = \sin\alpha = \text{const}$; thus, $dl^2 = dr^2 + r^2\sin^2\alpha d\varphi^2$.
2. Find the curve of the shortest length connecting the points $(a, 0, 0)$ и $(0, a, h)$ on the cylinder $x^2 + y^2 = a^2$, and the shortest distance between these points on the cylinder surface.
   Hint: *in the cylindrical coordinates*, use $dl^2 = a^2 d\theta^2 + dz^2$.

12.5   Functionals That Depend on Several Functions

12.5.1   Euler Equations

Consider a functional that depends on n functions $y_1(x), y_2(x), \ldots, y_n(x)$:

$$V[y_1(x), y_2(x), \ldots, y_n(x)] = \int_{x_0}^{x_1} F\left(x, y_1, y_1', y_2, y_2', \ldots, y_n, y_n'\right) dx; \qquad (12.58)$$

the boundary conditions at the endpoints of the interval (x_0, x_1) are

$$y_1(x_0) = y_{10}, \;\; y_2(x_0) = y_{20}, \;\; \ldots, \;\; y_n(x_0) = y_{n0},$$
$$y_1(x_1) = y_{11}, \;\; y_2(x_1) = y_{21}, \;\; \ldots, \;\; y_n(x_1) = y_{n1}.$$

We wish to find functions $y_i(x)$ on which the functional has an extremum.

A simple idea allows to generalize the Euler equation to this case. Let us fix all functions except $y_1(x)$ unchanged.

The functional now depends only on $y_1(x)$; therefore, the extremal has to satisfy the equation:

$$F_{y_1} - \frac{d}{dx} F_{y_1'} = 0.$$

Repeating the above consideration for each function $y_i(x)$ ($i = 1, \ldots, n$), we arrive to a system of the second order ODEs:

$$F_{y_i} - \frac{d}{dx} F_{y_i'} = 0 \quad (i = 1, \ldots, n). \qquad (12.59)$$

A solution of these equations with a subsequent application of a boundary conditions determines the extremals $y_i(x)$.

Example 12.11 Find the extremals of the functional:

$$V[y(x), z(x)] = \int\limits_0^{\pi/2} \left[(y')^2 + (z')^2 + 2yz \right] dx,$$

$$y(0) = 0, \quad z(0) = 0, \quad y\left(\frac{\pi}{2}\right) = 1, \quad z\left(\frac{\pi}{2}\right) = -1.$$

Solution The system of the Euler equations is

$$\begin{cases} F_y - \dfrac{d}{dx}F_{y'} = 2z - 2y'' = 0 \\ F_z - \dfrac{d}{dx}F_{z'} = 2y - 2z'' = 0 \end{cases} \quad \text{or} \quad \begin{cases} z - y'' = 0 \\ y - z'' = 0. \end{cases}$$

Eliminating $z(x)$, we obtain:

$$y^{(4)} - y = 0.$$

The roots of the characteristic equation $k^4 - 1 = 0$ are $k_{1,2} = \pm i$, $k_{3,4} = \pm 1$ and the general solution is

$$\begin{cases} y(x) = C_1 e^x + C_2 e^{-x} + C_3 \cos x + C_4 \sin x, \\ z(x) = y'' = C_1 e^x + C_2 e^{-x} - C_3 \cos x - C_4 \sin x. \end{cases}$$

Using the boundary conditions, we obtain $C_1 = C_2 = C_3 = 0$, $C_4 = 1$; thus, the extremal is given by the equations:

$$y(x) = \sin x, \quad z(x) = -\sin x.$$

The value of the functional on these functions is

$$V[y(x), z(x)] = \int\limits_0^{\pi/2} \left[2\cos^2 x - 2\sin^2 x \right] dx = 2 \int\limits_0^{\pi/2} \cos 2x\, dx = 0.$$

In order to check whether the obtained extremum is a minimum, it is sufficient to show that any deviation from the extremum gives the growth of the functional. Toward this end, let denote

$$y(x) = \sin x + Y(x), \qquad z(x) = -\sin x + Z(x),$$

where

$$Y(0) = Y(\pi/2) = 0, \qquad Z(0) = Z(\pi/2) = 0.$$

Expanding $Y(x)$ and $Z(x)$ into a sine series

$$Y(x) = \sum_{n=1}^{\infty} a_n \sin 2nx, \qquad Z(x) = \sum_{n=1}^{\infty} b_n \sin 2nx,$$

and using the orthogonality of the trigonometric functions, we find that

$$V[Y,Z] = \frac{\pi}{2} \sum_{n=1}^{\infty} \left[(a_n + b_n)^2/2 + (2n^2 - 1/2)(a_n^2 + b_n^2) \right],$$

which is obviously positive. Therefore, the obtained extremum is a minimum.

Example 12.12 Find the extremals of the functional:

$$V[y(x), z(x)] = \int_0^{\pi} \left(2yz - 2y^2 + y'^2 - z'^2 \right) dx,$$

$$y(0) = 0, \quad y(\pi) = 1, \quad z(0) = 0, \quad z(\pi) = 1.$$

Solution From equations

$$\begin{cases} F_y - \dfrac{d}{dx} F_{y'} = > y'' + 2y - z = 0 \\ F_z - \dfrac{d}{dx} F_{z'} = > z'' + y = 0, \end{cases}$$

we obtain $y^{(4)} + 2y'' + y = 0$; the general solution of this equation is

$$y(x) = C_1 \cos x + C_2 \sin x + x(C_3 \cos x + C_4 \sin x).$$

The boundary conditions give $C_1 = 0$, $C_3 = -1/\pi$; thus,

$$y(x) = C_2 \sin x + C_4 x \sin x - \frac{x}{\pi} \cos x.$$

Function $z(x)$ is

$$z(x) = y'' + 2y = C_2 \sin x + C_4(2\cos x + x \sin x) + \frac{1}{\pi}(2\sin x - x\cos x).$$

The boundary conditions give $C_4 = 0$ leaving C_2 arbitrary; thus, the extremals are

$$\begin{cases} y(x) = C_2 \sin x - \dfrac{x}{\pi} \cos x, \\[2mm] z(x) = C_2 \sin x + \dfrac{1}{\pi}(2\sin x - x\cos x). \end{cases}$$

Since C_2 is arbitrary, this problem has an infinite number of solutions.

Note that in the case of the boundary condition $z(\pi) = -1$, the solution $z(x)$ above cannot be satisfied and the functional does not have an extremal.

A system of the Euler equation (12.59) can be used to find the parametrically described curves, like in the following example.

Example 12.13 In the special relativity theory, the trajectory of a free particle corresponds to the extremal of the functional:

$$l[x(t), y(t), t(t)] = \int_{t_1}^{t_2} \sqrt{c^2 - (x')^2 - (y')^2 - (z')^2}\, dt,$$

which is the invariant length of the particle's world line in the Minkowski space (c is the speed of light). Find the trajectory of the particle moving between the point $(0, 0, 0)$ (at $t = t_1$) and the final point (x_f, y_f, z_f) (at $t = t_2$).

The first Euler equation

$$l_x - \frac{d}{dx} l'_x = 0,$$

gives $\dfrac{d}{dt}\left[\dfrac{x'}{\sqrt{c^2 - (x')^2 - (y')^2 - (z')^2}} \right] = 0$; hence, $\dfrac{x'}{\sqrt{c^2 - (x')^2 - (y')^2 - (z')^2}} = k_x = const.$

Similarly, other Euler equations give:

$$\frac{y'}{\sqrt{c^2 - (x')^2 - (y')^2 - (z')^2}} = k_y = const, \quad \frac{z'}{\sqrt{c^2 - (x')^2 - (y')^2 - (z')^2}} = k_z = const.$$

Evaluating $k_x^2 + k_y^2 + k_z^2$, we find that the expression

$$v^2 = (x')^2 + (y')^2 + (z')^2,$$

which is determined by the relation

$$\frac{v^2}{c^2 - v^2} = k_x^2 + k_y^2 + k_z^2, \tag{12.60}$$

is a constant that has to be smaller than c^2.

Therefore, quantities

$$x' = v_x, \quad y' = v_y, \quad z' = v_z$$

are also constants; hence,

$$x = v_x(t - t_1), \quad y = v_y(t - t_1), \quad z = v_z(t - t_1), \tag{12.61}$$

where

$$v_x = \frac{x_f}{t_2 - t_1}, \quad v_y = \frac{y_f}{t_2 - t_1}, \quad v_z = \frac{z_f}{t_2 - t_1}.$$

We come to the conclusion that, like in the case of the classical nonrelativistic mechanics, the trajectory of the free particle is the straight line (12.61). The only difference is the restriction

$$v^2 = v_x^2 + v_y^2 + v_z^2 < c^2$$

that follows from the relation (12.60). Therefore, the problem has a solution only if

$$x_f^2 + y_f^2 + z_f^2 < c^2(t_2 - t_1)^2.$$

Problems

Find the extremals of functionals:

1. $V[y, z] = \int\limits_0^{\pi/4} \left(2z - 4y^2 + y'^2 - z'^2\right) dx$, $y(0) = 0$, $y(\pi/4) = 1$, $z(0) = 0$, $z(\pi/4) = 1$.

2. $V[y, z] = \int\limits_{-1}^{1} \left(2xy - y'^2 + \frac{z'^3}{3}\right) dx$, $y(-1) = 2$, $y(1) = 0$, $z(-1) = -1$, $z(1) = 1$.

3. $V[y, z] = \int\limits_0^{\pi/2} \left(y'^2 + z'^2 - 2yz\right) dx$, $y(0) = 0$, $y(\pi/2) = 1$, $z(0) = 0$, $z(\pi/2) = 1$.

4. $V[y, z] = \int\limits_0^1 \left(y'^2 + z'^2 + 2y\right) dx$, $y(0) = 1$, $y(1) = \frac{3}{2}$, $z(0) = 0$, $z(1) = 1$.

5. $V[y, z] = \int\limits_{1/2}^1 \left(y'^2 - 2xyz'\right) dx$, $y\left(\frac{1}{2}\right) = 6$, $y(1) = 3$, $z\left(\frac{1}{2}\right) = 15$, $z(1) = 1$..

6. $V[y,z] = \int\limits_0^{\pi/2} \left(y'z + yz' + yz - \frac{1}{2}y'^2 - \frac{1}{2}z'^2 \right) dx$, $y(0) = 0$, $y(\pi/2) = 1$, $z(0) = 0$,

   $z(\pi/2) = 1$.

7. $V[y,z] = \int\limits_0^1 \left(x^2 + 2yz + y'^2 + z'^2 + z' \right) dx$,

   $y(0) = 1, y(1) = \frac{1}{e}, z(0) = 1, z(1) = \frac{1}{e}$.

8. $V[y,z] = \int\limits_0^1 [y'(z-y') + z'(y-z'+1) - 2yz] dx$,

   $y(0) = 1, y(1) = \frac{1}{e}, z(0) = 1, z(1) = \frac{1}{e}$.

9. $V[y,z] = \int\limits_0^{\pi/2} \left(x^2 + xy' + y + 4yz + 2y'^2 + 2z'^2 \right) dx$,

   $y(0) = 0, y\left(\frac{\pi}{2}\right) = 1, z(0) = 0, z\left(\frac{\pi}{2}\right) = -1$.

10. $V[y,z] = \int\limits_0^{\pi/2} \left[z' - 2x^2 + yz + \frac{1}{2}\left(y'^2 + z'^2 \right) \right] dx$,

    $y(0) = 1, y\left(\frac{\pi}{2}\right) = 0, z(0) = -1, z\left(\frac{\pi}{2}\right) = 0$.

11. $V[y,z] = \int\limits_0^{\pi/2} \left[1 + y'z + yz' - 2yz - y'^2 - z'^2 \right] dx$,

    $y(0) = 0, y\left(\frac{\pi}{2}\right) = 1, z(0) = 0, z\left(\frac{\pi}{2}\right) = -1$.

12. $V[y,z] = \int\limits_0^1 \left[y + xz' + z + \frac{1}{2}\left(y'^2 + z'^2 \right) \right] dx$,

    $y(0) = 0, y(1) = \frac{1}{2}, z(0) = 0, z(1) = -1$.

13. $V[y,z] = \int\limits_0^1 \left(x + 3y' - 2z' + y'^2 + 4z + z'^2 \right) dx$, $y(0) = 1$, $y(1) = 0$, $z(0) = 0$,

    $z(1) = 2$.

14. $V[y,z] = \int\limits_{-1}^1 \left(xy' + 3y + y'^2 + z'^2 - z' \right) dx$,

    $y(-1) = \frac{3}{2}, y(1) = \frac{3}{2}, z(-1) = 2, z(1) = 0$.

15. $V[y,z] = \int\limits_1^2 \left(y'^2 + z^2 + z'^2 \right) dx$, $y(1) = 1$, $y(2) = 2$, $z(1) = 0$, $z(2) = 1$.

12.5.2 Application: The Principle of the Least Action

The primary variational principle of the classical mechanics is *the principle of the least action*, which states that an object under the action of forces is moving along the trajectory on which the integral

$$I = \int\limits_{t_1}^{t_2} L dt, \tag{12.62}$$

which is called the *action*, has a *minimal* value.

Function L is called the *Lagrange function*, or the *Lagrangian*, and t_1 and t_2 are the initial and the final time moments. In most cases L depends on the time, the generalized coordinates and speeds, $L(t, q_i(t), \dot{q}_i(t))$. The independent variable in (12.62) is t instead of x in previous sections; also a dot is used as differentiation sign.

The functional I has the minimum (the least) value on the trajectories $q_i(t)$, obtained from the Euler-Lagrange equations:

$$\frac{\partial L}{\partial q_i} - \frac{d}{dt} \frac{\partial L}{\partial \dot{q}_i} = 0. \tag{12.63}$$

The number of these equations equals the number of the degrees of freedom, $i = 1, \ldots, N$. This system of equations along with the boundary conditions at t_1 and t_2, determines the actual trajectories $q_i(t)$.

In mechanics, $L = K - U$, where K and U are the kinetic and the potential energy of a system, respectively.

Example 12.14 Consider a mass m moving in the potential field $U(x, y, z)$. The system of Eq. (12.61) for the action $I = \int\limits_{t_1}^{t_2} (K - U) dt$ is

$$-\frac{\partial U}{\partial x} - \frac{d}{dt} \frac{\partial K}{\partial \dot{x}} = 0, \qquad -\frac{\partial U}{\partial y} - \frac{d}{dt} \frac{\partial K}{\partial \dot{y}} = 0, \qquad -\frac{\partial U}{\partial z} - \frac{d}{dt} \frac{\partial K}{\partial \dot{z}} = 0$$

and with the kinetic energy $K = \frac{m}{2} \left(\dot{x}^2 + \dot{y}^2 + \dot{z}^2 \right)$, it gives:

$$\frac{\partial U}{\partial x} + m\ddot{x} = 0, \qquad \frac{\partial U}{\partial y} + m\ddot{y} = 0, \qquad \frac{\partial U}{\partial x} + m\ddot{z} = 0.$$

Since $\vec{F} = -\nabla U$ (∇ is the gradient operator), these equations are equivalent to Newton's equation, $\vec{F} = m\vec{a}$.

Example 12.15 Consider the motion of an object under the action of the central force of attraction. Below we shall consider the potential $U = -\frac{a}{r}$, which is appropriate for two different physical problem, the electrostatic Coulomb interaction of charged particles and gravitational interaction of masses. Note that the interaction force $F = -\frac{dU}{dr}$ is inversely proportional to r^2. In the case of a Coulomb attraction of two point charges of the opposite sign, Q and q, $\alpha = k|qQ|$, where k is the Coulomb constant; in the case of a gravitational attraction of masses m and M, $\alpha = GmM$, G is the gravitational constant.

As generalized coordinates, we use the polar coordinates r and φ, which are related with Cartesian coordinates as follows:

$$x = r\cos\varphi, \quad y = r\sin\varphi,$$

and the kinetic energy is

$$K = \frac{m}{2}\left(\dot{x}^2 + \dot{y}^2\right) = \frac{m}{2}\left(\dot{r}^2 + r^2\dot{\varphi}^2\right).$$

Here dots are used to denote time derivatives. Then, for $L = K - U$ Eq. (12.63) give:

$$\frac{d}{dt}\frac{\partial L}{\partial \dot{r}} - \frac{\partial L}{\partial r} = \frac{d}{dt}(m\dot{r}) - mr\dot{\varphi}^2 + \frac{\alpha}{r^2} = 0 \tag{12.64}$$

and

$$\frac{d}{dt}\frac{\partial L}{\partial \dot{\varphi}} - \frac{\partial L}{\partial \varphi} = \frac{d}{dt}(mr^2\dot{\varphi}) = 0.$$

The last equation means that $mr^2\dot{\varphi} = \mathrm{const}$; this is the object's angular momentum, l, which, as we see, is a constant of motion. Then, Eq. (12.64) reads:

$$m\frac{d^2r}{dt^2} - \frac{l^2}{mr^3} + \frac{\alpha}{r^2} = 0. \tag{12.65}$$

Solution of Eq. (12.65) defines the object's trajectory, $r(t)$.
For a circular motion, $\dot{r} = 0$ and with $l = mr^2\dot{\varphi}$, we have:

$$\dot{\varphi}^2 = \frac{\alpha}{mr^3}.$$

Thus, the object's energy is

$$E = \frac{1}{2}mr^2\dot{\varphi}^2 - \frac{\alpha}{r} = -\frac{1}{2}\frac{\alpha}{r} = \frac{1}{2}U.$$

The orbital frequency is a constant:

$$\omega = \dot{\varphi} = \frac{l}{mr^2}.$$ (12.66)

As is mentioned in Sect. 12.5.2, the principle of least action is the basic principle in many areas of physics. For instance, the equations of the electromagnetic field (the Maxwell equations) can be derived by a minimization of the functional that depends on the fields of the electric potential $\varphi\left(\vec{r}, t\right)$ and the magnetic potential $\vec{A}\left(\vec{r}, t\right)$; the combinations of their derivatives form the electric field $\vec{E}\left(\vec{r}, t\right)$ and magnetic field $\vec{B}\left(\vec{r}, t\right)$, where $\vec{r} = (x, y, z)$ is a vector of the coordinates and t is the time. The action functional is the integral over all independent variables x, y, z and t.

Reading Exercise Generalize Example 12.16 to a situation, where the potential energy U depends both on r and θ. Is the angular momentum conserved in this situation?

12.6   Functionals Containing Higher-Order Derivatives

Consider a functional:

$$V[y(x)] = \int_{x_0}^{x_1} F\left[x, y(x), y'(x), y''(x), \ldots, y^{(n)}(x)\right] dx,$$ (12.67)

where a function $y(x)$ is subject to the following boundary conditions:

$$y(x_0) = y_0, \quad y'(x_0) = y'_0, \quad \ldots, y^{(n-1)}(x_0) = y_0^{(n-1)},$$
$$y(x_1) = y_1, \quad y'(x_1) = y'_1, \quad \ldots, y^{(n-1)}(x_1) = y_1^{(n-1)}.$$

Similar to the approach in Sect. 12.3, consider a family of functions

$$y(x, \alpha) = y(x) + \alpha \delta y,$$

which for $\alpha = 0$ contains the extremal $y(x)$.

For the functional $V[y(x, \alpha)]$ considered as a function of $\alpha\alpha$

$$V[y(x, \alpha)] \equiv \varphi(\alpha),$$

the necessary condition for the extremum is

$$\delta V = \varphi'(0) = 0.$$

For

$$\varphi(\alpha) = \int_{x_0}^{x_1} F\big[x, y(x, \alpha), y_x'(x, \alpha), y_x''(x, \alpha), \ldots\big] dx$$

assuming that the partial derivatives of the necessary order of function F exist, we have:

$$\varphi'(\alpha) = \frac{\partial}{\partial \alpha} \int_{x_0}^{x_1} F dx = \int_{x_0}^{x_1} \left[\frac{\partial F}{\partial y} \frac{\partial}{\partial \alpha} y(x, \alpha) + \frac{\partial F}{\partial y'} \frac{\partial}{\partial \alpha} y'(x, \alpha) + \frac{\partial F}{\partial y''} \frac{\partial}{\partial \alpha} y''(x, \alpha) + \ldots \right] dx.$$

With

$$\frac{\partial}{\partial \alpha} y(x, \alpha) = \frac{\partial}{\partial \alpha}[y(x) + \alpha \delta y(x)] = \delta y(x), \quad \ldots,$$
$$\frac{\partial}{\partial \alpha} y^{(n)}(x, \alpha) = \frac{\partial}{\partial \alpha}\left[y^{(n)}(x) + \alpha \delta y^{(n)}(x)\right] = \delta y^{(n)}(x)$$

we obtain:

$$\varphi'(0) = \int_{x_0}^{x_1} \big[F_y \delta y + F_{y'} \delta y' + F_{y''} \delta y'' + \ldots\big] dx = 0.$$

Applying the integration by parts and using $\delta y' = (\delta y)'$, we have:

$$\int_{x_0}^{x_1} F_{y'} (\delta y)' dx = F_{y'} \delta y \big|_{x_0}^{x_1} - \int_{x_0}^{x_1} \frac{dF_{y'}}{dx} \delta y dy.$$

The third term in the integral is integrated by parts two times and using $\delta'' y = (\delta y)''$, we obtain:

$$\int_{x_0}^{x_1} F_{y''}(\delta y)'' dx = F_{y''}(\delta y)' \big|_{x_0}^{x_1} - \frac{dF_{y''}}{dx} \delta y \bigg|_{x_0}^{x_1} + \int_{x_0}^{x_1} \frac{d}{dx^2} F_{y''} \delta y dx.$$

Repeating integration by parts with the remaining terms in $\varphi'(0)$ and taking into account that variations $\delta y = \delta' y = \delta y'' = \ldots = 0$ at $x = x_0$ and $x = x_1$, we find that the necessary condition for function $y(x)$ to be the extremal is that it must be a solution of equation:

$$F_y - \frac{d}{dx}F_{y'} - \frac{d^2}{dx^2}F_{y''} + \ldots + (-1)^n \frac{d^n}{dx^n}F_{y^{(n)}} = 0. \tag{12.68}$$

This is the differential equation of order $2n$. Its general solution contains $2n$ arbitrary constants, which should be determined from the boundary conditions (12.67).

For functionals that depend on several functions, such as

$$V[y(x), z(x)] = \int\limits_{x_0}^{x_1} F\left[x, y(x), y'(x), \ldots, y^{(n)}(x), z(x), z'(x), \ldots, z^{(n)}(x)\right] dx,$$

$$\tag{12.69}$$

with the boundary conditions for a function $z(x)$ similar to (12.67), we have the additional equation:

$$F_z - \frac{d}{dx}F_{z'} + \frac{d^2}{dx^2}F_{z''} + \ldots + (-1)^n \frac{d^n}{dx^n}F_{z^{(n)}} = 0. \tag{12.70}$$

Example 12.16 Find the extremals of the functional:

$$V[y(x)] = \int\limits_0^1 \left[1 + (y'')^2\right] dx, \quad y(0) = 0, \quad y'(0) = 1, \quad y(1) = 1, \quad y'(1) = 1.$$

Solution Equation (12.68) gives:

$$\frac{d^2}{dx^2}(2y'') = 0, \quad \text{or} \quad y^{(4)} = 0.$$

Its general solution is $y = C_1 x^3 + C_2 x^2 + C_3 x + C_4$. Using the boundary conditions, we obtain $C_1 = C_2 = C_3 = 0$, $C_4 = 1$. Thus, the extremum may be attained on the straight line $y = x$.

Let us define $y = x + Y$, where

$$Y(0) = Y'(0) = Y(1) = Y'(1) = 0 \tag{12.71}$$

and rewrite V as

$$V[Y(x)] = 1 + \int_0^1 [Y''(x)]^2 dx.$$

Obviously, $V[Y(x)] - 1 \geq 0$. The only function Y satisfying conditions (12.71) that gives $V[Y(x)] = 1$ is $Y(x) = 0$. Thus, the found extremum is a minimum.

Problems

Find the extremals:

1. $V[y] = \int_{-1}^{1} \left[x - x^2 + y^2 y' + (y'')^2 \right] dx$, $y(0) = y'(0) = 0$, $y(1) = 2$, $y'(1) = 5$.

2. $V[y] = \int_{-1}^{1} \left[(1 + x^2)^2 + (1 + y)y' + \frac{1}{2}(y'')^2 \right] dx$, $y(0) = 0$, $y'(0) = 1$, $y(1) = 1$, $y'(1) = 1$.

3. $V[y] = \int_{-1}^{1} \left[x^3 - x + (y^2 + 2y'')y' - 3(y'')^2 \right] dx$, $y(0) = 1$, $y'(0) = 0$, $y(1) = 0$, $y'(1) = -2$.

4. $V[y] = \int_0^2 [x^4 + 1 + y''(2y' - y'')] dx$, $y(0) = 1$, $y'(0) = 0$, $y(2) = 15$, $y'(2) = 17$.

5. $V[y] = \int_{-1}^{0} \left(240y + y'''^2 \right) dx$, $y(-1) = 1$, $y(0) = 0$, $y'(-1) = -4.5$, $y'(0) = 0$, $y''(-1) = 16$, $y''(0) = 0$.

6. $V[y] = \int_{x_0}^{x_1} (y + y'') dx$, $y(x_0) = y_0, y(x_1) = y_1, y'(x_0) = y_0', y'(x_1) = y_1'$.

7. $V[y] = \int_{x_0}^{x_1} \left(y'^2 + yy'' \right) dx$, $y(x_0) = y_0, y(x_1) = y_1, y'(x_0) = y_0', y'(x_1) = y_1'$.

8. $V[y] = \int_0^1 \left(y'^2 + y''^2 \right) dx$, $y(0) = 0$, $y(1) = \sinh 1$, $y'(0) = 1$, $y'(1) = \cosh 1_1$.

9. $V[y] = \int_{x_0}^{x_1} \left(y''^2 - 2y'^2 + y^2 - 2y \sin x \right) dx$.

10. $V[y] = \int_0^1 \left[x^2 + 12xy + (y'')^2 \right] dx$, $y(0) = 0, y'(0) = 0$, $y(1) = \frac{1}{20}$, $y'(1) = -\frac{3}{4}$.

11. $V[y] = \int_0^1 \left[\sin^2 x + 48xy - 16y^2 y' - (y'')^2 \right] dx$, $y(0) = 0, y'(0) = 0$, $y(1) = \frac{1}{5}$, $y'(1) = 0$.

12. $V[y] = \int_0^1 \left[x^2 + y'^2 + (y'')^2 \right] dx$, $y(0) = 1$, $y'(0) = 1$, $y(1) = e$, $y'(1) = e$.

13. $V[y] = \int_0^{\pi} \left[x^2 + y^2 y' + 4 \left(y' + y'^2 - (y'')^2 \right) \right] dx$, $y(0) = 0$, $y'(0) = 2$, $y(\pi) = 0$, $y'(\pi) = -2..$

14. $V[y] = \int\limits_{0}^{\pi/4} \left[x^2 - x^4 + 2y^3 y' + (y'')^2 - 4y'^2 \right] dx$, $y(0) = 0$, $y'(0) = 2$, $y(\pi/4) = 1$,
    $y'(\pi/4) = 0$.

15. $V[y] = \int\limits_{-1}^{0} \left[240y - (y''')^2 \right] dx$, $y(-1) = -1$, $y'(-1) = 4.5$, $y''(-1) = 16$,
    $y(0) = 0$, $y'(0) = 0$, $y''(0) = 0$.

16. $V[y(x)] = \int\limits_{0}^{1} \left(360x^2 y - y''^2 \right) dx$, $y(0) = 0$, $y'(0) = 1$, $y(1) = 0$, $y'(1) = 2.5$.

The material contained in the following sections can be used for deeper study of the Calculus of variation topics.

12.7   Moving Boundaries

Considering the functional

$$V[y] = \int\limits_{x_0}^{x_1} F(x, y, y') dx, \qquad (12.72)$$

previously we assumed that the endpoints $A(x_0, y_0)$ and $B(x_1, y_1)$ are fixed. Suppose that one or both endpoints, $A(x_0, y_0)$, $B(x_1, y_1)$, of the extremals $y(x)$ can move along the given smooth curves

$$y = \varphi(x) \quad \text{and} \quad y = \psi(x); \qquad (12.73)$$

thus, $y_0 = \varphi(x_0)$ and $y_1 = \psi(x_1)$. The problem is now formulated as follows: among all curves $y(x)$ which connect the curves (12.73), find the extremal of $V[y]$. The example of such problem is finding the shortest distance between two lines.

Let the curve $y = y(x)$ connecting two arbitrary points on the curves $y = \varphi(x)$ and $y = \psi(x)$ be the extremal. First, it is clear that if the extremum is attained on a curve when the endpoints can move, then it is also attained on a curve with fixed endpoints.

Therefore, the necessary condition for a function $y(x)$ to be an extremal is the same as in the case of fixed endpoints, that is, this condition is the Euler equation:

$$F_y - \frac{d}{dx} F_{y'} = 0. \qquad (12.74)$$

Let the coordinates of the endpoints of the curve $y(x)$ be (x_0, y_0) and (x_1, y_1) the coordinates of the endpoints of the curve $y + \delta y$ be $(x_0 + \delta x_0, y_0 + \delta y_0)$ and $(x_1 + \delta x_1, y_1 + \delta y_1)$. All functions and their first derivatives are assumed to be smooth. For small δx_1 and δx_0 (see Fig. 12.8)

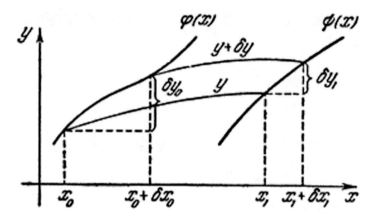

Fig. 12.8 Two close curves with moving right endpoint

$$\delta y_1 = \psi' \delta x_0, \quad \delta y_0 = \varphi' \delta x_0.$$

The increment of the functional is

$$\Delta V = \int_{x_0+\delta x_0}^{x_1+\delta x_1} F(x, y + \delta y, y' + \delta y')dx - \int_{x_0}^{x_1} F(x, y, y')dx =$$

$$= \int_{x_0}^{x_1} [F(x, y + \delta y, y' + \delta y') - F(x, y, y')]dx +$$

$$+ \int_{x_1}^{x_1+\delta x_1} F(x, y + \delta y, y' + \delta y')dx - \int_{x_0}^{x_0+\delta x_0} F(x, y + \delta y, y' + \delta y')dx.$$

Keeping the first-order terms in Taylor's expansion, we obtain the variation:

$$\delta V = \int_{x_0}^{x_1} \left[F_y - \frac{d}{dx} F_{y'} \right] \delta y \, dx + F_{y'} \big|_{x=x_1} \delta y_1 + \left(F - y' F_{y'} \right) \big|_{x=x_1} \delta x_1 -$$

$$- F_{y'} \big|_{x=x_0} \delta y_0 - \left(F - y' F_{y'} \right) \big|_{x=x_0} \delta x_0. \tag{12.75}$$

With (12.74) we have:

$$\delta V = F_{y'}\big|_{x=x_1}\delta y_1 + \left(F - y'F_{y'}\right)\big|_{x=x_1}\delta x_1 - F_{y'}\big|_{x=x_0}\delta y_0 - \left(F - y'F_{y'}\right)\big|_{x=x_0}\delta x_0$$

$$(12.76)$$

or

$$\delta V = \left(F_{y'}\psi' + F - y'F_{y'}\right)\big|_{x=x_1}\delta x_1 - \left(F_{y'}\varphi' + F - y'F_{y'}\right)\big|_{x=x_0}\delta x_0 = 0.$$

Because δx_0 and δx_1 are independent increments, we see that

$$\left[F + (\varphi' - y')F_{y'}\right]_{x=x_0} = 0, \quad \left[F + (\psi' - y')F_{y'}\right]_{x=x_1} = 0. \qquad (12.77)$$

The transversality conditions (12.77) give the relations between the tangents to the extremal and the curve $y = \varphi(x)$ at the endpoint A, and the tangents to the extremal and the curve $y = \psi(x)$ at the endpoint B. The curve $y = y(x)$ is *transversal* to the curves $\varphi(x)$ and $\psi(x)$.

When the endpoints are fixed, $\delta x_0 = \delta x_1 = 0$ and $\delta y_0 = \delta y_1 = 0$, we return to the standard Euler equation $F_y - \frac{d}{dx}F_{y'} = 0$ for the extremal.

To find four parameters: the integration constants of the Euler equation, C_1, C_2 and x_0, x_1 (the x-coordinates of the endpoints A and B), we have two conditions (12.77) and two conditions $y(x_0) = \varphi(x_0)$ and $y(x_0) = \psi(x_0)$. As always, the extremal may not exist.

When only one of the endpoints moves, let us say the endpoint (x_1, y_1) moves along the curve $y_1 = \psi(x_1)$ and the end $y(x_0) = y_0$ is fixed, we have only the second transversality relation (12.77).

In particular, if the point (x_1, y_1) moves along a horizontal line $y = y_1$ ($\psi' = 0$), then

$$\left[F + y'F_{y'}\right]_{x=x_1} = 0. \qquad (12.78)$$

If the point (x_1, y_1) moves along a vertical line $x = x_1$, i.e., $\psi' = \infty$ (a *free end*), then we have:

$$F_{y'}\big|_{x=x_1} = 0. \qquad (12.79)$$

In applications functionals often have the form:

$$\int_{x_0}^{x_1} f(x, y)\sqrt{1 + y'^2}\,dx. \qquad (12.80)$$

In this case

$$F_{y'} = f(x, y) \frac{y'}{\sqrt{1 + y'^2}} = \frac{Fy'}{1 + y'^2},$$

and the transversality relation becomes:

$$F + F_{y'}(\psi' - y') = \frac{F(1 + y'\psi')}{1 + y'^2} = 0, \qquad (12.81)$$

or

$$y' = -1/\psi' \text{ in the point } x = x_1.$$

Similarly, when both endpoints move

$$y' = -1/\varphi' \text{ in the point } x = x_0.$$

Thus, for functionals (12.80) the transversality relations lead to the orthogonality of the extremal and the curves $y = \varphi(x)$ and $y = \psi(x)$.

In the case of functionals that depends on several functions, for example,

$$\int_{x_0}^{x_1} F(x, y, z, y', z')dx, \qquad (12.82)$$

the problem can be formulated as follows: among all possible curves whose endpoints lie on two fixed surfaces $x = \varphi(y, z)$ and $x = \psi(y, z)$, find functions $y(x)$ и $z(x)$ that provide the extremum to the functional (6). For this problem, the Euler equations

$$F_y - \frac{d}{dx}F_{y'} = 0, \quad F_z - \frac{d}{dx}F_{z'} = 0,$$

are considered along with the transversality equations:

$$
\begin{aligned}
F_{y'} + \frac{\partial \varphi}{\partial y}(F - y'F_{y'} - z'F_{z'})\big|_{x=x_0} &= 0, \\
F_{z'} + \frac{\partial \varphi}{\partial y}(F - y'F_{y'} - z'F_{z'})\big|_{x=x_0} &= 0, \\
F_{y'} + \frac{\partial \psi}{\partial y}(F - y'F_{y'} - z'F_{z'})\big|_{x=x_1} &= 0, \\
F_{z'} + \frac{\partial \psi}{\partial y}(F - y'F_{y'} - z'F_{z'})\big|_{x=x_1} &= 0.
\end{aligned}
\qquad (12.83)
$$

Example 12.17 Find the extremal of the functional:

$$V[y(x)] = \int_0^1 y'(y' - x)dx,$$

when the endpoints are moving along the vertical lines $x_0 = 0$, $x_1 = 1$.

Solution The solution of the Euler equation is

$$y = \frac{x^2}{4} + C_1 x + C_2.$$

The transversality conditions are

$$F_{y'}\big|_{x=x_0} = 2y'(0) = 2C_1 = 0 \text{ and } F_{y'}\big|_{x=x_1} = 2y'(1) - 1 = 2C_1 = 0.$$

Thus, the extremal is $y(x) = \frac{x^2}{4} + C$.

Example 12.18 Find the extremum of the functional:

$$V[y(x)] = \int_0^{x_1} \frac{\sqrt{1 + (y')^2}}{y} dx,$$

with the fixed left endpoint $y(0) = 0$ and the right endpoint moving along $y_1 = x_1 - 5$.

Solution The integral curves of the Euler equation are the circles:

$$(x - C_1)^2 + y^2 = C_2^2.$$

The boundary condition $y(0) = 0$ gives $C_1 = C_2$. The functional is of the form (12.80); thus, the second condition (12.77) gives $y' = -1/\psi'$, meaning that the extremal $y = y(x)$ and $y_1 = \psi(x_1)$ are mutually orthogonal. Thus, $y_1 = x_1 - 5$ is the diameter of the circle with the center at the point $(0, 5)$ – the intersection of the line $y_1 = x_1 - 5$ and the x-axis. Thus, the extremal curve is the circle $(x - 5)^2 + y^2 = 25$ (Fig. 12.9).

Example 12.19 Find the shortest distance between the parabola $y = x^2$ and the straight line $y = x - 5$.

Solution We have to find the *extremal (minimal) value* of the functional:

Fig. 12.9 Graphs of functions $y = \sqrt{x(10-x)}$ and $y = x - 5$

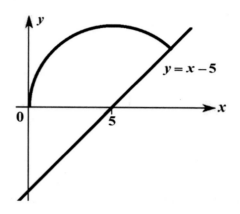

$$V[y] = \int_{x_0}^{x_1} \sqrt{1 + (y')^2}\, dx$$

under the condition that the left endpoint of the extremal can move along the curve $y = x^2$ and the right endpoint along the line $y = x - 5$. Thus, $\varphi(x) = x^2$, $\psi(x) = x - 5$, and at the endpoints $y(x_0) = \varphi(x_0)$, $y(x_1) = \psi(x_1)$.

The general solution of the Euler equation, $y'' = 0$, is $y = C_1 x + C_2$.
The condition (12.81) at the endpoints (x_0, y_0) and (x_1, y_1) reads

$$\sqrt{1 + y'^2(x_0)} + [\varphi'(x_0) - y'(x_0)] \frac{y'(x_0)}{\sqrt{1 + y'^2(x_0)}} = 0,$$

$$\sqrt{1 + y'^2(x_1)} + [\psi'(x_1) - y'(x_1)] \frac{y'(x_1)}{\sqrt{1 + y'^2(x_1)}} = 0,$$

where $y'(x_0) = C_1$, $y'(x_1) = C_1$. With $C_1 x_0 + C_2 = x_0^2$, $C_1 x_1 + C_2 = x_1 - 5$ and $\varphi'(x_0) = 2C_1$, $\psi'(x_1) = 1$, these equations determine C_1, C_2, x_0, x_1:

$$C_1 = -1, C_2 = 3/4, x_0 = 1/2, x_1 = 23/8.$$

Therefore, the extremal is $y = \frac{3}{4} - x$, and the distance between the curves is

$$l = \int_{1/2}^{23/8} \sqrt{1 + (-1)^2}\, dx = \frac{19\sqrt{2}}{8}.$$

Problems

Find the extremals of the functionals:

1. $V[y(x)] = \int\limits_{0}^{x_1} \frac{\sqrt{1+y'^2}}{x-2} dx$, the left end of the curve $y(x)$ is fixed at the point $(0,0)$, and the right end can move along the line $y = -4x+4$.

2. $V[y] = \int\limits_{0}^{\pi/4} \left(y'^2 - y^2\right) dx$, the fixed left end, $y(0) = 1$, and the right end can move along the line $x = \pi/4$.

3. $V[y(x)] = \int\limits_{-1}^{1} \left[y'^2(x) + 4y^2(x) - 8xy(x) + 2x^2\right] dx$, the fixed left end, $y(-1) = 3$, and the right end can move along the line $x = 1$.

4. $V[y] = \int\limits_{1}^{2} \left[x^3y'^2 - 8(x^2 - x)yy' + 4y^2 + 8x^2y'\right] dx$, $y(2) = -7$, and the left end can move along the line $x = 1$.

5. $V[y] = \int\limits_{0}^{\pi/4} \left(y'^2 - y^2 + 4y\cos x\right) dx$, $y(0) = 0$, and the right end can move along the line $x = \pi/4$.

6. $V[y] = \int\limits_{1}^{3} \left(8yy' \ln x - xy'^2 + 6xy'\right) dx$, $y(1) = 1$, and the right end can move along the line $x = 3$.

7. $V[y] = \int\limits_{0}^{x_1} \left(-2y + y'^2\right) dx$, $y(0) = 0$ and $y(x_1) = -2$.

8. $V[y] = \int\limits_{0}^{x_1} \left(yy' + y'^2 + 1\right) dx$, $y(0) = 1$ and $y(x_1) = -3$.

9. $V[y] = \int\limits_{0}^{x_1} \sqrt{1 + y'^2} dx$, $y(0) = 0$ and $y(x_1) = x_1 - 1$.

10. $V[y] = \int\limits_{0}^{x_1} y'^2 dx$, $y(0) = 0$ and $y(x_1) = x_1 - 1$.

In the following problems, find the shortest distance:

11. From the point $A(5; -1)$ and the parabola $y = x^2$
12. From the circle $x^2 + y^2 = 1$ and the line $y = 4 - x$
13. From the point $A(1; 0)$ and the ellipse $4x^2 + 9y^2 = 36$
14. From the point $A(-1; 5)$ and the parabola $x = y^2$
15. From the point $A(-1; 3)$ and the line $y = 1 - 3x$
16. Find the shortest distance from the point $O(0; 0)$ and the curve $x^2y = 1$
17. From the point $M(0; 0; 3)$ and the surface $z = x^2 + y^2$

12.8   Conditional Extremum: Isoperimetric Problems

Recall that in order to find an extremum of a *function* $z = f(x_1, x_2, \ldots, x_n)$ when there are additional *binding conditions* $\varphi_i(x_1, x_2, \ldots, x_n) = 0$ $(i = 1, 2, \ldots, m)$, $m < n$, the method of *Lagrange multipliers* is used.

This method is for finding the extremum of an auxiliary function:

$$z^* = f + \sum_{i=1}^{m} \lambda_i \varphi_i.$$

The necessary condition for extremum of a function z^* is determined from the system of equations:

$$\frac{\partial z^*}{\partial x_j} = 0 \quad (j = 1, 2, \ldots, n),$$

considered together with the binding relations $\varphi_i = 0$. From this, n unknowns $x_1, x_2, \ldots, x_n$ and the values of m parameters, $\lambda_1, \lambda_2, \ldots, \lambda_m$, can be found.

Similarly, if a functional and the boundary conditions

$$V[y_1, y_2, \ldots y_n] = \int_{x_0}^{x_1} F\left(x, y_1, y_2, \ldots y_n, y_1', y_2', \ldots y_n'\right) dx, \tag{12.84}$$

$$y_k(x_0) = y_{k0}, \quad y_k(x_1) = y_{k1} \quad (k = 1, 2, \ldots, n),$$

are accompanied by the binding conditions on the functions y_i,

$$\varphi_i(x, y_1, y_2, \ldots, y_n) = 0 \quad (i = 1, 2, \ldots, m < n), \tag{12.85}$$

an auxiliary functional is

$$V^* = \int_{x_0}^{x_1} \left(F + \sum_{i=1}^{m} \lambda_i(x)\varphi_i\right) dx \quad \text{or} \quad V^* = \int_{x_0}^{x_1} F^* dx, \tag{12.86}$$

where $F^* = F + \sum_{i=1}^{m} \lambda_i(x)\varphi_i$ and λ_i are the Lagrange multipliers. To find the extremum of the functional V^*, we should solve the Euler equations:

$$F_{y_j}^* - \frac{d}{dx} F_{y_j'}^* = 0 \quad (j = 1, 2, \ldots, n) \tag{12.87}$$

and then take into account the binding conditions (12.85).

Notice that the boundary conditions $y_j(x_0) = y_{j0}$, $y_j(x_1) = y_{j1}$ must not contradict the binding constraints. Also notice that the binding constraints in some situations may include the derivatives y'_j. (Such constraints are called non-holonomic constraints.)

A substantial and important class of variational problems with a conditional extremum are the *isoperimetric problems*.

The *isoperimetric conditions* are formulated as

$$\int_{x_0}^{x_1} \varphi_i\left(x, y_1, y_2, \ldots, y_n, y'_1, y'_2, \ldots, y'_n\right) dx = l_i, \quad l_i \text{ are constants}, \quad i = 1, 2, \ldots, m.$$

It can be shown that the method of Lagrange multipliers for isoperimetric problems can be reduced to the following:

$$V^* = \int_{x_0}^{x_1} \left(F + \sum_{i=1}^{m} \lambda_i \varphi_i \right) dx, \quad \text{or} \quad V^* = \int_{x_0}^{x_1} F^* dx; \lambda_i \text{ are constants}. \tag{12.88}$$

Let us present several examples.

Example 12.20 Find the extremals of the functional:

$$V[y] = \int_0^1 \left[(y')^2 + y^2 \right] dx, \text{ if } y(0) = 0, y(1) = e, \text{ and } \int_0^1 ye^x dx = \frac{e^2 + 1}{4}.$$

Solution For the Lagrange function $F^* = y'^2 + y^2 + \lambda y e^x$, the Euler equation is

$$2(-y'' + y) + \lambda e^x = 0, \text{ or } y'' - y = \lambda e^x / 2.$$

Its general solution is $y = C_1 e^x + C_2 e^{-x} + C_3 x e^x$. The boundary conditions give $C_2 = -C_1$ and $C_3 = 1 + C_1(1 - e^2)$. After that, the extremal equation contains only one constant, C_1. Substituting this extremal into the binding condition, one can obtain $C_1 = 0$; thus, the extremal is $y = xe^x$.

Example 12.21 Find the extremals of the functional:

$$V[y] = \int_0^1 (y')^2 dx, \text{ if } y(0) = 0, y(1) = 0, \text{ and } \int_0^1 y^2 dx = 2$$

Solution For the Lagrange function $F^* = y'^2 + \lambda y^2$, the general solution of the Euler equation, $y'' - \lambda y = 0$, depends on the sign of λ:

(a) $\lambda > 0 \Rightarrow y = C_1 e^{\sqrt{\lambda}x} + C_2 e^{-\sqrt{\lambda}x}$
(б) $\lambda = 0 \Rightarrow y = C_1 x + C_2$
(в) $\lambda < 0 \Rightarrow y = C_1 \sin\left(\sqrt{-\lambda}x\right) + C_2 \cos\left(\sqrt{-\lambda}x\right)$

For $\lambda > 0$ and $\lambda = 0$, the solution of the boundary value problem does not exist. For $\lambda < 0$ the extremals are $y = C_1 \sin(\pi n x)$, $n = \pm 1, \pm 2, \ldots$, and from the binding condition, one can obtain that $C_1 = 2$.

Thus, this problem has an *infinite number of extremals (solutions)*.

Note that the values of $V[y]$ are different at different extremals: only the solutions with $n = 1$ give a global minimum of the functional.

Shape of a Cable Hanged at the Ends

Consider a homogeneous cable of the length $2l$ and the linear density ρ, whose ends are hanged at the same height above the ground. If we take $y(-a) = y(a) = 0$, the potential energy of the cable, $U = \int \rho g y ds$ is

$$U = \rho g \int_{-a}^{a} y\sqrt{1 + y'^2}dx. \tag{12.89}$$

The cable length is

$$L = \int_{-a}^{a} \sqrt{1 + y'^2}dx = 2l. \tag{12.90}$$

The function $F^* = F + \lambda \varphi$ is $F^* = \rho g y \sqrt{1 + y'^2} + \lambda \sqrt{1 + y'^2}$. The first integral of the Euler equation is

$$\rho g y - \lambda = C_1 \sqrt{1 + y'^2}.$$

Integrating, we have:

$$x = C_1 \int \frac{dy}{\sqrt{(\rho g y - \lambda)^2 - C_1^2}} + C_2.$$

Substitution $\rho g y - \lambda = C_1 \cosh z$ gives:

$$x = \int \frac{C_1 \sinh z dz}{\rho g \sqrt{\cosh^2 z - 1}} = \frac{C_1}{\rho g} z + C_2; \tag{12.91}$$

thus,

$$y = \frac{\lambda}{\rho g} + \frac{C_1}{\rho g} \cosh \frac{\rho g(x - C_2)}{C_1}. \tag{12.92}$$

Using $y(-a) = y(a) = 0$ and (9.72), we obtain:

$$y(x) = \frac{\alpha}{\rho g} \left(\cosh \frac{\rho g x}{\alpha} - \cosh \frac{\rho g a}{\alpha} \right), \tag{12.93}$$

where $\alpha \sinh (\rho g a / \alpha) = \rho g l$.

Equation (12.93) is the *equation of a chain line*.

Dido's Problem
Find a curve of the fixed length that bounds a plane region of a largest area.

The solution of this problem, named after Dido, the legendary founder and the first Queen of Carthage (around 814 BC), is a circle. By means of a calculus of variations, it was solved by Euler. This problem was one of his largest motivations in this area.

Note that a curve in question must be convex; otherwise, there exists a straight line, such that, if you mirrored a segment of the border in it, you would get a region of larger area than the original one, with the same border length. We can split the region of the largest area in two parts by a straight line; thus, it is sufficient to consider the region bounded by some straight line. To be more specific, let us find the largest area under an arc $y = y(x)$ of length $l > 2a$, whose endpoints are on the straight line. If we take $-a \le x \le a$ and $y(-a) = y(a) = 0$, we have the following isoperimetric problem: find the extremal of the functional:

$$A[y] = \int_{-a}^{a} y dx, \quad y(-a) = y(a) = 0, \tag{12.94}$$

subject to the additional condition

$$\int_{-a}^{a} \sqrt{1 + (y')^2} dx = l. \tag{12.95}$$

In the functional

$$A^* = \int_{-a}^{a} \left[y + \lambda \sqrt{1 + (y')^2} \right] dx \tag{12.96}$$

the integrand does not contain variable x; thus, the first integral of Euler equation is $F - y' F_{y'} = C_1$, or

$$y + \lambda\sqrt{1 + y'^2} - \frac{\lambda y'^2}{\sqrt{1 + y'^2}} = C_1,$$

which gives

$$y - C_1 = -\frac{\lambda}{\sqrt{1 + y'^2}}.$$

With the substitution $y' = \tan t$, we obtain:

$$y - C_1 = -\lambda\cos t. \tag{12.97}$$

From $\frac{dy}{dx} = \tan t$ it follows $dx = \frac{dy}{\tan t} = \frac{\lambda\sin t\, dt}{\tan t} = \lambda\cos t\, dt$; thus,

$$x - C_2 = -\lambda\sin t. \tag{12.98}$$

Eliminating the parameter t from the parametric Eqs. (12.97) and (12.98), we obtain *circles*:

$$(y - C_1)^2 + (x - C_2)^2 = \lambda^2.$$

Constants C_1, C_2, λ can be obtained from the condition (12.95) and $y(-a) = y(a) = 0$.

Among the physical applications of 3D Dido's problem, one can mention the spherical shape of unsupported droplets in the absence of gravity, or the spherical shape of droplets in a free fall.

The Geodesics

Let us return to the problem of finding the shortest lines connecting two given points on the surface (geodesics) that we considered in Sect. 12.4.5. In the general case, to tackle geodesic problems, it is convenient to apply an approach similar to the method of Lagrange multipliers.

The shortest line connecting two given points on the surface is called geodesics.

To tackle geodesic problems, one can develop an approach similar to the one used for isoperimetric problems.

Here we need to find the minimum of the functional:

$$l = \int_{x_0}^{x_1} \sqrt{1 + y'^2 + z'^2}\, dx \tag{12.99}$$

subject to the condition $\psi(x, y, z) = 0$, which is the surface equation.

Introduce the functional

$$I^* = \int\limits_{x_0}^{x_1} \left[\sqrt{1 + y'^2 + z'^2} + \lambda \psi(x, y, z) \right] dx \qquad (12.100)$$

and write the Euler equations for I^*:

$$\lambda \psi_y - \frac{d}{dx} \frac{y'}{\sqrt{1 + y'^2 + z'^2}} = 0, \quad \lambda \psi_z - \frac{d}{dx} \frac{z'}{\sqrt{1 + y'^2 + z'^2}} = 0. \qquad (12.101)$$

From these equations and $\psi(x, y, z) = 0$, one can find the geodesic equations $y = y(x)$ and $z = z(x)$.

When the curve is defined parametrically, $L : \{x = x(t), y = y(t), z = x(t)\}$; then,

$$I = \int\limits_{t_0}^{t_1} \left[\sqrt{x'^2 + y'^2 + z'^2} \right] dt \text{ and } I^* = \int\limits_{t_0}^{t_1} \left[\sqrt{x'^2 + y'^2 + z'^2} + \lambda \varphi(x, y, z) \right] dt.$$

The Euler equations for the functional I^* are

$$\begin{cases} \lambda \varphi_x - \dfrac{d}{dt} \dfrac{x'}{\sqrt{x'^2 + y'^2 + z'^2}} = 0, \\[3mm] \lambda \varphi_y - \dfrac{d}{dt} \dfrac{y'}{\sqrt{x'^2 + y'^2 + z'^2}} = 0, \\[3mm] \lambda \varphi_z - \dfrac{d}{dt} \dfrac{z'}{\sqrt{x'^2 + y'^2 + z'^2}} = 0. \end{cases}$$

Example 12.22 Find the shortest distance between points $A(1, -1, 0)$ and $B(2, 1, -1)$ lying on the plane $15x - 7y + z - 22 = 0$.

Solution For the functional

$$I^* = \int\limits_{1}^{2} \left[\sqrt{1 + y'^2 + z'^2} + \lambda(15x - 7y + z - 22) \right] dx,$$

the Euler equations are

$$-7\lambda - \frac{d}{dx} \left(\frac{y'}{\sqrt{1 + y'^2 + z'^2}} \right) = 0, \lambda - \frac{d}{dx} \left(\frac{z'}{\sqrt{1 + y'^2 + z'^2}} \right) = 0.$$

Multiplying the second equation by 7 and adding with the first one, we obtain:

$$\frac{d}{dx}\left(\frac{y' + 7z'}{\sqrt{1 + y'^2 + z'^2}}\right) = 0, \text{ or } \frac{y' + 7z'}{\sqrt{1 + y'^2 + z'^2}} = C.$$

This DE, with $z' = 7y' - 15$ (which follows from $15x - 7y + z - 22 = 0$), has the solution $y = C_1x + C_2$. With the boundary conditions $y(1) = -1$, $y(2) = 1$, we obtain $y(x) = 2x - 3$, and then with $z(1) = 0$, $z(2) = -1$, we have $z(x) = 1 - x$. From the either the Euler equations, we obtain $\lambda = 0$.

The shortest distance is

$$l = \int\limits_{1}^{2} \sqrt{1 + y'^2 + z'^2}\,dx = \sqrt{6}.$$

The same result is easy to obtain from simple geometric consideration.

Example 12.23 Find the geodesic lines on the cylinder surface.

Solution In cylindrical coordinates, r, φ, z, the length of the curve connecting points P_0 and P_1 on some surface is

$$l = \int\limits_{P_0}^{P_1} \sqrt{dr^2 + r^2 d\varphi^2 + dz^2} = \int\limits_{\varphi_0}^{\varphi_1} \sqrt{r^2 + r'^2_\varphi + z'^2_\varphi}\,d\varphi,$$

where φ_0 and φ_1 are the polar angles for points P_0 and P_1.

For the functional

$$l^* = \int\limits_{\varphi_0}^{\varphi_1} \left[\sqrt{r^2 + r'^2_\varphi + z'^2_\varphi} + \lambda\psi(\varphi, r, z)\right]d\varphi = \int\limits_{\varphi_0}^{\varphi_1} F^*\left(\varphi, r, z, r'_\varphi, z'_\varphi\right)d\varphi,$$

the Euler equations are

$$\frac{\partial F^*}{\partial r} - \frac{d}{d\varphi}\frac{\partial F^*}{\partial r'_\varphi} = 0, \quad \frac{\partial F^*}{\partial z} - \frac{d}{d\varphi}\frac{\partial F^*}{\partial z'_\varphi} = 0.$$

In case of a cylinder of radius a,

$$l^* = \int\limits_{\varphi_0}^{\varphi_1} \left[\sqrt{r^2 + r'^2_\varphi + z'^2_\varphi} + \lambda(r - a)\right]d\varphi = \int\limits_{\varphi_0}^{\varphi_1} F^*\left(\varphi, r, z, r'_\varphi, z'_\varphi\right)d\varphi,$$

and the Euler equations give:

Fig. 12.10 Geodesic line
on a cylinder surface

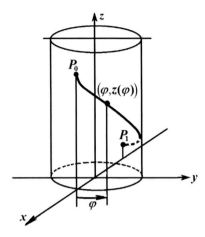

$$\frac{r}{\sqrt{r^2 + r'^2_\varphi + z'^2_\varphi}} + \lambda - \frac{d}{d\varphi}\left(\frac{r'_\varphi}{\sqrt{r^2 + r'^2_\varphi + z'^2_\varphi}}\right) = 0, \quad \frac{d}{d\varphi}\left(\frac{z'_\varphi}{\sqrt{r^2 + r'^2_\varphi + z'^2_\varphi}}\right) = 0.$$

Because $r = a = const.$ and $r'_\varphi = 0$, from the second equation, we have:

$$\frac{z'_\varphi}{\sqrt{a^2 + z'^2_\varphi}} = c, z'_\varphi = \pm \frac{ca}{\sqrt{1 - c^2}} = c_1 \ (c_1 \neq 1), \text{thus } z = c_1\varphi + c_2$$

From the first equation, we now obtain that $\lambda = -1$.

The extremals $z = c_1\varphi + c_2$ are the *helix lines*. Through points P_0 and P_1 (if $z_0 \neq z_1$), one can draw the infinite number of helix lines with a different number of rotations around the cylinder (Fig. 12.10). Only one of those curves has indeed the shortest length. Other solutions correspond to local minima of the functional, i.e., they have a shorter length than any close curves, but they are not global minima. But they are the shortest curves among the curves with the given number of rotations around the cylinder. Note that the number of rotations is a topological invariant, i.e., we cannot continuously transform a curve into another curve with a different number of rotations.

Problems

Find the extremals in isoperimetric problems:

1. $V[y] = \int\limits_0^1 (y')^2 dx$, $y(0) = 1$, $y(1) = 0$, subject to the condition $\int\limits_0^1 y \, dx = 0$.

2. $V[y] = \int\limits_0^1 (y')^2 dx$, $y(0) = 0$, $y(1) = 1$, subject to the condition $\int\limits_0^1 xy \, dx = 0$.

3. $V[y] = \int\limits_0^\pi (y')^2 dx$, $y(0) = 1$, $y(\pi) = -1$, subject to the condition $\int\limits_0^\pi y \cos x \, dx = \frac{\pi}{2}$.

4. $V[y] = \int\limits_0^\pi (y')^2 dx$, $y(0) = 1$, $y(\pi) = 1$, subject to the condition $\int\limits_0^\pi y \cos x \, dx = 0$.

5. $V[y] = \int\limits_0^\pi y \sin x \, dx$, $y(0) = 0$, $y(\pi) = \pi$ subject to the condition $\int\limits_0^\pi (y')^2 dx = \frac{3\pi}{2}$.

6. $V[y] = \int\limits_0^1 (y')^2 dx$, $y(0) = 2e+1$, $y(1) = 2$ subject to the condition $\int\limits_0^1 y e^{-x} dx = e$.

7. $V[y] = \int\limits_0^1 (y')^2 dx$, $y(0) = 0$, $y(1) = 1$, subject to the condition $\int\limits_0^1 y e^x dx = 0$.

8. $V[y] = \int\limits_1^2 x^2 (y')^2 dx$, $y(1) = 1$, $y(2) = 2$, subject to the condition $\int\limits_1^2 xy \, dx = \frac{7}{3}$.

9. $V[y] = \int\limits_1^2 x^3 (y')^2 dx$, $y(1) = 4$, $y(2) = 1$, subject to the condition $\int\limits_1^2 y \, dx = 2$.

10. $V[y] = \int\limits_0^1 \left(x^2 + y'^2 \right) dx$, $y(0) = 0$, $y(1) = 0$, subject to the condition $\int\limits_0^1 y^2 dx = 2$.

11. $V[y] = \int\limits_0^1 y'^2 dx$, $y(0) = 0$, $y(1) = \frac{1}{4}$, subject to the condition $\int\limits_0^1 \left(y - y'^2 \right) dx = \frac{1}{12}$.

12. Find the shortest distance between points $A(1, 0, -1)$ и $B(0, -1, 1)$ lying on the plane $x+y+z = 0$.

13. Find a geodesics on a sphere. This problem was solved in Sect. 12.4.5 by direct substitution of coordinates of a point on the sphere in the functional. Here use the method of Lagrange multipliers and substitute $r = a$ into the solution of the Euler equation.

14. Without using the method of Lagrange multipliers, find the curve of the shortest length connecting the points $(a, 0, 0)$ and $(0, a, h)$ on the cylinder $x^2 + y^2 = a^2$, and the shortest distance between these points on the cylinder surface. Solve this problem in the Cartesian coordinates determining line in parametric form $x = x(t)$, $y = y(t)$, $z = x(t)$.

12.9   Functionals That Depend on a Multivariable Function

In many problems, functionals depend on a multivariable function, for instance:

$$V[u(x_1, \ldots, x_m)] = \int\limits_D F(x_1, \ldots, x_m, u, u_{x_1}, u_{x_2}, \ldots, u_{x_m}) dx_1 \ldots dx_m. \qquad (12.102)$$

D is the domain of the independent variables $x_1, x_2, \ldots, x_m$; the function $u = u(x_1, x_2, \ldots, x_m)$ is assumed to be fixed on the boundary Γ of this domain; $u_{x_m} \equiv \partial u / \partial x_m$.

As always, we assume that the derivatives of the necessary order of the functions F and u exist and are continuous in D.

Let us find the variation of the functional $V[u]$ using Eq. (12.15):

$$\delta V = \left(\frac{d}{d\alpha} V[u + \alpha \delta u]\right)\Big|_{\alpha=0} =$$

$$= \left\{\frac{d}{d\alpha}\left[\int_D F(x_1, \ldots, x_m, u + \alpha\delta u, u_{x_1} + \alpha\delta u_{x_1}, \ldots, u_{x_m} + \alpha\delta u_{x_m})dx_1 \ldots dx_m\right]\right\}\Big|_{\alpha=0}$$

$$= \int_D \left(\frac{dF}{du}\delta u + \frac{dF}{du_{x_1}}\delta u_{x_1} + \ldots + \frac{dF}{du_{x_m}}\delta u_{x_m}\right)dx_1 \ldots dx_m.$$

The integrals of $\frac{\partial F}{\partial u_{x_k}}\delta u_{x_k}$ can be evaluated by means of integrating by parts. With $\delta u_{x_k} = \frac{\partial}{\partial x_k}\delta u$ and $\delta u|_\Gamma = 0$, we obtain:

$$\int_D \left(\frac{\partial F}{\partial u_{x_k}}\delta u_{x_k}\right)dx_1 \ldots dx_m = \int dx_1 \ldots dx_{k-1}dx_{k+1} \ldots dx_m \int \left(\frac{\partial F}{\partial u_{x_k}}\delta u_{x_k}\right)dx_k =$$

$$= \int dx_1 \ldots dx_{k-1}dx_{k+1} \ldots dx_m\left\{\frac{\partial F}{\partial u_{x_k}}\delta u\Big|_\Gamma - \int\left(\frac{\partial}{\partial x_k}\frac{\partial F}{\partial u_{x_k}}\right)\delta u dx_k\right\} =$$

$$= -\int_D \left(\frac{\partial}{\partial x_k}\frac{\partial F}{\partial u_{x_k}}\right)\delta u dx_1 \ldots dx_m.$$

Then,

$$\delta V = \int_D \left(\frac{\partial F}{\partial u}\delta u + \frac{\partial F}{\partial u_{x_1}}\delta u_{x_1} + \ldots + \frac{\partial F}{\partial u_{x_m}}\delta u_{x_m}\right)dx_1 \ldots dx_m =$$

$$= \int_D \left(\frac{\partial F}{\partial u} - \frac{\partial}{\partial x_1}\frac{\partial F}{\partial u_{x_1}} - \ldots - \frac{\partial F}{\partial x_m}\frac{\partial F}{\partial u_{x_m}}\right)\delta u dx_1 \ldots dx_m = 0.$$

Because δu is arbitrary and continuous, and equals zero on Γ, the extremals $u(x_1, x_2, \ldots, x_m)$ of the functional $V[u]$ are defined by the Euler equation of the form:

$$\frac{\partial F}{\partial u} - \sum_{k=1}^{m} \frac{\partial}{\partial x_k}\frac{\partial F}{\partial u_{x_k}} = 0. \tag{12.103}$$

If a functional $V[u(x_1, \ldots, x_m), v(x_1, \ldots, x_m), \ldots]$ depends on several functions, then the extremals $u, v, \ldots$ can be found from the system of equations:

$$\frac{\partial F}{\partial u} - \sum_{k=1}^{m} \frac{\partial}{\partial x_k} \frac{\partial F}{\partial u_{x_k}} = 0, \qquad \frac{\partial F}{\partial v} - \sum_{k=1}^{m} \frac{\partial}{\partial x_k} \frac{\partial F}{\partial v_{x_k}} = 0. \qquad (12.104)$$

We leave it to the reader to generalize Eqs. (12.103) and (12.104) to the cases when the functionals contain higher-order derivatives.

Example 12.24 Let

$$V[u] = \int_D \left[\left(\frac{\partial u}{\partial x}\right)^2 + \left(\frac{\partial u}{\partial y}\right)^2 - 2uf(x,y) \right] dxdy, \quad u|_\Gamma = \varphi(x,y).$$

Find the equation for the extremals.

Solution Equation (12.103) gives:

$$F'_u - \frac{\partial}{\partial x} F'_{u_x} - \frac{\partial}{\partial y} F'_{u_y} = 2\left[f - \frac{\partial^2 u}{\partial x^2} - \frac{\partial^2 u}{\partial y^2} \right] = 0,$$

or

$$\nabla^2 u = f(x,y).$$

The obtained equation is called the *Poisson equation* (see Chap. 13).

Problems

Obtain the Euler equation for the following functionals:

1. $V[u(x,y)] = \iint_D \left[\left(\frac{\partial u}{\partial x}\right)^2 - \left(\frac{\partial u}{\partial y}\right)^2 \right] dxdy.$

2. $V[u(x,y)] = \int_D \left[\left(\frac{\partial u}{\partial x}\right)^2 + \left(\frac{\partial u}{\partial y}\right)^2 + \left(\frac{\partial u}{\partial z}\right)^2 + 2uf(x,y,z) \right] dxdydz.$

3. $V[u(x,y)] = \iint_D \left[\left(\frac{\partial u}{\partial x}\right)^4 + \left(\frac{\partial u}{\partial y}\right)^4 + 24ux\sin y \right] dxdy.$

4. $V[u(x,y)] = \iint_D \left[\left(\frac{\partial u}{\partial x}\right)^4 + \left(\frac{\partial u}{\partial y}\right)^4 + 12uf(x,y) \right] dxdy.$

5. $V[u(x,y)] = \iint_D \left[\left(\frac{\partial u}{\partial x}\right)^2 + \left(\frac{\partial u}{\partial y}\right)^2 + 2\left(\frac{\partial^2 u}{\partial x\partial y}\right)^2 - 2uf(x,y) \right] dxdy.$

6. The Lagrangian for a nonrelativistic particle of the mass m and charge e, moving through the electric potential $\varphi\left(\vec{r}, t\right)$ is $L = \frac{mv^2}{2} - e\varphi$.

   Using the Lagrange-Euler equation $\frac{\partial L}{\partial \vec{r}} - \frac{d}{dt}\frac{\partial L}{\partial \vec{v}} = 0$, show that the equation of motion of the particle is $\frac{d\vec{p}}{dt} = e\vec{E}$, where $\vec{E} = -\nabla\varphi$ is the electric field and $\vec{p} = m\vec{v}$ is the particle's momentum.

7. In the relativistic field theory, a field function, $\phi(x)$, is the function of the 4-*coordinate*, $x_\alpha = (x, y, z, ict)$. The Lagrangian L depends on the field and its derivatives, and the action is

$$A = \int L\left(\phi(t, x), \frac{\partial \phi(t, x)}{\partial x_\alpha}\right) dV dt. \qquad (12.105)$$

The principle of a least action, $\delta A = 0$, leads to the Lagrange-Euler equation:

$$\frac{\partial L}{\partial \phi} - \frac{\partial}{\partial x_\alpha} \frac{\partial L}{\partial \frac{\partial \phi}{\partial x_\alpha}} = 0 \qquad (12.106)$$

(a summation over the repeating index α is assumed). For *scalar particles* with the mass m and zero spin, the Lagrangian is

$$L(\phi, \partial_\alpha \phi) = -\frac{1}{2}\left[\left(\frac{\partial \phi}{\partial x_\alpha}\right)^2 + m^2 \phi^2\right]. \qquad (12.107)$$

Check that Eqs. (12.106) and (12.107) give the *Klein-Gordon equation* for the scalar field:

$$\left(-\frac{\partial}{\partial x_\alpha} \frac{\partial}{\partial x_\alpha} + m^2\right)\phi(x) = 0. \qquad (12.108)$$

12.10 Introduction to the Direct Methods for Variational Problems

In the previous sections, the Euler equation and its modifications are used to solve variational problems. In the so-called direct methods, a variational problem is considered as the problem of finding an *extremum of a certain function*, and then one constructs a sequence of such functions which converges to the extremal of a given functional.

Among analytical approaches one of the most popular is *Ritz's method*, named after a Swiss theoretical physicist Walther Ritz (1878–1909).

12.10.1   Ritz's Method

To find a *minimum* of a functional

$$V[y(x)] = \int_a^b F(x, y, y')dx, \quad y(a) = y_a, \ y(b) = y_b \qquad (12.109)$$

using Ritz's method, linear combinations are considered:

$$y_n(x) = \sum_{k=1}^n \alpha_k \varphi_k(x), \qquad (12.110)$$

where $\varphi_k(x)$ is a system of linearly independent basis functions that satisfy the boundary conditions in (12.109) and α_k are constants.

On functions $y_n(x)$, the functional $V[y_n(x)]$ is a function $\Phi(\alpha_1, \alpha_2, \ldots, \alpha_n)$. The necessary conditions for the extremum are

$$\frac{\partial \Phi}{\partial \alpha} = 0, \quad k = 1, 2, \ldots, n. \qquad (12.111)$$

This results in a system of equations to find the coefficients α_k. Thus, $y_n(x)$ can be determined that give the approximation to the solution of the variational problem. Taking different values of n, we have the sequence $y_1(x)$, $y_2(x)$, $\ldots$, $y_n(x)$. If it converges

$$y(x) = \lim_{n \to \infty} y_n(x),$$

then the function $y(x)$ is the solution of the variational problem. In practice, the functions y_n and y_{n+1} should be close with some imposed precision; if not, we should find y_{n+2} and compare with y_{n+1}, etc.

Often for $\{\varphi_n\}$ the monomials 1, x, $\ldots$, x^n, or the trigonometric functions, $\sin x$, $\sin 2x$, $\ldots$, $\sin nx$ are taken.

When the boundary conditions are zero, one can take

$$\varphi_k(x) = (x - a)^{k-1}(x - b)\psi_k(x) \quad (k = 1, 2, \ldots), \qquad (12.112)$$

where $\psi_k(x)$ are some continuous functions; another choice is

$$\varphi_k(x) = \sin\left(\frac{k\pi}{2} \frac{x - a}{b - a}\right) \quad (k = 1, 2, \ldots). \qquad (12.113)$$

When the boundary conditions are not zero, one can take

$$y_n(x) = \varphi_0(x) + \sum_{k=1}^{n} \alpha_k \varphi_k(x) \tag{12.114}$$

with $\varphi_0(x)$ satisfying $\varphi_0(a) = y_a$, $\varphi_0(b) = y_b$, and all other $\varphi_k(x)$ satisfying zero boundary conditions. $\varphi_0(x)$ can be chosen as a linear function:

$$\varphi_0(x) = y_a + \frac{y_b - y_a}{b - a}(x - a), \quad \text{or} \quad \varphi_0(x) = y_a + \sin\left(\frac{k\pi}{2}\frac{x - a}{b - a}\right)(y_b - y_a). \tag{12.115}$$

In the case of the functionals $V[z(x_1, x_2, \ldots, x_n)]$, the functions φ_k depend on the variables x_1, x_2, $\ldots$, x_n. Functionals that depend on multiple functions can be considered similarly.

Example 12.25 Find the minimum of the functional:

$$V[y] = \frac{1}{2} \int_0^\pi \left(y'^2 + y^2 + 2y \sin x\right) dx, \quad y(0) = y(\pi) = 0$$

using Ritz's method and compare with the exact solution.

Solution Let us take the linearly independent functions:

$$\varphi_k(x) = (1 - x)x^k \quad (k = 1, 2, \ldots),$$

which satisfy the boundary conditions $\varphi_k(0) = \varphi_k(1) = 0$.

Choosing the first term in (12.112) as

$$y_1(x) = \alpha_1 x(\pi - x),$$

we obtain:

$$\Phi[y_1] = \frac{1}{2} \int_0^\pi \left[\alpha_1^2(\pi - 2x)^2 + \alpha_1^2 x^2(\pi - x)^2 + 2\alpha_1 x(\pi - x) \sin x\right] dx.$$

Equation $\frac{\partial \Phi}{\partial \alpha_1} = 0$ gives:

$$\alpha_1 \int_0^\pi \left[(\pi - 2x)^2 + x^2(\pi - x)^2\right] dx = \int_0^\pi x(\pi - x) \sin x \, dx.$$

Integrating, we find:

Fig. 12.11 Solid line, the approximate solution $y_1(x)$; dashed line, the exact solution $y(x) = -0.5\sin x$

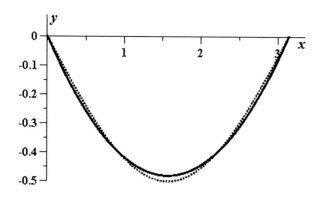

$$\frac{\pi^3(\pi^2 + 10)}{30}\alpha_1 = -4, \text{ and } y_1(x) = -0.1948x(\pi - x).$$

In Sect. 12.3.3, Example 12.11, we obtained the exact solution of this problem, $y(x) = -0.5\sin x$, and we also found that the functional has the minimum on this function.

From Fig. 12.11 one can see that the maximum difference between the approximate and exact solutions is ~ 0.02, about 4%.

Example 12.26 Find the minimum of the functional:

$$V[y] = \int\limits_0^1 \left(y'^2 - y^2 + 2xy\right)dx, \quad y(0) = y(1) = 0$$

using Ritz's method and compare with the exact solution.

Solution Let us choose the linearly independent functions:

$$\varphi_k(x) = (1-x)x^k \qquad (k = 1, 2, \ldots).$$

For $n = 1$, $y_1(x) = \alpha_1(x - x^2)$ and

$$V[y_1(x)] = \int\limits_0^1 \left[\alpha_1^2(1 - 2x)^2 - \alpha_1^2\left(x - x^2\right)^2 + 2\alpha_1 x\left(x - x^2\right)\right]dx =$$

$$= \frac{3}{10}\alpha_1^2 + \frac{1}{6}\alpha_1 = \Phi(\alpha_1).$$

Then the condition

$$\frac{\partial\Phi}{\partial\alpha_1} = \frac{3}{5}\alpha_1 + \frac{1}{6} = 0$$

gives $\alpha_1 = -\frac{5}{18}$; thus, $y_1(x) = -\frac{5}{18}x(1-x)$.

Fig. 12.12 Solid line, the
approximate solution $y_1(x)$;
dashed line, the exact
solution

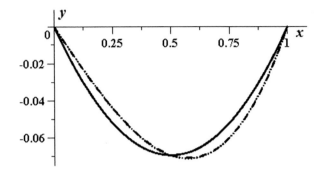

The Euler equation gives $y'' + y = x$, and its solution using the boundary conditions is $y(x) = x - \dfrac{\sin x}{\sin 1}$.

The graphs of $y_1(x)$ and the exact solutions are shown in Fig. 12.12.

To obtain better agreement, let us add one more basis function, $y_2(x) = \alpha_1 x(1-x) + \alpha_2 x^2(1-x)$; then,

$$V[y_2(x)] = \int\limits_0^1 \left\{ \left[\alpha_1(1-2x) + \alpha_2(2x - 3x^2) \right]^2 - \left[\alpha_1(x - x^2) + \alpha_2(x^2 - x^3) \right]^2 + \right.$$
$$\left. + 2x \left[\alpha_1(x - x^2) + \alpha_2(x^2 - x^3) \right] \right\} dx = \Phi(\alpha_1, \alpha_2).$$

The system of equations

$$\begin{cases} \dfrac{\partial \Phi}{\partial \alpha_1} = \dfrac{3}{5}\alpha_1 + \dfrac{3}{10}\alpha_2 + \dfrac{1}{6} = 0, \\[3mm] \dfrac{\partial \Phi}{\partial \alpha_2} = \dfrac{3}{10}\alpha_1 + \dfrac{26}{105}\alpha_2 + \dfrac{1}{10} = 0 \end{cases}$$

gives $\alpha_1 = -\dfrac{71}{369}$, $\alpha_2 = -\dfrac{7}{41}$; thus,

$$y_2(x) = -x(1-x)\left(\dfrac{71}{369} + \dfrac{7}{41}x \right).$$

In Fig. 12.12 we see that with two basic functions, we obtain reasonable
approximation.

Example 12.27 Find the minimum of the functional:

$$V[y] = \int_{-1}^{1} \left(y'^2 - 4y^2 + 2xy - x^2 \right) dx, \quad y(-1) = 2, y(1) = 4$$

using Ritz's method and compare with the exact solution.

Solution The boundary conditions are not zero; thus, we consider:

$$y_n(x) = \varphi_0(x) + \sum_{k=1}^{n} \alpha_k \varphi_k(x),$$

with $\varphi_0(x)$ satisfying these conditions, $\varphi_0(-1) = 2$, $\varphi_0(1) = 4$, and all other functions $\varphi_k(x)$ satisfying zero boundary conditions.

For $n = 1$

$$y_1(x) = \varphi_0(x) + \varphi_1(x)$$

take $\varphi_0(x) = 2 + \frac{4-2}{1+1}(x+1) = 3 + x$, $\varphi_1(x) = \alpha_1(x+1)(x-1) = \alpha_1(x^2 - 1)$.
For the function $y_1(x) = 3 + x + \alpha_1(x^2 - 1)$,

$$V[y_1(x)] = \int_{-1}^{1} \left\{ [1 + 2\alpha_1 x]^2 - 4[3 + x + \alpha_1(x^2 - 1)]^2 + \right.$$

$$\left. + 2x[3 + x + \alpha_1(x^2 - 1)] - x^2 \right\} dx = \Phi(\alpha_1).$$

From $\frac{\partial \Phi}{\partial \alpha_1} = 0$ we have $\alpha_1 = 10$, thus, $y_1(x) = 10x^2 + x - 7$.
The solution of the Euler equation

$$y'' + 4y = x$$

with the given boundary conditions is $y(x) = \frac{3}{\cos 2} \cos 2x + \frac{3}{4\sin 2} \sin 2x + \frac{x}{4}$.
The graphs of this exact solution and the first Ritz's approximation $y_1(x)$ are shown in Fig. 12.13.

Example 12.28 Find the extremum of the functional:

$$V[z(x,y)] = \iint_{D} \left[\left(\frac{\partial z}{\partial x} \right)^2 + \left(\frac{\partial z}{\partial y} \right)^2 - 2z \right] dx dy,$$

where D is $-a \le x \le a$, $-a \le y \le a$, and $z = 0$ on the boundaries of the square.

Solution Let us search for an extremal in the form:

$$z_0(x, y) = \alpha_0 (x^2 - a^2)(y^2 - a^2).$$

This $z_0(x, y)$ satisfies the boundary conditions.
    Integrating, we obtain:

$$V[z_0] = \frac{256}{45} \alpha_0^2 a^8 - \frac{32}{9} \alpha_0 a^6 = \Phi(\alpha_0).$$

Then,

$$\frac{\partial \Phi}{\partial \alpha_0} = \frac{512}{45} \alpha_0 a^8 - \frac{32}{9} a^6 = 0 \text{ and } \alpha_0 = \frac{5}{16 a^2}.$$

12.10.2 Ritz's Method for Quantum Systems

The energy of a quantum mechanical system is given by the integral

$$E = \int \psi^* \widehat{H} \psi \, dq, \tag{12.116}$$

where operator $\widehat{H}$ is the system's *Hamiltonian* and $\psi(q)$ is the *wave function* of the generalized coordinates. For discrete energy states, the wave functions can be normalized:

$$\int \psi^* \psi \, dq = 1. \tag{12.117}$$

The ground state energy E_0 is

$$E_0 = \min \int \psi^* \widehat{H} \psi \, dq. \tag{12.118}$$

First, we should choose a trial function and to evaluate the integral

$$\Phi(\alpha, \beta, \ldots) = \int \psi^*(q; \alpha, \beta, \ldots) H \psi(q; \alpha, \beta, \ldots) dq.$$

The necessary conditions for the minimum

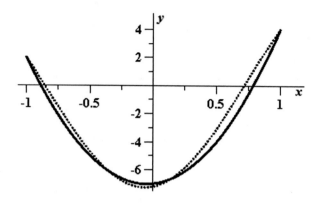

Fig. 12.13 Solid line, the approximate solution $y_1(x)$; dashed line, the exact solution

$$\frac{\partial \Phi}{\partial \alpha} = \frac{\partial \Phi}{\partial \beta} = \ldots = 0$$

give the values of the parameters $\alpha_0, \beta_0, \ldots$. The corresponding approximate ground state energy and the wave function are $E_0 = \Phi(\alpha_0, \beta_0, \ldots)$ and $\psi_0(q; \alpha_0, \beta_0, \ldots)$.

The energy E_1 of the first excited state is given by

$$E_1 = \min \int \psi_1^* \widehat{H} \psi_1 \, dq, \tag{12.119}$$

considering the additional conditions of orthogonality

$$\int \psi_1^* \psi_1 \, dq = 1, \quad \int \psi_1^* \psi_0 \, dq = 0. \tag{12.120}$$

For the energy of the second excited state, the variational problem is

$$E_2 = \min \int \psi_2^* \widehat{H} \psi_2 \, dq \tag{12.121}$$

with the additional conditions

$$\int \psi_2^* \psi_2 \, dq = 1, \quad \int \psi_2^* \psi_1 \, dq = \int \psi_2^* \psi_0 \, dq = 0. \tag{12.122}$$

A choice of the trial functions is based on the qualitative analysis of a problem and its symmetry properties.

For example, let us find the ground state of the *hydrogen atom*. The orbital momentum in this state is zero and the Hamiltonian is

$$\widehat{H} = -\frac{\hbar^2}{2m}\nabla^2 - \frac{e^2}{r}. \tag{12.123}$$

Here m is the electron's mass; e is the nucleus charge. The wave function must approach zero as $r \to \infty$, and let us take the trial function in the form:

$$\psi = A\exp(-\beta). \tag{12.124}$$

In the spherical coordinates, the normalization condition is

$$\int_0^\infty \psi^2 r^2 dr = 1, \text{ which gives } A^2 = \beta^2/\pi.$$

Then,

$$\Phi(\beta) = \int_0^\infty \psi\widehat{H}\psi r^2 dr = \frac{2\beta^3\hbar^2}{m}\int_0^\infty e^{-\beta r}\left(\nabla^2 e^{-\beta r}\right)r^2 dr - 4\beta^3 e^2 \int_0^\infty e^{-2\beta r} r dr. \tag{12.125}$$

Evaluation of the first integral gives:

$$\int_0^\infty e^{-\beta r}\left(\nabla^2 e^{-\beta r}\right)r^2 dr = -\int_0^\infty \left(\frac{\partial}{\partial r}e^{-\beta r}\right)^2 r^2 dr = -(4\beta)^{-1}.$$

Second integral in (12.125) is

$$\int_0^\infty e^{-2\beta r} r dr = (2\beta)^{-2};$$

thus,

$$\Phi(\beta) = \frac{\hbar^2\beta^2}{2m} - e^2\beta. \tag{12.126}$$

From the condition for the minimum, $\Phi'(\beta) = 0$, we obtain the variational parameter $\beta_0 = 1/a$, where $a = \hbar^2/(me^2) \approx 0.53 \times 10^{-8}$cm is called the *Bohr radius*. This is the most probable distance between the nucleus and the electron in the hydrogen atom in its ground state. Substituting β_0 in (12.126) and (12.124), we obtain the energy and the wave function of the hydrogen ground state:

$$E_{1s} = \Phi(\beta_0) = -\frac{e^2}{2a}, \quad \psi_{1s} = \frac{1}{\sqrt{\pi a^3}} \exp\left(-\frac{r}{a}\right). \tag{12.127}$$

(The atomic physics notation is $1s$, 1 stays for the principle quantum number, $n = 1$, s for zero angular momentum, $l = 0$.)

Next, we can describe the first excited state, $2s$ with $n = 2$ and $l = 0$; thus, the wave function also depends only on r. Choose the trial function in the form containing two parameters, α and γ:

$$\psi_{2s} = B(1 + \gamma r) \exp(-\alpha r/a). \tag{12.128}$$

From the orthogonality condition, $\int \psi_{2s} \psi_{1s} r^2 dr = 0$, one can find:

$$-\gamma = \frac{1}{3}(1 + \alpha).$$

Then, the normalization of the function ψ_{2s} gives:

$$B^2 = \frac{3a^5}{\pi a^3 (1 - \alpha + \alpha^2)}.$$

Next, evaluate the integral:

$$\Phi(\alpha) = \int_0^{\infty} \psi_{2s} \hat{H} \psi_{2s} dq = \frac{e^2}{a} \left[-\frac{\alpha}{2} + \frac{7}{6}\alpha^2 - \frac{\alpha^2}{2(\alpha^2 - \alpha + 1)} \right]. \tag{12.129}$$

The minimum condition, $\Phi'(\alpha) = 0$, gives $\alpha = 1/2$. Finally, with (12.129) and (12.128), we have:

$$E_{2s} = -\frac{e^2}{8a}, \quad \psi_{2s} = \frac{1}{\sqrt{8\pi a^3}} \left(1 - \frac{r}{2a}\right) \exp\left(-\frac{r}{2a}\right).. \tag{12.130}$$

Problems

Find the extremum of the functional using Ritz's method and compare with the exact solution obtained using the Euler equation:

1. $V[y] = \int_{-1}^{1} \left(y'^2 + 4y^2 - 8xy + 2x^2\right) dx, \quad y(-1) = 3, y(1) = 1.$

2. $V[y] = \int_{-1}^{1} \left(y'^2 + 4y^2 + 4x^2y + x \cos x\right) dx, \quad y(-1) = 2, y(1) = 0.5.$

3. $V[y] = \int_{0}^{2} \left(y'^2 - 4y' \sin 2x - x^2\right) dx, \quad y(0) = 1, y(2) = -1.$

4. $V[y] = \int\limits_{0.5}^{1.5} \left(y' + y'^2 \sin 2x - \cos 2x \right) dx, \quad y(0.5) = 1, y(1.5) = 2.$

5. $V[y] = \int\limits_{1}^{2} \left(y' + xy'^2 - x^2 y' \right) dx, \quad y(1) = 2, y(2) = -1.$

6. $V[y] = \int\limits_{0}^{2} \left(y' + y'^2 + x^2 y'^2 \right) dx, \quad y(0) = 2, y(2) = -2.$

7. $V[y] = \int\limits_{-1}^{1} \left(y'^2 + 2y' e^x \sin x - e^x \cos x \right) dx, \quad y(-1) = 2, y(1) = 3.$

8. $V[y] = \int\limits_{0}^{2} \left(y'^2 + 2yy' - 16y^2 \right) dx, \quad y(0) = 0, y(2) = 0.$

9. $V[y] = \int\limits_{0}^{2} \left(xy' + y'^2 \right) dx, \quad y(0) = 1, y(2) = 0.$

10. $V[y] = \int\limits_{-1}^{1} \left(y^2 + y'^2 - 2y \sin x \right) dx, \quad y(-1) = 2, y(1) = 3.$

Chapter 13
Partial Differential Equations

13.1 Introduction

In the previous chapters of this book, we have seen that the ODEs provide a mathematical formulation for various problems in mechanics, physics, and biology. Also, the ODEs serve as a powerful tool for solving mathematical problems, even when their initial formulation does not contain ODEs. In Chap. 12, the application of ODEs in the variational analysis is demonstrated. In the present chapter, we consider a new class of mathematical problems, the partial differential equations (PDEs). Let us emphasize that this chapter cannot replace a comprehensive textbook on PDEs (see, e.g., [9]). Here, we are going to explain how definite classes of PDEs can be solved be their reduction to ODEs.

Until now, we considered problems where the unknown functions were functions of one variable. However, the processes in the real world take place in the three-dimensional space and in time. Therefore, as the rule, functions of several variables are needed for their description. For instance, the electromagnetic phenomena are governed by functions $\vec{E}\left(\vec{r},t\right)$ and $\vec{H}\left(\vec{r},t\right)$ that determine the electric and magnetic fields at the point with the coordinates $\vec{r} = (x_1, x_2, x_3)$, at the time instance t.

Partial differential equation (PDE) is the differential equation for unknown function of several variables, $u(x_1, x_2, \ldots x_n)$. The highest order of derivatives of the function u is called the order of equation.

For instance, the general form of a second-order PDE for the function $u(x_1, x_2, \ldots x_n)$ is

$$F\left(x_1, \ldots x_n, u, \frac{\partial u}{\partial x_1}, \ldots, \frac{\partial u}{\partial x_n}, \frac{\partial^2 u}{\partial x_1{}^2}, \frac{\partial^2 u}{\partial x_1 \partial x_2}, \ldots, \frac{\partial^2 u}{\partial x_n{}^2}\right) = 0, \qquad (13.1)$$

where F is a given (linear or nonlinear) function.

© The Author(s), under exclusive license to Springer Nature Switzerland AG 2023
V. Henner et al., *Ordinary Differential Equations*,
https://doi.org/10.1007/978-3-031-25130-6_13

The PDE itself is not sufficient for the determination of its solution describing a particular natural phenomenon; additional information, which takes the form of the initial and boundary conditions, is required. This gives rise to the existence and necessity of the initial-boundary value problems (or boundary value problems) for PDEs.

Many natural phenomena – such as sound, electromagnetic and other types of waves, fluid flow, diffusion, propagation of heat, and electrons' motion in atom – are described by PDEs. Also, countless models and applications in many areas of pure and applied science and technology are largely based on PDEs. Several important PDEs are presented below.

Maxwell's equations for the electric, $\vec{E}\left(\vec{r},t\right)$, and magnetic, $\vec{H}\left(\vec{r},t\right)$, fields can be written as equations for a vector potential, $\vec{A}\left(\vec{r},t\right)$, and an electric (scalar) potential, $\varphi\left(\vec{r},t\right)$. For an electromagnetic wave, these potentials are governed by the equation:

$$\nabla^2 u\left(\vec{r},t\right) - \frac{1}{c^2}\frac{\partial^2 u\left(\vec{r},t\right)}{\partial t^2} = 0, \tag{13.2}$$

where ∇^2 is the *Laplace operator or Laplacian* (in Cartesian coordinates $\nabla^2 = \frac{\partial^2}{\partial x^2} + \frac{\partial^2}{\partial x^2} + \frac{\partial^2}{\partial x^2}$), c is the speed of light, and function $u\left(\vec{r},t\right)$ represents either potential $\vec{A}\left(\vec{r},t\right)$ or $\varphi\left(\vec{r},t\right)$.

Equation (13.2) is the particular case of the *wave equation*, which describes a wave propagating with speed c. Similar equations describe sound waves, waves propagating in strings, membranes, and many other types of waves.

The simplest one-dimensional wave equation is

$$\frac{\partial^2 u(x,t)}{\partial t^2} = a^2 \frac{\partial^2 u(x,t)}{\partial x^2}, \tag{13.3}$$

where the coefficient $a^2 \equiv 1/v^2$ (a is the wave speed). Equation (13.3) has to be supplemented by some initial and boundary conditions.

Stationary (time-independent) electric and magnetic fields are governed by the following equations:

$$\nabla^2 \varphi\left(\vec{r}\right) = -4\pi\rho\left(\vec{r}\right), \qquad \nabla^2\vec{A}\left(\vec{r}\right) = -\frac{4\pi}{c}\vec{j}\left(\vec{r}\right), \tag{13.4}$$

where $\rho\left(\vec{r}\right)$ is the charge density distribution and $-\vec{j}\left(\vec{r}\right)/c$ is the current density distribution. Equation (1.4) are the particular cases of the *Poisson equation*:

$$\nabla^2 u\left(\vec{r}\right) = f\left(\vec{r}\right), \tag{13.5}$$

where $f\left(\vec{r}\right)$ is a known function.

In the absence of charges and currents, Eq. (1.4) have the form:

$$\nabla^2 u\left(\vec{r}\right) = 0, \tag{13.6}$$

which is called the *Laplace equation*.

Equations (13.5) and (13.6) are time-independent (static); thus, the complete formulation of a problem requires only boundary conditions without initial conditions.

One more basic equation is the *heat equation*, the simplest form of which in one-dimensional case can be presented as

$$\frac{\partial u(x,t)}{\partial t} = a\frac{\partial^2 u(x,t)}{\partial x^2}, \tag{13.7}$$

where the parameter a is called the heat conductivity. The heat equation describes a process with a loss (dissipation) of energy, such as heat flow and diffusion. The solutions of Eq. (1.7) decay exponentially in time, in a contradistinction to time-periodic solutions of Eq. (13.3).

The *Helmholtz equation* has the form:

$$\nabla^2 u\left(\vec{r}\right) + k^2 u\left(\vec{r}\right) = 0, \tag{13.8}$$

where k^2 can be a function of coordinates. The stationary Schrödinger equation of quantum mechanics also has a form of Helmholtz equation.

Equations shown above are classified as either *hyperbolic (wave equation)*, *parabolic (heat equation)*, or *elliptic (Laplace, Poisson, Helmholtz equations)*. This terminology comes from the classification based on conic sections. The nontrivial solutions of a second-order algebraic equation

$$ax^2 + bxy + cy^2 + dx + ey + f = 0 \tag{13.9}$$

describe hyperbolas, parabolas, or ellipses depending on the sign of the discriminant $b^2 - 4ac$. Similarly, considering for simplicity functions of two variables, $u(x, y)$, the *second-order general linear PDE*

$$au''_{xx}(x,y) + bu''_{xy}(x,y) + cu''_{yy}(x,y) + du'_x(x,y) + eu'_y(x,y) + fu = g(x,y) \tag{13.10}$$

is called hyperbolic if $b^2 - 4ac > 0$, elliptic if $b^2 - 4ac < 0$, and parabolic if $b^2 - 4ac = 0$. The sign of the discriminant for Eqs. (13.2, 13.3, 13.4, 13.5, 13.6, 13.7, and 13.8) confirms their classification as stated previously. This classification is useful and important because, as we mentioned earlier, the solutions of these different types of equations exhibit very different physical behaviors, but within a given class, the solutions have many similar features. For instance, solutions of all hyperbolic equations describing very different natural phenomena are oscillatory.

In the following sections, we shall demonstrate how the methods for solving ODEs described in the previous chapters can be used for finding the solutions of initial-boundary value problems and boundary value problems for PDEs.

13.2   The Heat Equation

13.2.1   *Physical Problems Described by the Heat Equation*

The first universal equation of mathematical physics that we consider is the heat equation that describes irreversible processes in nature. Below we present some physical examples.

Heat Conduction
The process that gives the name to the heat equation is the heat transfer, i.e., the flow of energy through a body due to a difference in temperature. The energy transfer takes place by molecular collisions.

Heat flow through a solid due to the spatial temperature inhomogeneity is characterized by the linear dependence (established by Fourier) between the heat flux $\vec{q}$ (the energy flow through a unit cross-sectional surface area per unit time) and the temperature gradient ∇T:

$$\vec{q} = -\kappa \nabla T. \tag{13.11}$$

The coefficient κ is called the thermal conductivity.
The energy conservation leads to the *continuity equation*:

$$\rho c \frac{\partial T}{\partial t} + \text{div } \vec{q} = 0, \tag{13.12}$$

where ρ is the mass density of medium and c is its heat capacity. Assuming that the thermal conductivity is constant, we find:

$$\text{div } \vec{q} = \text{div } [-\kappa \nabla T] = -\kappa \nabla^2 T.$$

Hence, Eqs. (13.11) and (13.12) lead to the *heat conduction equation* given by

$$\frac{\partial T}{\partial t} = a\nabla^2 T. \tag{13.13}$$

The positive coefficient $a = \kappa/c\rho$ is called *the thermal diffusivity*.

Note that in a steady state, when the temperature in the solid body is time independent, the heat equation reduces to *Laplace equation* given by

$$\nabla^2 T = 0, \tag{13.14}$$

that will be considered in Sect. 13.4.

If the heat is added (or removed) with the rate Q per unit time and unit volume, the heat conduction is described by the inhomogeneous heat equation:

$$\frac{\partial T}{\partial t} = a\nabla^2 T + \frac{Q}{\rho c}. \tag{13.15}$$

Diffusion

Diffusion is a mixing process which occurs when one substance is introduced into a second substance. The introduced substance, by means of molecular or atomic transport, spreads from locations where its concentration is higher to locations where the concentration is lower. Examples include the diffusion of perfume molecules into a room when the bottle is opened and the diffusion of neutrons in a nuclear reactor.

For low concentration gradients, diffusion obeys *Fick's law* in which the diffusion flux I of a substance in the mixture is proportional to the concentration gradient ∇c of that substance:

$$\vec{I} = -D\nabla c. \tag{13.16}$$

Here D is the diffusion coefficient.

Similarly to the case of the heat transfer, the mass transfer of the substance is described by the continuity equation:

$$\frac{\partial c}{\partial t} + \text{div } \vec{I} = 0, \tag{13.17}$$

which follows from the conservation of mass. Substituting $\vec{I}$ from Eq. (13.16), we obtain the diffusion equation:

$$\frac{\partial c}{\partial t} = D\nabla^2 c, \tag{13.18}$$

which is mathematically equivalent to the heat Eq. (13.13).

If the introduced substance is being created (or destroyed) – for example, in chemical reactions – we may describe this action by some function f of time and coordinates on the right side of the continuity equation. The diffusion equation then takes the form:

$$\frac{\partial c}{\partial t} = D\nabla^2 c + f, \tag{13.19}$$

which is equivalent to (13.15), up to notations.

13.2.2   The Method of Separation of Variables for One-Dimensional Heat Equation

The heat equation

$$\frac{\partial u}{\partial t} = a\frac{\partial^2 u}{\partial x^2} + f(x,t), \quad 0 < x < l, \ t > 0, \tag{13.20}$$

describes the temperature, $u(x,t)$, of a bar of length l with insulated lateral surface.

The boundary conditions depend on the physical situation on the ends of the bar. The so-called the first kind conditions correspond to controlled temperature, the *second kind* to controlled heat flow, and the *third kind* to the Newton's law of heat transfer.

These three kinds of the boundary conditions can be combined in the following way:

$$\alpha_1\frac{\partial u}{\partial x} + \beta_1 u\bigg|_{x=0} = g_1(t), \quad \alpha_2\frac{\partial u}{\partial x} + \beta_2 u\bigg|_{x=l} = g_2(t). \tag{13.21}$$

The condition of the first kind, or *Dirichlet condition* (the temperature at the boundary j, $j = 1, 2$, is fixed), corresponds to $\alpha_j = 0$; the condition of the second kind, or *Neumann condition* (heat insulated boundary j), corresponds to $\beta_j = 0$. For the condition of third kind (also called *Robin condition*), both α_j and β_j are nonzero. The physical meaning is the proportionality of the heat flux on the bar ends to the temperature on the boundary. Because the heat flux proportional to $-du/dx$ is always directed toward the colder region, the coefficients α_1 and β_1 cannot be of the same sign, and coefficients α_2 and β_2 cannot be of different signs. Equation (13.20) can have different kinds of boundary conditions on different boundaries; in that case, one says that mixed conditions are applied.

In the case of the boundary conditions of the first kind, the temperatures on the left and right ends of the bar are $g_1(t)/\beta_1$ and $g_2(t)/\beta_2$. In the case of the boundary conditions of the second kind, $g_1(t)$ and $g_2(t)$ are heat fluxes on the boundaries.

The boundary conditions of the third kind correspond to the Newton's law of the heat transfer, $-\alpha \partial u/\partial x = \beta(u - u_0)$, where u_0 is the external temperature (i.e., that of the medium outside the bar ends). In the latter case, $g_1(t)/\beta_1$ and are external temperatures, rather than temperatures of the bar ends.

Function $g_j(t) = 0$ in the boundary conditions of the first and third kind function means that the environment temperature is zero; in the boundary conditions of the second kind, it means thermal insulations of the ends.

The initial distribution of the temperature within the bar is described by the initial condition:

$$u(x, t)|_{t=0} = \varphi(x). \tag{13.22}$$

Let us first consider the *homogeneous equation* corresponding to the situation with no heat sources (or removal) within the bar:

$$\frac{\partial u}{\partial t} = a \frac{\partial^2 u}{\partial x^2}, \quad 0 < x < l, \ t > 0, \tag{13.23}$$

and *homogeneous boundary conditions*

$$\alpha_1 \frac{\partial u}{\partial x} + \beta_1 u \bigg|_{x=0} = 0, \quad \alpha_2 \frac{\partial u}{\partial x} + \beta_2 u \bigg|_{x=l} = 0. \tag{13.24}$$

Let us solve the initial-boundary value problem (13.22, 13.23, and 13.24) using the method of separation of variables (also known as Fourier method) that is described below.

The method starts with the construction of a special kind of solutions of Eq. (13.23) that can be presented as a product of two functions, the first one depending only on x and the second one depending only on t:

$$u(x, t) = X(x) \, T(t). \tag{13.25}$$

Substituting Eq. (13.25) into Eq. (13.23), we obtain:

$$X(x)T'(t) - aX''(x)T(t) = 0,$$

where primes indicate derivatives with respect to x or t.
The variables can be separated:

$$\frac{T'(t)}{aT(t)} = \frac{X''(x)}{X(x)},$$

and because the left-hand side of the equation can depend only on t and the right-hand side can depend only on x, each of them is just constant. Denoting that *separation constant as* $-\lambda$, we obtain:

$$\frac{T'(t)}{aT(t)} \equiv \frac{X''(x)}{X(x)} = -\lambda. \qquad (13.26)$$

Equation (13.26) gives two ordinary linear homogeneous differential equations, a first-order equation for function $T(t)$:

$$T'(t) + a\lambda T(t) = 0 \qquad (13.27)$$

and a second-order equation for function $X(x)$:

$$X''(x) + \lambda X(x) = 0. \qquad (13.28)$$

To find the allowed values of λ, we apply the boundary conditions. Homogenous boundary conditions (13.24) imposed on $u(x, t)$ yield homogeneous boundary conditions on the function $X(x)$ given by

$$\alpha_1 X' + \beta_1 X|_{x=0} = 0, \qquad \alpha_2 X' + \beta_2 X|_{x=l} = 0. \qquad (13.29)$$

Thus, we obtain the *Sturm-Liouville boundary value problem* (13.28), (13.29) for eigenvalues, λ, and the corresponding eigenfunctions, $X(x)$.

Let us briefly remind the main properties of eigenvalues and eigenfunctions of the Sturm-Liouville problem that are considered in Chap. 9 in more detail.

1.	There exists an *infinite set* (discrete spectrum) of real nonnegative discrete eigenvalues $\{\lambda_n\}$ and corresponding eigenfunctions $\{X_n(x)\}$. The eigenvalues increase as the index n increases: $\qquad \lambda_1 < \lambda_2 < \lambda_3 < \ldots < \lambda_n < \ldots$ ($\lim_{n \to \infty} \lambda_n = +\infty$).
2.	Eigenfunctions corresponding to different eigenvalues are *linearly independent and orthogonal*: $$\int_0^l X_i(x)X_j(x)dx = 0, \ i \neq j \qquad (13.30)$$
3.	*The completeness property* states that any function $f(x)$ which is twice differentiable on $(0, l)$ and satisfies the homogeneous boundary conditions in Eq. (13.39) can be expanded in an absolutely and uniformly converging series with eigenfunctions of the boundary value problem given in Eqs. (13.38) and (13.39) $$f(x) = \sum_{n=1}^{\infty} f_n X_n(x), \qquad f_n = \frac{1}{\|X_n\|^2} \int_0^l f(x)X_n(x)dx, \qquad (13.31)$$

where $\|X_n\|^2 = \int_0^1 X_n^2 dx$

Recall that coefficients α_j and β_j cannot vanish simultaneously. If coefficients α_k and β_k satisfy the conditions $\beta_1/\alpha_1 < 0$ and $\beta_2/\alpha_2 > 0$ (the physical meaning of those conditions is discussed above), then all the eigenvalues are positive. The eigenvalues are positive also if one of or both coefficients α_j vanish. If $\beta_1 = \beta_2 = 0$, then there exists an eigenvalue equal to zero.

Eigenvalues and eigenfunctions of a boundary value problem depend on the kind of the boundary conditions: Dirichlet (the first kind), Neumann (the second kind), Robin (the third kind), or mixed ones.

The general solution of Eq. (13.28) is

$$X(x) = C_1 \cos \sqrt{\lambda} x + C_2 \sin \sqrt{\lambda} x. \tag{13.32}$$

The coefficients C_1 and C_2 are determined from the boundary conditions (13.29):

$$\begin{cases} C_1 \beta_1 + C_2 \alpha_1 \sqrt{\lambda} = 0, \\ C_1 \left[-\alpha_2 \sqrt{\lambda} \sin \sqrt{\lambda} l + \beta_2 \cos \sqrt{\lambda} l \right] + C_2 \left[\alpha_2 \sqrt{\lambda} \cos \sqrt{\lambda} l + \beta_2 \sin \sqrt{\lambda} l \right] = 0. \end{cases} \tag{13.33}$$

This system of linear homogeneous algebraic equations has a nontrivial solution only when its determinant equals zero:

$$(\alpha_1 \alpha_2 \lambda + \beta_1 \beta_2) \tan \sqrt{\lambda} l - \sqrt{\lambda} (\alpha_1 \beta_2 - \beta_1 \alpha_2) = 0. \tag{13.34}$$

It is easy to determine (for instance by using graphical methods) that Eq. (13.34) has an infinite number of roots $\{\lambda_n\}$, which conforms to the general Sturm-Liouville theory. For each root λ_n, we obtain a nonzero solution of Eq. (13.33).

It is often convenient to present the solution in a form that allows us to consider in a unified way a mixed boundary condition and two other kinds of boundary conditions, when some of the constants α_i or β_i may be equal to zero. We will do this in the following way. Using the first expression in Eqs. (13.33), we represent C_1 and C_2 as

$$C_1 = C \alpha_1 \sqrt{\lambda_n}, \qquad C_2 = -C \beta_1,$$

where $C \neq 0$ is an arbitrary constant (because the determinant is equal to zero, the same C_1 and C_2 satisfy the second equation of the system of Eq. (13.33)). For these constraints the choice

$$C = 1 / \sqrt{\lambda_n \alpha_1^2 + \beta_1^2}$$

(with positive square root and $\alpha_1^2 + \beta_1^2 \neq 0$) allows us to obtain a simple set of coefficients C_1 and C_2. For Dirichlet boundary conditions, we assign $\alpha_1 = 0$, $\beta_1 = -1$, so that $C_1 = 0$ and $C_2 = 1$. For Neumann boundary conditions $\beta_1 = 0$, $\alpha_1 = 1$, and we have $C_1 = 1$, $C_2 = 0$. With this choice of C, the functions $X_n(x)$ are *bounded by the values ± 1.*

The alternative often used for the coefficients C_1 and C_2 corresponds to the normalizations $\|X_n\|^2 = 1$. Here and elsewhere in the book, we will use the first choice because in this case the graphs for $X_n(x)$ are easier to plot.

From the above discussion, the eigenfunctions of the Sturm-Liouville problem given by the equations:

$$X'' + \lambda X = 0,$$

$$\alpha_1 X' + \beta_1 X|_{x=0} = 0, \; \alpha_2 X' + \beta_2 X|_{x=l} = 0$$

can be written as

$$X_n(x) = \frac{1}{\sqrt{\alpha_1^2 \lambda_n + \beta_1^2}} \left[\alpha_1 \sqrt{\lambda_n} \cos \sqrt{\lambda_n} x - \beta_1 \sin \sqrt{\lambda_n} x \right]. \tag{13.35}$$

The orthogonality property (13.30) of these functions can be verified by the reader as a *Reading Exercise*. The square norms of eigenfunctions are

$$\|X_n\|^2 = \int_0^l X_n^2(x)dx = \frac{1}{2} \left[l + \frac{(\beta_2 \alpha_1 - \beta_1 \alpha_2)(\lambda_n \alpha_1 \alpha_2 - \beta_1 \beta_2)}{(\lambda_n \alpha_1^2 + \beta_1^2)(\lambda_n \alpha_2^2 + \beta_2^2)} \right]. \tag{13.36}$$

The eigenvalues are $\lambda_n = \left(\frac{\mu_n}{l} \right)^2$, where μ_n is the n-th root of the equation:

$$\tan \mu = \frac{(\alpha_1 \beta_2 - \alpha_2 \beta_1) l \mu}{\mu^2 \alpha_1 \alpha_2 + l^2 \beta_1 \beta_2}. \tag{13.37}$$

This equation remains unchanged when the sign of μ changes, which indicates that positive and negative roots are placed symmetrically on the μ axis. Because the eigenvalues λ_n do not depend on the sign of μ, it is *enough to find only positive roots* μ_n of Eq. (13.37) since negative roots do not give new values of λ_n. Clearly, μ is dimensionless.

In the cases when either both boundary conditions are of Dirichlet type, or Neumann type, or when one is Dirichlet type and the other is Neumann type, in a graphical solution of Eq. (13.37) is not needed, and we leave it to the reader as the *Reading Exercises* to obtain analytical expressions for $X_n(x)$.

As we mentioned in Chap. 9, normally there are physical restrictions on the signs of the coefficients in Eq. (13.97) so that we have $\alpha_1/\beta_1 < 0$ and $\alpha_2/\beta_2 > 0$. Obviously, only these ratios are significant, but to formulate a solution of the Sturm-Liouville problem for functions $X(x)$, it is more convenient to keep all the constants α_1, β_1, α_2, and β_2.

Now consider Eq. (13.27). It is a linear first-order differential equation and the general solution with $\lambda = \lambda_n$ is

$$T_n(t) = C_n e^{-a\lambda_n t}, \tag{13.38}$$

where C_n is an arbitrary constant. Note that negative values of λ create instability, i.e., unbounded growth of solutions. Now we have that each function

$$u_n(x,t) = T_n(t)X_n(x) = C_n e^{-a\lambda_n t} X_n(x)$$

is a solution of Eq. (13.23) satisfying boundary conditions (13.24). To satisfy the initial conditions (13.22), we compose the series

$$u(x,t) = \sum_{n=1}^{\infty} C_n e^{-a\lambda_n t} X_n(x). \tag{13.39}$$

If this series converges uniformly and so does the series obtained by differentiation twice in x and once in t, the sum gives a solution to Eq. (13.23) and satisfies the boundary conditions (13.24). The initial condition (13.22) gives:

$$u|_{t=0} = \varphi(x) = \sum_{k=1}^{\infty} C_k X_k(x), \tag{13.40}$$

which is the expansion of the function $\varphi(x)$ in a series of the eigenfunctions of the boundary value problem given by Eqs. (13.28) and (13.29).

Assuming uniform convergence of the series (13.39), we can find the coefficients C_n. Multiplying both sides of Eq. (13.40) by $X_n(x)$, integrating from 0 to l, and imposing the orthogonality condition of the functions $X_n(x)$, we obtain:

$$C_n = \frac{1}{\|X_n\|^2} \int_0^l \varphi(x)X_n(x)dx. \tag{13.41}$$

If the series (13.39) and the series obtained by differentiating it in t and twice differentiating in x are uniformly convergent then by substituting these values of coefficients, C_n, into the series (13.39), we obtain the solution of the problem stated in Eqs. (13.22) through (13.24).

Equation (13.39) gives a solution for *free heat exchange* (heat exchange without sources of heat within the body). It can be considered as the decomposition of an unknown function, $u(x,t)$, into a Fourier series over an orthogonal set of functions $\{X_n(x)\}$.

Example 13.1 Let zero temperature be maintained on both ends, $x = 0$ and $x = l$, of a uniform isotropic bar of length l with a heat-insulated lateral surface. Initially the temperature distribution inside the bar is given by

$$u(x,0) = \varphi(x) = \begin{cases} \dfrac{x}{l}u_0 & \text{for } 0 < x \le \dfrac{l}{2}, \\ \dfrac{l-x}{l}u_0 & \text{for } \dfrac{l}{2} < x < l, \end{cases}$$

where $u_0 = $ const. There are no sources of heat inside the bar. Find the temperature distribution in the interior of the bar for time $t > 0$.

Solution The problem is described by the equation:

$$\frac{\partial u}{\partial t} = a\frac{\partial^2 u}{\partial x^2} \ (0 < x < l)$$

with initial and boundary conditions

$$u(x,0) = \varphi(x), u(0,t) = u(l,t) = 0.$$

These conditions are consistent, if $\varphi(0) = \varphi(l) = 0$. The boundary conditions of the problem are Dirichlet homogeneous boundary conditions; therefore, eigenvalues and eigenfunctions of the problem are

$$\lambda_n = \left(\frac{n\pi}{l}\right)^2, X_n(x) = \sin\frac{n\pi x}{l}, \|X_n\|^2 = \frac{l}{2}, n = 1,2,3,\ldots$$

Equation (13.41) gives:

$$C_n = \frac{2}{l}\int_0^l \varphi(x)\sin\frac{n\pi x}{l}dx = \frac{4u_0}{n^2\pi^2}\sin\frac{n\pi}{2}$$

$$= \begin{cases} 0, & n = 2k, \\ \dfrac{4u_0}{(2k-1)^2\pi^2}(-1)^k, & n = 2k-1. \end{cases} \quad k = 1,2,3,\ldots$$

Hence, the distribution of temperature inside the bar at certain time is described by the series:

$$u(x,t) = \frac{4u_0}{\pi^2}\sum_{k=1}^{\infty}\frac{(-1)^k}{(2k-1)^2}e^{-\frac{a(2k-1)^2\pi^2}{l^2}t}\sin\frac{(2k-1)\pi x}{l}.$$

We have obtained the analytical solution of the problem, but, of course, we cannot visualize it without numerical computations. For that task, in this chapter we shall widely use the software described in Appendix B. Figure 13.1 shows the spatial-time-dependent solution $u(x,t)$ obtained with program *Heat* for the case when $l = 10$, $u_0 = 5$, and $a = 0.25$. All parameters are dimensionless. The dark gray line

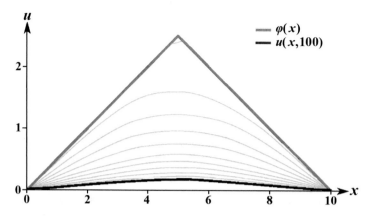

Fig. 13.1 Solution $u(x, t)$ for Example 13.1

represents the initial temperature; the black line is the temperature at time $t = 100$. The gray lines in between show the temperature evolution within the period of time from 0 until 100, with a step $\Delta t = 2$. The series for the solution converges rather rapidly because of the exponential factor.

In the case of nonhomogeneous boundary condition, an auxiliary function, $w(x, t)$, satisfying the given boundary condition, could be introduced, and the solution to the problem can be expressed as the sum of two functions:

$$u(x, t) = v(x, t) + w(x, t),$$

where $v(x, t)$ is a new, unknown function with *zero boundary condition*. Function $w(x, t)$ should be sought in a way that allows adjustment to satisfy the boundary conditions. For instance, for Dirichlet conditions $u(0, t) = g_1(t)$, $u(l, t) = g_2(t)$, such function is

$$w(x, t) = \left[1 - \frac{x}{l}\right] \cdot g_1(t) + \frac{x}{l} \cdot g_2(t). \tag{13.42}$$

Functions $w(x, t)$ for all possible situations in the nonhomogeneous boundary conditions are listed in [9].

Example 13.2 Let the pressure of air in a cylinder be equal to the atmospheric pressure. One end of the cylinder at $x = 0$ is opened at $t = 0$, and the other, at $x = l$, remains closed. The concentration of some gas in the external environment is constant ($u_0 = $ const). Find the concentration of gas in the cylinder for $t > 0$, if at $t = 0$ the gas begins to diffuse into the cylinder through the opened end.

Solution This problem can be represented by the equation:

$$\frac{\partial u}{\partial t} = D\frac{\partial^2 u}{\partial x^2},$$

subject to conditions

$$u(x,0) = 0, \quad u(0,t) = u_0, \quad DS\frac{\partial u}{\partial x}(l,t) = 0,$$

where D is the diffusion coefficient.

In order to reduce that problem to the problem with homogeneous boundary conditions, let us replace the unknown function $u(x,t)$ by another function $v(x,t)$ such that $u(x,t) = v(x,t) + w(x,t)$, where $w(x,t)$ satisfies the equation

$$\frac{\partial w}{\partial t} = D\frac{\partial^2 w}{\partial x^2}$$

and the nonhomogeneous boundary conditions $w(0,t) = u_0$, $\frac{\partial w}{\partial x}(l,t) = 0$, but not necessarily satisfies the given initial conditions. It is easy to guess that $w(x,t) = u_0$ is appropriate. Then, we obtain the problem with homogeneous boundary conditions:

$$\frac{\partial v}{\partial t} = D\frac{\partial^2 v}{\partial x^2}, v(0,t) = 0, \frac{\partial v}{\partial x}(l,t) = 0, v(x,0) = -u_0$$

which can be solved by means of the method of separation of variables.

The eigenvalues and eigenfunctions of the problem for $X_n(x)$ can be easily found (see examples in Chap. 9):

$$\lambda_n = \left[\frac{(2n-1)\pi}{2l}\right]^2, X_n(x) = \sin\frac{(2n-1)\pi x}{2l}, \|X_n\|^2 = \frac{l}{2}, n = 1,2,3,\ldots$$

Applying formula (13.31), we obtain:

$$C_n = -\frac{2u_0}{l}\int\limits_0^l \sin\frac{(2n-1)\pi x}{2l}dx = -\frac{4u_0}{(2n-1)\pi}.$$

Substituting the expression for C_n into the general formula, we obtain the final solution:

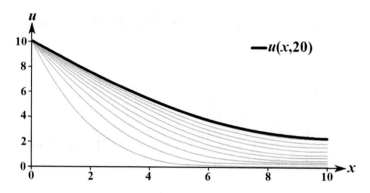

Fig. 13.2 Solution $u(x, t)$ for Example 13.2

$$u(x,t) = w(x,t) + \sum_{n=1}^{\infty} C_n e^{-\frac{D(2n-1)^2\pi^2}{4l^2}t} \sin\frac{(2n-1)\pi x}{2l} =$$

$$= u_0 - \frac{4u_0}{\pi}\sum_{n=1}^{\infty}\frac{1}{2n-1}e^{-\frac{D(2n-1)^2\pi^2}{4l^2}t} \sin\frac{(2n-1)\pi x}{2l}.$$

Figure 13.2 shows the concentration evolution from $u(x, 0) = 0$ to $u(x, 20)$ (black line) for $D = 1$, $l = 10$, $u_0 = 10$ (all dimensionless). The time step is $\Delta t = 2$. The number of harmonics in the solution is $N = 15$. The solution was obtained using the program *Heat*.

Now consider the *nonhomogeneous equation*:

$$\frac{\partial u}{\partial t} = a\frac{\partial^2 u}{\partial x^2} + f(x,t) \qquad (13.43)$$

with the initial condition (13.22) and nonhomogeneous boundary conditions (13.21).

We cannot apply the Fourier method directly to obtain a solution of the problem, because the boundary conditions are nonhomogeneous. As we discussed, the problem can be reduced to a problem with zero boundary conditions by introducing the auxiliary function, $w(x,t)$, chosen so that it satisfies the given nonhomogeneous boundary conditions.

For the function $v(x, t)$, we obtain the following boundary value problem:

$$\frac{\partial v}{\partial t} = a\frac{\partial^2 v}{\partial x^2} + \tilde{f}(x,t),$$

$$v(x, t)|_{t=0} = \tilde{\varphi}(x),$$

$$\alpha_1 v_x + \beta_1 v|_{x=0} = 0, \qquad \alpha_2 v_x + \beta_2 v|_{x=l} = 0,$$

where

$$\widetilde{f}(x,t) = f(x,t) - \frac{\partial w}{\partial t} + a\frac{\partial^2 w}{\partial x^2},$$
$$\widetilde{\varphi}(x) = \varphi(x) - w(x,0).$$

Reading Exercise check these results above.

The solution of such a problem with homogeneous boundary conditions has been considered in the previous section.

Let us express the function $v(x,t)$ (or $u(x,t)$ in case of homogeneous boundary conditions) as the sum of two functions:

$$v(x,t) = u_1(x,t) + u_2(x,t),$$

where $u_1(x,t)$ satisfies the *homogeneous equation* (free heat exchange) with the homogeneous boundary conditions and the initial condition:

$$\frac{\partial u_1}{\partial t} = a\frac{\partial^2 u_1}{\partial x^2},$$
$$u_1(x,t)|_{t=0} = \varphi(x),$$
$$\alpha_1 u_{1x} + \beta_1 u_1|_{x=0} = 0, \quad \alpha_2 u_{1x} + \beta_2 u_1|_{x=l} = 0.$$

The function $u_2(x,t)$ satisfies the *nonhomogeneous* equation (heat exchange involving internal sources) with *zero boundary and initial conditions*:

$$\frac{\partial u_2}{\partial t} = a\frac{\partial^2 u_2}{\partial x^2} + f(x,t), \tag{13.44}$$

$$u_2(x,t)|_{t=0} = 0, \tag{13.45}$$

$$\alpha_1 u_{2x} + \beta_1 u_2|_{x=0} = 0, \quad \alpha_2 u_{2x} + \beta_2 u_2|_{x=l} = 0. \tag{13.46}$$

The methods for finding $u_1(x,t)$ have been discussed in the previous section; therefore, here we concentrate our attention on finding the solutions $u_2(x,t)$. As in the case of free heat exchange inside the bar, let us expand function $u_2(x,t)$ in a series:

$$u_2(x,t) = \sum_{n=1}^{\infty} T_n(t)X_n(x), \tag{13.47}$$

where $X_n(x)$ are eigenfunctions of the corresponding homogeneous boundary value problem and $T_n(t)$ are unknown functions of t.

The boundary conditions in Eq. (13.46) for $u_2(x,t)$ are valid for any choice of functions $T_n(t)$ (when the series converge uniformly) because they are valid for the functions $X_n(x)$. Substituting the series (13.47) into Eq. (13.44), we obtain:

$$\sum_{n=1}^{\infty} \left[T_n'(t) + (a^2\lambda_n + \gamma)T_n(t)\right]X_n(x) = f(x,t). \tag{13.48}$$

Using the completeness property, we can expand the function $f(x,t)$, as function of x, into a Fourier series of the functions $X_n(x)$ on the interval $(0,l)$ such that

$$f(x,t) = \sum_{n=1}^{\infty} f_n(t)X_n(x). \tag{13.49}$$

Using the orthogonality property of the functions $X_n(x)$, we find that

$$f_n(t) = \frac{1}{\|X_n\|^2} \int_0^l f(x,t)X_n(x)dx. \tag{13.50}$$

Comparing the two expansions in Eqs. (13.48) and (13.49) for the same function $f(x,t)$, we obtain a differential equations for the functions $T_n(t)$:

$$T_n'(t) + a\lambda_n T_n(t) = f_n(t). \tag{13.51}$$

To ensure that $u_2(x,t)$ given by Eq. (13.47) satisfies the initial condition (13.45), it is necessary that the functions $T_n(t)$ obey the condition:

$$T_n(0) = 0. \tag{13.52}$$

The solution to the first-order ordinary differential equation (13.51) with initial condition (13.52) can be presented in the integral form:

$$T_n(t) = \int_0^t f_n(\tau)e^{-a\lambda_n(t-\tau)}d\tau, \tag{13.53}$$

or

$$T_n(t) = \int_0^t f_n(\tau)Y_n(t-\tau)d\tau, \text{ where } Y_n(t-\tau) = e^{-a\lambda_n(t-\tau)}.$$

Thus, the solution of the nonhomogeneous heat conduction problem for a bar with zero boundary conditions has the form:

$$u(x,t) = u_1(x,t) + u_2(x,t) = \sum_{n=1}^{\infty} \left[T_n(t) + C_ne^{-a\lambda_n t}\right]X_n(x), \tag{13.54}$$

where functions $T_n(t)$ are defined by Eq. (13.53) and coefficients C_n have been found earlier when we considered the homogeneous heat equation.

Example 13.3 A point-like heat source with power $Q = $ const is located at x_0 $(0 < x_0 < l)$ in a uniform isotropic bar with insulated lateral surfaces. The initial temperature of the bar is zero. Temperatures at the ends of the bar are maintained at zero. Find the temperature inside the bar for $t > 0$.

Solution The boundary value problem modeling heat propagation for this example is

$$\frac{\partial u}{\partial t} = a\frac{\partial^2 u}{\partial x^2} + \frac{Q}{c\rho}\delta(x - x_0),$$

$$u(x, 0) = 0, \quad u(0, t) = u(l, t) = 0,$$

where $\delta(x - x_0)$ is the delta function.

The boundary conditions are Dirichlet homogeneous boundary conditions, so the eigenvalues and eigenfunctions of the problem are

$$\lambda_n = \left(\frac{n\pi}{l}\right)^2, \quad X_n(x) = \sin\frac{n\pi x}{l}, \quad \|X_n\|^2 = \frac{l}{2} \quad (n = 1, 2, 3, \ldots).$$

In the case of homogeneous initial conditions, $\varphi(x) = 0$, we have $C_n = 0$, and the solution $u(x, t)$ is defined by the series:

$$u(x, t) = \sum_{n=1}^{\infty} T_n(t) \sin\frac{n\pi x}{l},$$

where f_n and T_n are defined by Eqs. (2.40) and (2.44):

$$f_n(t) = \frac{2}{l}\int_0^l \frac{Q}{c\rho}\delta(x - x_0)\sin\frac{n\pi x}{l}dx = \frac{2Q}{lc\rho}\sin\frac{n\pi x_0}{l}dx,$$

$$T_n(t) = \frac{2Ql}{c\rho a^2 n^2 \pi^2}\left(1 - e^{-n^2 a\pi^2 t/l^2}\right)\sin\frac{n\pi x_0}{l}.$$

Substituting the expression for $T_n(t)$ into the general formulas, we obtain the solution of the problem:

$$u(x, t) = \frac{2Ql}{c\rho a^2 \pi^2}\sum_{n=1}^{\infty}\frac{1}{n^2}\left(1 - e^{-n^2 a\pi^2 t/l^2}\right)\sin\frac{n\pi x_0}{l}\sin\frac{n\pi x}{l}.$$

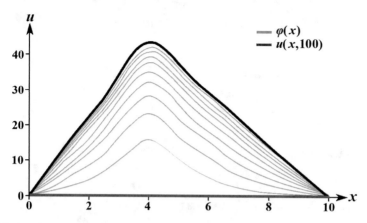

Fig. 13.3 Solution $u(x, t)$ for Example 13.3

Figure 13.3 shows the spatial-time-dependent solution $u(x, t)$ for Example 2.3. This solution was obtained using the program *Heat* for the case when $l = 10$, $Q/cp = 5$, $x_0 = 4$, and $a = 0.25$. All parameters are dimensionless. The dark gray line represents the initial temperature; the black line is the temperature at time $t = 100$. The gray lines in between show the temperature evolution within the period of time from 0 until 100, with a step $\Delta t = 2$.

Example 13.4 Consider a rod ($0 \leq x \leq l$) that is thermally insulated over its lateral surface. The left end is kept at the temperature u_0 and the right end at zero temperature. The initial temperature of the rod is $u(x, 0) = u_0(1 - x/l)$. Internal heat source is active in the rod, and its intensity (per unit length of the rod) is given by $f(x, t) = B$. Find the distribution of temperature in the rod.

Solution The problem is described by the equation:

$$\frac{\partial u}{\partial t} = a \frac{\partial^2 u}{\partial x^2} + B$$

with initial and boundary conditions

$$u(l, t) = 0, \quad u(x, 0) = u_0(1 - x/l), \quad u(0, t) = u_0.$$

The boundary conditions are nonhomogeneous, so the solution to the problem should be sought in the form:

$$u(x, t) = v(x, t) + w(x, t),$$

where a function $v(x, t)$ is the solution to the boundary value problem with zero boundary conditions and $w(x, t)$ is an auxiliary function satisfying the given boundary conditions; clearly,

$$w(x, t) = \left(1 - \frac{x}{l}\right)u_0.$$

The boundary conditions are *Dirichlet* type, and the eigenvalues and eigenfunctions of the problem are

$$\lambda_n = \left(\frac{n\pi}{l}\right)^2, X_n(x) = \sin\frac{n\pi x}{l}, \|X_n\|^2 = \frac{l}{2}, n = 1, 2, 3, \quad \ldots$$

The coefficients C_n

$$C_n = \frac{2}{l}\int_0^l 0 \cdot \sin\frac{n\pi x}{l}dx = 0, \text{ for all } n.$$

Following the algorithm described above, we have $\widetilde{f} = f(x, t) = B$; then, we find:

$$f_0(t) = \frac{1}{l}\int_0^l f(x)dx = \frac{1}{l}\int_0^l Bdx = B,$$

$$T_0(t) = \int_0^t f_0(\tau)\,d\tau = B\int_0^t d\tau = Bt.$$

For $n > 0$

$$f_n(t) = \frac{2}{l}\int_0^l f(x)\sin\frac{n\pi x}{l}dx = \frac{2B}{l}\int_0^l \sin\frac{n\pi x}{l}dx = \frac{2B}{n\pi}[1 - (-1)^n],$$

$$T_n(t) = \int_0^t f_n(p)\cdot e^{-\frac{an^2\pi^2}{l^2}(t-p)}dp = \frac{2B}{n\pi}[1 - (-1)^n]e^{-\frac{an^2\pi^2}{l^2}t}\int_0^t e^{\frac{an^2\pi^2}{l^2}p}dp =$$

$$= \frac{2B}{n\pi}[1 - (-1)^n]\cdot\frac{l^2}{an^2\pi^2}\left(1 - e^{-\frac{an^2\pi^2}{l^2}t}\right).$$

Finally, we obtain the solution of the problem:

$$u(x, t) = v(x, t) + w(x, t) =$$

$$= B + Bt + \frac{2Bl^2}{a\pi^3}\sum_{n=1}^{\infty}\left[\frac{1 - (-1)^n}{n^3}\left(1 - e^{-\frac{an^2\pi^2}{l^2}t}\right)\right]\sin\frac{n\pi x}{l} + \left(1 - \frac{x}{l}\right)u_0.$$

The temporal evolution of $u(x, t)$ is shown in Fig. 13.4.

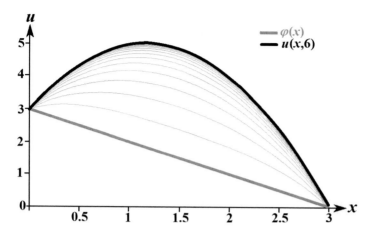

Fig. 13.4 Solution $u(x, t)$ obtained using the program *Heat* for Example 13.4

Problems

In all problems consider rods that are thermally insulated over their lateral surfaces. In the initial time, $t = 0$, the temperature distribution is given by $u(x, 0) = \varphi(x)$, $0 < x < l$. There are no heat sources or absorbers inside the rod. Find the temperature evolution inside the rod for the following cases:

1. (a) $\varphi_1(x) = 1$, (b) $\varphi_2(x) = x$, and (c) $\varphi_3(x) = x(l - x)$. The ends of the rod are kept at zero temperature.
2. (a) $\varphi_1(x) = x^2$, (b) $\varphi_2(x) = x$, and (c) $\varphi_3(x) = x(l - x/2)$. The left end of the rod is kept at zero temperature, and the right end is thermally insulated from the environment.
3. (a) $\varphi_1(x) = x$, (b) $\varphi_2(x) = 1$, and (c) $\varphi_3(x) = l^2 - x^2$. The left end of the rod is thermally insulated, and the right end is kept at zero temperature.
4. (a) $\varphi_1(x) = x$, (b) $\varphi_2(x) = l^2 - x^2$, and (c) $\varphi_3(x) = x^2\left(1 - \frac{2x}{3}\right)$. Both ends of the rod are thermally insulated.

In problems 5 and 6, consider rods whose ends are subject to heat transfer according to Newton's law.

5. The left end of the rod is kept at a constant temperature $u(0, t) = u_1$, and a constant heat flow is supplied to the right end of the rod. The initial temperature of the rod is (a) $\varphi_1(x) = x^2$, (b) $\varphi_2(x) = x$, and (c) $\varphi_3(x) = x\left(l - \frac{x}{2}\right)$.
6. Constant heat flows are supplied to both ends of the rod. The initial temperature of the rod is (a) $\varphi_1(x) = x$, (b) $\varphi_2(x) = l^2 - x^2$, and (c) $\varphi_3(x) = x^2(1 - 2x/3)$.

In the following problems, the initial temperature of the rod is zero. Find the temperature change inside the rod for the following cases:

7. The left end of the rod is kept at the constant temperature $u(0, t) = u_1$, and the temperature of the right end changes according to $g_2(t) = A \cos \omega t$.

8. The heat flow $g_1(t) = A \cos \omega t$ is supplied to the left end of the rod from outside, while the right end of the rod is kept at the constant temperature $u(l, t) = u_2$.
9. The left end of the rod is kept at the constant temperature $u(0, t) = u_1$, and the right end is subjected to a convective heat transfer with the environment that has a temperature that varies as $u_{md}(t) = A \sin \omega t$.
10. The left end of the rod is subjected to a convective heat transfer with the environment that has a temperatures that varies as $u_{md}(t) = A \cos \omega t$, and the right end is kept at the constant temperature $u(l, t) = u_2$.

13.2.3   Heat Conduction Within a Circular Domain

Let a uniform circular plate be placed in the horizontal xy plane and bounded at the circular periphery by a radius of length l. The plate is assumed to be thin enough so that the temperature is the same at all points with the same x, y coordinates.

For a circular domain, it is convenient to write the heat conduction equation in polar coordinates, where the Laplace operator is

$$\nabla^2 u = \frac{\partial^2 u}{\partial r^2} + \frac{1}{r}\frac{\partial u}{\partial r} + \frac{1}{r^2}\frac{\partial^2 u}{\partial \varphi^2}, \tag{13.55}$$

and the heat conduction is described by the equation:

$$\frac{\partial u}{\partial t} = a\left(\frac{\partial^2 u}{\partial r^2} + \frac{1}{r}\frac{\partial u}{\partial r} + \frac{1}{r^2}\frac{\partial^2 u}{\partial \varphi^2}\right) + f(r, \varphi, t), \tag{13.56}$$

$$0 \le r < l, \quad 0 \le \varphi < 2\pi, \quad t > 0.$$

Here $u(r, \varphi, t)$ is the temperature of the plate at a point (r, φ) at time t.

The *initial condition* defines the temperature distribution within the membrane at time zero and may be expressed in generic form as

$$u(r, \varphi, 0) = \xi(r, \varphi).$$

The *boundary condition* describes the thermal condition around the boundary at any time t. In general, the boundary condition can be written as follows:

$$\alpha\frac{\partial u}{\partial r} + \beta u\bigg|_{r=l} = g(\varphi, t). \tag{13.57}$$

It is obvious that a function $g(\varphi, t)$ must be a singled-valued periodic function of φ of period 2π, that is,

$$g(\varphi + 2\pi, t) = g(\varphi, t).$$

Again, we consider three main types of boundary conditions:

(i)	Boundary condition of the first kind (*Dirichlet condition*), $u(l, \varphi, t) = g(\varphi, t)$, where the temperature at the boundary is given; specifically, it can be taken as zero, $g(\varphi, t) \equiv 0$.
(ii)	Boundary condition of the second kind (*Neumann condition*), $u_r(l, \varphi, t) = g(\varphi, t)$, in which case the heat flux at the boundary is given; specifically, the boundary can be thermally insulated, $g(\varphi, t) \equiv 0$.
(iii)	Boundary condition of the third kind (*Robin condition*), $u_r(l, \varphi, t) + hu(l, \varphi, t) = g(\varphi, t)$, where the conditions of heat exchange with a medium are specified (here $h =$ const).

In the case of nonhomogeneous boundary condition, we introduce an auxiliary function, $w(r, \varphi, t)$, satisfying the given boundary condition, and express the solution to the problem as the sum of two functions:

$$u(r, \varphi, t) = v(r, \varphi, t) + w(r, \varphi, t),$$

where $v(r, \varphi, t)$ is a new, unknown function with *zero boundary condition*. Function $w(r, \varphi, t)$ can be sought in the form:

$$w(r, \varphi, t) = \left(c_0 + c_2 r^2\right) g(\varphi, t)$$

with constants c_0 and c_2 to be adjusted to satisfy the boundary conditions.

In the case of *nonhomogeneous equation* and *zero boundary conditions*, we may express the solution $v(r, \varphi, t)$ as the sum of two functions:

$$v(r, \varphi, t) = u_1(r, \varphi, t) + u_2(r, \varphi, t),$$

where $u_1(r, \varphi, t)$ is the solution of the homogeneous equation (free heat exchange) with the given initial conditions and $u_2(r, \varphi, t)$ is the solution of the nonhomogeneous equation (heat exchange involving internal sources) with zero initial and boundary conditions. In some situations a function $u_2(x, t)$ is easy to guess. A standard procedure to find a function $u_2(x, t)$ can be found in [9].

Let us find the solution of the homogeneous equation:

$$\frac{\partial u_1}{\partial t} = a \left(\frac{\partial^2 u_1}{\partial r^2} + \frac{1}{r} \frac{\partial u_1}{\partial r} + \frac{1}{r^2} \frac{\partial^2 u_1}{\partial \varphi^2} \right) \tag{13.58}$$

with the initial condition

$$u_1(r, \varphi, 0) = \xi(r, \varphi) \tag{13.59}$$

and homogeneous boundary condition

$$\alpha\frac{\partial u_1}{\partial r}+\beta u_1\bigg|_{r=l}=0. \tag{13.60}$$

As was done previously, we can separate the variables, so that

$$u_1(r,\varphi,t)=T(t)V(r,\varphi). \tag{13.61}$$

Substituting this in Eq. (13.58) and separating the variables, we obtain:

$$\frac{T''(t)}{aT(t)}=\frac{1}{V(r,\varphi)}\left[\frac{\partial^2 V}{\partial r^2}+\frac{1}{r}\frac{\partial V}{\partial r}+\frac{1}{r^2}\frac{\partial^2 V}{\partial\varphi^2}\right]=-\lambda,$$

where λ is a separation of variables constant (recall that a choice of negative sign before λ is convenient). Thus, the function $T(t)$ is the solution of the first-order ordinary linear homogeneous differential equation:

$$T'(t)+a\lambda T(t)=0 \tag{13.62}$$

and $V(r,\varphi)$ is the solution to the following boundary value problem:

$$\frac{\partial^2 V}{\partial r^2}+\frac{1}{r}\frac{\partial V}{\partial r}+\frac{1}{r^2}\frac{\partial^2 V}{\partial\varphi^2}+\lambda V=0, \tag{13.63}$$

$$\alpha\frac{\partial V}{\partial r}+\beta V\bigg|_{r=l}=0. \tag{13.64}$$

Two restrictions on $V(r,\varphi)$ are that it be bounded $|V(r,\varphi)|<\infty$ and that it be periodic in φ: $V(r,\varphi)=V(r,\varphi+2\pi)$.

Let us represent the function in the form:

$$V(r,\varphi)=R(r)\Phi(\varphi). \tag{13.65}$$

Substituting (13.65) into (13.63) and separating the variables, we have:

$$\frac{r^2\nabla_r R}{R}=-\frac{\nabla_\varphi^2\Phi}{\Phi}=\lambda.$$

Because the first term does not depend on the angular variable φ and the second does not depend on r, each term must equal a constant which we denoted as λ. From here we obtain two separate equations for $R(r)$ and $\Phi(\varphi)$:

$$\Phi''+\lambda\Phi=0, \tag{13.66}$$

$$r^2 R'' + r R' - \lambda R = 0. \tag{13.67}$$

Clearly, the solution of Eq. (13.66) must obey the periodicity condition:

$$\Phi(\varphi + 2\pi) = \Phi(\varphi), \tag{13.68}$$

which can be true only for positive values, $\lambda = n^2$ and can be written as $\Phi = e^{in\varphi}$. Then, from (13.68) we obtain $e^{2in\pi} = 1$; thus, n is *integer*: $n = 0, \pm 1, \pm 2, \ldots$.
The eigenfunctions are

$$\Phi = \Phi_n(\varphi) = \begin{cases} \cos n\varphi, \\ \sin n\varphi. \end{cases} \tag{13.69}$$

Negative values of n correspond to the same eigenfunctions and therefore need not be included in the list of eigenvalues.
The boundary condition for the function $R(r)$ is

$$\left. \alpha \frac{\partial R}{\partial r} + \beta R \right|_{r=l} = 0. \tag{13.70}$$

Equation (13.167) with $\lambda = n^2$ is the Bessel equation. In Chap. 10 we obtained the following eigenvalues and eigenfunctions for the BVP problem (13.167):

$$\lambda_{nm} = \left(\frac{\mu_m^{(n)}}{l} \right)^2, \quad R_{nm}(r) = J_n\left(\frac{\mu_m^{(n)}}{l} r \right), \quad n, m = 0, 1, 2, \ldots, \tag{13.71}$$

where $\mu_m^{(n)}$ is the m-th *positive* root of the equation

$$\alpha \mu J_n'(\mu) + \beta l J_n(\mu) = 0, \tag{13.72}$$

and $J_n(\mu)$ is the Bessel function of the first kind.
Collecting the above results, we may write the eigenfunctions of the given boundary value problem as

$$V_{nm}^{(1)}(r, \varphi) = J_n\left(\frac{\mu_m^{(n)}}{l} r \right) \cos n\varphi \quad \text{and} \quad V_{nm}^{(2)}(r, \varphi) = J_n\left(\frac{\mu_m^{(n)}}{l} r \right) \sin n\varphi. \tag{13.73}$$

We will see in Sect. 13.4.3 that the functions $V_{nm}^{(1,2)}(r, \varphi)$ are the eigenfunctions of the Laplace operator in polar coordinates in the domain $0 \leq r \leq l, 0 \leq \varphi < 2\pi$, and λ_{nm} are the corresponding eigenvalues.
The squared norms of eigenfunctions, $\left\| V_{nm}^{(1,2)}(r, \varphi) \right\|^2 = \int_0^{2\pi} d\varphi \int_0^\infty \left| V_{nm}^{(1,2)}(r, \varphi) \right|^2 r^2 dr$, are

$$\left\|V_{nm}^{(1,2)}(r,\varphi)\right\|^2 = \pi\|R_{nm}\|^2 \text{ for } n>0, \text{ and } \left\|V_{0m}^{(1)}(r,\varphi)\right\|^2 = 2\pi\|R_{nm}\|^2;$$

the norms $\|R_{nm}\|^2$ are presented in Sect. 10.2.7.

With the eigenvalues, λ_{nm}, we can write the solution of the differential equation:

$$T'_{nm}(t) + a\lambda_{nm}T_{nm}(t) = 0 \qquad (13.74)$$

as

$$T_{nm}(t) = C_{nm}e^{-a\lambda_{nm}t}. \qquad (13.75)$$

From this we see that the general solution for the function u_1 is

$$u_1(r,\varphi,t) = \sum_{n=0}^{\infty}\sum_{m=0}^{\infty}\left[a_{nm}V_{nm}^{(1)}(r,\varphi) + b_{nm}V_{nm}^{(2)}(r,\varphi)\right]e^{-a\lambda_{nm}t}. \qquad (13.76)$$

The coefficients a_{nm} and b_{nm} are defined using the function which expresses the initial condition and the orthogonality property of functions $V_{nm}^{(1)}(r,\varphi)$ and $V_{nm}^{(2)}(r,\varphi)$:

$$a_{nm} = \frac{1}{\left\|V_{nm}^{(1)}\right\|^2}\int_0^l\int_0^{2\pi}\xi(r,\varphi)\,V_{nm}^{(1)}(r,\varphi)\,rdrd\varphi, \qquad (13.77)$$

$$b_{nm} = \frac{1}{\left\|V_{nm}^{(2)}\right\|^2}\int_0^l\int_0^{2\pi}\xi(r,\varphi)\,V_{nm}^{(2)}(r,\varphi)\,rdrd\varphi. \qquad (13.78)$$

Example 13.5 The initial temperature distribution within a very long (infinite) cylinder of radius l is

$$\xi(r,\varphi) = u_0\left(1 - \frac{r^2}{l^2}\right), u_0 = \text{const.}$$

Find the distribution of temperature within the cylinder if its surface is kept at constant zero temperature. Generation (or absorption) of heat by internal sources is absent.

Solution The boundary value problem modeling the process of the cooling of an infinite cylinder is

$$\frac{\partial u}{\partial t} = a\left[\frac{\partial^2 u}{\partial r^2} + \frac{1}{r}\frac{\partial u}{\partial r}\right],$$

$$u(r, \varphi, 0) = u_0 \left(1 - \frac{r^2}{l^2} \right), u(l, \varphi, t) = 0.$$

The initial temperature does not depend on the polar angle φ; thus, only terms with $n = 0$ are not zero, and solution includes only functions $V_{0m}^{(1)}(r, \varphi)$ given by the first of Eq. (13.68). The solution $u(r, \phi, t)$ is therefore given by the series:

$$u(r, \varphi, t) = \sum_{m=0}^{\infty} a_{0m} e^{-a\lambda_{0m}t} J_0 \left(\frac{\mu_m^{(0)}}{l} r \right).$$

The boundary condition of the problem is of Dirichlet type, so eigenvalues are given by the equation:

$$J_n(\mu) = 0;$$

therefore, $\mu_m^{(n)}$ are the roots of this equation.

The coefficients a_{0m} are given by Eq. (13.72):

$$a_{0m} = \frac{2\pi}{\left\| V_{0m}^{(1)} \right\|^2} \int_0^l u_0 \left(1 - \frac{r^2}{l^2} \right) J_0 \left(\mu_m^{(0)} r/l \right) r dr.$$

Using the formulas

$$\int x^n J_{n-1}(x) dx = x^n J_n(x), \quad J_{n+1}(x) = \frac{2n}{x} J_n(x) - J_{n-1}(x),$$

and taking into account that $J_0 \left(\mu_m^{(0)} \right) = 0$, we can arrive to the result:

$$a_{0m} = \frac{8A}{\left(\mu_m^{(0)} \right)^3 J_1 \left(\mu_m^{(0)} \right)}.$$

Thus, the distribution of temperature within the cylinder is given by the series in Bessel functions of zero-th order:

$$u(r, \varphi, t) = 8u_0 \sum_{m=0}^{\infty} \frac{e^{-a\lambda_{0m}t}}{\left(\mu_m^{(0)} \right)^3 J_1 \left(\mu_m^{(0)} \right)} J_0 \left(\frac{\mu_m^{(0)}}{l} r \right).$$

On examination of the solution, we see that, due to the exponential nature of the coefficients, the final temperature of the cylinder after a long time will be zero. This is due to dissipation of energy to the surrounding space and could have been anticipated from the physical configuration of the problem.

Solutions of the heat equation on a rectangular domain can be found in [9].

Problems
In problems below we consider a circular plate (or very long cylinder) of radius l thermally insulated over its lateral surface. In the initial time, $t = 0$, the temperature distribution is given by $u(r, \varphi, 0) = \xi(r, \varphi)$. There are no heat sources or absorbers inside the plate. Find the distribution of temperature within the plate at any later time:

1. $\xi(r, \varphi) = Ar(l^2 - r^2) \sin \varphi$. The boundary of the plate is kept at constant zero temperature.
2. $\xi(r, \varphi) = u_0 r \cos 2\varphi$. The boundary of the plate is thermally insulated.
3. $\xi(r, \varphi) = u_0(1 - r^2/l^2)$. The boundary of the plate is subjected to convective heat transfer with the environment that has zero temperature.
4. $\xi(r, \varphi) = Ar(l^2 - r^2) \sin \varphi$. The surface of the cylinder is kept at constant temperature $u = u_0$.
5. $\xi(r, \varphi) = u_0 \sin 4\varphi$. The surface of the cylinder is subjected to convective heat transfer with the environment that has a temperature $u = u_0$.

In the following problems $\xi(r, \varphi) = 0$ and the edge of the plate is kept at zero temperature. A heat is generated uniformly throughout the plate; the intensity of internal sources (per unit area of the plate) is $Q(t)$. Find the temperature distribution within the plate:

6. $Q(t) = A \cos \omega t$.
7. $Q(t) = A \sin \omega t$.
8. $Q(t) = A(l - r) \sin \omega t$.

13.3 The Wave Equation

13.3.1 Physical Problems Described by the Wave Equation

Transverse Oscillations of a String
Let us start with the following physical example. Consider the problem of small transverse oscillations of a thin, stretched string. The transverse oscillations mean that the movement of each point of the string is perpendicular to the x-axis with no displacements or velocities along this axis. Let $u(x, t)$ represent displacements of the points of the string from the equilibrium so that a graph of the function $u(x, t)$ gives the string's amplitude at the location x and time t (u plays the role of the y coordinate; see Fig. 13.5). Small oscillations mean that the displacement amplitude $u(x, t)$ is small relative to the string length and that the partial derivative $u_x(x, t)$ is small for all values of x and t (i.e., the slope is small everywhere during the string's motion); therefore, the square of the slope can be neglected: $(u_x)^2 << 1$ (u_x is dimensionless). With these assumptions, which are justified in many applications, the equations that we will derive will be *linear* partial differential equations.

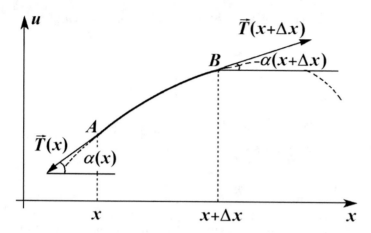

Fig. 13.5 Small oscillations of a string

Consider an interval $(x, x + \Delta x)$ in Fig. 13.5. Points on the string move perpendicular to the x direction; thus, the sum of the x components of the tension forces at points x and $x + \Delta x$ equals zero. Because the tension forces are directed along the tangent lines, we have:

$$- T(x) \cos \alpha(x) + T(x + \Delta x) \cos \alpha(x + \Delta x) = 0.$$

Clearly, $\tan \alpha = u_x$ and for small oscillations $\cos \alpha = 1/\sqrt{1 + \tan^2 \alpha} = 1/\sqrt{1 + u_x^2} \approx 1$; thus,

$$T(x) \approx T(x + \Delta x),$$

that is, the value of the tension T does not depend on x, and for all x and t it approximately equals its value in the equilibrium state.

In the situation shown in Fig. 13.5, at a point x the vertical component of the tension force is

$$T_{vert} = - T \sin \alpha(x).$$

The same expression with a positive sign holds at the point $x + \Delta x$: $T_{vert} = T \sin \alpha(x + \Delta x)$. The signs of T_{vert} at x and $x + \Delta x$ depend on the orientation of the string segment and are opposite for the two ends of the segment.

For small oscillations $\sin \alpha = \tan \alpha/\sqrt{1 + \tan^2 \alpha} = u_x/\sqrt{1 + u_x^2} \approx u_x$, so that

$$T_{vert} = - T u_x(x, t)$$

and the sum of vertical components of tension forces at points x and $x + \Delta x$ is

$$T^{net}_{vert} = T[\sin \alpha(x + \Delta x) - \sin \alpha(x)] = T[u_x(x + \Delta x, t) - u_x(x, t)].$$

As $\Delta x \to 0$ we arrive at

$$T^{net}_{vert} = T \frac{\partial^2 u}{\partial x^2} dx.$$

On the other hand, the force T^{net}_{vert} acting on segment Δx is equal to the mass of this segment, $\rho(x)dx$ (where $\rho(x)$ is a linear mass density of the string) times acceleration:

$$T^{net}_{vert} = \rho(x) \frac{\partial^2 u}{\partial t^2} dx.$$

If there is also an additional external force $F(x, t)$ per unit length acting on the string perpendicular to the x-axis (for small oscillations the force should be small with respect to tension, T), we obtain the equation for *forced transverse oscillations of a string*:

$$\rho(x) \frac{\partial^2 u}{\partial t^2} = T \frac{\partial^2 u}{\partial x^2} + F(x, t). \tag{13.79}$$

For the case of a constant mass density, $\rho = $ const, i.e., for a uniform string, this equation can be written as

$$\frac{\partial^2 u}{\partial t^2} = a^2 \frac{\partial^2 u}{\partial x^2} + f(x, t), \tag{13.80}$$

where $a = \sqrt{T/\rho} = $ const, $f(x, t) = F(x, t)/\rho$. For instance, if the weight of the string cannot be neglected and the force of gravity is directed down perpendicular to the x-axis, we have $f(x, t) = -mg/l\rho = -g$. If there is no external force, $F(x, t) \equiv 0$, we have the equation for free oscillations of a string:

$$\frac{\partial^2 u}{\partial t^2} = a^2 \frac{\partial^2 u}{\partial x^2}, \tag{13.81}$$

which is referred to as the *homogeneous wave equation*; Eq. (13.80) is called *nonhomogeneous wave equation*. With subscripts for derivatives, Eq. (13.81) is

$$u_{tt}(x, t) = a^2 u_{xx}(x, t).$$

The general solution of that equation is the sum of two solutions:

$$u(x, t) = f_1(x - at) \quad \text{and} \quad u(x, t) = f_2(x + at),$$

where f_1 and f_2 are arbitrary, twice differentiable functions. Each of these solutions has a simple physical interpretation. In the first case, the displacement $u = f_1$ at point x and time t is the same as that at point $x + a\Delta t$ at time $t + \Delta t$. Thus, the disturbance moves in the direction of increasing x with velocity a. The quantity a is therefore the "velocity of propagation" of the disturbance, or the *wave speed*. In the second case, the displacement $u = f_2$ at point x at time t is found at the point with a coordinate $x - a\Delta t$ at a later time $t + \Delta t$. This disturbance therefore travels in the direction of decreasing x with velocity a. The general solution of Eq. (3.4) can be written as the sum of f_1 and f_2:

$$u(x, t) = f_1(x - at) + f_2(x + at). \tag{13.82}$$

Equation (13.81) describes the simplest situation with no external forces (including string's weight) and no dissipation. For a string vibrating in an elastic medium when the force on the string from the medium is proportional to string's deflection, $F = -\alpha u$ (that is Hooke's law; α is a coefficient with the dimension of force per length squared; F is a force per unit length), we have the wave equation in the form:

$$\rho \frac{\partial^2 u}{\partial t^2} = T \frac{\partial^2 u}{\partial x^2} - \alpha u. \tag{13.83}$$

When a string oscillates in a medium with a force of friction proportional to the speed, the force per unit length, F, is given by $F = -k u_t$, where k is a coefficient of friction. In that case, the equation contains the time derivative $u_t(x, t)$:

$$\frac{\partial^2 u}{\partial t^2} = a^2 \frac{\partial^2 u}{\partial x^2} - 2\kappa \frac{\partial u}{\partial t}, \tag{13.84}$$

where $2\kappa = k/\rho$ (κ has dimension of inverse time).

Equations (13.79), (13.83), and (13.84) are of hyperbolic type (as we discussed in Sect. 13.1).

Next, consider the homogeneous wave equation with constant coefficients in the general form:

$$u_{tt} - a^2 u_{xx} + b_1 u_t + b_2 u_x + c u = 0. \tag{13.85}$$

If we introduce a new function, $v(x, t)$, using the substitution

$$u(x, t) = e^{\lambda x + \mu t} v(x, t), \tag{13.86}$$

with $\lambda = b_1/2$ and $\mu = b_2/2$, then Eq. (13.85) reduces to a substantially simpler form:

$$v_{tt} - a^2 v_{xx} + cv = 0. \tag{13.87}$$

Exercise Prove the statement above.

It is immediately seen that if $c \neq 0$, Eq. (13.87) does not allow a solution in the form $f(x \pm at)$. Physically this result is related with the phenomena of wave dispersion.

One-dimensional wave equations describe many other phenomena of periodic nature, such as electric oscillations, acoustic longitudinal waves propagating in different materials, and transverse waves in fluids in shallow channels – for details see [7].

Rod Oscillations

Consider a thin elastic rod of cylindrical, rectangular, or other uniform cross section. In this case, a force applied along the axis, perpendicular to the (rigid) cross section, will cause changes in the length of the rod. We will assume that a force acts along the rod axis and each cross-sectional area can move only along the rod axis. Such assumptions can be justified if the transverse dimensions are substantially smaller compared to the length of the rod and the force acting on the rod is comparatively weak.

If a force compresses the rod along its axis and next the rod is released, it will begin to vibrate along this axis. Contrary to transverse string oscillations, such considered rod oscillations are *longitudinal*. Let the ends of the rod be located at the points $x = 0$ and $x = l$ when it is at rest. The location of some cross section at rest will be given by x (Fig. 13.6). Let the function $u(x, t)$ be the longitudinal shift of this cross section from equilibrium at time t. The force of tension T is proportional to $u_x(x, t)$, the relative length change at location x, and the cross-sectional area of the rod (Hooke's law); thus, $T(x, t) = EAu_x(x, t)$, where E is the elasticity modulus.

Consider the element of the rod between two cross sections A and A_1 with coordinates at rest x and $x + dx$. For small deflections, the resultant of two forces of tension at these cross sections is

$$T_{x+dx} - T_x = EA\frac{\partial u}{\partial x}\bigg|_{x+dx} - EA\frac{\partial u}{\partial x}\bigg|_x \approx EA\frac{\partial^2 u}{\partial x^2}dx.$$

The acceleration of this element is $\partial^2 u/\partial t^2$ in the direction of the resultant force. Together, these two equations give the equation of longitudinal motion of a cross-sectional element as

$$\rho A dx \frac{\partial^2 u}{\partial t^2} = EA\frac{\partial^2 u}{\partial x^2}dx,$$

where ρ is the rod density. Using the notation

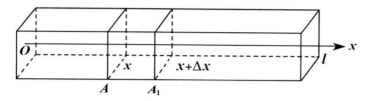

Fig. 13.6 Arbitrary segment of a rod of length Δx

$$a = \sqrt{E/\rho}, \qquad (13.88)$$

we obtain the differential equation for *longitudinal free oscillations of a uniform rod*:

$$\frac{\partial^2 u}{\partial t^2} = a^2 \frac{\partial^2 u}{\partial x^2}. \qquad (13.89)$$

As we already know, solutions of such hyperbolic equations have a wave character with the speed of wave propagation, a, given by (13.88).

If there is also an external force $F(x, t)$ per unit volume, we obtain instead the equation:

$$\rho A dx \frac{\partial^2 u}{\partial t^2} = EA \frac{\partial^2 u}{\partial x^2} dx + F(x, t)A dx,$$

or

$$\frac{\partial^2 u}{\partial t^2} = a^2 \frac{\partial^2 u}{\partial x^2} + \frac{1}{\rho} F(x, t). \qquad (13.90)$$

This is the equation for *forced oscillations of a uniform rod*. Note that Eq. (13.90) is equivalent to Eq. (13.80) for a string under forced oscillations.

For longitudinal waves there is an important physical restriction: the derivative $u_t(x, t)$ must be small in comparison with the speed of wave propagation: $u_t < < a$. The initial conditions are similar to those for a string:

$$u(x, 0) = \varphi(x), \quad \frac{\partial u}{\partial t}(x, 0) = \psi(x),$$

which are initial deflection and initial speed of points of a rod, respectively.

Electrical Oscillations in a Circuit

Let us briefly set up the boundary value problem for a current and a voltage in a circuit which contains the resistance R, capacitance C, and inductance L as well as the possibility of leakage, G. For simplicity, these quantities are considered uniformly distributed in a wire placed along the x-axis and defined to be per unit length. The functions $i(x, t)$ and $V(x, t)$ represent the current and voltage at a location x along the wire and at a time t.

Applying *Ohm's law*, introduced by the German physicist Georg Ohm in 1827, to a circuit with nonzero self-inductance and using charge conservation, the so-called telegraph equations can be obtained (for more details see [7]):

$$I_{xx}(x, t) = LCI_{tt}(x, t) + (RC + LG)I_t(x, t) + RGI(x, t) \tag{13.91}$$

and

$$V_{xx}(x, t) = LCV_{tt}(x, t) + (RC + LG)V_t(x, t) + RGV(x, t). \tag{13.92}$$

These equations are similar to the equations of string oscillations; they describe electrical oscillations in RCL circuit, which can be considered as a longitudinal wave along a conductor. When $R = 0$ and $G = 0$, the equations have the simplest form. For instance, for current

$$\frac{\partial^2 I}{\partial t^2} = a^2 \frac{\partial^2 I}{\partial x^2},$$

where $a^2 = 1/LC$, and once again we got the wave Eq. (13.89). If $G \neq 0$, the equation is similar to Eq. (13.90) describing oscillations in a medium with the force of resistance proportional to speed:

$$\frac{\partial^2 I}{\partial t^2} = a^2 \frac{\partial^2 I}{\partial x^2} - 2\kappa \frac{\partial I}{\partial t},$$

where $a^2 = 1/LC$, $2\kappa = (R/L + G/C)$.

Consider initial and boundary conditions for a current and voltage. Let initially, at $t = 0$, the current be $\varphi(x)$, and the voltage be $\psi(x)$ along the wire.

The equation for current contains the second-order time derivative; thus, we need two initial conditions. One initial condition is the initial current in the wire:

$$I(x, 0) = \varphi(x). \tag{13.93}$$

The second initial condition for the current is

$$I_t(x, 0) = -\frac{1}{L}[\psi'(x) - R\varphi(x)]. \tag{13.94}$$

Two initial conditions for $V(x, t)$ are

$$V(x,0) = \psi(x), \tag{13.95}$$

$$V_t(x,0) = -\frac{1}{C}[\varphi'(x) - G\psi(x)]. \tag{13.96}$$

For the details how to obtain conditions (13.94) and (13.96), see [7].
Let us give two examples of the boundary conditions.

1.	One end of the wire of length l is attached to a source of electromotive force (*emf*) $E(t)$, and the other end is grounded. The boundary conditions are $V(0,t) = E(t), \ V(l,t) = 0.$
2.	A sinusoidal voltage with frequency ω is applied to the left end of the wire, and the other end is insulated. The boundary conditions are $V(0,t) = V_0 \sin \omega t, \ I(l,t) = 0.$

13.3.2   Separation of Variables for One-Dimensional Equation

Let us apply the Fourier method for the general one-dimensional equation:

$$\frac{\partial^2 u}{\partial t^2} + 2\kappa \frac{\partial u}{\partial t} - a^2 \frac{\partial^2 u}{\partial x^2} + \gamma u + f(x,t) = 0, \quad 0 \leq x \leq l, t \geq 0, \tag{13.97}$$

where a, κ, γ are constants. As we saw when discussed Eqs. (13.83) and (13.84) in Sect. 13.3.1, in physical situations $\kappa \geq 0$ and $\gamma \geq 0$. Typical wave problems do not contain the term like bu_x, but if included into the equation, the substitution $u(x,t) = e^{bx/2a^2} v(x,t)$ leads to the equation for function $v(x,t)$ without v_x term.
The initial conditions are

$$u(x,t)|_{t=0} = \varphi(x), \quad u_t(x,t)|_{t=0} = \psi(x), \tag{13.98}$$

where $\varphi(x)$ and $\psi(x)$ are given functions.
The boundary conditions can be formulated as

$$\alpha_1 \frac{\partial u}{\partial x} + \beta_1 u \bigg|_{x=0} = g_1(t), \quad \alpha_2 \frac{\partial u}{\partial x} + \beta_2 u \bigg|_{x=l} = g_2(t). \tag{13.99}$$

In the case of a nonhomogeneous boundary condition, one can use an auxiliary function, $w(x,t)$, which satisfies the boundary condition:

$$u(x,t) = v(x,t) + w(x,t),$$

where $v(x, t)$ is a function with *zero boundary condition*. In Sect. 13.2 we discussed how to obtain function $w(x, t)$.

In the case of *nonhomogeneous equation* and *zero boundary conditions*, we may express the solution $v(x, t)$ as the sum of two functions:

$$v(x, t) = u_1(x, t) + u_2(x, t),$$

where $u_1(x, t)$ is the solution of the homogeneous equation with the given initial conditions and $u_2(x, t)$ is the solution of the nonhomogeneous equation with zero initial and boundary conditions. In simple cases a function $u_2(x, t)$ can be just guessed.

Let us search for a solution of a *homogeneous* equation:

$$\frac{\partial^2 u}{\partial t^2} + 2\kappa \frac{\partial u}{\partial t} - a^2 \frac{\partial^2 u}{\partial x^2} + \gamma u = 0 \qquad (13.100)$$

satisfying the *homogeneous* boundary conditions

$$\alpha_1 u_x + \beta_1 u|_{x=0} = 0, \qquad \alpha_2 u_x + \beta_2 u|_{x=l} = 0 \qquad (13.101)$$

where α_1, β_1, α_2, and β_2 are constants.

Clearly, $u(0, t) = 0$, or $u(l, t) = 0$, correspond to a fixed end of the string, and $u'(0, t) = 0$, or $u'(l, t) = 0$, correspond to a free end of the string.

We begin by assuming that a nontrivial (nonzero) solution of Eq. (13.100) can be found that is a product of two functions, one depending only on x and another depending only on t:

$$u(x, t) = X(x)T(t). \qquad (13.102)$$

Substituting (13.102) into (13.100), we obtain:

$$X(x)T''(t) + 2\kappa X(x)T'(t) - a^2 X''(x)T(t) + \gamma X(x)T(t) = 0$$

or, by rearranging terms

$$\frac{T''(t) + 2\kappa T'(t) + \gamma T(t)}{a^2 T(t)} = \frac{X''(x)}{X(x)}.$$

The left-hand side of this equation depends only on t, and the right-hand side only on x, which is possible only if each side equals a constant. By using the notation $-\lambda$ for this constant, we obtain:

$$\frac{T''(t) + 2\kappa T'(t) + \gamma T(t)}{a^2 T(t)} \equiv \frac{X''(x)}{X(x)} = -\lambda$$

(it is seen from this relation that λ has the dimension of inversed length squared).

Thus, Eq. (13.99) gives two ordinary second-order linear homogeneous differential equations:

$$T''(t) + 2\kappa T'(t) + \left(a^2\lambda + \gamma\right)T(t) = 0 \tag{13.102}$$

and

$$X''(x) + \lambda X(x) = 0. \tag{13.103}$$

Therefore, we see that we have successfully separated the variables, resulting in separate equations for functions $X(x)$ and $T(t)$. These equations share the common parameter λ.

To find λ, we apply the boundary conditions. Homogenous boundary condition of Eq. (13.101), imposed on $u(x, y)$, gives the homogeneous boundary conditions on the function $X(x)$:

$$\alpha_1 X' + \beta_1 X|_{x=0} = 0, \qquad \alpha_2 X' + \beta_2 X|_{x=l} = 0 \tag{13.104}$$

with restrictions $\alpha_1/\beta_1 < 0$ and $\alpha_2/\beta_2 > 0$.

Eigenvalues and eigenfunctions of this Sturm-Liouville boundary value problem depend on the type of the boundary conditions: Dirichlet, Neumann, fRobin, or mixed ones.

The next step of the solution is to *find the function* $T(t)$. Equation (13.102) is an ordinary linear second-order homogeneous differential equation. For $\lambda = \lambda_n$

$$T_n''(t) + 2\kappa T_n''(t) + \left(a^2\lambda_n + \gamma\right)T_n(t) = 0 \tag{13.105}$$

and a general solution of this equation is

$$T_n(t) = a_n y_n^{(1)}(t) + b_n y_n^{(2)}(t), \tag{13.106}$$

where a_n and b_n are arbitrary constants. Two particular solutions, $y_n^{(1)}(t)$ and $y_n^{(2)}(t)$, are

$$y_n^{(1)}(t) = \begin{cases} e^{-\kappa t}\cos\omega_n t, & \text{if } \kappa^2 < a^2\lambda_n + \gamma, \\ e^{-\kappa t}\cosh\omega_n t, & \text{if } \kappa^2 > a^2\lambda_n + \gamma, \\ e^{-\kappa t}, & \text{if } \kappa^2 = a^2\lambda_n + \gamma, \end{cases} \tag{13.107}$$

and

$$y_n^{(2)}(t) = \begin{cases} e^{-\kappa t} \sin \omega_n t, & \text{if } \kappa^2 < a^2 \lambda_n + \gamma, \\ e^{-\kappa t} \sinh \omega_n t, & \text{if } \kappa^2 > a^2 \lambda_n + \gamma, \\ te^{-\kappa t}, & \text{if } \kappa^2 = a^2 \lambda_n + \gamma, \end{cases} \tag{13.108}$$

where

$$\omega_n = \sqrt{|a^2 \lambda_n + \gamma - \kappa^2|}$$

(obviously, ω_n has the dimension of inversed time).

It is clear that each function

$$u_n(x,t) = T_n(t)X_n(x) = \left[a_n y_n^{(1)}(t) + b_n y_n^{(2)}(t)\right]X_n(x) \tag{13.109}$$

is a solution of Eq. (13.100) and satisfies boundary conditions in (13.101).

Then, we compose the series:

$$u(x,t) = \sum_{n=1}^{\infty} \left[a_n y_n^{(1)}(t) + b_n y_n^{(2)}(t)\right]X_n(x), \tag{13.110}$$

which can be considered as the expansion of the unknown function $u(x,t)$ into a Fourier series using an orthogonal system of functions $\{X_n(x)\}$.

The superposition (13.110) allows to satisfy the initial conditions (13.94). The first one gives:

$$u|_{t=0} = \varphi(x) = \sum_{n=1}^{\infty} a_n X_n(x). \tag{13.111}$$

If the series (13.110) converges uniformly, it can be differentiating with t and the second initial condition gives:

$$\frac{\partial u}{\partial t}\bigg|_{t=0} = \psi(x) = \sum_{n=1}^{\infty} [\omega_n b_n - \kappa a_n]X_n(x). \tag{13.112}$$

For initial conditions that are "reasonably smooth," like piecewise continuous, the series (13.111) and (13.112) converge uniformly, and they allow to find coefficients a_n and b_n. By multiplying both sides of these equations by $X_n(x)$, integrating from 0 to l, and using the orthogonality condition, we obtain:

$$a_n = \frac{1}{\|X_n\|^2} \int_0^l \varphi(x)X_n(x)dx \tag{13.113}$$

and

$$b_n = \frac{1}{\omega_n} \left[\frac{1}{\|X_n\|^2} \int_0^l \psi(x) X_n(x) dx + \kappa a_n \right]. \qquad (13.114)$$

Note that if in formulas (13.112) and (13.114) we replace ω_n by 1 when $\kappa^2 = a^2\lambda_n + \gamma$, then we can consider all three cases for κ^2 in (13.107) and (13.108) simultaneously.

The series (13.110) with these coefficients gives the solution of the boundary value problem given in Eqs. (13.97), (13.98), and (13.99).

Recall that the success of this method is based on the following conditions: the functions $\{X_n(x)\}$ are orthogonal to each other and form a complete set (i.e., a basis for an expansion of $u(x, t)$); the functions $\{X_n(x)\}$ satisfy the same boundary conditions as the solutions, $u(x, t)$; and solutions to linear equations obey the superposition principle (i.e., the sums of solutions are also solutions).

The obtained solution describes *free oscillations*. Processes with not very large damping, i.e., $\kappa^2 < a^2\lambda_n + \gamma$, are periodic (or quasiperiodic) and have a special physical interest. For $\kappa = 0$, the motion is purely periodic; thus, $\omega_n = a\sqrt{\lambda_n}$ are the frequencies. The partial solutions $u_n(x, t) = T_n(t)X_n(x)$ are called *normal modes*. The first term, $u_1(x, t)$, is called the *first* (or *fundamental*) *harmonic* and has time dependence with frequency ω_1 and period $2\pi/\omega_1$. The *second harmonic* (or *first overtone*), $u_2(x, t)$, oscillates with greater frequency ω_2 (for Dirichlet or Neumann type boundary conditions and $\kappa = \gamma = 0$, it is twice ω_1); the *third harmonic* is called *second overtone*, etc. The points where $X_n(x) = 0$ are not moving are called *nodes* of the harmonic $u_n(x, t)$. Between the nodes the string oscillates up and down. The waves $u_n(x, t)$ are also called *standing waves* because the positions of the nodes are fixed in time. The general solution, $u(x, t)$, is a *superposition of standing waves*; thus, any oscillation can be presented this way.

Clearly, $\omega_n = a\sqrt{\lambda_n}$ increase with the tension and decrease with the length and density: tuning any stringed instrument is based on changing the tension, and the bass strings are longer and heavier. The loudness of a sound is characterized by the energy or amplitude of the oscillations; tone by the period of oscillations; and timbre by the ratio of energies of the main mode and overtones. The presence of high overtones destroys the harmony of a sound producing dissonance. Low overtones, in contrast, give a sense of completeness to a sound.

Example 13.6 The ends of a uniform string of length l are fixed, and all external forces including the gravitational force can be neglected. Displace the string from equilibrium by shifting the point $x = x_0$ by distance A at time $t = 0$, and then release it with zero initial speed. Find the displacements $u(x, t)$ of the string for times $t > 0$.

Solution The boundary value problem is

$$u_{tt} - a^2 u_{xx} = 0, 0 < x < l, t > 0$$

$$u(x,0) = \begin{cases} \dfrac{A}{x_0}x, & 0 < x \le x_0, \\[2ex] \dfrac{A(l-x)}{l-x_0}, & x_0 < x < l, \end{cases} \qquad \frac{\partial u}{\partial t}(x,0) = 0, \quad u(0,t) = u(l,t) = 0.$$

This initial condition is not differentiable in the point $x = x_0$; it means we are searching for a generalized solution.

The eigenvalues and eigenfunctions are those of the Dirichlet problem on $0 < x < l$:

$$\lambda_n = \left(\frac{n\pi}{l}\right)^2, X_n(x) = \sin\frac{n\pi x}{l}, \|X_n\|^2 = \frac{l}{2}, n = 1, 2, 3, \ldots.$$

Using Eqs. (13.113), we obtain:

$$a_n = \frac{2Al^2}{\pi^2 x_0(l-x_0)n^2}\sin\frac{n\pi x_0}{l}, b_n = 0.$$

Therefore, string vibrations (13.110) are given by the series:

$$u(x,t) = \frac{2Al^2}{\pi^2 x_0(l-x_0)}\sum_{n=1}^{\infty}\frac{1}{n^2}\sin\frac{n\pi x_0}{l}\sin\frac{n\pi x}{l}\cos\frac{n\pi at}{l}.$$

Figure 13.7 shows the space-time-dependent solution $u(x,t)$ for Example 13.6 (for the case when $a^2 = 1$, $l = 100$, $A = 6$, and $x_0 = 25$). The sequence in Fig. 13.7 obtained using the program *Waves* shows the snapshots of the animation at times $t = 0, 1, \ldots, 12$. Because there is no dissipation, it is sufficient to run the simulation until the time is equal to the period of the main harmonic (until $2l/a = 200$ in this example).

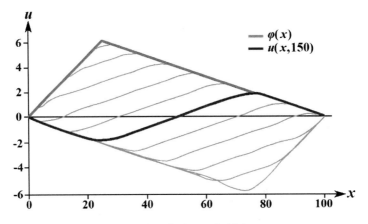

Fig. 13.7 Solutions, $u(x,t)$, at various times for Example 13.6

For $t = 0$ the obtained solution gives:

$$u(x, 0) = \sum_{n-1}^{\infty} a_n \sin \frac{n\pi x}{l}$$

– this smooth function is the Fourier expansion of the initial condition function given above.

In the solution $u(x, t)$, the terms for which $\sin(n\pi x_0/l) = 0$ vanish from the series, that is, the solution does not contain overtones for which $x = x_0$ is a node. For instance, if x_0 is at the middle of the string, the solution does not contain harmonics with even numbers.

The formula for $u(x, t)$ can be presented in a more physically intuitive form. Let us denote $\sqrt{\lambda_n}$ as k_n, where $k_n = \frac{n\pi}{l}$ which are called *wave numbers*. The frequencies $\omega_n = \frac{n\pi a}{l}$ give *the frequency spectrum* of the problem. The solution to the problem now can be written as

$$u(x, t) = \sum_{n=1}^{\infty} A_n \cos \omega_n t \cdot \sin k_n x,$$

where the amplitude $A_n = \frac{D}{n^2} \sin \frac{n\pi x_0}{l}$ with $D \equiv \frac{2Al^2}{\pi^2 x_0(l - x_0)}$.

The first harmonic (or mode), $u_1(x, t)$, is zero for all x when $\omega_1 t = \pi/2$, that is, when $t = l/2a$. It becomes zero again at $t = 3l/2a$, $t = 5l/2a$, etc. The second harmonic $u_2(x, t)$ is zero for all x at $t = l/4a$, $t = 3l/4a$, etc.

Problems

In the following problems, the initial shape of a string with fixed ends is $u(x, 0) = \varphi(-x)$; the initial speed is $u_t(x, 0) = \psi(x)$. External forces are absent. Find the vibrations of the string:

1. $\varphi(x) = Ax(1 - \frac{x}{l})$, $\psi(x) = 0$.
2. $\varphi(x) = A \sin \frac{\pi x}{l}$, $\psi(x) = 0$.
3. $\varphi(x) = 0$, $\psi(x) = v_0 = \text{const.}$
4. $\varphi(x) = 0$, $\psi(x) = \frac{B}{l} x(l - x)$.
5. $\varphi(x) = Ax(1 - \frac{x}{l})$, $\psi(x) = \frac{B}{l} x(l - x)$.

In the following problems, the ends of a string are rigidly fixed. The initial conditions are zero. Starting at time $t = 0$, a uniformly distributed force with linear density $f(x, t)$ acts on the string. Find oscillations of the string:

6. $f(x) = Ax(x/l - 1)$.
7. $f(x, t) = Axe^{-Bt}$.
8. $f(x, t) = Ax(x/l - 1) \sin \omega t$.
9. $f(x, t) = Ax(x/l - 1) \cos \omega t$.

13.3.3 Transverse Oscillations of a Circular Membrane

Consider oscillations of a circular membrane (rectangular membrane oscillations can be found in [9])) when all its points move perpendicular to the equilibrium plane (the transverse oscillations). In polar coordinates the displacement of points on the membrane will be a function of r, φ, and time t: $u = u(r, \varphi, t)$.

Using the expression for the Laplace operator in polar coordinates

$$\nabla^2 u = \frac{\partial^2 u}{\partial r^2} + \frac{1}{r}\frac{\partial u}{\partial r} + \frac{1}{r^2}\frac{\partial^2 u}{\partial \varphi^2},$$

the equation of oscillations of a circular membrane is

$$\frac{\partial^2 u}{\partial t^2} + 2\kappa \frac{\partial u}{\partial t} - a^2\left(\frac{\partial^2 u}{\partial r^2} + \frac{1}{r}\frac{\partial u}{\partial r} + \frac{1}{r^2}\frac{\partial^2 u}{\partial \varphi^2}\right) + \gamma u = f(r, \varphi, t). \tag{13.115}$$

The domains of the independent variables are $0 \leq r \leq l$, $0 \leq \varphi < 2\pi$, and $0 \leq t < \infty$, respectively.

Boundary conditions in polar coordinate in general from can be written as

$$\alpha \frac{\partial u}{\partial r} + \beta u \bigg|_{r=l} = g(\varphi, t). \tag{13.116}$$

If $\alpha = 0$ we have Dirichlet boundary condition, if $\beta = 0$ we have Neumann boundary condition, and if $\alpha \neq 0$ and $\beta \neq 0$, then we have Robin boundary conditions. From physical arguments it is normally the case that $\beta/\alpha > 0$.

The initial deviation of the points of a membrane and the initial velocities of these points give the initial conditions:

$$u|_{t=0} = \xi(r, \varphi), \quad \frac{\partial u}{\partial t}\bigg|_{t=0} = \psi(r, \varphi). \tag{13.117}$$

Below we shall consider only the *homogeneous equation* of oscillations with *homogeneous* boundary conditions, i.e., functions $f(r, \varphi, t) = 0$ and $g(\varphi, t) = 0$.

Let us represent the function $u(r, \varphi, t)$ as a product of two functions: $V(r, \varphi)$ and $T(t)$:

$$u(r, \varphi, t) = V(r, \varphi)T(t). \tag{13.118}$$

Substituting (13.116) into (13.113) and separating variables, we obtain:

$$\frac{T''(t) + 2\kappa T'(t) + \gamma T(t)}{a^2 T(t)} \equiv \frac{1}{V(r,\varphi)} \left[\frac{\partial^2 V}{\partial r^2} + \frac{1}{r} \frac{\partial V}{\partial r} + \frac{1}{r^2} \frac{\partial^2 V}{\partial \varphi^2} \right] = -\lambda,$$

where λ is a separation of variables constant. Thus, the function $T(t)$ satisfies the ordinary linear homogeneous differential equation of second order:

$$T''(t) + 2\kappa T'(t) + (a^2 \lambda + \gamma) T(t) = 0, \tag{13.119}$$

and the function $V(r, \varphi)$ satisfies the equation:

$$\frac{\partial^2 V}{\partial r^2} + \frac{1}{r} \frac{\partial V}{\partial r} + \frac{1}{r^2} \frac{\partial^2 V}{\partial \varphi^2} + \lambda V = 0 \tag{13.120}$$

with a boundary condition

$$\alpha \frac{\partial V(l, \varphi)}{\partial r} + \beta V(l, \varphi) = 0. \tag{13.121}$$

Using physical arguments we also require that the solutions remain finite (everywhere, including point $r = 0$) so that

$$|V(r, \varphi)| < \infty \tag{13.122}$$

and require the solutions to be periodic which may be defined as

$$V(r, \varphi) = V(r, \varphi + 2\pi). \tag{13.123}$$

The solution of the boundary value problems (13.120), (13.121), (13.122), and (13.123) was already obtained in Sect. 13.2.3:

$$V_{nm}^{(1)}(r, \varphi) = J_n\left(\frac{\mu_m^{(n)}}{l} r\right) \cos n\varphi \quad \text{and} \quad V_{nm}^{(2)}(r, \varphi) = J_n\left(\frac{\mu_m^{(n)}}{l} r\right) \sin n\varphi, \tag{13.124}$$

where $\mu_m^{(n)}$ is the m-th *positive* root of the equation

$$\alpha \mu J_n'(\mu) + \beta l J_n(\mu) = 0. \tag{13.125}$$

Substituting the eigenvalues, λ_{nm}, $\lambda = \lambda_{nm} = \left(\frac{\mu_m^{(n)}}{l}\right)^2$ in (13.119) and denoting the corresponding solution of this equation as $T_{nm}(t)$, we have:

$$T_{nm}''(t) + 2\kappa T_{nm}'(t) + (a^2 \lambda_{nm} + \gamma) T_{nm}(t) = 0. \tag{13.126}$$

   Two linearly independent solutions of this equation were already obtained when we derived formulas (13.103, 13.104, 13.105, and 13.106) (the only difference is that now there are two indices, n and m):

$$
y_{nm}^{(1)}(t) = \begin{cases} e^{-\kappa t}\cos\omega_{nm}t, & \kappa^2 < a^2\lambda_{nm} + \gamma, \\ e^{-\kappa t}\cosh\omega_{nm}t, & \kappa^2 > a^2\lambda_{nm} + \gamma, \\ e^{-\kappa t}, & \kappa^2 = a^2\lambda_{nm} + \gamma, \end{cases}
$$

$$
y_{nm}^{(2)}(t) = \begin{cases} e^{-\kappa t}\sin\omega_{nm}t, & \kappa^2 < a^2\lambda_{nm} + \gamma, \\ e^{-\kappa t}\sinh\omega_{nm}t, & \kappa^2 > a^2\lambda_{nm} + \gamma, \\ te^{-\kappa t}, & \kappa^2 = a^2\lambda_{nm} + \gamma, \end{cases} \tag{13.127}
$$

with $\omega_{nm} = \sqrt{|\,a^2\lambda_{nm} + \gamma - \kappa^2\,|}$.

   A general solution of Eq. (13.102) is a linear combination of $y_{nm}^{(1)}(t)$ and $y_{nm}^{(2)}(t)$. Thus, we have the solutions to Eq. (13.77) satisfying the given boundary conditions:

$$
u_{nm}^{(1)}(r,\varphi,t) = T_{nm}^{(1)}V_{nm}^{(1)}(r,\varphi) = \left[a_{nm}y_{nm}^{(1)} + b_{nm}y_{nm}^{(2)}\right]V_{nm}^{(1)}(r,\varphi),
$$

$$
u_{nm}^{(2)}(r,\varphi,t) = T_{nm}^{(2)}V_{nm}^{(2)}(r,\varphi) = \left[c_{nm}y_{nm}^{(1)} + d_{nm}y_{nm}^{(2)}\right]V_{nm}^{(2)}(r,\varphi).
$$

   To find solutions satisfying not only the boundary conditions above but also the initial conditions, let us sum these functions as a series, superposing all $u_{nm}^{(1)}(r,\varphi,t)$ and $u_{nm}^{(2)}(r,\varphi,t)$:

$$
u(r,\varphi,t) = \sum_{n=0}^{\infty}\sum_{m=0}^{\infty}\left\{\left[a_{nm}y_{nm}^{(1)}(t) + b_{nm}y_{nm}^{(2)}(t)\right]V_{nm}^{(1)}(r,\varphi) + \right.
$$
$$
\left. + \left[c_{nm}y_{nm}^{(1)}(t) + d_{nm}y_{nm}^{(2)}(t)\right]V_{nm}^{(2)}(r,\varphi)\right\}. \tag{13.128}
$$

   If this series and the series obtained from it by twice differentiating term by term with respect to the variables r, φ, and t converges uniformly, then its sum will be a solution to Eq. (13.115), satisfying the boundary condition (13.116).

   To satisfy the initial conditions given in Eq. (13.117), we require that

$$
u|_{t=0} = \xi(r,\varphi) = \sum_{n=0}^{\infty}\sum_{m=0}^{\infty}\left[a_{nm}V_{nm}^{(1)}(r,\varphi) + c_{nm}V_{nm}^{(2)}(r,\varphi)\right] \tag{13.129}
$$

and

$$\left.\frac{\partial u}{\partial t}\right|_{t=0} = \psi(r,\varphi) = \sum_{n=0}^{\infty} \sum_{m=0}^{\infty} \left\{ [\omega_{mn} b_{mn} - \kappa a_{mn}] V_{nm}^{(1)}(r,\varphi) + \right.$$
$$\left. + [\omega_{mn} d_{mn} - \kappa c_{mn}] V_{nm}^{(2)}(r,\varphi) \right\}. \tag{13.130}$$

From here the coefficients a_{nm}, b_{mn}, c_{nm}, and d_{nm} can be obtained:

$$a_{nm} = \frac{1}{\|V_{nm}^{(1)}\|^2} \int_0^l \int_0^{2\pi} \xi(r,\varphi)\, V_{nm}^{(1)}(r,\varphi)\, r dr d\varphi,$$

$$b_{nm} = \frac{1}{\omega_{nm}} \left[\frac{1}{\|V_{nm}^{(1)}\|^2} \int_0^l \int_0^{2\pi} \psi(r,\varphi)\, V_{nm}^{(1)}(r,\varphi)\, r dr d\varphi + \kappa a_{nm} \right],$$

$$c_{nm} = \frac{1}{\|V_{nm}^{(2)}\|^2} \int_0^l \int_0^{2\pi} \xi(r,\varphi)\, V_{nm}^{(2)}(r,\varphi)\, r dr d\varphi, \tag{13.131}$$

$$d_{nm} = \frac{1}{\omega_{nm}} \left[\frac{1}{\|V_{nm}^{(2)}\|^2} \int_0^l \int_0^{2\pi} \psi(r,\varphi)\, V_{nm}^{(2)}(r,\varphi)\, r dr d\varphi + \kappa c_{nm} \right].$$

Equation (13.128) gives the evolution of *free oscillations* of a circular membrane when boundary conditions are homogeneous. It can be considered as the expansion of the (unknown) function $u(r,\varphi,t)$ in a Fourier series using the orthogonal system of functions $\{V_{nm}(r,\varphi)\}$. This series converges under sufficiently reasonable assumptions about initial and boundary conditions.

The particular solutions $u_{nm}(r,\varphi,t) = T_{nm}^{(1)}(t) V_{nm}^{(1)}(r,\varphi) + T_{nm}^{(2)}(t) V_{nm}^{(2)}(r,\varphi)$ are called *standing wave* solutions. From this we see that the profile of a standing wave depends on the functions $V_{nm}(r,\varphi)$; the functions $T_{nm}^{(1)}(t)$ and $T_{nm}^{(2)}(t)$ only change the amplitude of the standing wave over time, as was the case for standing waves on a string. Lines on the membrane defined by $V_{nm}(r,\varphi) = 0$ remain at rest for all times and are called *nodal lines* of the standing wave $V_{nm}(r,\varphi)$. Points, where $V_{nm}(r,\varphi)$ reaches a relative maximum or minimum for all times, are called *antinodes* of this standing wave. From the above discussion of the Fourier expansion, we see that an arbitrary motion of the membrane may be thought of as an infinite sum of these standing waves.

Each mode $u_{nm}(r,\varphi,t)$ possesses a characteristic pattern of nodal lines. The first few of these normal vibration modes for $V_{nm}^{(1)}(r,\varphi)$ are sketched in Fig. 13.8 with similar pictures for the modes $V_{nm}^{(2)}(r,\varphi)$. In the fundamental mode of vibration corresponding to $\mu_0^{(0)}$, the membrane vibrates as the whole. In the mode corresponding to $\mu_1^{(0)}$, the membrane vibrates in two parts as shown with the part labeled with a plus sign initially above the equilibrium level and the part labeled with a minus sign initially below the equilibrium. The nodal line in this case is a circle

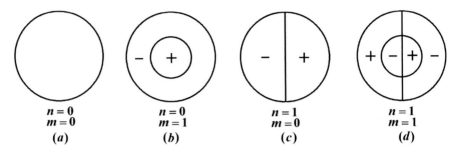

Fig. 13.8 Sketch of the first few modes of vibration for the mode $V_{nm}^{(1)}(r, \varphi)$

which remains at rest as the two sections reverse location. The mode characterized by $\mu_0^{(1)}$ is equal to zero when $\varphi = \pm \pi/2$ and is positive and negative as shown.

Let us consider oscillations that do not depend on the polar angle φ – such oscillations are called *axisymmetric (or radial)*. Physically, radial oscillations will occur when initial displacements and initial velocities do not depend on φ, but rather are functions only of r:

$$u(r, \varphi, t)|_{t=0} = \xi(r), \quad \frac{\partial u}{\partial t}(r, \varphi, t)\bigg|_{t=0} = \psi(r). \tag{13.132}$$

In this case all coefficients, a_{nm}, b_{nm}, c_{nm}, and d_{nm} with $n \geq 1$ equal zero. We may easily verify this, for example, for a_{nm}:

$$a_{nm} = \frac{1}{\left\| V_{nm}^{(1)} \right\|^2} \int\limits_0^l \int\limits_0^{2\pi} \xi(r) J_n\left(\mu_m^{(n)} r/l\right) \cos n\varphi \, r dr d\varphi.$$

Because $\int\limits_0^{2\pi} \cos n\varphi \, d\varphi = 0$ for any integer $n \geq 1$, we have $a_{nm} = 0$.

Similarity, $b_{nm} = 0$ for $n \geq 1$, and $c_{nm} = 0$, and $d_{nm} = 0$ for all n. Thus, *the solution does not contain the functions* $V_{nm}^{(2)}(r, \varphi)$ and the nonzero coefficients are

$$a_{0m} = \frac{2\pi}{\left\| V_{0m}^{(1)} \right\|^2} \int\limits_0^l \xi(r) J_0\left(\mu_m^{(0)} r/l\right) r dr, \tag{13.133}$$

$$b_{0m} = \frac{2\pi}{\left\| V_{0m}^{(1)} \right\|^2} \int\limits_0^l \psi(r) J_0\left(\mu_m^{(0)} r/l\right) r dr. \tag{13.134}$$

Substituting these coefficients into the series in Eq. (13.126), we notice that the series reduces from a double series to a single one since all terms in the second sum of this series disappear. Only those terms in the first sum remain for which $n = 0$, making it necessary to sum only on m but not on n. The final result is

$$u(r, \varphi, t) = \sum_{m=0}^{\infty} \left[a_{0m} y_{0m}^{(1)}(t) + b_{0m} y_{0m}^{(2)}(t) \right] J_0 \left(\frac{\mu_m^{(0)}}{l} r \right). \tag{13.135}$$

Thus, for radial oscillations the solution contains only Bessel functions of zero order.

Example 13.7 Find the transverse oscillations of a circular membrane with a fixed edge. Assume the initial displacement has the form of a paraboloid of rotation, initial velocities are zero and the reaction of the environment can be neglected.

Solution We have the following boundary value problem:

$$\frac{\partial^2 u}{\partial t^2} - a^2 \left[\frac{\partial^2 u}{\partial r^2} + \frac{1}{r} \frac{\partial u}{\partial r} \right] = 0 \ (0 \leq r < l, 0 \leq \varphi < 2\pi, t > 0).$$

$$u(r, 0) = A \left(1 - \frac{r^2}{l^2} \right), \ \frac{\partial u}{\partial t}(r, \varphi, 0) = 0, u(l, \varphi, t) = 0.$$

The oscillations of the membrane are radial since the initial displacement and the initial velocities do not depend on the polar angle φ. Thus, only terms with $n = 0$ are not zero.

Boundary conditions of the problem are of *Dirichlet type*; thus, eigenvalues $\mu_m^{(0)}$ are the solutions of the equation $J_0(\mu) = 0$, and the eigenfunctions are

$$V_{0m}^{(1)}(r, \varphi) = J_0 \left(\frac{\mu_m^{(0)}}{l} r \right).$$

The three-dimensional picture shown in Fig. 13.9 depicts the two eigenfunctions (chosen as examples), $V_{00}^{(1)}(r, \varphi) = J_0 \left(\mu_0^{(0)} r/l \right)$ and $V_{02}^{(1)}(r, \varphi) = J_0 \left(\mu_2^{(0)} r/l \right)$.
The solution $u(r, \varphi, t)$ is given by the series:

$$u(r, \varphi, t) = \sum_{m=0}^{\infty} a_{0m} \cos \frac{a \mu_m^{(0)} t}{l} J_0 \left(\mu_m^{(0)} r/l \right).$$

The coefficients a_{0m} can be obtained using Eq. (13.133):

$$a_{0m} = \frac{2\pi}{\left\| V_{0m}^{(1)} \right\|^2} \int_0^l A \left(1 - \frac{r^2}{l^2} \right) J_0 \left(\mu_m^{(0)} r/l \right) r dr = \frac{8A}{\left(\mu_m^{(0)} \right)^3 J_1 \left(\mu_m^{(0)} \right)}$$

(the integral was already calculated in Example 13.5).

(a)   (b)

Fig. 13.9 Two eigenfunctions, (**a**) $V_{00}^{(1)}(r, \varphi)$ and (**b**) $V_{02}^{(1)}(r, \varphi)$ for the Dirichlet boundary conditions in Example 13.7. The solution was obtained using the program *Wave*

Thus,

$$u(r, \varphi, t) = 8A \sum_{m=0}^{\infty} \frac{1}{\left(\mu_m^{(0)}\right)^3 J_1\left(\mu_m^{(0)}\right)} \cos \frac{a\mu_m^{(0)} t}{l} J_0\left(\mu_m^{(0)} r/l\right).$$

Problems

In the following problems, consider transverse oscillations of a homogeneous circular membrane of radius l fixed along its contour. For all problems the displacement of the membrane, $u(r, \varphi, 0) = \xi(r, \varphi)$ is given at some initial moment of time, $t = 0$, and the membrane is released with the initial velocity $u_t(r, \varphi, 0) = \psi(r, \varphi)$:

1. $\xi(r, \varphi) = Ar(l^2 - r^2) \sin \varphi$, $\psi(r, \varphi) = 0$.
2. $\xi(r, \varphi) = 0$, $\psi(r, \varphi) = Ar(l^2 - r^2) \cos 4\varphi$.
3. $\xi(r, \varphi) = Ar(1 - r) \sin \varphi$, $\psi(r, \varphi) = 0$.
4. $\xi(r, \varphi) = 0$, $\psi(r, \varphi) = Ar(1 - r) \cos \varphi$.
5. $\xi(r, \varphi) = 0$, $\psi(r, \varphi) = Ar\left(l - \frac{r}{2}\right) \cos 4\varphi$.

13.4   The Laplace Equation

13.4.1   Physical Problems Described by the Laplace Equation

When studying different stationary (time-independent) processes, very often we encounter the *Laplace equation*:

$$\nabla^2 u = 0. \qquad (13.136)$$

The nonhomogeneous equation

$$\nabla^2 u = -f \qquad (13.137)$$

with a given function f of the coordinates is called the *Poisson equation*.

The Laplace and Poisson partial differential equations are of *Elliptic type*.

Among fundamental physical problems described by these equations are the following:

1.	If a temperature distribution created by an external heating does not change with time, $\partial T/\partial t = 0$, the homogeneous heat Eq. (13.13) reduces to the Laplace equation: $$\nabla^2 T = 0. \qquad (13.138)$$ If a medium contains a heat source or heat absorber, Q, and the temperature is time independent, the heat conduction Eq. (13.15) becomes: $$\nabla^2 T = -\frac{Q}{\rho c a^2}, \qquad (13.139)$$ which is a particular example of the Poisson equation (13.137).
2.	The diffusion equation (13.19) for stationary diffusion, $\partial c/\partial t = 0$, becomes: $$\nabla^2 c = -f/D, \qquad (13.140)$$ where c is the concentration of a material diffusing through a medium, D is the diffusion coefficient and f is a source or sink of the diffusing material. This is also the Poisson equation (or the Laplace equation when $f = 0$).
3.	The electrostatic potential due to a point charge q is $$\varphi = \frac{q}{r}, \qquad (13.141)$$ where r is the distance from the charge to the point where the electric field is measured. For continuous charge distribution with charge density ρ, the potential φ is related with ρ as $$\nabla^2 \varphi = -4\pi\rho. \qquad (13.142)$$ Equation (13.142) is the Poisson equation for electrostatic potential. In regions that do not contain electric charges, for instance at points outside of a charged body, $\rho = 0$ and the potential that the body creates obeys the Laplace equation: $$\nabla^2 \varphi = 0. \qquad (13.143)$$

13.4.2 BVP for the Laplace Equation in a Rectangular Domain

The Laplace operator in Cartesian coordinates is

$$\nabla^2 = \frac{\partial^2}{\partial x^2} + \frac{\partial^2}{\partial y^2}. \qquad (13.144)$$

Boundary value problems for the Laplace equation in a rectangular domain can be solved using the method of separation of variables. Here we consider the *Dirichlet problem* defined by

$$\nabla^2 u = 0 \quad \left(0 < x < l_x, 0 < y < l_y\right), \qquad (13.145)$$

$$u(x,y)|_{x=0} = g_1(y), \quad u(x,y)|_{x=l_x} = g_2(y),$$
$$u(x,y)|_{y=0} = g_3(x), \quad u(x,y)|_{y=l_y} = g_4(x). \tag{13.146}$$

Let us split the problem in Eqs. (13.145) through (13.146) into two parts, each of which has *homogeneous (zero) boundary conditions in one variable*. To proceed we introduce

$$u(x,y) = u_1(x,y) + u_2(x,y), \tag{13.147}$$

where $u_1(x,y)$ and $u_2(x,y)$ are the solutions to the following problems on a rectangular boundary:

$$\nabla^2 u_1 = 0, \tag{13.148}$$

$$u_1(x,y)|_{x=0} = u_1(x,y)|_{x=l_x} = 0, \tag{13.149}$$

$$u_1(x,y)|_{y=0} = g_3(x), \quad u_1(x,y)|_{y=0} = g_3(x), \tag{13.150}$$

and

$$\nabla^2 u_2 = 0, \tag{13.151}$$

$$u_2(x,y)|_{y=0} = u_2(x,y)|_{y=l_y} = 0, \tag{13.152}$$

$$u_2(x,y)|_{x=0} = g_1(y), \quad u_2(x,y)|_{x=l_x} = g_2(y). \tag{13.153}$$

First, we consider the problem for the function $u_1(x,y)$ and search for the solution in the form:

$$u_1(x,y) = X(x)Y(y). \tag{13.154}$$

Substituting (13.154) into the Laplace equation and separating the variables yields

$$\frac{X''(x)}{X(x)} \equiv -\frac{Y''(y)}{Y(y)} = -\lambda, \tag{13.155}$$

where we assume $\lambda > 0$.

Next we obtain equations for $X(x)$ and $Y(y)$. With the homogeneous boundary conditions in Eq. (13.149), we obtain the one-dimensional Sturm-Liouville problem for $X(x)$ given by

$$X'' + \lambda X = 0, 0 < x < l_x$$

$$X(0) = X(l_x) = 0.$$

The solution to this problem is

$$X_n = \sin \sqrt{\lambda_{xn}} x, \quad \lambda_{xn} = \left(\frac{\pi n}{l_x} \right)^2, \quad n = 1, 2, 3, \ldots \tag{13.156}$$

With these eigenvalues, λ_{xn}, we obtain an equation for $Y(y)$ from (13.155):

$$Y'' - \lambda_{xn} Y = 0, \quad 0 < y < l_y. \tag{13.157}$$

A general solution to this equation can be written as

$$Y_n = C_n^{(1)} \exp\left(\sqrt{\lambda_{xn}} y \right) + C_n^{(2)} \exp\left(-\sqrt{\lambda_{xn}} y \right). \tag{13.158}$$

Such form of a solution does not fit well the purposes of the further analysis. It is more suitable to take a fundamental system of solution $\{Y_n^{(1)}, Y_n^{(2)}\}$ of Eq. (13.157) such that the functions $Y_n^{(1)}$ and $Y_n^{(2)}$ satisfy the homogeneous boundary condition, the former at $y = 0$ and the latter at $y = l_y$:

$$Y_n^{(1)}(0) = 0, \quad Y_n^{(2)}(l_y) = 0.$$

It is convenient to choose the following conditions at two other boundaries:

$$Y_n^{(1)}(l_y) = 1, \quad Y_n^{(2)}(0) = 1.$$

As the result the proper fundamental solutions of Eq. (13.155) are

$$Y_n^{(1)} = \frac{\sinh \sqrt{\lambda_{xn}} y}{\sinh \sqrt{\lambda_{xn}} l_y} \quad \text{and} \quad Y_n^{(2)} = \frac{\sinh \sqrt{\lambda_{xn}} (l_y - y)}{\sinh \sqrt{\lambda_{xn}} l_y}. \tag{13.159}$$

It is easily verified that they both satisfy Eq. (13.157) and are linearly independent; thus, they can serve as a fundamental set of particular solutions for this equation.

Using the above relations, we may write a general solution of the Laplace equation satisfying the homogeneous boundary conditions at the boundaries $x = 0$ and $x = l_x$ in Eq. (13.149), as a series in the functions $Y_n^{(1)}(y)$ and $Y_n^{(2)}(y)$:

$$u_1 = \sum_{n=1}^{\infty} \left[A_n Y_n^{(1)}(y) + B_n Y_n^{(2)}(y) \right] \sin \sqrt{\lambda_{xn}} x. \tag{13.160}$$

The coefficients of this series are determined from the boundary conditions (13.150):

$$u_1(x,y)|_{y=0} = \sum_{n=1}^{\infty} B_n \sin \sqrt{\lambda_{xn}}x = g_3(x),$$

$$u_1(x,y)|_{y=l_y} = \sum_{n=1}^{\infty} A_n \sin \sqrt{\lambda_{xn}}x = g_4(x).$$

(13.161)

Therefore, we see that B_n and A_n are Fourier coefficients of functions $g_3(x)$ and $g_4(x)$ in the system of eigenfunctions $\{\sin \sqrt{\lambda_{xn}}x\}_1^{\infty}$:

$$B_n = \frac{2}{l_x} \int_0^{l_x} g_3(\xi) \sin \sqrt{\lambda_{xn}}\xi d\xi,$$

$$A_n = \frac{2}{l_x} \int_0^{l_x} g_4(\xi) \sin \sqrt{\lambda_{xn}}\xi d\xi.$$

(13.162)

This completes the solution of the problem given in Eqs. (13.148) through (13.150).

Obviously, the solution of the similar problem given in Eqs. (13.151) through (13.153) can be obtained from that of Eqs. (13.160) and (13.162) by replacing y for x, l_y for l_x, and $g_3(x)$, $g_4(x)$ for $g_1(y)$ and $g_2(y)$. Carrying out this procedure yields

$$u_2 = \sum_{n=1}^{\infty} \left[C_n X_n^{(1)}(x) + D_n X_n^{(2)}(x) \right] \sin \sqrt{\lambda_{yn}}y,$$

(13.163)

where

$$X_n^{(1)}(x) = \frac{\sinh \sqrt{\lambda_{yn}}x}{\sinh \sqrt{\lambda_{yn}}l_x}, X_n^{(2)}(x) = \frac{\sinh \sqrt{\lambda_{yn}}(l_x - x)}{\sinh \sqrt{\lambda_{yn}}l_x}, \lambda_{yn} = \left(\frac{\pi n}{l_y}\right)^2$$

(13.164)

and

$$C_n = \frac{2}{l_y} \int_0^{l_y} g_2(\xi) \sin \sqrt{\lambda_{yn}}\xi d\xi, \quad D_n = \frac{2}{l_y} \int_0^{l_y} g_1(\xi) \sin \sqrt{\lambda_{yn}}\xi d\xi.$$

(13.165)

Finally, the solution to Problems (13.145) and (13.146) has the form:

$$u(x, y) = u_1(x, y) + u_2(x, y), \tag{13.166}$$

where functions $u_1(x, y)$ and $u_2(x, y)$ are defined by formulas (13.160) and (13.163), respectively.

In the same way, a BVP for the Laplace equation in a rectangular domain with *other types of boundary conditions* can be solved. The only difference is that the other fundamental solutions should be used.

Example 13.8 Find a steady-state temperature distribution inside a rectangular material which has boundaries maintained under the following conditions:

$$T(x, y)|_{x=0} = T_0 + (T_3 - T_0)\frac{y}{l_y}, T(x, y)|_{x=l_x} = T_1 + (T_2 - T_1)\frac{y}{l_y},$$

and

$$T(x, y)|_{y=0} = T_0 + (T_1 - T_0)\frac{x}{l_x}, T(x, y)|_{y=l_y} = T_3 + (T_2 - T_3)\frac{x}{l_x},$$

i.e., at the corners of the rectangle, the temperatures are T_0, T_1, T_2, T_3, and on the boundaries the temperature is a linear function.

Solution Introduce the function $u = T - T_0$ so that we measure the temperature relative to T_0. Then, $g_1(y) = (T_3 - T_1)\frac{y}{l_y}$, etc., and evaluating the integrals in Eqs. (13.162) and (13.165), we obtain:

$$A_n = \frac{2}{\pi n}[T_3 - T_0 - (-1)^n T_2], B_n = -2\frac{(-1)^n}{\pi n}(T_1 - T_0),$$

$$C_n = \frac{2}{\pi n}[T_1 - T_0 - (-1)^n T_2], D_n = -2\frac{(-1)^n}{\pi n}(T_3 - T_0).$$

These coefficients decay only as $1/n$; thus, the series in Eqs. (13.162) and (13.164) converge rather slowly.

Problems

In the following problems, consider rectangular plates ($0 \le x \le l_x, 0 \le y \le l_y$) which are thermally insulated over their lateral surfaces. There are no heat sources or absorbers inside the plates. Find the steady-state temperature distribution in the plates:

1. The sides $x = 0$, $y = 0$, and $y = l_y$ have a fixed zero temperature, and the side $x = l_x$ has the temperature $u(l_x, y) = \sin^2(\pi y/l_y)$.
2. The sides $x = 0$, $x = l_x$, and $y = 0$ have a fixed zero temperature, and the side $y = l_y$ has the temperature $u(x, l_y) = \sin^2(\pi x/l_x)$.
3. The sides $x = 0$ and $y = 0$ have a fixed zero temperature, the side $x = l_x$ is thermally insulated, and the side $y = l_y$ has the temperature $u(x, l_y) = \sin(5\pi x/l_x)$.

4. The sides $x = 0$ and $x = l_x$ have a fixed zero temperature, and the sides $y = 0$ and $y = l_y$ have the temperature $u(x, 0) = \sin \frac{\pi x}{l_x}$ and $u(x, l_y) = \sin (3\pi x, l_x)$.

In the following problems, an infinitely long rectangular cylinder has its axis along the z-axis, and its cross section is a rectangle with sides of length π. The sides of the cylinder are kept at an electric potential described by functions $u(x, y)|_\Gamma$ given below. Find the electric potential within the cylinder:

5. $u|_{x = 0} = u|_{x = \pi} = y^2, \quad u|_{y = 0} = x, \quad u|_{y = \pi} = 0.$
6. $u|_{x = 0} = 0, \quad u|_{x = \pi} = y^2, \quad u|_{y = 0} = 0, \quad u|_{y = \pi} = \cos x.$
7. $u|_{x = 0} = u|_{x = \pi} = \cos 2y, \quad u|_{y = 0} = u|_{y = \pi} = 0.$
8. $u|_{x = 0} = \cos 3y, \quad u|_{x = \pi} = 0, \quad u|_{y = 0} = x^2, \quad u|_{y = \pi} = 0.$

13.4.3   The Laplace Equation in Polar Coordinates

In this section we consider two-dimensional problems with a symmetry allowing the use of polar coordinates. Solutions to corresponding problems contain simple trigonometric functions of the polar angle as well as power and logarithmic functions of the radius.

In polar coordinates (r, φ), the Laplacian has the form:

$$\nabla^2 = \frac{\partial^2}{\partial r^2} + \frac{1}{r}\frac{\partial}{\partial r} + \frac{1}{r^2}\frac{\partial^2}{\partial \varphi^2} \equiv \nabla_r^2 + \frac{1}{r^2}\nabla_\varphi^2. \tag{13.167}$$

We begin by solving the Laplace equation $\nabla^2 u = 0$ using the Fourier method of separation of variables. First, we represent $u(r, \varphi)$ in the form:

$$u(r, \varphi) = R(r)\Phi(\varphi). \tag{13.168}$$

Substituting Eq. (13.168) into $\nabla^2 u = 0$ and separating the variables, we have:

$$\frac{r^2 \nabla_r R}{R} = -\frac{\nabla_\varphi^2 \Phi}{\Phi} = \lambda.$$

Because the first term does not depend on the angular variable φ and the second does not depend on r, each term must equal a constant which we denoted as λ. Thus, we obtain two separate equations for $R(r)$ and $\Phi(\varphi)$:

$$\Phi'' + \lambda\Phi = 0, \tag{13.169}$$

$$r^2 R'' + rR' - \lambda R = 0. \tag{13.170}$$

As we have discussed in Sect. 13.2.3, the solution of Eq. (13.169) must obey the periodicity condition, $\Phi(\varphi + 2\pi) = \Phi(\varphi)$, which is possible only if $\lambda = n^2$, $n = 0, \pm 1, \pm 2, \ldots$, and it is

$$\Phi = \Phi_n(\varphi) = \begin{cases} \cos n\varphi, \\ \sin n\varphi. \end{cases} \tag{13.171}$$

Equation for $R(r)$

$$r^2 R'' + r R' - n^2 R = 0 \tag{13.172}$$

is known as the *Euler equation* (see Sect. 3.8). The general solution to this equation is

$$\begin{aligned} R = R_n(r) &= C_1 r^n + C_2 r^{-n}, & n \neq 0, \\ R_0(r) &= C_1 + C_2 \ln r, & n = 0. \end{aligned} \tag{13.173}$$

Combining the above results for $\Phi_n(\varphi)$ and $R_n(r)$, we obtain the following particular and general solutions of Laplace equation:

(a) Under the condition that the solution be finite at $r = 0$, we have:

$$u_n(r, \varphi) = r^n \begin{cases} \cos n\varphi \\ \sin n\varphi \end{cases}, \quad n = 0, 1, \ldots$$

We can write a general solution for the Laplace problem for an *interior boundary value problem*, $0 \leq r \leq l$, via the expansion with these particular solutions $u_n(r, \varphi)$ as

$$u(r, \varphi) = \sum_{n=0}^{\infty} r^n (A_n \cos n\varphi + B_n \sin n\varphi).$$

The term with $n = 0$ is more conveniently written as $A_0/2$; thus, we have:

$$u(r, \varphi) = \frac{A_0}{2} + \sum_{n=1}^{\infty} r^n (A_n \cos n\varphi + B_n \sin n\varphi). \tag{13.174}$$

(b) For the case that the solution is finite at $r \to \infty$, we have:

$$u_n(r, \varphi) = \frac{1}{r^n} \begin{cases} \cos n\varphi \\ \sin n\varphi \end{cases}, \quad n = 0, 1, \ldots$$

These functions may be used as solutions to the Laplace problem for regions outside of a circle. The general solution of the Laplace equation for such an *exterior boundary value problem*, $r \geq l$, limited (i.e., bounded) at infinity, can be written as

$$u(r, \varphi) = \frac{A_0}{2} + \sum_{n=1}^{\infty} \frac{1}{r^n} (A_n \cos n\varphi + B_n \sin n\varphi). \tag{13.175}$$

(c) We also have a third set of solutions:

$$u_n(r, \varphi) = 1, \quad \ln r, \quad r^n \left\{ \begin{array}{c} \cos n\varphi \\ \sin n\varphi \end{array} \right\}, \quad \frac{1}{r^n} \left\{ \begin{array}{c} \cos n\varphi \\ \sin n\varphi \end{array} \right\}, \quad n = 1, 2, \ldots$$

for the cases where the solution is unbounded as $r \to 0$, as well as $r \to \infty$. This set of functions is used to solve Laplace equation for regions which form a circular ring, $l_1 \leq r \leq l_2$.

Let us solve the boundary value problem for a disk:

$$\nabla^2 u = 0 \quad \text{in} \quad 0 \leq r < l, \tag{13.176}$$

with Dirichlet boundary condition

$$u(r, \varphi)|_{r=l} = f(\varphi). \tag{13.177}$$

Applying (13.177) to formula (13.174), we obtain:

$$\frac{A_0}{2} + \sum_{n=1}^{\infty} l^n (A_n \cos n\varphi + B_n \sin n\varphi) = f(\varphi). \tag{13.178}$$

From this we see that $l^n A_n$ and $l^n B_n$ are the Fourier coefficients of expansion of the function $f(\varphi)$ in the system (or basis) of trigonometric functions $\{\cos n\varphi, \sin n\varphi\}$. We may evaluate the coefficients using the formulas:

$$A_n l^n = \frac{1}{\pi} \int_0^{2\pi} f(\varphi) \cos n\varphi \, d\varphi,$$

$$\tag{13.179}$$

$$B_n l^n = \frac{1}{\pi} \int_0^{2\pi} f(\varphi) \sin n\varphi \, d\varphi, \quad n = 0, 1, 2, \ldots$$

Thus, the solution of the *interior Dirichlet problem* for the Laplace equation is

$$u(r, \varphi) = \frac{A_0}{2} + \sum_{n=1}^{\infty} \left(\frac{r}{l}\right)^n [A_n \cos n\varphi + B_n \sin n\varphi]. \tag{13.180}$$

Example 13.8 Find the temperature distribution inside a circle if the boundary is kept at the temperature $T_0 = C_1 + C_2 \cos \varphi + C_3 \sin 2\varphi$.

Solution It is obvious that for this particular case, the series given by Eq. (13.178) reduces to three nonzero terms:

$$A_0 = 2C_1, \quad lA_1 = C_2, \quad l^2 B_2 = C_3.$$

In this case the solution is

$$T = C_1 + C_2 \frac{r}{l} \cos \varphi + C_3 \left(\frac{r}{l}\right)^2 \sin 2\varphi.$$

Example 13.9 Find the temperature distribution outside a circle with the same distribution of the temperature on its boundary $r = l$ as in Example 13.8.

Solution Substituting the general solution (13.175) into the boundary condition, we find that nonzero Fourier coefficients A_0, A_1, and B_2 satisfy the following relations:

$$A_0 = 2C_1, A_1 = lC_2, B_2 = l^2 C_3.$$

Thus, the solution is

$$T = C_1 + C_2 \frac{l}{r} \cos \varphi + C_3 \left(\frac{l}{r}\right)^2 \sin 2\varphi.$$

The boundary value problems with other boundary condition can be solved in a similar way. Note that in the case of the Neumann boundary condition $\frac{\partial u}{\partial r}|_{r=l} = g(\varphi)$ (i.e., the heat flux is fixed on the boundary), the solution exists only if integral $\int_0^{2\pi} g(\varphi) d\varphi = 0$ (the temperature field can be steady only if there is no total heating or cooling of the disk). If the condition given above is satisfied, the problem has infinitely many solutions: if $u(r, \varphi)$ satisfies the Laplace equation and the Neumann boundary condition, then $u(r, \varphi) + cont.$ satisfies them as well.

Problems

Consider a circular plate of radius l that is thermally insulated over its lateral surfaces. The circular periphery of the plate is kept at the temperature given by the function of the polar angle $u(l, \varphi)$, $0 \le \varphi \le 2\pi$. Find the steady-state temperature distribution in the plate.

1. $u(l, \varphi) = \varphi$.
2. $u(l, \varphi) = \cos 3\varphi$.
3. $u(l, \varphi) = \sin (\varphi/4)$.
4. $u(l, \varphi) = \cos (\varphi/2)$.
5. $u(l, \varphi) = \sin 3\varphi$.

Solve the boundary value problem for the Laplace equation in the disk, $0 \leq r \leq l$, $0 \leq \varphi \leq 2\pi$, with the following boundary conditions:

6. $u(l, \varphi) = 2\sin^2\varphi + 4\cos^3\varphi$.
7. $u(l, \varphi) = 32(\sin^6\varphi + \cos^6\varphi)$.

Solve the boundary value problem for the Laplace equation outside the disk, $r \geq l$, $0 \leq \varphi \leq 2\pi$, with the following boundary conditions:

8. $u(l, \varphi) = 8\cos^4\varphi$.
9. $u(l, \varphi) = \begin{cases} 1, & 0 < \varphi < \pi, \\ -1, & \pi < \varphi < 2\pi. \end{cases}$

13.4.4   The Laplace Equation in a Sphere

The next important example is the three-dimensional Laplace equation:

$$\nabla^2 u = 0$$

inside a sphere of radius a (internal problem) or outside that sphere (external problem). Both kinds of problems can appear in electrostatics. In the case of gravitation, the external problem is typical.

Using the *spherical coordinates*, we obtain the following equation for function $u = u(r, \theta, \varphi)$:

$$\nabla^2 u = \frac{1}{r^2} \frac{\partial}{\partial r} \left(r^2 \frac{\partial u}{\partial r} \right) + \frac{1}{r^2 \sin \theta} \frac{\partial}{\partial \theta} \left(\sin \theta \frac{\partial u}{\partial \theta} \right) + \frac{1}{r^2 \sin^2 \theta} \frac{\partial^2 u}{\partial \varphi^2} = 0, \quad (13.181)$$

$r < a$ (for internal problem) or $r > a$ (for external problem), $0 < \theta < \pi, 0 \leq \varphi \leq 2\pi$. We shall discuss only the *Dirichlet boundary value problem*:

$$u(a, \theta, \varphi) = F(\theta, \varphi), \quad 0 \leq \theta \leq \pi, \quad 0 \leq \varphi \leq 2\pi. \quad (13.182)$$

We shall only consider the case when the boundary condition does not depend on the azimuthal angle φ:

$$u(a, \theta, \varphi) = F(\theta), \quad 0 \leq \theta \leq \pi; \quad (13.183)$$

therefore, the solution also does not depend on φ:

$$u = u(r, \theta). \qquad (13.184)$$

Equation (13.181) is reduced to

$$\nabla^2 u = \frac{1}{r^2} \frac{\partial}{\partial r} \left(r^2 \frac{\partial u}{\partial r} \right) + \frac{1}{r^2 \sin \theta} \frac{\partial}{\partial \theta} \left(\sin \theta \frac{\partial u}{\partial \theta} \right) = 0. \qquad (13.185)$$

As usual, we apply the method of separation of variables and find particular solutions in the form:

$$u(r, \theta) = R(r) \Theta(\theta). \qquad (13.186)$$

Substituting (13.186) into (13.185), we get:

$$\left(r^2 R'(r) \right)' \Theta(\theta) + \frac{R(r)}{\sin \theta} [\sin \theta \Theta'(\theta)]' = 0; \qquad (13.187)$$

hence,

$$\frac{(r^2 R'(r))'}{R(r)} = -\frac{[\sin \theta \Theta'(\theta)]'}{\sin \theta \Theta(\theta)} = \lambda, \qquad (13.188)$$

where λ is constant.

It is convenient to introduce the variable $x = \cos \theta$:

$$\frac{d}{dx} = -\frac{1}{\sin \theta} \frac{d}{d\theta},$$

which allows to rewrite the equation for $\Theta(\theta) \equiv X(x)$ as

$$\frac{d}{dx} \left[(1 - x^2) \frac{dX}{dx} \right] + \lambda X = 0, \quad 1 \leq x \leq 1.$$

Thus, function $X(x)$ is determined by *the Legendre equation*.

The properties of solutions of that equation are described in Chap. 10. It is shown that the bounded solutions of that equations, *the Legendre polynomials*, $X_n(x) \equiv P_n(x)$, exist only for $\lambda = \lambda_n = n(n + 1)$, where $n = 0, 1, 2, \ldots$

Equation for $R_n(r)$

$$\left[r^2 R_n'(r) \right]' - n(n + 1) R_n = 0 \qquad (13.189)$$

has two particular solutions

$$r^n \quad \text{and} \quad r^{-(n+1)}.$$

In the case of an internal problem $(r < a)$, the solution $r^{-(n + 1)}$ has to be dropped because it is unbounded when $r \to 0$. In the case of an external problem $(r > a)$, the solution r^n has to be dropped because it is unbounded when $r \to \infty$.

Thus, in the case of an internal problem, the bounded solution of Eq. (13.183) can be presented in the form:

$$u(r, \theta) = \sum_{n=0}^{\infty} A_n \left(\frac{r}{a}\right)^n P_n(\cos \theta). \tag{13.190}$$

The boundary condition (13.183) prescribes

$$u(a, \theta) = \sum_{n=0}^{\infty} A_n P_n(\cos \theta) = F(\theta). \tag{13.191}$$

Using the orthogonality property of the Legendre polynomials

$$\int_{-1}^{1} P_n(x) P_{n'}(x) dx = \frac{2}{2n + 1} \delta_{nn'},$$

we find that

$$A_n = \frac{2n + 1}{2} \int_{0}^{\pi} F(\theta) P_n(\cos \theta) \sin \theta d\theta.$$

The solution of an external problem can be found in a similar way.

Some topics that involve the Poisson equation are discussed in [9].

Example 13.10 The top surface of the sphere $(0 \le \theta \le \pi/2)$ with the radius a has the electric potential U_0, while its bottom surface $(\pi/2 < \theta \le \pi)$ is grounded, i.e., has zero potential. Find the electric potential inside the sphere.

Solution The potential u inside the sphere is governed by the following boundary value problem:

$$\nabla^2 u = 0 \quad \text{in} \quad 0 \le r < a,$$

$$u|_{r=a} = \left\{ \begin{array}{ll} U_0, & 0 \le \theta \le \pi/2 \\ 0, & \pi/2 \le \theta \le \pi \end{array} \right\} = F(\theta).$$

Its solution is axially symmetric (i.e., it does not depend on φ). Hence, it can be written as

$$u(r, \theta) = \sum_{n=0}^{\infty} \left(\frac{r}{a}\right)^n A_n P_n(\cos \theta).$$

The coefficients A_n are determined from the boundary condition using the formula:

$$A_n = \frac{2n+1}{2} \int_0^\pi F(\theta) P_n(\cos \theta) \sin \theta d\theta = U_0 \frac{2n+1}{2} \int_0^1 P_n(x) dx.$$

First, note that for even $n \neq 0$

$$\int_0^1 P_n(x) dx = \frac{1}{2} \int_{-1}^1 P_n(x) dx = 0.$$

For $n = 0$

$$\int_0^1 P_n(x) dx = \int_0^1 dx = 1.$$

Now, let $n = 2k + 1$. Using the formula

$$P_n(x) = \frac{1}{2n+1} \left[P'_{n+1}(x) - P'_{n-1}(x) \right],$$

we find that

$$\int_0^1 P_{2k+1}(x) dx = \frac{1}{4k+3} \int_0^1 \left[P'_{2k+2}(x) - P'_{2k}(x) \right] dx = = \frac{1}{4k+3} \left[P'_{2k}(0) - P'_{2k+2}(0) \right].$$

Because

$$P_{2k}(0) = (-1)^k \frac{(2k)!}{2^{2k}(k!)^2},$$

We obtain:

$$\int_0^1 P_{2k+1}(x)dx = (-1)^k \frac{(2k)!}{2^{(2k+1)(k+1)}(k!)^2}.$$

Thus,

$$A_{2k} = 0, \quad A_{2k+1} = (-1)^k \frac{4k+3}{k+1} \frac{(2k)!}{2^{2k+1}(k!)^2} U_0,$$

and the solution is

$$u(r,\theta) = \frac{U_0}{2} + U_0 \sum_{k=0}^{\infty} (-1)^k \frac{4k+3}{k+1} \cdot \frac{(2k)!}{2^{2k+1}(k!)^2} \left(\frac{r}{a}\right)^{2k+1} P_{2k+1}(\cos\theta).$$

Problems

Solve the internal Dirichlet problems for the ball $0 \le r \le a$ with the boundary conditions:

1. $u|_{r=a} = \cos\theta$.
2. $u|_{r=a} = \sin\theta$.
3. $u|_{r=a} = \cos 2\theta$.
4. $u|_{r=a} = \sin 2\theta$.
5. $u|_{r=a} = \cos\theta + \cos^2\theta$. *Hint*: $\cos^2\theta = (1 + \cos 2\theta)/2$.

Solve the external Dirichlet problem for the ball $r \ge a$ with the boundary conditions:

6. $u|_{r=a} = \cos\theta$.
7. $u|_{r=a} = \sin\theta$.
8. $u|_{r=a} = \cos 2\theta$.
9. $u|_{r=a} = \sin 2\theta$.

13.5   Three-Dimensional Helmholtz Equation and Spherical Functions

Let us consider the Helmholtz equation for function $\psi(r, \theta, \varphi)$, where r, θ, and φ are spherical coordinates ($r \ge 0$, $0 \le \theta \le \pi$, and $0 \le \varphi \le 2\pi$):

$$\nabla^2\psi + k^2\psi = 0. \tag{13.192}$$

If k^2 is a function of only r or it is a constant, this equation allows a separation of variables:

$$\psi(r,\theta,\varphi) = R(r)Y(\theta,\varphi). \tag{13.193}$$

Using the Laplacian in spherical coordinates

$$\nabla^2 = \nabla_r^2 + \frac{1}{r^2}\nabla_{\theta\varphi}^2 \tag{13.194}$$

where

$$\nabla_r^2 = \frac{\partial^2}{\partial r^2} + \frac{2}{r}\frac{\partial}{\partial r}, \quad \nabla_{\theta\varphi}^2 = \nabla_\theta^2 + \frac{1}{\sin^2\theta}\nabla_\varphi^2$$

$$\nabla_\theta^2 = \frac{1}{\sin\theta}\frac{\partial}{\partial\theta}\left(\sin\theta\frac{\partial}{\partial\theta}\right), \quad \nabla_\varphi^2 = \frac{\partial^2}{\partial\varphi^2}.$$

Equation (13.192) becomes:

$$\left(\nabla_r^2 + \frac{1}{r^2}\nabla_{\theta\varphi}^2\right)RY + k^2RY = 0,$$

which leads to

$$Y\nabla_r^2R + \frac{R}{r^2}\nabla_{\theta\varphi}^2Y + k^2RY = 0.$$

Since $k = k^2(r)$, dividing this equation by RY separates the functions that depend on r, θ, and φ:

$$r^2\frac{\nabla_r^2R}{R} + k^2r^2 = -\frac{\nabla_{\theta\varphi}^2Y}{Y} = \lambda, \tag{13.195}$$

where λ is the separation constant. Thus, we obtain two equations, the equation for *the radial part*, $R(r)$:

$$\nabla_r^2R + \left(k^2 - \frac{\lambda}{r^2}\right)R = 0 \tag{13.196}$$

and the equation for *the angular function*, $Y(\theta,\varphi)$:

$$\nabla_{\theta\varphi}^2Y + \lambda Y = 0. \tag{13.197}$$

Next, we can continue the separation of variables by representing $Y(\theta,\varphi)$ as

$$Y(\theta, \varphi) = \Theta(\theta)\Phi(\varphi). \tag{13.198}$$

Substitution of Eq. (13.198) into (13.197) gives:

$$\Phi\nabla_\theta^2\Theta + \frac{\Theta}{\sin^2\theta}\nabla_\varphi^2\Phi + \lambda\Theta\Phi = 0.$$

Dividing by $\Theta\Phi$ and separating the functions of θ and φ, we obtain:

$$\left(\frac{\nabla_\theta^2\Theta}{\Theta} + \lambda\right)\sin^2\theta = -\frac{\nabla_\varphi^2\Phi}{\Phi} \equiv m^2, \tag{13.199}$$

where m^2 is another separation constant. Thus,

$$\nabla_\theta^2\Theta + \left(\lambda - \frac{m^2}{\sin^2\theta}\right)\Theta = 0, \tag{13.200}$$

and

$$\nabla_\varphi^2\Phi + m^2\Phi = 0 \quad \text{or} \quad \frac{d^2\Phi(\varphi)}{d\varphi^2} + m^2\Phi(\varphi) = 0. \tag{13.201}$$

The solution of Eq. (13.201), $\Phi = e^{im\varphi}$, must obey a periodicity condition in the angle φ, e.g., $\Phi(\varphi + 2\pi) = \Phi(\varphi)$; thus, $e^{2im\pi} = 1$ and therefore, m is an *integer*: $m = 0, \pm 1, \pm 2, \ldots$

Let us consider the equation for $\Theta(\theta)$. We may introduce a new variable, $x \equiv \cos\theta$, which has the range $-1 \le x \le 1$ (this follows from $0 \le \theta \le \pi$), and change the notation to the conventional one: $\Theta(\theta) \Rightarrow P(x)$.

Reading Exercise Show that $\nabla_\theta^2\Theta = \frac{d}{dx}\left[(1-x^2)\frac{d}{dx}P(x)\right]$.

Thus, Eq. (13.200) is equivalent to

$$\frac{d}{dx}\left[(1-x^2)\frac{dP}{dx}\right] + \left(\lambda - \frac{m^2}{1-x^2}\right)P = 0. \tag{13.202}$$

We discussed this equation in Chap. 10. Its solutions are called the *associated Legendre functions*. We obtained there that the parameter λ can be presented as $\lambda = l(l+1)$, $l = 0, 1, 2, \ldots$, and the values of the parameter m are $m = 0, \pm 1, \pm 2, \ldots$, with $|m| \le l$. With these notations, the solutions of Eq. (13.200) are $P_l^m(x)$, or $P_l^m(\cos\theta)$.

Thus, we obtain the set of *spherical harmonics* as the solutions of the Helmholtz equation:

$$Y_l^m(\theta, \varphi) = N_l^m P_l^m(\cos\theta) e^{im\varphi} \quad (|m| < l; l = 0, 1, \ldots). \tag{13.203}$$

The coefficients N_l^m can be obtained by a normalization of functions $Y_l^m(\theta, \varphi)$:

$$\int |Y_l^m|^2 \sin\theta d\theta d\varphi = 1. \tag{13.204}$$

This gives:

$$N_l^m = (-1)^{\frac{1}{2}(m+|m|)} \left[\frac{2l+1}{4\pi} \frac{(l-|m|)!}{(l+|m|)!} \right]^{\frac{1}{2}}. \tag{13.205}$$

The first four spherical harmonics are

$$Y_0^0 = \frac{1}{\sqrt{4\pi}}, \quad Y_1^0 = \left(\frac{3}{4\pi}\right)^{1/2} \cos\theta, \quad Y_1^{\pm 1} = \mp \left(\frac{3}{8\pi}\right)^{1/2} \sin\theta e^{\pm i\varphi}. \tag{13.206}$$

In quantum mechanics, Eq. (13.192) is called the Schrödinger equation for the *wave function* ψ. In the case of a *spherical symmetry*, k^2 does not depend on the angles, and as we obtained, $\psi(r, \theta, \varphi) = R(r) Y_l^m(\theta, \varphi)$, with $l = 0, 1, \ldots$, $m = -l, -l+1, \ldots, l$. The parameter l is called *the angular (or orbital) quantum number*; m is called *the magnetic quantum number*. Thus, from a purely mathematical consideration, we obtain a fundamental result: the angular momentum and the magnetic moment are *quantized*, i.e., only certain values are allowed.

Next, consider Eq. (13.196) for the function $R(r)$. As we obtained, $\lambda = l(l+1)$ in that equation.

It is not difficult to solve Eq. (13.196) in a simple case $k^2 = const$. For application, consider a quantum particle trapped in a spherically symmetric impenetrable potential well of radius a. In the Schrödinger equation for a particle of mass μ

$$k^2 = \frac{2\mu}{\hbar^2} (E - U),$$

where E is the particle's total energy, U is its potential energy and $\hbar$ is Planck's constant.

Inside the well, where $r < a$, the electron moves freely and $k = \sqrt{2\mu E}/\hbar$ does not depend on r. Using the substitution $x = kr$, we have $\frac{d}{dr} = k\frac{d}{dx}$, $\frac{d^2}{dr^2} = k^2 \frac{d^2}{dx^2}$, and Eq. (13.196) becomes:

$$\frac{d^2 R(x)}{dx^2} + \frac{2}{x}\frac{dR(x)}{dx} + \left[1 - \frac{l(l+1)}{x^2}\right]R(x) = 0. \tag{13.207}$$

This is the *spherical Bessel equation* (see Sect. 10.2.8). Its solutions that are finite at $x = 0$ are the *spherical Bessel* functions, $j_l(x)$. Thus, $R(x) = Cj_l(x)$, or

$$R_{kl}(kr) = Cj_l(kr). \tag{13.208}$$

The coefficient C can be found from the normalization condition:

$$\int_0^\infty |R_{kl}|^2 r^2 dr = 1. \tag{13.209}$$

Inside the well

$$\psi_{klm}(r, \theta, \varphi) = Cj_l(kr)Y_l^m(\theta, \varphi). \tag{13.210}$$

The wave function vanishes for $r > a$ because the particle cannot penetrate into a region, where $U = \infty$. The continuity of the wave function leads to the boundary condition at $r = a$:

$$j_l(ka) = 0. \tag{13.211}$$

Denoting the roots of the Bessel spherical function $j_l(x)$ as x_{nl}, Eq. (13.209) gives:

$$x_{nl} = ak_{nl}, \tag{13.212}$$

where $n = 1, 2, 3, ..$ numbers the roots; thus,

$$k_{nl} = x_{nl}/a. \tag{13.213}$$

Using the relationship $k = \sqrt{2\mu E}/\hbar$, we obtain the energy spectrum of a particle inside the well:

$$E_{nl} = \frac{\hbar^2 k_{nl}^2}{2\mu} = \frac{\hbar^2}{2\mu a^2}x_{nl}^2. \tag{13.214}$$

This formula shows that the energy does not depend on the value of the magnetic quantum number m, which is why each eigenvalue of the boundary value problem is $(2l + 1)$-fold degenerated, i.e., the same energy E_{nl} corresponds to $(2l + 1)$ different functions ψ_{nlm}.

For $l = 0$ the solutions of equation $j_0(x) = 0$ are $x = n\pi$; thus,

$$E_{n0} = \frac{1}{2\mu}\left(\frac{n\pi\hbar}{a}\right)^2. \tag{13.215}$$

For the *ground state* $n = 1$, and the lowest energy level is $E_{10} = \frac{1}{2\mu}\left(\frac{\pi\hbar}{a}\right)^2$. The wave function for the state with $l = 0$ is

$$\psi_{n00}(r,\theta,\varphi) = Cj_0(kr)Y_0^0(\theta,\varphi) = \frac{1}{\sqrt{2\pi a}}\frac{1}{\sqrt{4\pi}}\frac{\sin(n\pi r/a)}{n\pi r/a}, \tag{13.216}$$

where the coefficient $C = 1/\sqrt{2\pi a}$ was obtained from the normalization (13.207). For $l = 0$ the solution of the Schrödinger equation does not depend on the angles θ and φ; in other words, it is spherically isotropic.

The quantum mechanical interpretation of the wave function $\psi_{nlm}(r,\theta,\varphi)$ is that its modulus squared, $|\psi_{nlm}(r,\theta,\varphi)|^2$, determines the probability of finding a particle at the point with coordinates r, θ, φ. Equation (13.216) for ψ_{n00} shows that this probability changes periodically with the distance from the center and there are some distances where the particle cannot be located at all, such as $r = a/2$ for the state with $n = 2$. Such "nodes" are separated by the maxima where a probability of locating a particle is the largest. These maxima correspond to the "orbits" in a classical interpretation. For the states with $l \neq 0$, such orbits have more complex geometry, which is determined by the shape of $\left|Y_l^m(\theta,\varphi)\right|^2$.

Reading Exercise For $n = 1, 2, 3$, plot graphs of ψ_{n00} as a function of r.

Reading Exercise For $l = 1$, plot a three-dimensional graphs of this function for $m = 0, \pm 1$.

Chapter 14
Introduction to Numerical Methods for ODEs

Numerical methods for solution of ODEs are employed when exact analytical or approximate solution either does not exist, or obtaining it is difficult.

Prior to applying a numerical method, equations and systems of equations of order two or higher, it can be convenient to reduce them to a first-order system using a change of variable(s) (see Sect. 4.1). For example, a second-order equation

$$\frac{d^2 y(x)}{dx^2} = f(x, y(x), y'(x)), \tag{14.1}$$

is reduced to a first-order system

$$\begin{cases} \dfrac{dz(x)}{dx} = f(x, y(x), z(x)), \\ \dfrac{dy(x)}{dx} = z(x), \end{cases} \tag{14.2}$$

by changing a variable, viz.,

$$\frac{dy(x)}{dx} = z(x). \tag{14.3}$$

Below we give some examples of numerical methods for initial and boundary value problems (IVP and BVP, respectively). A general description of numerical methods for solving ODEs can be found in [10].

© The Author(s), under exclusive license to Springer Nature Switzerland AG 2023
V. Henner et al., *Ordinary Differential Equations*,
https://doi.org/10.1007/978-3-031-25130-6_14

14.1 Two Numerical Methods for IVP

IVP for first-order equation reads:

$$y'(x) = f(x, y)$$
$$y(x_0) = y_0. \tag{14.4}$$

Euler's Method

Euler's method is the simplest, but least accurate, numerical method for IVP. Suppose that we seek a solution of (14.4) on the interval $[x_0, b]$. First, this interval is partitioned into n subintervals $[x_0, x_1]$, $[x_1, x_2]$, ..., $[x_{n-1}, b]$ of equal width $h = \frac{b - x_0}{n}$. Then, the solution $y(x)$ at the points $x_i = x_0 + ih$, $i = 1, \ldots, n$ can be approximated by Euler's formula:

$$y_i = y_{i-1} + hf(x_{i-1}, y_{i-1}), \quad i = 1, \ldots, n, \tag{14.5}$$

(see Sect. 2.1). According to (14.5), in order to calculate a numerical solution on the interval $[x_0, b]$, it is necessary to take n steps from x_0 to b. Each step requires one evaluation of a function $f(x, y)$ and incurs a small error. According to Taylor's Theorem, the difference between the approximate value y_i and the exact value $y(x_i)$ is $O(h^2)$. Because the errors accumulate, the mean difference between the exact and approximate values of $y(x)$ is $O(h)$. Thus, Euler's method is a method of the first-order accuracy. When using Euler's method, one has to use a very small value of h.

Euler's method is easily extended to an IVP for a first-order system:

$$y'_1 = f_1(x, y_1, y_1, \ldots, y_n),$$
$$y'_2 = f_2(x, y_1, y_1, \ldots, y_n),$$
$$\ldots\ldots\ldots\ldots\ldots\ldots\ldots\ldots\ldots\ldots\ldots\ldots\ldots\ldots \tag{14.6}$$
$$y'_n = f_n(x, y_1, y_1, \ldots, y_n),$$

subject to the initial conditions

$$y_i(x_0) = y_{i0}.$$

Approximate values $y_k(x_i)$ are calculated as

$$y_{k(i)} = y_{k(i-1)} + hf_k\left(x_{i-1}, y_{1(i-1)}, y_{2(i-1)}, \ldots, y_{n(i-1)}\right), \quad k = 1, \ldots, n; \ i = 1, \ldots, n.$$

Runge-Kutta Method of Order 4

For IVP (14.4) a fourth order Runge-Kutta method (RK4) is [11,12]

$$y_i = y_{i-1} + \frac{1}{6}[k_1 + 2k_2 + 2k_3 + k_4], i = 1, \ldots, n,$$

where

$$
\begin{aligned}
k_1 &= hf(x_{i-1}, y_{i-1}), \\
k_2 &= hf\left(x_{i-1} + \frac{h}{2}, y_{i-1} + \frac{k_1}{2}\right), \\
k_3 &= hf\left(x_{i-1} + \frac{h}{2}, y_{i-1} + \frac{k_2}{2}\right), \\
k_4 &= hf(x_{i-1} + h, y_{i-1} + k_3).
\end{aligned}
\tag{14.7}
$$

RK4 is more accurate than Euler's method. From (14.7) it is apparent that this method requires four evaluations of $f(x, y)$ at each step. A single-step error is $O(h^5)$, and a total error on the interval $[x_0, b]$ is $O(h^4)$. Thus, RK4 is a method of the fourth-order accuracy. Due to its higher accuracy, when using RK4 one can choose larger step size than in Euler's method (for given input error tolerance). Method (14.7) was developed by German mathematicians Carl Runge (1856–1927) and Wilhelm Kutta (1867–1944) around 1900.

RK4 for system (14.6) is a straightforward generalization of (14.7). After casting (14.6) in the form

$$\vec{y} = \vec{f}\left(x, \vec{y}\right), \vec{y}(x_0) = \vec{y}_0,$$

one calculates at each step the quantities:

$$
\begin{aligned}
\vec{k}_1 &= h\vec{f}\left(x_{i-1}, \vec{y}_{i-1}\right), \\
\vec{k}_2 &= h\vec{f}\left(x_{i-1} + \frac{1}{2}h, \vec{y}_{i-1} + \frac{1}{2}\vec{k}_1\right), \\
\vec{k}_3 &= h\vec{f}\left(x_{i-1} + \frac{1}{2}h, \vec{y}_{i-1} + \frac{1}{2}\vec{k}_2\right), \\
\vec{k}_4 &= h\vec{f}\left(x_{i-1} + h, \vec{y}_{i-1} + \vec{k}_3\right).
\end{aligned}
\tag{14.8}
$$

Then, $\vec{y}_{i+1}$ is calculated as weighted average of $\vec{k}_i$:

$$\vec{y}_i = \vec{y}_{i-1} + \frac{1}{6}\left[\vec{k}_1 + 2\vec{k}_2 + 2\vec{k}_3 + \vec{k}_4\right], i = 1, \ldots, n. \tag{14.9}$$

14.2   A Finite-Difference Method for Second-Order BVP

In a basic *finite-difference* (FD) *method*, the derivatives of a continuous unknown function $y(x)$ are approximated by *finite differences* on a *uniform grid* $x_i = a + ih$, $i = 0, \ldots, n$ covering the interval $[a, b]$. Here $h = (b - a)/n$.

Let us consider applying this method to a linear DE:

$$y''(x) + p(x)y'(x) + q(x)y(x) = f(x), \tag{14.10}$$

where the functions $p(x)$, $q(x)$, and $f(x)$ are continuous on $[a, b]$. The boundary conditions at the endpoints $x = a$ and $x = b$ read:

$$\alpha_1 y' + \beta_1 y|_{x=a} = A, \quad \alpha_2 y' + \beta_2 y|_{x=b} = B. \tag{14.11}$$

We choose to approximate the first and second derivatives in (14.10) as

$$y_i' \approx \frac{y_{i+1} - y_i}{h}, \quad y_i'' \approx \frac{1}{h}\left[\frac{y_{i+1} - y_i}{h} - \frac{y_i - y_{i-1}}{h}\right] = \frac{y_{i+1} - 2y_i + y_{i-1}}{h^2}. \tag{14.12}$$

Then, (14.10) turns into $n - 1$ linear algebraic equations:

$$\frac{y_{i+1} - 2y_i + y_{i-1}}{h^2} + p_i \frac{y_{i+1} - y_i}{h} + q_i y_i = f_i, \quad i = 1, \ldots, n - 1. \tag{14.13}$$

Boundary conditions (14.11) give two additional equations:

$$\alpha_1 \frac{y_1 - y_0}{h} + \beta_1 y_0 = A, \quad \alpha_2 \frac{y_n - y_{n-1}}{h} + \beta_2 y_n = B. \tag{14.14}$$

Equations (14.13) and (14.14) form a linear algebraic system of $n + 1$ equations for $n + 1$ unknowns $y_0, y_1, \ldots y_n$. This system can be easily written in the form [13]:

$$\begin{cases} y_{i+2} + m_i y_{i+1} + n_i y_i = f_i h^2, & i = 0, \ldots, n - 2, \\ \alpha_1 \dfrac{y_1 - y_0}{h} + \beta_1 y_0 = A, \\ \alpha_2 \dfrac{y_n - y_{n-1}}{h} + \beta_2 y_n = B, \end{cases} \tag{14.15}$$

where $m_i = -2 + hp_i$, $n_i = 1 - hp_i + h^2 q_i$. Notice that the matrix of system (14.15) is tridiagonal and diagonally dominant. The most efficient solution method for (14.15) is *Thomas method*, named after a British physicist and applied mathematician Llewellyn Thomas (1903–1992). The details of the method can be found in [14,15]. This method consists of two stages, a forward one and a backward one. In the forward stage, one determines the coefficients that are necessary for the backward stage, where the approximate solution values y_i, $i = 0, \ldots, n$ are computed.

Forward stage: Compute

$$
\begin{cases}
c_0 = \dfrac{\alpha_1 - \beta_1 h}{m_0(\alpha_1 - \beta_1 h) + n_0 \alpha_1} \\[2ex]
d_0 = \dfrac{n_0 A h}{\alpha_1 - \beta_1 h} + f_0 h^2
\end{cases}
\text{ and }
\begin{cases}
c_i = \dfrac{1}{m_i - n_i c_{i-1}} \\[2ex]
d_i = f_i h^2 - n_i c_{i-1} d_{i-1},
\end{cases}
\tag{14.16}
$$

$i = 1, \ldots, n-2.$

Backward stage: The last solution of (14.15) is computed as

$$
y_n = \frac{\alpha_2 c_{n-2} d_{n-2} + Bh}{\alpha_2(1 + c_{n-2}) + \beta_2 h}.
\tag{14.17}
$$

Other solutions are computed recursively:

$$
\begin{aligned}
y_i &= c_{i-1}\left(d_{i-1} - y_{i+1}\right), \quad i = n-1, \ldots 1, \\
y_0 &= \frac{\alpha_1 y_1 - Ah}{\alpha_1 - \beta_1 h}.
\end{aligned}
\tag{14.18}
$$

The accuracy of the above finite-difference method is $O(h)$.

In order to reach a given precision goal, one may consecutively double the number of grid points. When approximation values at a grid point corresponding to same x for computations with n and $2n$ grid points agree within given precision, the solutions is declared complete.

14.3 Applications

Magnetic Resonance
(the example of IVP numerical solution using RK4 method)

In Sect. 4.7.2 we discussed a precession of a magnetic moment $\vec{\mu}$ of a particle in a magnetic field $\vec{B}$ which is governed by equation:

$$
\frac{d\vec{\mu}}{dt} = \gamma \vec{\mu} \times \vec{B},
\tag{14.19}
$$

where γ is the gyromagnetic ratio. A physically interesting situation occurs when a particle with a magnetic moment is placed in a combination of a strong constant magnetic field B_0, whose direction we can chose along the z-axis, and a much weaker periodic magnetic field in the xy-plane:

$$B_x = B_1 \cos \omega t, \quad B_y = -B_1 \sin \omega t, \quad B_1 << B_0. \tag{14.20}$$

We had seen in Sect. 4.7.2 that in the presence of only the magnetic field B_0, the magnetic moment is rotating about the direction of vector B_0 with *the Larmor frequency* $\omega_0 = \gamma B_0$. The result that we want to demonstrate now is that in the case of the resonance, $\omega = \omega_0$, the vector $\vec{\mu}$ periodically changes its orientation along the z-axis. This phenomenon is called the *magnetic resonance*.

The component form of Eq. (14.19) is the system of ODEs:

$$\begin{cases} \dfrac{d\mu_x}{dt} = \gamma\left(\mu_y B_z - \mu_z B_y\right), \\[2mm] \dfrac{d\mu_y}{dt} = \gamma\left(\mu_z B_x - \mu_x B_z\right), \\[2mm] \dfrac{d\mu_z}{dt} = \gamma\left(\mu_x B_y - \mu_y B_x\right). \end{cases} \tag{14.21}$$

It is convenient to cast this system in the dimensionless form using the dimensionless time, $\tilde{t} = t\omega_0$, and the unit magnetic moment, $\vec{e} \equiv \vec{\mu}/\mu$. The system of equations for vector $\vec{e}$ in the resonant case, $\omega = \omega_0$, reads (here $a \equiv B_1/B_0$):

$$\begin{cases} \dfrac{de_x}{d\tilde{t}} = e_y + ae_z \sin\tilde{t} \\[2mm] \dfrac{de_y}{d\tilde{t}} = ae_z \cos\tilde{t} - e_x \\[2mm] \dfrac{de_z}{d\tilde{t}} = -ae_x \sin\tilde{t} - ae_y \cos\tilde{t}. \end{cases} \tag{14.22}$$

The numerical solution of this system obtained with RK4 method describes the time variation of the vector $\vec{e}(\tilde{t})$ with the period $2\pi/a$ (see Fig. 14.1).

The direction of $\vec{e}(\tilde{t})$ along the z-axis reverses when each period ends, and also the vector $\vec{e}(\tilde{t})$ is rotating in the *xy*-plane with the period 2π (in variable $\tilde{t}$). Flips in

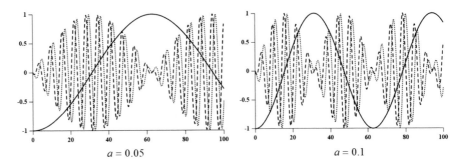

$a = 0.05$ $a = 0.1$

Fig. 14.1 Solution of system (14.22) with the initial conditions $e_z(0) = -1$, $e_{x,y}(0) = 0$. Solid line, $e_z(\tilde{t})$; dotted line, $e_x(\tilde{t})$; dashed line, $e_y(\tilde{t})$

the direction of the field B_0 are the manifestation of the magnetic resonance. The flips happen only when the frequency of the transverse field is equal to (or is very close to) the Larmor frequency. Thus, scanning the system of magnetic moments (spin system) with the periodic external field allows to very precisely find the frequency $\omega = \omega_0$ and, therefore, to determine the particle's gyromagnetic ratio, in other words, the particle spins. Because different atoms or molecules can be distinguished by their spins, this method provides a very powerful tool to find the compositions of different materials, for instance, tissues.

In a spin system, *the dipole interactions* of individual spins lead to their precession dephasing, which causes an exponential decay of the transverse components of the *total magnetic moment* (the magnetization of a sample), while the amplitude of the longitudinal component of the total magnetization diminishes much slower. A decay of the transverse magnetization can be taken into account by adding the corresponding terms in the first two equations of system (14.22):

$$\begin{cases} \dot{e}_x = e_y + e_z a \sin \tilde{t} - b e_x \\ \dot{e}_y = - e_x + e_z a \cos \tilde{t} - b e_y \\ \dot{e}_z = - e_x a \sin \tilde{t} - e_y a \cos \tilde{t}, \end{cases} \qquad (14.23)$$

where the parameter b gives the scale of the transverse relaxation time.

The numerical solution of the system (14.23) obtained with RK4 method demonstrates the so-called Rabi oscillations after Isidor Isaac Rabi (1898–1988), an American physicist, awarder the 1944 Nobel Prize for discovery of nuclear magnetic resonance; see Fig. 14.2.

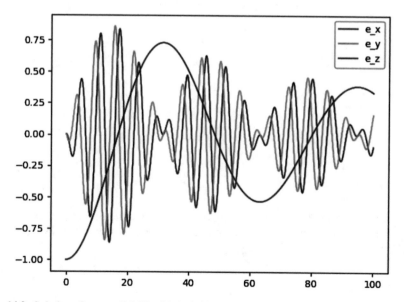

Fig. 14.2   Solution of system (14.23) with the initial conditions $e_z(0) = -1$, $e_x(0) = 0$, $e_y(0) = 0$. Parameters: $a = 0.1, b = 0.02$

The Code in Python for the System (14.23)

```python
from numpy import*
import matplotlib.pyplot as plt

def rungeKutta(f, to, yo, tEnd, dtau):
  def DeltaY(f, t, y, tau): # to find approximate solution using RK4 method
    k1 = dtau * f(t, y)
    k2 = dtau * f(t + 0.5 * dtau, y + 0.5 * k1)
    k3 = dtau * f(t + 0.5 * dtau, y + 0.5 * k2)
    k4 = dtau * f(t + dtau, y + k3)
    return (k1 + 2 * k2 + 2 * k3 + k4) / 6.

  t = [] # initialization (create empty array)
  y = [] # initialization (create empty array)
  t.append(to) # include the initial value to in the array t
  y.append(yo) # include the initial value yo in the array y
  while to < tEnd: # include compuation results in arrays t,y
    #tau = min(tau, tEnd - to) # determine the step size tau
    yo = yo + DeltaY(f, to, yo, dtau) # compute solution at (to,yo)
    to = to + dtau # increase the time
    t.append(to) # include new time value in array t
    y.append(yo) # include new solution value in array y
  return array(t), array(y)

def f(t, y): # the right-hand side of the ODE system
    f = zeros([3])
    f[0] = y[1] + y[2] * a * sin(t) - b * y[0]
    f[1] = -y[0] + y[2] * a * cos(t) - b * y[1]
    f[2] = -y[0] * a * sin(t) - y[1] * a * cos(t)
    return f
#===================================================
a = 0.1
b = 0.02
to = 0.    # the initial time
tEnd = 100. # the final time
yo = array([0., 0., -1.]) # the initial conditions
dtau = 0.01 # шаг

t, y = rungeKutta(f, to, yo, tEnd, dtau)

plt.plot(t,y)
plt.legend(['e_x', 'e_y', 'e_z'])
plt.show()
```

Besides the *transverse relaxation*, there is also the *longitudinal relaxation* caused by spin-lattice interactions; this relaxation can be taken into account by an additional term in the third equation in (14.23)

$$\begin{cases} \dot{e}_x = e_y + e_z a \sin t - b e_x \\ \dot{e}_y = - e_x + e_z a \cos t - b e_y \\ \dot{e}_z = - e_x a \sin t - e_y a \cos t - c\left(e_z - e_z^{(0)}\right). \end{cases} \qquad (14.24)$$

Here $e_z^{(0)}$ is the steady-state nuclear magnetization in an external magnetic field B_0; constant c determines the time scale of slow ($c < < b$) longitudinal relaxation. Equations (14.24) are called the *Bloch equations*, after Felix Bloch (1905–1983), Swiss-American physicist, awarder the 1952 Nobel Prize for his works on nuclear magnetic resonance.

Reading Exercise modify the above Python code to include longitudinal relaxation in Eq. (14.24), generate the graphs of $e_x(t)$, $e_y(t)$, and $e_z(t)$ for different values of parameters a, b, c, and $e_z^{(0)}$ (start with $e_z^{(0)} = 0.08$ and $c = 0.01b$).

Stationary Heat Equation
(*the example of BVP numerical solution using FD method*)

In Chap. 13 we studied heat equation, which is an evolution partial differential equation for the temperature of certain object. Here we solve a stationary one-dimensional heat equation for a solid rod. The rod extends from $x = -1$ to $x = 1$. This equation describes the rod's temperature in the limit $\partial T/\partial t = 0$:

$$T'' + Q(x) = 0. \qquad (14.25)$$

Here $T(x)$ is the temperature and $Q(x)$ is the spatial density of heat sources inside the rod. Let $Q(x) = A(1 - x^2)$, $-1 \leq x \leq 1$, and let the boundary conditions be

$$T(-1) = T(1) = T_0.$$

The exact solution is

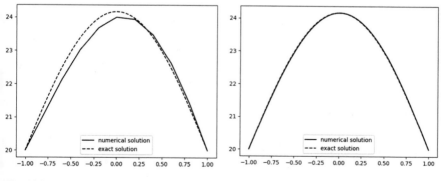

Fig. 14.3 Stationary temperature distribution in the rod at $A = 10$, $T_0 = 20$ °C. Left panel, $N = 10$; right panel $N = 100$

$$T(x) = T_0 + \frac{A}{12}\left[x^4 - 6x^2 + 5\right].$$

In Fig. 14.3 this solution is compared with FD solution employing N grid points.

The Code in Python

```
#  solution of BVP using FD method
import numpy as np
from matplotlib import pyplot as plt
import math

N = 100
A = 10;  T0 = 20
xFirst = -1; xLast = 1;
al0 = 1; al1 = 0; Ap = 20;
bt0 = 1; bt1 = 0; Bp = 20;

def p(x):
   return 0
def q(x):
   return 0
def f(x):
   return -10 * (1 - x*x)

def y_exact(x):
   return T0 + (A/12) * (x**4 - 6 * x**2 + 5);

xFirst = 0; xLast = 10.0;
al0 = 1; al1 = 0; Ap = 20;
bt0 = 1; bt1 = 0; Bp = 20;

def p(x):
   return 0
def q(x):
   return 0
def f(x):
   return 0.3

def y_exact(x):
   return 20 - (0.3/2) * x * (10 - x);
#=================================================

x0 = xFirst
h = (xLast - xFirst)/N
#MM, NN, C, D, Y
MM = np.zeros(N+1); NN = np.zeros(N+1)
C = np.zeros(N+1); D = np.zeros(N+1)
X = np.zeros(N+1); Y = np.zeros(N+1)
yE = np.zeros(N+1);
```

```
for i in range (N + 1):
  X[i] = xFirst + i * h
  yE[i] = y_exact (X[i])
# right-hand sides of the linear algebraic system (LAS)
for i in range (N - 1):
  MM[i] = -2 + h * p(X[i])
  NN[i] = 1 - h * p(X[i]) + h * h * q(X[i])
# Solution of LAS using Thomas algorithm: Forward stage
C[0] = (al1 - al0*h) / (MM[0] * (al1 - al0*h) + NN[0] * al1)
D[0] = (NN[0] * Ap * h) / (al1 - al0*h) + f(X[0]) * (h*h)
for i in range (1, N - 1):
  C[i] = 1 / (MM[i] - NN[i] * C[i - 1])
  D[i] = f(X[i]) * (h*h) - NN[i] * C[i - 1] * D[i- 1]
# Solution of LAS using Thomas algorithm: Backward stage
Y[N] = (bt1 * C[N - 2] * D[N-2] + Bp*h) / (bt1 * (1 + C[N - 2]) + bt0 * h)
i = N
while i > 0:
  i = i - 1
  Y[i] = C[i - 1] * (D[i - 1] - Y[i + 1])
Y[0] = (al1 * Y[1] - Ap*h) / (al1 - al0 * h)

print (" i    X        Y        YE")
for i in range (N + 1):
  print (f"{i:4} {X[i]:10.4f} {Y[i]:10.4f}  {yE[i]:10.4f}")


plt.plot (X, Y, 'k', label='numerical solution')
plt.plot (X, yE, 'k--', label='exact solution')
plt.legend ()
plt.show ()
```

Problems

Solve numerically and analytically the following IVP with (a) Euler's method and
(b) fourth-order Runge-Kutta method. Use different values of h starting with $h = 0.1$
and then taking smaller values of h.

For numerical solution you can write your own code or use program *ODE 1st
order*. With this program you can compare the results of your numerical and
analytical solutions:

1. $y' = 2(x^2 + y)$, $y(0) = 1$, $x \in [0, 1]$.
2. $y' = 3x^2 y$, $y(0) = 1$, $x \in [0, 1]$.
3. $y'x^2 + y^2 = 0$, $y(1) = 1$, $x \in [1, 3]$.
4. $y'y + x = 1$, $y(2) = 1$, $x \in [2, 5]$.
5. $y'x^2 y^2 - y + 1 = 0$, $y(1) = 1$, $x \in [1, 3]$.
6. $y' + 2y = y^2 e^x$, $y(0) = 1$, $x \in [0, 1]$.

Solve the IVP for the following systems of first-order ODE with fourth-order
Runge-Kutta method. You can write your own code, or use Python code presented
above, or use program *ODE 1st order*. Also solve the IVP analytically. *ODE 1st
order* program allows to compare the results of your numerical and analytical
solutions:

7. $\begin{cases} x' = 2x - y, \\ y' = x + 2y. \end{cases}$ $x(0) = 1,\ y(0) = 0,\ t \in [0, 1].$

8. $\begin{cases} x' = x + 2y, \\ y' = x + e^{-t}. \end{cases}$ $x(0) = 0,\ y(0) = 0,\ t \in [0, 2].$

9. $\begin{cases} x' = 4y - 2z - 3x, \\ y' = z + x, \\ z' = 6x - 6y + 5z. \end{cases}$ $x(0) = 0,\ y(0) = 0,\ z(0) = 1,\ t \in [0, 2].$

10. $\begin{cases} x' = x - y - z, \\ y' = x + y, \\ z' = 3x + z. \end{cases}$ $x(0) = 1,\ y(0) = 0,\ z(0) = 0,\ t \in [0, 1].$

11. $\begin{cases} x' + 4y = \cos 2t, \\ y' + 4x = \sin 2t. \end{cases}$ $x(0) = 0,\ y(0) = 1,\ t \in [0, \pi].$

12. $\begin{cases} x' = 2x - y, \\ y' = 2y - x - 5e^t \sin t. \end{cases}$ $x(0) = 1,\ y(0) = 2,\ t \in [0, 1].$

Solve numerically and analytically the following BVP. For numerical solution use the finite-difference method. Write your own code (the Python code above can give some guidelines) or use any program packages. Take a different number of grid points starting with $N = 10$ and then increase N. Compare the results of your analytical and numerical solutions:

13. $y'' + 4y = x,\ y(0) = 0,\ y(\pi/4) = 1,\ x \in [0, \pi/4].$
14. $y'' + y = 1,\ y(0) = 0,\ y'(\pi/2) = 1,\ x \in [0, \pi/2].$
15. $y'' - y' = 0,\ y(0) = 3,\ y(1) - y'(1) = 1,\ x \in [0, 1].$
16. $y'' - 4y' - 2 = -4x,\ y(0) - y'(0) = 0,\quad y(3) = 1,\quad x \in [0, 3].$
17. $y'' + \frac{y'}{x} + 2y = x,\ y(0.7) = 0.5,\quad 2y(1) + 3y'(1) = 1.2,\quad x \in [0.7, 1].$
18. $y'' - xy' - y = 0,\ y(0) = 1,\ y'(1) + 2y(1) = 0x \in [0, 1].$

Appendix A: Existence and Uniqueness of a Solution of IVP for First-Order ODE

The initial value problem (IVP) for first-order equation consists of the equation:

$$y' = f(x, y)$$

and an initial condition

$$y(x_0) = y_0.$$

The question that we now should be asking is the following one: under what conditions on a function $f(x, y)$ does there exist a unique solution of the IVP? The following Theorem (*Picard–Lindelöf theorem*) provides the answer.

Theorem

Let a function $f(x, y)$ be:

1. *Continuous in a neighborhood of a point (x_0, y_0)*
2. *In that neighborhood satisfies the Lipschitz condition*

$$|f(x, y) - f(x_0, y_0)| \leq N|y - y_0|$$

where N is a constant.

Then, in some neighborhood $(x_0 - \delta, x_0 + \delta)$ of x_0, there exists a unique solution of equation $y' = f(x, y)$, which satisfies the initial condition $y(x_0) = y_0$.

A neighborhood of a point (x_0, y_0) is assumed to be a rectangle D: $x_0 - a \leq x \leq x_0 + a$, $y_0 - b \leq y \leq y_0 + b$, in which a solution of an IVP $y' = f(x, y)$, $y(x_0) = y_0$ is sought (Fig. A.1).

We will prove the Theorem in three stages:

1. *We will construct the sequence of functions $y = y_n(x)$ that uniformly converges to a certain function $\bar{y}(x)$.*

V. Henner et al., *Ordinary Differential Equations*,
https://doi.org/10.1007/978-3-031-25130-6

Fig. A.1 Solution $\bar{y}(x)$ in a domain D

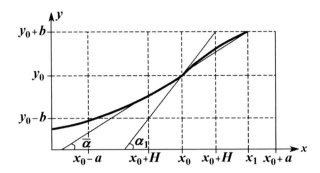

2. *We will show that a function* $\bar{y}(x) = \lim\limits_{n \to \infty} y_n(x)$ *is a solution of an IVP.*

3. *We will show that a solution* $\bar{y}(x)$ *is unique.*

Note that a Lipschitz condition can be replaced by a stronger condition that is easier to check. This condition is the existence in D of a bounded absolute value of a partial derivative:

$$\left| f'_y(x, y) \right| \leq N.$$

Then, according to the mean value theorem

$$\left| f(x, y_1) - f(x, y_2) \right| = f'_y(x, \xi) |y_1 - y_2|,$$

where $y_1 \leq \xi \leq y_2$. This gives the Lipschitz condition.

Let us first determine the interval of values of x, such that the integral curve $\bar{y}(x)$ is confined to D.

Function $f(x, y)$ is uniformly bounded, since it is continuous in D. Therefore, there exists a constant M, such that $|f(x, y)| \leq M$. Let us show that within the interval

$$x \in [x_0 - H, x_0 + H], \quad \text{where } 0 < H < \min\left(a, \frac{b}{M}, \frac{1}{N}\right)$$

the integral curve $\bar{y}(x)$ is confined to D.

Let us assume that the curve $y = \bar{y}(x)$ crosses the upper horizontal boundary of D at $x = x_1$, $x_0 < x_1 < x_0 + a$. The angle α of a tangent line to the curve $\bar{y}(x)$ satisfies the inequality $-M < \tan \alpha < M$. Since

$$\tan \bar{\alpha} = \frac{b}{x_1 - x_0} < \tan \alpha_1 = \frac{b}{H} = M,$$

it is obvious that at $x_0 < x < x_0 + H$, the curve $\bar{y}(x)$ does not leave the domain D via crossing of the upper horizontal boundary (Fig. A.1).

In a similar fashion, it can be shown that the curve $\bar{y}(x)$ does not cross the lower horizontal boundary of D.

Proof of Stage 1

Recall that the Cauchy problem is equivalent to the integral equation:

$$y(x) = y_0 + \int_{x_0}^{x} f(t, y(t))dt. \qquad (A.1)$$

That equivalence can be verified by the integration of the equation $y' = f(x, y)$ from x_0 to x.

Let us construct the Euler's broken line $y = y_n(x)$ on the interval $x \in [x_0, x_0 + H]$. Here the step size $h = H/n$, where n is a positive integer. The line passes through the point (x_0, y_0); see Fig. 2.2 in Chap. 2.

In the interval $x_k \leq x \leq x_{k+1}$, $k = 0, 1, \ldots, n-1$

$$y'_n(x) = f(x_k, y_n(x)) + \mu_n(x), \qquad (A.2)$$

where

$$\mu_n(x) = f(x_k, y_k) - f(x_k, y_n(x)).$$

Since a function $f(x, y)$ is continuous on a bounded region D, $f(x, y)$ is uniformly continuous. Then,

$$|\mu_n(x)| = |f(x_k, y_k) - f(x_k, y_n(x))| < \varepsilon \qquad (A.3)$$

at $n > N(\varepsilon_n)$, where $\varepsilon_n \to 0$ at $n \to \infty$, since $|x - x_k| \leq h$, $|y - y_k| \leq Mh$ and $h = H/n \to 0$ at $n \to \infty$.

Integrating (A.2) with respect to x from x_0 to x, and taking into account that $y_n(x_0) = y_0$, we arrive to

$$y_n(x) = y_0 + \int_{x_0}^{x} f(t, y_n(t))dt + \int_{x_0}^{x} \mu_n(t)dt. \qquad (A.4)$$

For any positive integer $m > 0$

$$y_{n+m}(x) = y_0 + \int_{x_0}^{x} f(t, y_{n+m}(t))dt + \int_{x_0}^{x} \mu_{n+m}(t)dt. \qquad (A.5)$$

Subtracting (A.4) from (A.5) and taking the absolute value of the difference gives

$$\left| y_{n+m}(x) - y_n(x) \right| = \left| \int\limits_{x_0}^{x} \left[f\left(t, y_{n+m}(t)\right) - f\left(t, y_n(t)\right) \right] dt + \int\limits_{x_0}^{x} \mu_{n+m}(t) dt - \int\limits_{x_0}^{x} \mu_n(t) dt \right|$$

$$\leq \int\limits_{x_0}^{x} \left| f\left(t, y_{n+m}(t)\right) - f\left(t, y_n(t)\right) \right| dt + \int\limits_{x_0}^{x} \left| \mu_{n+m}(t) \right| dt + \int\limits_{x_0}^{x} \left| \mu_n(t) \right| dt$$

at $x_0 < x < x_0 + H$. Taking into account the Lipschitz condition and the inequality (A.3), we get:

$$\left| y_{n+m}(x) - y_n(x) \right| \leq N \int\limits_{x_0}^{x} \left| y_{n+m}(t) - y_n(t) \right| dt + (\varepsilon_{n+m} + \varepsilon_n) \cdot H.$$

Correspondingly,

$$\max_{x_0 \leq x \leq x_0 + H} \left| y_{n+m}(x) - y_n(x) \right| \leq N \max_{x_0 \leq x \leq x_0 + H} \int\limits_{x_0}^{x} \left| y_{n+m}(t) - y_n(t) \right| dt + (\varepsilon_{n+m} + \varepsilon_n) \cdot H.$$

Then, we obtain:

$$\max_{x_0 \leq x \leq x_0 + H} \left| y_{n+m}(x) - y_n(x) \right| \leq \frac{(\varepsilon_{n+m} + \varepsilon_n) \cdot H}{1 - NH} \leq \frac{2\varepsilon_n H}{1 - NH}.$$

To derive the last relation, we used the monotonicity of a sequence ε_n: $\varepsilon_{n+m} < \varepsilon_n$, i.e., $\varepsilon_{n+m} + \varepsilon_n \leq 2\varepsilon_n$. It is obvious that this condition holds at $NH < 1$. Since $\lim\limits_{n \to \infty} \varepsilon_n = 0$, $2\varepsilon_n H / (1 - NH)$ can be made less than any $\varepsilon > 0$ when n is sufficiently large, $n > K(\varepsilon)$. Therefore, we get:

$$\max_{x_0 \leq x \leq x_0 + H} \left| y_{n+m}(x) - y_n(x) \right| < \varepsilon$$

at $n > K(\varepsilon)$, i.e., the sequence of a continuous function $y_n(x)$ at $x_0 < x < x_0 + H$ converges uniformly to a certain function $\bar{y}(x)$:

$$y_n(x) \Rightarrow \bar{y}(x).$$

Proof of Stage 2

Let us take the limit of (A.4) at $n \rightarrow \infty$:

$$\lim_{n \rightarrow \infty} y_n(x) = y_0 + \lim_{n \rightarrow \infty} \int_{x_0}^{x} f(t, y_n(t)) dt + \lim_{n \rightarrow \infty} \int_{x_0}^{x} \mu_n(t) dt,$$

or

$$\bar{y}(x) = y_0 + \lim_{n \rightarrow \infty} \int_{x_0}^{x} f(t, y_n(t)) dt + \lim_{n \rightarrow \infty} \int_{x_0}^{x} \mu_n(t) dt. \qquad (A.6)$$

Since $y_n(x)$ uniformly converges to $\bar{y}(x)$, and $f(x, y)$ is uniformly continuous, the sequence $f(x, y_n(x))$ converges:

$$f(x, y_n(x)) \Rightarrow f(x, \bar{y}(x)).$$

Indeed,

$$|f(x, \bar{y}(x)) - f(x, y_n(x))| < \varepsilon,$$

if $|\bar{y}(x) - y_n(x)| < \delta(\varepsilon)$ at $n > K(\varepsilon)$ for all x.

Since the sequence $f(x, y_n(x))$ uniformly converges to $f(x, \bar{y}(x))$, in (A.6) the limit calculation and the integration can be interchanged. Considering that $|\mu_n(\varepsilon)| < \varepsilon_n$, where $\varepsilon_n \rightarrow 0$ at $n \rightarrow \infty$, (A.6) gives

$$\bar{y}(x) = y_0 + \int_{x_0}^{x} f(x, \bar{y}(x)) dx.$$

Therefore, $\bar{y}(x)$ satisfies Eq. (A.1).

Proof of Stage 3

Let us assume the existence of two different solutions, $y_1(x)$ and $y_2(x)$, of Eq. (A.1). Then,

$$\max_{x_0 \leq x \leq x_0 + H} |y_1(x) - y_2(x)| \neq 0.$$

Subtracting $y_2(x) \equiv y_0 + \int_{x_0}^{x} f(x, y_2(x)) dx$ from $y_1(x) \equiv y_0 + \int_{x_0}^{x} f(x, y_1(x)) dx$ gives

$$y_1(x) - y_2(x) \equiv \int_{x_0}^{x} [f(x, y_1(x)) - f(x, y_2(x))]dx.$$

Therefore,

$$\max_{x_0 \le x \le x_0+H} |y_1(x) - y_2(x)| = \max_{x_0 \le x \le x_0+H} \left| \int_{x_0}^{x} [f(x, y_1(x)) - f(x, y_2(x))]dx \right|$$

$$\le \max_{x_0 \le x \le x_0+H} \int_{x_0}^{x} |f(x, y_1(x)) - f(x, y_2(x))|dx.$$

Using the Lipschitz condition, we arrive to

$$\max_{x_0 \le x \le x_0+H} |y_1(x) - y_2(x)| = N \max_{x_0 \le x \le x_0+H} \left| \int_{x_0}^{x} |y_1(x) - y_2(x)|dx \right|$$

$$\le N \max_{x_0 \le x \le x_0+H} |y_1(x) - y_2(x)| \max_{x_0 \le x \le x_0+H} \left| \int_{x_0}^{x} dx \right|$$

$$= NH \max_{x_0 \le x \le x_0+H} |y_1(x) - y_2(x)|.$$

If $\max_{x_0 \le x \le x_0+H} |y_1(x) - y_2(x)| \ne 0$ that inequality

$$\max_{x_0 \le x \le x_0+H} |y_1(x) - y_2(x)| \le NH \max_{x_0 \le x \le x_0+H} |y_1(x) - y_2(x)|, \tag{A.7}$$

does not hold, since according to the conditions of the Theorem $H < 1/N$, but it follows from (A.7) that $NH \ge 1$. The contradiction can be eliminated only when

$$\max_{x_0 \le x \le x_0+H} |y_1(x) - y_2(x)| = 0,$$

i.e., $y_1(x) \equiv y_2(x)$ at $x_0 < x < x_0 + H$.

This concludes the proof of Theorem.

Appendix B: How to Use the Accompanying Software

The book includes a user-friendly laboratory-type interactive software. Using software is completely optional, but choosing to use it would make understanding and studying the material more efficient and visual. Each program has an integrated help system where detailed explanation of mathematical background can be found.

The programs included are *ODE first order*, *ODE second order*, *Heat*, *Waves*, and *FourierSeries*. (Other programs to work with different kinds of PDE can be found in Ref. [9].)

Let us start with the *ODE first* order and *ODE second* order programs which are designed to solve equations and systems of equations of the first and second order (Chaps. 2, 3, and 4). These programs are mainly used for self-checking – the analytical solution obtained by the user is compared here with the numerical solution obtained by the program.

B.1 Examples Using the Program *ODE First Order*

The *ODE first order* program provides a suitable and simple laboratory environment for solving first-order ordinary differential equations.

This program is designed for the following problem types:

- *Plot slope fields and isoclines.*
- *Plot solutions of differential equations for different values of arbitrary constant.*
- *Solve Cauchy problems for first-order ODEs.*
- *Solve Cauchy problems for systems of two first-order ODEs.*
- *Solve Cauchy problems for systems of three first-order ODEs.*

Interaction with the rest of the package programs (input of functions and parameters, output of graphs and tables, analysis of results, etc.) is performed in the same way.

© The Editor(s) (if applicable) and The Author(s), under exclusive license to
Springer Nature Switzerland AG 2023
V. Henner et al., *Ordinary Differential Equations*,
https://doi.org/10.1007/978-3-031-25130-6

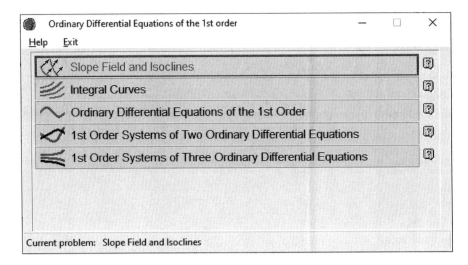

Fig. B.1 Starting interface of the program ODE first order

Figure B.1 shows the starting interface of the program ODE first order.

Suppose, for example, that you choose the second option, *Slope fields and isoclines*. To solve a particular problem, you must next enter the values of the parameters. Starting from the *Data* menu, there are two basic ways to proceed, by choosing *Library Example* or *New Problem*. If you choose to load a library example, you will be asked to select one of the available problems (see Fig. B.2).

A description of each of the examples can be accessed by clicking the *Text + Hint* button. Once you select an example, you will see the parameters of the problem in a dialog box with parameters already entered according to the selected example. If you choose *New problem* in the previous step, the spaces for input functions and parameters on the previous screen will be left blank.

Example B.1 (*Slope Field and Isoclines*)

Plot isoclines and the direction field for differential equation:

$$y' = x + y - 1$$

in the domain $[-2, 2; -2, 2]$. With the help of isoclines, plot (approximately) two particular solutions of this equation passing through the points: (*a*) $y(0) = 0.5$; (*b*) $y(0) = -0.5$.

Solution This problem is not in the library set. To solve it with the program, select a *New problem* from the *Data* menu on the problem screen.

Then enter the right-hand side $f(x)$ of a differential equation and the values of the domain boundaries for drawing slope field and isoclines ($x_{min} = -2$, $x_{max} = 2$, $y_{min} = -2$, $y_{max} = 2$) (see Fig. B.3).

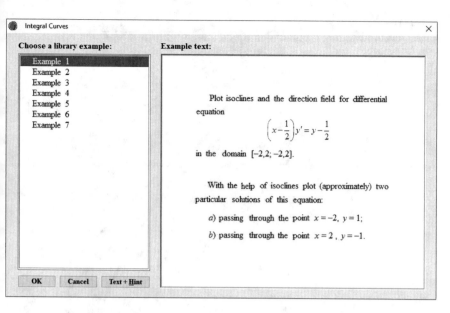

Fig. B.2 Starting interface of the program ODE first order

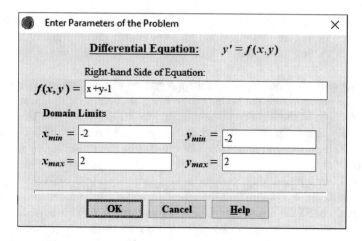

Fig. B.3 Parameters of the problem dialog box for Example B.1

The *Execute* command displays the plot of a slope field and a set of isoclines for the successive angles $\alpha_i = \alpha_0 + i\Delta\alpha$, where i runs from zero to N, N is the assigned number of angles and $\Delta\alpha$ is a distance between the successive angles. By default, the values of these parameters are $\alpha_0 = \arctan(f_{min})$, $\Delta\alpha = (\mathrm{arctg}f_{max} - \mathrm{arctg}f_{min})/N$, and $N = 20$. The arrows along isoclines show the slope field of the given differential equation.

Fig. B.4 Direction field,
isoclines, and two integral
curves for equation
$y' = x + y - 1$

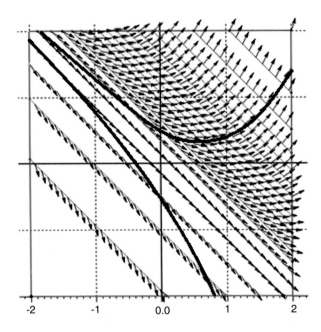

Let us plot the particular solution of this equation, passing through the point
$(0; 0.5)$. Click the button *Draw Solution Polyline* (screen cursor converts to *Cross*),
and then click the left mouse button at the point $(0; 0.5)$ (*initial point* on the integral
curve).

Build the segments, *one by one*, of the integral curve by clicking the left mouse
button at the points where the nodes of the integral curve should appear. Choose
these nodes to be at the locations where the slopes on different close isoclines are
approximately equal. Get up to $y = 2$, click the right mouse button, and in the context
menu, choose the command *Finish line plotting*. Screen cursor will return to the form
of an arrow.

Now build the left part of integral curve – on the interval $[-2, 0]$. Again, click the
button *Draw Solution Polyline* and then the left mouse button at the point $(0; 0.5)$.
Consecutively build the segments of the integral curve by clicking the left mouse
button at the points where the nodes of the left branch of the integral curve should
appear. Get up to $y = -2$, click the right mouse button and in the context menu
choose the command *Finish line plotting*.

Similarly build the integral curve passing through the point $(0; -0.5)$ (Fig. B.4).

Example B.2 (*Ordinary Differential Equations of the First Order*)
Solve the differential equation $y' + 5y = e^x$ on the interval $[-0.3; 3]$; the initial
condition is $y(0) = 1$. Present the IVP solution as a graph.

Solution The given problem is from the library set. To do this example, start with
Ordinary Differential Equations of the 1st Order and click on "*Data*" – "*Library
example*" – "*Example 5*". Then, as shown in Fig. B.5, you should enter the equation,

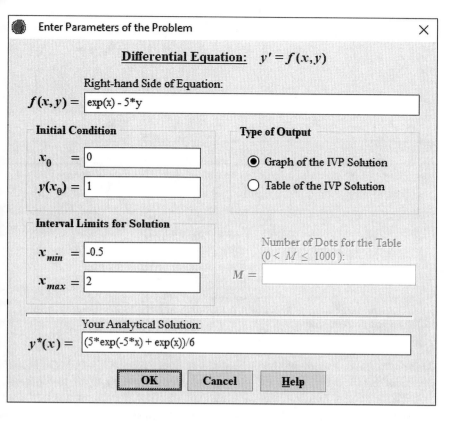

Fig. B.5 Parameters of the problem dialog box for Example B.2

initial conditions and interval where you want to obtain the solution of IVP. Also, you can enter the analytical solution of IVP to compare with the numerical one.

The program solves equation numerically (the solution can be presented in the form of a graph, Fig. B.6a, or table, Fig. B.6b) and compares with the reader's analytical solution, $y = (5e^{-5x} + e^x)/6$ (in the interface it is denoted as $y^*(x)$).

Example B.3 (*First Order System of Two Ordinary Differential Equations*)
Find the general solution of the system of two first-order equation:

$$\begin{cases} \dfrac{dx}{dt} = -x + 8y + 1, \\ \dfrac{dy}{dt} = x + y + t. \end{cases}$$

Solve IVP with the initial conditions $x(0) = 1$, $y(0) = -1$. Compare your analytical solution with the numerical one obtained with the program on the interval $[-2, 2]$.

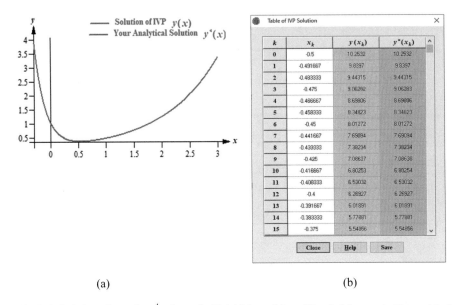

(a)   (b)

Fig. B.6 Solution of equation $y' + 5y = e^x$ with initial condition $y(0) = 1$; (a) presented in graphical form; (b) presented in table form

Solution To do this example, begin with *First Order System of Two Ordinary Differential Equations* in the starting interface of the program ODE first order; then click on *Data – Library example – Example 6*. Then, as shown in Fig. B.7, you should enter the equations, initial conditions, and interval where you want to obtain the solution of IVP.

Also, you can enter the analytical solution of IVP to compare with the numerical.

The program solves the system numerically (the solution can be presented in the form of a graph or a table) and compares with the reader's analytical solution (in the interface it is denoted as $x^*(t)$ and $y^*(t)$, and should be input as shown in the interface) (Fig. B.8).

B.2 Examples Using the Program *ODE Second Order*

The *ODE second order* program provides a suitable and simple laboratory environment for experimenting with second-order ordinary differential equations.

This program is designed for the following problem types:

- *Solve Cauchy problems for second-order ODEs.*
- *Solve Cauchy problems for second-order ODEs with constant coefficients.*
- *Solve a problem of forced one-dimensional oscillations damped by viscous friction.*

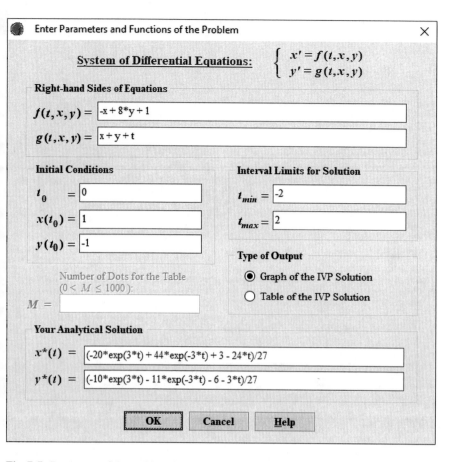

Fig. B.7 Parameters of the problem dialog box for Example B.3

Figure B.9 shows the starting interface of the program ODE second order.

Example B.4 (*Second-Order ODE*)

Find the particular solution of equation $yy'' - 2y'^2 = 0$ for the initial conditions $y(1) = -1$, $y'(1) = 2$. Present the IVP solution as a graph on the interval $[1;3]$.

Solution To solve this example, start with *second order ODE*; then, click on *Data – Library example – Example 6*. Then, as shown in Fig. B.10, you should enter the equation, initial conditions, and interval where you want to obtain the solution of IVP. Also, you can enter the analytical solution of IVP to compare with the numerical one. The solution can be presented in graphical or table form.

The program solves it numerically and compares with the reader's analytical solution $y = 1/(1 - 2x)$. In the interface this solution is denoted as y^* and should be input in the form shown in the interface (Fig. B.11).

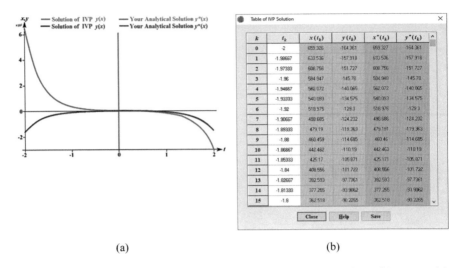

Fig. B.8 Solution of IVP of Example B.3; (**a**) presented in graphical form; (**b**) presented in table form

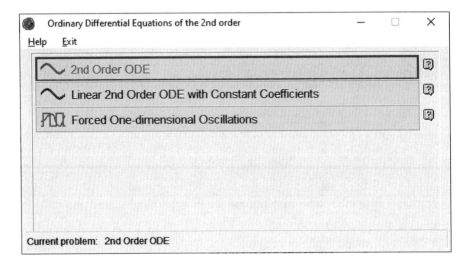

Fig. B.9 Starting interface of the program ODE second order

B.3 Examples Using the Program *Heat*

The *Heat* program provides a suitable laboratory environment for experimenting with the problem of heat conduction within rods and membranes.

This program is designed to solve the following types of problems:

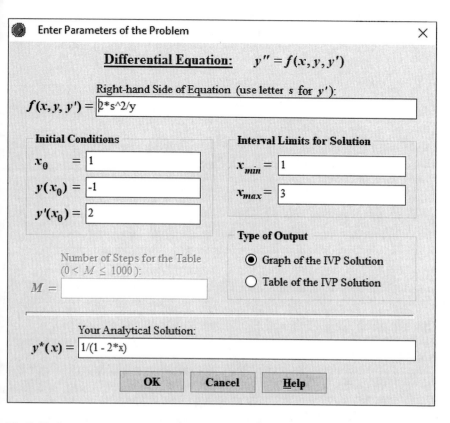

Enter Parameters of the Problem   ✕

Differential Equation: $y'' = f(x, y, y')$

Right-hand Side of Equation (use letter s for y'):

$f(x, y, y') =$ 2*s^2/y

Initial Conditions   **Interval Limits for Solution**

x_0 = 1 $x_{min} =$ 1

$y(x_0) =$ -1 $x_{max} =$ 3

$y'(x_0) =$ 2

Number of Steps for the Table   **Type of Output**
(0 < M ≤ 1000):   ⦿ Graph of the IVP Solution

$M =$ ○ Table of the IVP Solution

Your Analytical Solution:

$y^*(x) =$ 1/(1 - 2*x)

| OK | Cancel | **Help** |

Fig. B.10 Parameters of the problem dialog box for Example B.4

- *Heat conduction within an infinite or semi-infinite rod*
- *Heat conduction within a finite uniform rod*
- *Heat conduction within a thin uniform rectangular membrane*
- *Heat conduction within a thin uniform circular membrane*

Figure B.12 shows the starting interface of the program Heat.

Example B.5 *(Heat Conduction Within a Finite Uniform Rod)*

The initial temperature of a slender wire of length l lying along the x-axis is $u(x, 0) = u_0$, where $u_0 =$ const, and the ends of the wire are kept under the constant temperatures $u(0, t) = u_1 =$ const, $u(l, t) = u_2 =$ const. The wire is experiencing a heat loss through the lateral surface, proportional to the difference between the wire temperature and the surrounding temperature. The temperature of the surrounding medium is $u_{md} =$ const. Let γ be a heat loss parameter.

Find the distribution of temperature $u(x, t)$ in the wire for $t > 0$ if it is free of internal heat sources. The assigned values of the parameters are $a^2 = 0.25$ (the thermal diffusivity of the material), $l = 10$, $\gamma = h/c\rho = 0.05$, $u_{md} = 15$ (the temperature of the medium), and $u_0 = 5$ (the initial temperature of the wire).

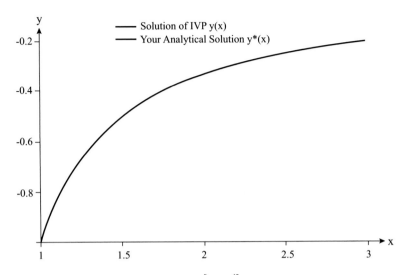

Fig. B.11 Graph of the solution of equation $yy'' - 2y'^2 = 0$ with initial conditions $y(1) = -1$, $y'(1) = 2$

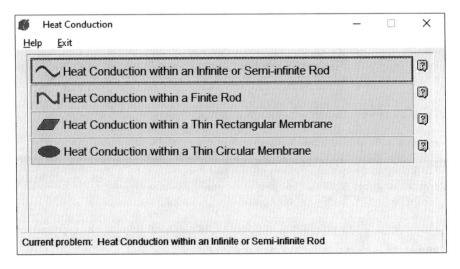

Fig. B.12 Starting interface of the program Heat

Solution The problem is an example of the equation (Fig. B.13):

$$\frac{\partial u}{\partial t} = a^2 \frac{\partial^2 u}{\partial x^2} - \gamma \cdot (u - u_{md})$$

subject to the conditions $u(x, 0) = u_0$, $u(0, t) = u_1$, $u(l, t) = u_2$.

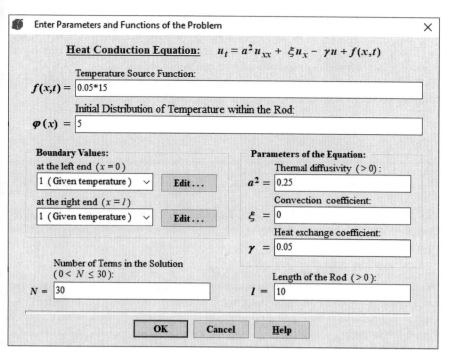

Fig. B.13 Parameters of the problem dialog box for Example B.5

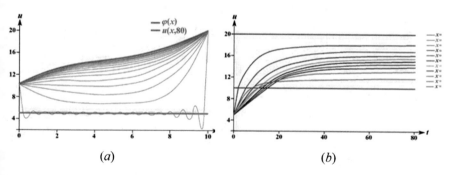

(a)   (b)

Fig. B.14 (a) The animated solution, $u(x, t)$; (b) time traces of temperature at rod point, (for Example B.5)

This problem is also given in the library set. To load it, select *Data – Library example – Example 6*.

Since all parameters for the problem have been entered, you may run the solver by choosing the *Execute* command from the main menu and then selecting *Results – Evolution of Rod Temperature*. The animation sequence in Fig. B.14 shows the solution $u(x, t_k)$ at times $t_k = k\Delta t$.

The menu option *Time Traces of Temperature at Rod Points* allows one to investigate the significance of the changes in the magnitudes of the diffusivity, a^2, and of the coefficient of lateral heat exchange with a medium, γ, by modifying them and noting their effect on the solution to the problem (Fig. B.14).

More results can be displayed with the *Bar charts of* $|V_n(t)|$ command. This command produces a sequence of bars of the values:

$$V_n(t) = T_n(t) + C_n e^{-\left(a^2 \lambda_n + \gamma\right)t}$$

at successive time points. Each bar chart gives a graphic picture of the contribution of the individual terms to the solution $u(x, t)$ at a certain time. Bar charts display data by drawing a rectangle of the height equal to $|V_n(t)|$ (measured along the vertical axis) at each n value along the graph. Therefore, the height of the nth rectangle corresponds to the absolute value of the amplitude of the nth harmonic at the respective time t.

Example B.6 (*Heat Conduction Within a Thin Uniform Circular Membrane*)

Consider a very long (infinite) cylinder of radius l. At time $t = 0$ let a magnetic field, parallel to cylinder axis, be instantly established outside of the cylinder. The external magnetic field strength is $H = H_0 \sin \omega t$, $H_0 = $ const $(0 < t < \infty)$.

Find the magnetic field strength within the cylinder if the initial field is zero. To start, take the following values of the parameters: $a^2 = 0.25$, $l = 4$, $H_0 = 5$, and $\omega = 2$.

Solution The problem depends on the solution of the equation:

$$\frac{\partial u}{\partial t} = a^2 \left[\frac{\partial^2 u}{\partial r^2} + \frac{1}{r} \frac{\partial u}{\partial r} \right]$$

subject to the conditions $u(r, \varphi, 0) = 0$, $u(l, \varphi, t) = H_0 \sin \omega t$.

This problem is mathematically identical to a heat conduction problem found in the program library of problems. Select the problem type from the main menu by clicking *Heat Conduction within a Thin Circular Membrane*. To load the appropriate problem, select *Data – Library example – Example 6* (Fig. B.15).

Eigenvalues of the given boundary value problem have the form $\lambda_{nm} = \left(\mu_m^{(n)}/l\right)^2$, $n, m = 0, 1, 2, \ldots$, where $\mu_m^{(n)}$ are the positive roots of the equation $J_n(\mu) = 0$. Figure B.16 shows the graph of the function $J_n(\mu)$ and associated eigenvalues for $n = 2$.

Eigenfunctions of the given boundary value problem are

$$V_{nm}^{(1)} = J_n\left(\frac{\mu_m^{(n)}}{l} r\right) \cos n\varphi, \quad V_{nm}^{(2)} = J_n\left(\frac{\mu_m^{(n)}}{l} r\right) \sin n\varphi.$$

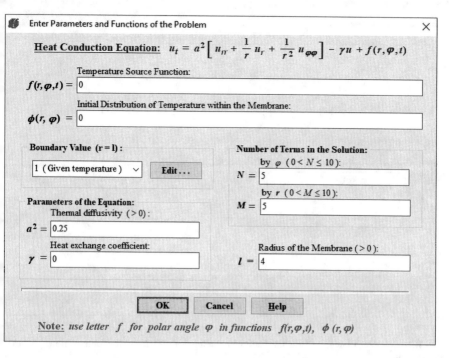

Fig. B.15 Parameters of the problem dialog box for Example B.6

The three-dimensional picture shown in Fig. B.17 depicts the one of eigenfunction for the given problem ($\lambda_{21} = 4.42817$, $V_{21}^{(1)} = J_2\left(\mu_1^{(2)}r/4\right)\cos 2\varphi$).

To solve the problem with the help of the program, follow the same steps as in Example B.6. Results of the solution can be displayed by selecting the menu option *Evolution of Membrane Temperature*. The three-dimensional surface shown in Fig. B.17b depicts the animated solution $u(r, \varphi, t)$ at time $t = 2$.

By changing the values of thermal diffusivity, a^2, radius, l, and parameters H_0 and ω of the boundary value function, different pictures of a magnetic field distribution can be obtained and studied.

B.4 Examples Using the Program Waves

The *Waves* program provides a suitable laboratory environment for experimenting with the problems of forced vibrations of strings, membranes, and rods. This program is designed to solve the following types of problems:

- *Vibrations of infinite or semi-infinite string*
- *Vibrations of finite string*

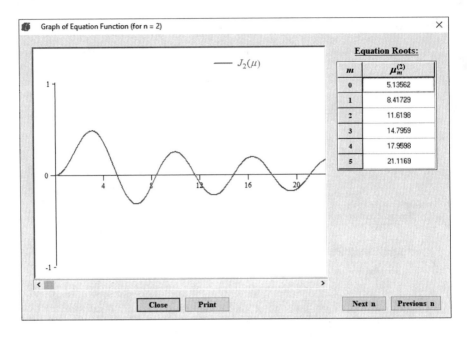

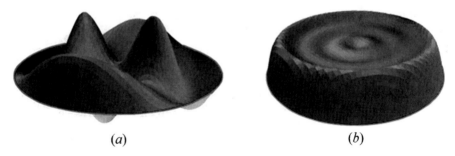

Fig. B.16 Graph of equation for evaluating eigenvalues for $n = 2$ and the table of roots $\mu_m^{(2)}$ for Example B.6

Fig. B.17 (a) Eigenfunction $V_{21}^{(1)}(r, \varphi)$; (b) distribution of magnetic field, $u(r, \varphi, t)$, at $t = 2$ (for Example B.6)

- *Vibrations of a flexible rectangular membrane*
- *Vibrations of a flexible circular membrane*

Example B.7 (*Vibrations of a Finite String*)

   An elastic uniform string is fixed at the ends $x = 0$ and $x = l$. At the point $x = x_0$, the string is deflected a small distance, h, from the equilibrium position, and at time $t = 0$, it is released with zero speed. The string is oscillating in a medium, where a resistance is proportional to speed.

Find the vibrations of the string. Choose a value of a coefficient that corresponds to the medium resistance so that the oscillations decay (with precision of about 5%) during the following: (a) two periods; (b) three periods; and (c) four periods of the main mode.

Find a point x_0 which will eliminate all even harmonics. To start, the following values of the parameters are assigned: $l = 100$, $h = 6$, $\rho = 1$, $\kappa = R/2\rho = 0.001$ (damping coefficient), $a^2 = T/\rho = 1$, and $x_0 = 35$.

Solution This problem involves the solution of the homogeneous wave equation:

$$\frac{\partial^2 u}{\partial t^2} + 2\kappa \frac{\partial u}{\partial t} - a^2 \frac{\partial^2 u}{\partial x^2} = 0, \quad a = \sqrt{\frac{T}{\rho}}.$$

The initial and boundary conditions are

$$u(x,0) = \begin{cases} \dfrac{h}{x_0}x, & \text{if } 0 \le x \le x_0, \\ \dfrac{h}{l-x_0}(l-x), & \text{if } x_0 < x \le l, \end{cases} \qquad \frac{\partial u}{\partial t}(x,0) = 0,$$

$$u(0,t) = u(l,t) = 0.$$

To load the problem, select *Data* from the main menu and click *Library example*. From the given list of problems, select *Example 4*.

The initial deflection of the string can be entered using the *Impulse function* defined by $\text{Imp}(x, a, b) = 1$ if $a \le x < b$ and 0 otherwise. For the given parameters, the following should be entered in the dialog box:

$$\varphi(x) = \text{Imp}(x, 0, 35) * 6^*x/35 + \text{Imp}(x, 35, 100) * 6^*(100 - x)/(100 - 35).$$

Since all parameters for the problem have been entered, you may run the solver by choosing the *Execute* command from the main menu and then selecting *Results – Evolution of String Profile*. The animation sequence in Fig. B.18 shows the profile of the string at times $t_k = k\Delta t$.

To change the decay time of the oscillations due to friction, use the menu option *Results – Time Traces of String Points*. In choosing different values of the damping coefficient κ, trace the time evolution of the string until the free oscillations have decayed (with some reasonable precision). For example, for a damping coefficient $\kappa = 0.004$, the oscillations amplitude decays to approximately zero during five periods of the main mode (Fig. B.19).

In the expansion of the solution $u(x,t)$ the terms for which $\sin \frac{n\pi x_0}{l} = 0$ vanish, that is, the solution does not contain harmonics for which the point $x = x_0$ is a node. If x_0 is at the middle of the string, the solution does not contain even-numbered harmonics. To eliminate all even harmonics, select the menu *Data – Change Current*

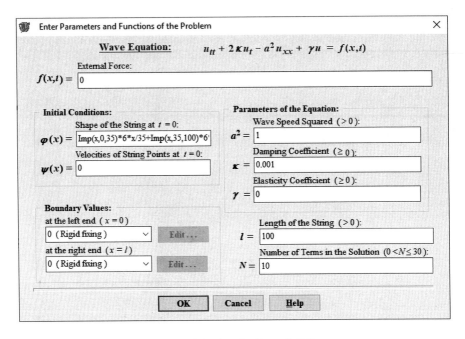

Fig. B.18 Parameters of the problem dialog box for Example B.7

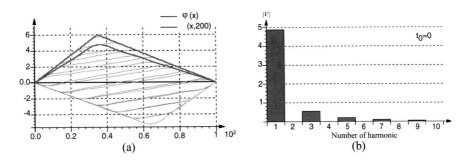

Fig. B.19 (a) Eigenfunction $V_{21}^{(1)}(r, \varphi)$; (b) bar charts of $|V_n(x)|$ (for Example B.7)

Problem, and in the parameters of the problem dialog box, enter the new initial function:

$$\varphi(x) = \mathrm{Imp}(x, 0, 50) * 6^* x/50 + \mathrm{Imp}(x, 50, 100) * 6^* (100 - x)/(100 - 50).$$

Run the problem solver and then select menu *Results – Free Vibrations* or *Results – Bar Charts of |V(t)|* to ensure that even harmonic have been eliminated.

Example B.8 (*Vibrations of a Flexible Rectangular Membrane*)
A flexible rectangular membrane is clamped at the edges $x = 0$, $y = 0$, the edge $y = l_y$ is attached elastically, and the edge $x = l_x$ is free. Initially the membrane is at rest in a horizontal plane. Starting at the time $t = 0$, a uniformly distributed transversal force

$$f(x, y, t) = Ae^{-0.5t}x\sin\frac{4\pi y}{l_y}$$

acts on the membrane.

Find the transverse vibrations of the membrane if the resistance of the surrounding medium is proportional to the speed of the membrane. Choose the coefficient of medium resistance so that oscillations decay to zero (with precision of about 5%) during the time (a) $t = 20$; (b) $t = 30$; and (c) $t = 40$. The assigned parameter values are $A = 0.5$, $a^2 = 1$, $\kappa = 0.01$, $l_x = 4$, and $l_y = 6$.

Solution The problem involves solving the wave equation:

$$\frac{\partial^2 u}{\partial t^2} + 2\kappa\frac{\partial u}{\partial t} - a^2\left[\frac{\partial^2 u}{\partial x^2} + \frac{\partial^2 u}{\partial y^2}\right] = Ae^{-0.5t}x\sin\frac{4\pi y}{l_y},$$

under the conditions $u(x, y, 0) = 0$, $\frac{\partial u}{\partial t}(x, y, 0) = 0$, $u(0, y, t) = 0$, $\frac{\partial u}{\partial x}(l_x, y, t) = 0$, $u(x, 0, t) = 0$, $\frac{\partial u}{\partial y}(x, l_y, t) + h_4 u(x, l_y, t) = 0$.

The given problem is not from the library set. To solve it, instead of selecting a library example as we did before, select a *New problem* from the *Data* menu on the problem screen (Fig. B.20).

To specify the boundary condition at the edge $y = l_y$, choose option "3 (elastic fixing)" by selecting this item from the dialog box. Enter the parameters by clicking the *Edit* button and typing $h_4 = 0.1$ and $g(x, t) = 0$.

The following menus may be used to investigate other properties of the solution:

Results → *Free Vibrations*
Results → *Surface Graphs of Eigenfunctions*
Results → *Evolution of Membrane Profile at y = const*
Results → *Evolution of Membrane Profile at x = const*

More information on each option can be found by selecting the *Help* menu on the corresponding screen.

Eigenvalues of the given boundary value problem have the form:

$$\lambda_{nm} = \lambda_{xn} + \lambda_{ym} \text{ where } \lambda_{xn} = \left[\frac{\pi(2n-1)}{2l_x}\right]^2, \quad \lambda_{ym} = \left[\frac{\mu_{ym}}{l_y}\right]^2, \quad n, m = 1, 2, 3, \ldots.$$

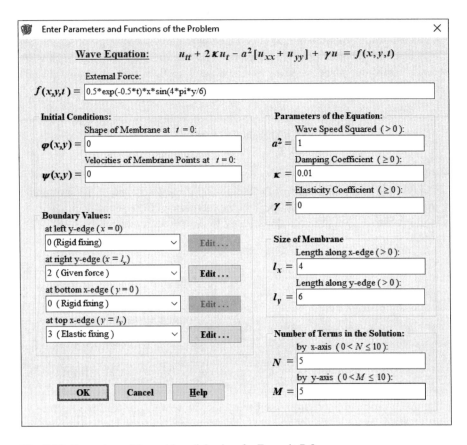

Fig. B.20 Parameters of the problem dialog box for Example B.8

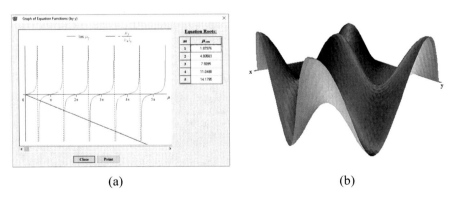

(a)   (b)

Fig. B.21 (**a**) Graphs of functions for evaluating eigenvalues and the table of roots μ_{ym}; (**b**) eigenfunction $V_{33}(x, y)$ (for Example B.8)

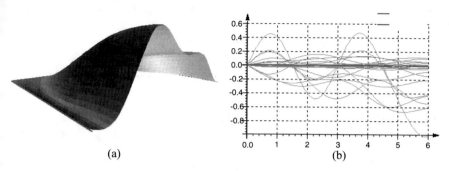

Fig. B.22 (a) Surface graph of the membrane at $t = 10.5$; (b) evolution of the membrane profile at $x = l_x$, $\kappa = 0.175$ (for Example B.8)

and μ_{ym} is the m-th root of the equation $\tan\mu_y = -\mu_y/(h_4 l_y)$. Figure B.21a shows graphs of the functions $\tan\mu_y$ and $-\frac{\mu_y}{h_4 l_y}$ used for evaluating eigenvalues. Eigenfunctions of the given boundary value problem are

$$V_{nm}(x, y) = X_n(x)Y_m(y) = \sin \frac{(2n-1)\pi x}{2l_x} \sin \sqrt{\lambda_{ym}} y.$$

The three-dimensional picture shown in Fig. B.21b depicts one of the eigenfunctions for the problem ($\lambda_{33} = 5.6019$, $V_{33}(x, y) = \sin \frac{5\pi x}{2l_x} \sin \sqrt{\lambda_{y3}} y$).

To solve the problem with the help of the program, follow the same steps as in Example B.8. Results can be displayed by selecting the menu option *Surface Graphs of Membrane*. The three-dimensional surface shown in Fig. B.22a depicts the animated solution, $u(x, y, t)$, at $t = 10.5$. Note that the edges of the surface adhere to the given boundary conditions.

To change the decay time of oscillations due to friction, use the menu option *Surface Graphs of Membrane, Evolution of Membrane Profile at $y = const$*, or *Evolution of Membrane Profile at $y = const$*. Choosing different values of the damping coefficient, κ, trace the time evolution of the membrane until the free oscillations decay (with some reasonable precision). For example, for the damping coefficient $\kappa = 0.175$, the amplitude of oscillations vanishes during time $t = 30$ (Fig. B.22b).

B.5 Examples Using the Program *FourierSeries*

The *FourierSeries* program provides solutions to three kinds of problems: expansion in a generalized Fourier series in terms of classical orthogonal polynomials (Legendre, Chebyshev of the first and second kind, Jacoby, Laguerre, Hermite), expansion in Fourier-Bessel series, and expansion in Fourier series in terms of associated Legendre functions.

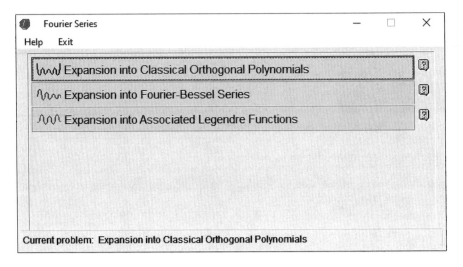

Fig. B.23 Starting interface of the program FourierSeries

Program *FourierSeries* offers solutions via Fourier series expansions to three problem types:

- *Expansion in a generalized Fourier series in terms of classical orthogonal polynomials*
- *Expansion in Fourier-Bessel series*
- *Expansion in Fourier series in terms of associated Legendre functions*

Figure B.23 shows the starting interface of the program FourierSeries.

Example B.9 (*Expansion into Fourier-Bessel Series*)
Expand the function:

$$f(x) = \begin{cases} x^2, & \text{if} \quad 0 \le x < 1 \\ x, & \text{if} \quad 1 \le x \le 2 \end{cases}$$

using the system of Bessel functions of the first kind of order $m = 2$

$$X_k(x) = J_2\left(\frac{\mu_k^{(2)}}{l}x\right),$$

where $\mu_k^{(2)}$ are the positive roots of the equation $\mu J_2'(\mu) + hlJ_2(\mu) = 0$.
Solve this problem for $h = 0.1, 1, 10$.

Solution Select the problem type from the main menu by clicking *Expansion into Fourier-Bessel Series*. The given problem is from the library set. To load it select *Data – Library problem – Problem* 6. To proceed with the selected problem, click

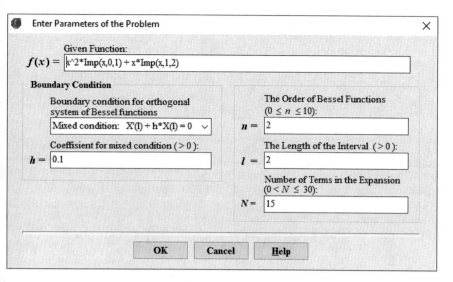

Fig. B.24 Parameters of the problem dialog box for Example B.9

the OK button; the parameters for the current problem will be displayed in the dialog window (Fig. B.24).

In the expression for the given function

$$f(x) = x^{2^*} \operatorname{Imp}(x, 0, 1) + x^* \operatorname{Imp}(x, 1, 2),$$

$\operatorname{Imp}(x, a, b)$ is an impulse function, such that $\operatorname{Imp}(x, a, b) = 1$ if $a \le x < b$ and 0 otherwise.

Since all parameters for the problem have been entered, we can proceed by clicking OK. To see the given function, select *View* from the main menu, and then click *Return* to go back to the problem solution screen. To see values of the coefficients c_k in the expansion, select *Results – Fourier Coefficients*.

To run the solver, click the *Execution* option; results can be seen by selecting *Results – Graph of the Partial Sum* (Fig. B.25a).

Figure B.25b depicts the bar chart of Fourier coefficients, c_k, (option *Bar Chart of Fourier Coefficients*). The bar chart gives a graphic picture of the contribution of the individual terms to the Fourier-Bessel expansion of $f(x)$. The bar chart displays data by drawing a rectangle of the height equal to the kth Fourier coefficient (measured along the vertical axis) at each k value along the graph. The height of the k-th rectangle corresponds to the value of the Fourier coefficient c_k of the general expansion of $f(x)$.

Menu option *Results – Graphs of Bessel Functions* allows to see the plots of Bessel functions of the first kind of the order $m = 2$ (the functions of the orthonormal system) for $k = 0, 1, \dots N$. One of the Bessel functions for the problem

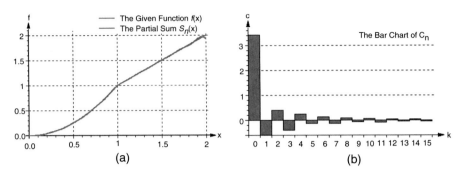

Fig. B.25 (a) Graph of the partial sum $S_{15}(x)$, $\delta^2 = 0.00021$; (b) bar chart of Fourier coefficients, c_k (for Example B.9)

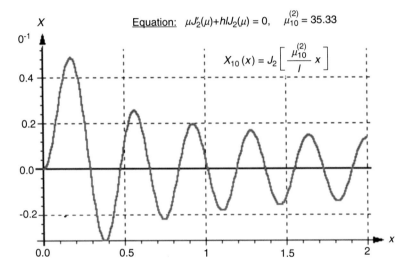

Fig. B.26 Graph of Bessel function $X_{10}(x)$ for Example B.9

$$X_{10}(x) = J_2\left(\mu_{10}^{(2)}x/l\right),$$

is presented in Fig. B.26, where $\mu_{10}^{(2)}$ is the tenth root of the equation $\mu J_2'(\mu) + hlJ_2(\mu) = 0$.

Example B.10 (*Expansion into Associated Legendre Functions*)
Expand the function:

$$f(x) = \begin{cases} -1, & -1 < x < 0, \\ 1, & 0 < x < 1, \end{cases}$$

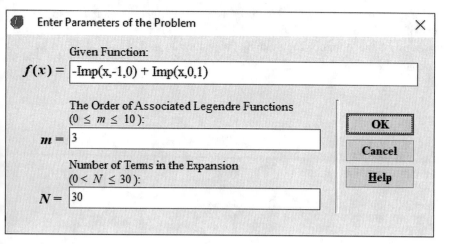

Fig. B.27 Parameters of the problem dialog box for Example B.10

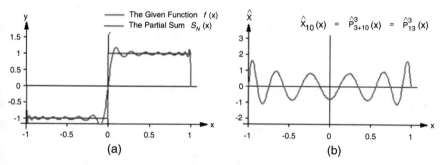

Fig. B.28 Graphs for Example B.10. (**a**) The partial sum $S_{30}(x)$, $\delta^2 = 0.0036$; (**b**) associated Legendre function, $X_{10}(x)$

defined in the interval $(-1,1)$ using the system of associated Legendre functions of order $m = 3$, which in normalized form are

$$\widehat{X}_k(x) = \widehat{P}^3_{3+k}(x) = \sqrt{(k+3) \cdot \frac{k!}{(k+6)!}} \, P^3_{3+k}(x).$$

Solution This problem is not in the library set. To solve it with the program, select a *New problem – Formula for f(x)* from the *Data* menu on the problem screen, and then enter the given function and parameters (Fig. B.27).

To run the solver, select the *Execute* command. Figure B.28a depicts the partial sum of the expansion.

Use menu option *Results – Graphs of Associated Legendre Functions* to see the plots of associated Legendre functions of the first kind of the order $m = 3$ (the functions of the orthonormal system) for values $k = 0, 1, \ldots, N$. One of the associated Legendre functions, $\widehat{X}_{10}(x) = \widehat{P}_{13}^{3}(x)$, is shown in Fig. B.28b.

Appendix C: Suggested Syllabi

To put the following syllabi in perspective, we note that many universities offer introductory (primary) courses in ODE and an advanced course. For instance, at the Technion (Israel Institute of Technology), the introductory course contains the standard material, with some variations at different departments (mathematics, physics, and engineering). The standard materials are Chapters 2–4, 6, 7, 9 of this textbook. The advanced course titled "Analytical Methods in DE" is required for applied mathematics majors, and it is often taken by physics majors as the elective. This course includes the material in Chapters 5, 8, and 10–13 of this textbook. (A number of universities offer a proof-based advanced course only for mathematics/applied mathematics majors.)

Suggested Syllabus for a One-Semester Course

Chapter 1: Introduction

Chapter 2: First-order differential equations
    2.1. Existence and uniqueness of solution
    2.2. Integral curves and isoclines
    2.3. Separable equations
    2.4. Linear first-order differential equations
        2.4.1. Homogeneous linear equations
        2.4.2. Nonhomogeneous linear equations: Method of a parameter variation
        2.4.3. Nonhomogeneous linear equations: Method of integrating factor
        2.4.4. Nonlinear equations that can be transformed into linear equations
    2.5. Exact equations

2.6. Equations unresolved with respect to a derivative (*optional*)
    2.6.1. Regular and irregular solutions
    2.6.2. Lagrange's equation
    2.6.3. Clairaut's equation
2.7. Qualitative approach for autonomous first-order equations: Equilibrium solutions and phase lines
2.8. Examples of problems leading to first-order differential equations

Chapter 3: Differential equations of order $n > 1$
    3.1. General considerations
    3.2. Second-order differential equations
    3.3. Reduction of order
    3.4. Linear second-order differential equations
        3.4.1. Homogeneous equations
        3.4.2. Reduction of order for a linear homogeneous equation
        3.4.3. Nonhomogeneous equations
    3.5. Linear second-order equations with constant coefficients
        3.5.1. Homogeneous equations
        3.5.2. Nonhomogeneous equations: Method of undetermined coefficients
    3.6. Linear equations of order $n > 2$
    3.7. Linear equations of order $n > 2$ with constant coefficients
    3.8. Euler equation
    3.9. Applications
        3.9.1. Mechanical oscillations
        3.9.2. RLC circuit
        3.9.3. Floating body oscillations

Chapter 4: Systems of differential equations
    4.1. General considerations
    4.2. Systems of first-order differential equations
    4.3. Systems of first-order linear differential equations
    4.4. Systems of linear homogeneous differential equations with constant coefficients
    4.5. Systems of linear nonhomogeneous differential equations
    4.6. Matrix approach (*optional*)
        4.6.1. Homogeneous systems of equations
        4.6.2. Nonhomogeneous systems of equations
    4.7. Applications
        4.7.1. Charged particle in a magnetic field
        4.7.2. Precession of a magnetic moment in a magnetic field
        4.7.3. Spring-mass system

Suggested Syllabus for Semester Two

Appendix D: Biographical Notes

The modern science begins with the fundamental works of Sir *Isaac Newton* (1643–1727), the great British mathematician and physicist. He invented the differential calculus, and he was the first person who wrote some ordinary and partial differential equations and solved them using the method of infinite power series that he developed. Everybody who studied the elementary algebra knows about Newton's binomial theorem for positive integer exponents; actually, he proved the much more general binomial theorem for arbitrary real exponents, as well as the corresponding multinomial theorem. He is also the founder of the numerical analysis, and his method for finding zeros of a function is widely used even nowadays.

In physics, Newton formulated the basic laws of mechanics (Newton's laws) and discovered the famous universal law of gravitation. Also, he was the founder of physical optics.

The scientific way of Newton is inextricably linked with the Trinity College of Cambridge University (admitted in 1661, BSc degree in 1664, MSc degree in 1668, professor from 1669 to 1701). Newton became a fellow of the Royal Society in 1672, and its president since 1703 till his death.

Amazingly, many of Newton's discoveries have been made during the Great Plague (1665–1667) when Newton continued his studies privately at his home. His fundamental work on mechanics was published only in 1687, and his work on optics only in 1704, since justification of the assumptions took a lot of time.

Jacob Bernoulli (1654–1705), the first in a line of Bernoulli mathematicians, was the professor of mathematics at the University of Basel. Jacob and his brother Johann developed the calculus of variations. Jacob Bernoulli made contributions also in many other areas of mathematics. One of his most important contributions was in the theory of probability; specifically, he found the law of large numbers. Also, he did an important work on infinite series.

In 1690 Jacob Bernoulli showed that the problem of determining the tautochrone is equivalent to solving a first-order nonlinear differential equation. Recall that the tautochrone, or the curve of constant descent, is the curve along which a particle will

V. Henner et al., *Ordinary Differential Equations*,
https://doi.org/10.1007/978-3-031-25130-6

descend under gravity from any point to the bottom in exactly the same time, no matter where the starting point is (see Sect. 12.4.2 of this book). Formerly, that problem was studied by Huygens in 1687 and Leibniz in 1689. After finding the differential equation, Bernoulli then solved it by what we now call separation of variables. Jacob Bernoulli's paper of 1690 is important for the history of calculus, since the term integral appears there for the first time with its integration meaning.

The younger brother of Jacob, *Johann Bernoulli* (1667–1748), the professor of mathematics at the University of Groningen, formulated and solved in 1696 another famous variational problem, the brachistochrone problem (see Sect. 12.4.1). Among the mathematical achievement of Johann Bernoulli are the development of the isocline approach and the technique for calculation of limits of indeterminate ratios. That technique is known as l'Hopital's rule: Johann had financial problems, and he signed a secret contract with Guillaume de l'Hopital who paid him a yearly pension. According to that contract, Johann had to deliver to l'Hopital all his inventions. In 1696, l'Hopital published, under his own name, the very first, extremely popular textbook on calculus based on Bernoulli's lecture notes. In 1705, after the death of l'Hopital, Johann Bernoulli returned to Basel, first as a professor of Greek. After the death of his brother Jacob, Johann inherited his chair of mathematics.

Daniel Bernoulli (1700–1782), the son of Johann Bernoulli, is most known as the discoverer of the famous Bernoulli's law in fluid dynamics. His achievements include the explanation of Boyle's law on the basis of kinetic analysis, the development of the Euler-Bernoulli beam equation in the elasticity theory, and the solution of the so-called St. Petersburg paradox in probability theory (nonexistence of the mathematical expectation for a certain simple game) invented by his relative, *Nicolas Bernoulli*. Initially, Daniel Bernoulli studied medicine, and he had a PhD degree in anatomy and botany (1721). His career as a professor of mathematics started in St. Petersburg (1724–1733). Then, he returned to Basel, where he taught medicine and later metaphysics. Only in 1750 he was appointed the chair in natural philosophy (i.e., in physics).

Leonhard Euler (1707–1783), one of the greatest mathematicians in history, is the author of more than 850 papers (including 2 dozen fundamental monographs) on mathematical analysis, differential geometry, number theory, approximate calculations, celestial mechanics, mathematical physics, optics, music theory, and other areas. He studied medicine, chemistry, botany, and many European and ancient languages.

Euler was admitted to Basel University, Switzerland, at age 13 and first studied to become a minister, but soon switched to mathematics. He remained a devoted believer throughout his life. After finishing the university in 1726, he was invited to the Saint Petersburg Academy of Science in Russia. Surrounded there by gifted scientists, he was able to study all branches of mathematics. After a year of stay in Russia, he knew Russian well and published part of his works (especially textbooks) in Russian. In 1741 he accepted a position at the Berlin Academy of Science (remaining an honorary member of the Saint Petersburg Academy).

Euler's "Letters on various physical and philosophical matters written to a certain German princess" (1768) gained the immense popularity. This science encyclopedia had over 40 editions in 10 languages.

In 1766 Euler returned to St. Petersburg. In 1771 he became blind, but remained prolific. He once noticed that now he would be less distracted from doing mathematics. The number of his published works even increased – during his second stay in Russia, Euler dictated more than 400 articles and 10 books, which is more than half of his creative legacy. After his death, the Academy at Saint Petersburg continued to publish his unpublished works for almost 40 years.

"He stopped calculating and living" was said at the morning meeting of the Paris Academy of Sciences (*Il cessa de calculer et de vivre*). He was buried at the Lutheran cemetery in Saint Petersburg. The inscription on the monument read: "Here lie the remains of the world-famous Leonhard Euler, a sage and a righteous man. Born in Basel April 4, 1707, died September 7, 1783."

Joseph Louis Lagrange (1736–1813) was the French mathematician of Italian origin. Initially, he taught mathematics in Turin (since 1755), where he founded the society named *Miscellannea Taurinensia* (1758) (Turin Miscellaneous) that was a kind of local Academy of Sciences.

In 1766, he received a personal invitation of the King of Prussia where he was called "the greatest mathematician in Europe," and he worked in Berlin till 1786. Then, he accepted the invitation of Louis XVI to move to Paris, where he lived till his death and taught at Ecole Normale (since 1795) and Ecole Polytechnique (since 1794).

The greatest scientific achievements of Lagrange are in the field of variational calculus (Euler-Lagrange equations, Lagrange multipliers), mathematical reformulation of Newtonian mechanics (Lagrangian function, Lagrange's equations), differential equations (method of variation of parameters, Lagrange's ordinary and partial differential equations), and probability theory (Lagrangian probability distributions).

French mathematician *Pierre-Simon Laplace* (1749–1827) is ranked as one of the greatest scientists of all time.

The achievements of Laplace in pure and applied mathematics, and especially in astronomy, are enormous, as he improved almost all areas. In four volumes of Laplace's main work, *Mécanique Céleste*, written over a period of 26 years (1799–1825), he formulated laws of mechanics using methods of calculus. He also summarized both his own research and the works of his predecessors, starting with Newton. He gave an analysis of the planetary motion using Newton's law of universal gravitation and proved the stability of the solar system. Back in 1695, Edmond Halley, an English astronomer, discovered that Jupiter gradually accelerates its motion and approaches the Sun, while Saturn slows down and moves away from the Sun. Laplace discovered that the causes of these displacements are the planet's interactions, and showed that these are nothing more than periodic perturbations of their orbits. Laplace also discovered that the periodic change in the eccentricity of the lunar orbit, which perplexed all astronomers, occurs due to the attraction of large planets.

Along with the results concerning the movements of the planets, satellites and comets, the shapes of the planets, the theory of tides, etc., the most important was the general conclusion that refuted the opinion (which was shared by Newton) that maintaining the stability of the solar system requires the intervention of some extraneous supernatural forces.

Laplace proposed the first mathematically substantiated nebular hypothesis of the origin of the solar system (previously suggested by the philosopher Immanuel Kant). Ahead of his time, he was one of the first scientists to suggest an idea similar to that of a black hole.

In solving applied problems, Laplace developed methods of mathematical physics, a field that he took a leading role in forming. His particularly important results relate to potential theory and special functions. The Laplace transform and the Laplace equation are named after him.

Laplace is one of the founders of the probability theory. In 1812, the 63-year-old Laplace published the grandiose *Théorie Analytique des Probabilités*. One of the fundamental theorems of the probability theory bears his name – the de Moivre-Laplace limit theorem. In the area of applications of the probability theory, he developed the theory of errors and the method of least squares.

Laplace was the adherent of absolute determinism. He argued that if it were possible to know the positions and speeds of all particles in the world at some time, then we would be able to predict their future locations. The erroneousness of Laplace's determinism was shown even before the advent of quantum mechanics – at the beginning of the twentieth century, another French genius, Henri Poincaré, discovered a fundamentally unpredictable process in which an insignificant change in the initial state causes arbitrarily large deviations in the final state over time.

Jean-Baptiste Joseph Fourier (1768–1830), the French physicist and mathematician, is known by his discovery of the expansion of functions into series of trigonometric functions (Fourier series) that he applied for solution of the problem of heat conduction. His paper on that subject presented in 1807 was rejected by the Academy of Sciences in Paris for lack of rigor. Indeed, Fourier did not find the conditions for the convergence of those series that were obtained by Dirichlet much later, in 1829. Nevertheless, in 1822, Fourier published his fundamental work *The Analytical Theory of Heat* that contained the formulation of the heat equation and its solution by means of Fourier series. Among the scientific achievements of Fourier are the formulation of the concept of dimensional analysis and the discovery of the greenhouse effect.

The life trajectory of Fourier was quite unusual for a scientist. He was a son of a tailor, and he was orphaned at the age of nine. He promoted the French revolution in his district and was imprisoned during the Terror. Fourier accompanied Napoleon in his Egyptian campaign (1798) and organized production of munitions for the French army. He taught in Ecole Normale and Ecole Polytechnique in Paris and oversaw road construction as a Prefect at Grenoble. Since 1822, he served as Permanent Secretary of the French Academy of Sciences. His name is engraved on the Eiffel Tower among 72 names of the most prominent French scientists, mathematicians, and engineers.

Henri Poincaré (1854–1912), the French mathematician, physicist, and engineer, was one of the greatest mathematicians of all time and the last universalist in mathematics. His achievements include the creation of the qualitative theory of differential equations, the discovery of chaos in deterministic dynamical systems, the development of the theory of automorphic functions and the theory of analytic functions of many variables, and the creation of topology. His works in group theory, especially his analysis of the symmetry properties of the Maxwell equations, formed the mathematical basis for the creation of the relativity theory by Albert Einstein. A simple statement of his achievements in physics and pure and applied mathematics, and even of theorems and notions named after him, would be too long for this short Appendix.

Poincaré graduated from Ecole Polytechnique in 1875 and from Ecole of Mines in 1879. Even being a professor of Sorbonne (since 1881), he continued to work as a mining engineer, and later as a chief engineer and the inspector general at the Corps of Mines. His activities included serving as president of the French Academy of Sciences and French Astronomical Society, working for the French Bureau of Longitudes, and the legal defense of Alfred Dreyfus, the French officer of Jewish descent falsely convicted of treason and sentenced to life imprisonment.

Aleksandr Lyapunov (1857–1918), the great Russian mathematician, is one of the founders of the stability theory and the theory of dynamical systems. He entered the University of St. Petersburg in 1876, graduated in 1880, and submitted his Master's thesis in 1884. He worked at Kharkiv University (1885–1902) and then at the University of Saint Petersburg (1902–1917). In 1918, he started teaching in Odessa, but he did not finish his course: he committed suicide after the death of his wife.

Numerous mathematical notions are named after Lyapunov. The notions of Lyapunov stability and Lyapunov function are explained in Chap. 5 of this book. The notions of Lyapunov exponents and Lyapunov vectors are now the fundamental notions of the theory of dynamical systems, specifically the chaos theory. The Lyapunov equation is an important equation of the control theory. The Lyapunov central limit theorem is a significant extension of the classical central limit theorem of the probability theory.

The textbook presents a rather unique combination of topics in ODEs, examples, and presentation style. The primary intended audience is undergraduate (second, third, or fourth year) students in engineering and science (physics, biology, economics). The needed prerequisite is a mastery of single-variable calculus. A wealth of included topics allows using the textbook in up to three sequential, one-semester ODE courses. Presentation emphasizes the development of practical solution skills by including a very large number of in-text examples and end-of-section exercises. All in-text examples, be they of a mathematical nature or a real-world examples, are fully solved, and the solution logic and flow are explained. Even advanced topics are presented in the same undergraduate-friendly style as the rest of the textbook. Completely optional interactive laboratory-type software is included with the textbook.

Bibliography

1. Boyce, W.E., DiPrima, R.C.: Elementary Differential Equations and Boundary Value Problems. Wiley (2008)
2. Coddington, E.A.: An Introduction to Ordinary Differential Equations. Prentice Hall, Englewood Cliffs (1961); Dover, New York (1989)
3. Braun, M.: Differential Equations and Their Applications. Springer (1978)
4. Abramowitz, M., Stegun, I.A.: Handbook of Mathematical Functions, Tenth Printing. National Bureau of Standards (1972)
5. Kevorkian, J., Cole, J.D.: Multiple Scales and Singular Perturbation Methods. Springer (1996)
6. Pinkus, A., Zafrany, S.: Fourier Series and Integral Transforms. Cambridge University Press (1997)
7. Henner, V., Belozerova, T., Forinash, K.: Mathematical Methods in Physics. AK Peters, Ltd. (2009)
8. Butkov, E.: Mathematical Physics. Addison-Wesley (1968)
9. Henner, V., Belozerova, T., Nepomnyashchy, A.: Partial Differential Equations – Analytical Methods and Applications. CRC Press, London/New York (2020)
10. Sauer, T.: Numerical Analysis. Pearson (2017)
11. Butcher, J.C.: Numerical Analysis of Ordinary Differential Equations. Wiley, London (1987)
12. Lambert, J.D.: Numerical Methods for Ordinary Differential Systems. Wiley, Chichester (1991)
13. Ortega, J.M.: Numerical Analysis: A Second Course. Academic Press, New York (1972)
14. Conte, S.D., de Boor, C.: Elementary Numerical Analysis. McGraw-Hill, New York (1972)
15. Higham, N.J.: Accuracy and Stability of Numerical Algorithms, 2nd edn, p. 175. SIAM (2002)

Index

Printed in the United States
by Baker & Taylor Publisher Services